SUPERCOLLIDER 3

SUPERCOLLIDER 3

Edited by
John Nonte

Superconducting Super Collider Laboratory
Dallas, Texas

Springer Science+Business Media, LLC

Library of Congress Cataloging-in-Publication Data

International Industrial Symposium on the Super Collider (3rd : 1991 :
Atlanta, Ga.)
    Supercollider 3 / edited by John Nonte.
        p.   cm.
    "Proceedings of the third International Industrial Symposium on
the Super Collider, held March 13-15, 1991, in Atlanta, Georgia"-
-T.p. verso.
    Includes bibliographical references and index.
    ISBN 978-1-4613-6668-3
    1. Superconducting Super Collider--Congresses.   I. Nonte, John.
II. Title.   III. Title: Supercollider three.
QC787.P7I57   1991
539.7'3--dc20   91-27634
                                                                CIP

Proceedings of the Third International Industrial Symposium on The Super Collider,
held March 13-15, 1991, in Atlanta, Georgia

ISBN 978-1-4613-6668-3   ISBN 978-1-4615-3746-5 (eBook)
DOI 10.1007/978-1-4615-3746-5

© 1991 Springer Science+Business Media New York
Originally published by Plenum Press, New York in 1991
Softcover reprint of the hardcover 1st edition 1991

Preface

The third annual International Industrialization Symposium on the SuperCollider, IISSC—held March 13–15, 1991, in Atlanta, Ga.—was an enormous success. The number of attendees, exhibitors, and representatives from foreign countries surpassed the totals of previous years. There were 740 attendees, representing more than 2 dozen universities and colleges, 32 states, 9 national labs, 6 research centers, several government entities at the local, state, and federal level, 182 businesses & industry and 14 countries. More than 100 exhibits, sponsored by 85 organizations, added to the excitement.

"Getting Down to Business" was the theme of this year's Symposium. The fact that the Superconducting SuperCollider (SSC) is indeed underway was the message delivered by the Symposium's keynote speaker, Dr. Roy Schwitters, and expanded upon by the opening plenary speakers. The project is moving from the planning stage to actual construction, to development and procurement of equipment, and to resolution of the technical issues involved in advancing the state-of-the-art in areas such as theory, controls, systems, metallurgy, quality control, management, cryogenics, power systems, detectors, interagency cooperation and funding.

Plenary speakers included:

Paul Gilbert, Chairman of Parsons Brinckerhoff Quade & Douglas, Inc.

Dr. Thomas Kirk, Director of High Energy Physics at Argonne National Laboratory

Joseph Cipriano, DOE Project Manager for the SSC

Edward Siskin, General Manager of the SSCL

Robert Johnson, Program Director of Energy Programs, General Dynamics Space Systems

Bruce Boswell, General Manager of Collider Magnet Division, Westinghouse Electric Corporation

Kent Higgins, Project Director, Koch Processing Systems

Dr. Raphael Kasper, Chief of the Directorate for the SSCL

Edward Donley, Chairman of the Executive Committee of the Board of Directors for Air Products and Chemicals, Inc.

Dr. John Marburger, President of State University of New York at Stony Brook

Hon. Joe Barton, Congressman from Texas

Fred Bucy, Chairman of the Texas National Research Laboratory Commission

Hon. John Meyers, Congressman from Indiana

Hon. Jim Chapman, Congressman from Texas

Dr. Jack Sandweiss, Professor of Physics at Yale University

Dr. Thomas Appelquist, Professor of Physics at Yale University

Dr. John Peoples, Jr., Director of Fermi National Accelerator Laboratory

Dr. Victor Yarba, First Vice-Director of the Institute of High Energy Physics at Serpukhov, USSR

John Clendenin, Chairman of the Board and Chief Executive Officer of BellSouth Corporation

Dr. Gary Gibbs, Acting Director, Office of the SSC, DOE

The symposium included a series of technical sessions offering discussions and technical papers on a variety of topics: accelerator systems, computers and controls, conventional construction, cryogenics, detectors, education, and various aspects of project management and systems engineering.

The program committee for this symposium included:

Dr. Thomas Kirk, Chairman

Dr. Edward Bingler

Dr. Thomas Bush

Dr. Paul Mantsch

Dr. Satoshi Ozaki

Dr. Giuseppe Scarfi

Dr. Clyde Taylor

Dr. William Wallenmeyer

Equally important was the contribution made by Pamela Patterson, who served as Conference Manager for the third year and tended to the numerous details necessary to coordinate such a large-scale, international conference.

Two special awards were presented at the meeting. The IISSC award was presented to Dr. Alvin Trivelpiece by the IISSC Board of Directors for his substantial contribution to the SSC program in its formative stages. The U.S. Department of Energy's Distinguished Associate Award was presented to E. Parke Rohrer by the DOE in recognition of his outstanding contributions to the SSC magnet research and development program.

The 1991 International Industrial Symposium on the Super Collider is a private, nonprofit corporation, staffed totally by dedicated volunteers from business and industry; local, state, and federal government; university research; and national labs. The organization exists solely to conduct an annual symposium on the SSC in order to educate people about the project and thereby increase understanding and support.

The IISSC Board of Directors for 1990/91 were:

Member	Affiliation
Mr. Charles Anderson (*)	Air Products and Chemicals, Inc.
Mr. Owen Anglum	Armco, Inc.
Mr. Robert Baldi	General Dynamics Space Systems Div.
Dr. David Berley	National Science Foundation
Dr. Edward Bingler	Texas National Research Laboratory

Ms. Regina Borchard	Martin Marietta Strategic Systems
Dr. Tom Bush	SSC Laboratory
Mr. Tony Favale	Grumman Space Systems
Mr. Paul Gilbert (*)	Parsons Brinckerhoff Quade & Douglas, Inc.
Dr. Leonard Goldman	Bechtel National, Inc.
Dr. Eric Gregory	IGC Advanced Superconductors, Inc.
Ms. Phyllis Hale (*)	SSC Laboratory
Mr. Andrew Jarabak (*)	Westinghouse Electric Corp.
Dr. Thomas Kirk	Argonne National Laboratory
Dr. Paul Mantsch (*)	Fermi National Accelerator Laboratory
Mr. Robert Marsh	Teledyne Wah Chang Albany
Dr. Michael McAshan	SSC Laboratory
Mr. John Nonte	Lockheed Engineering & Sciences Co.
Dr. Satoshi Ozaki	Brookhaven National Laboratory
Dr. W. Arthur Porter	Houston Area Research Center
Mr. Maurice Sabado	Science Applications International Corp.
Dr. Giuseppe Scarfi	Ansaldo Componenti S.p.A.
Dr. Clyde Taylor	Lawrence Berkeley Laboratory
Dr. Robert Tener	Sverdrup Corporation
Dr. William Wallenmeyer	Southeastern Universities Research Association

(*) Designates IISSC Officers

The following companies, organizations, societies, and agencies assisted us in producing the IISSC 1991:

Air Products and Chemicals, Inc.

American Physical Society

American Society of Civil Engineers

Ansaldo Componeneti S.p.A.

Argonne National Laboratory

Armco, Inc.

Babcock & Wilcox

Bechtel National, Inc.

Brookhaven National Laboratory

CRSS Inc.

Daniel, Mann, Johnson & Mendenhall

EG&G, Inc.

E.I. du Pont de Nemours & Company

Fermi National Accelerator Laboratory

General Dynamics Space Systems

Grumman Space Systems

Hitachi, Ltd.

Houston Area Research Center

IGC Advanced Superconductors, Inc.

Institute of Electrical and Electronics Engineers--
Nuclear and Plasma Sciences Society

Intermagnetics General Corporation

Koch Process Systems, Inc.

Lake Shore Cyrotronics, Inc.

Lawrence Berkeley Laboratory

Lockheed Engineering & Sciences Company

Maxwell Laboratories, Inc.

Martin Marietta Corporation

Morrison Knudsen Company, Inc.

National Science Foundation

National Society of Professional Engineersv

Noell, Inc.

Parsons Brinckerhoff Quade & Douglas, Inc.

Plainfield Stamping

RDS Manufacturing, Inc.

Science Applications International Corporation

Southeastern Universities Research Association

SSC Laboratory

Structural Composites Industries

Sulzer Bros., Inc.

Sverdrup Corporation

Sumitomo Electric Industries, Ltd.

Teledyne Japan K.K.

Teledyne Wah Chang Albany

Tesla Engineering Limited

Texas National Research Laboratory Commission

U.S. Department of Energy

Vector Fields, Inc.

Wang NMR, Inc.

Welded Ring Products, Inc.

Westinghouse Electric Corporation

Those listed provided support in many ways, including financial or volunteer staff assistance, or both. A total of $15,550 in direct funds contributions were raised this year for operating funds for the IISSC.

Continuing solid support is vital to the success of the SSC. Support is very much like a brick wall. Each brick is important to the whole and, in the wall, mortar gives all the individual bricks connection, continuity, and purpose. The SSC enjoys many good quality bricks of support. By learning more about the SSC, renewing our commitment, and working together, we refresh the mortar and provide the connectivity, the continuity, and quality of solid support necessary to assure that the SSC will become a reality.

The Board of Directors met in Atlanta on March 13, 1991, prior to the start of the symposium, to discuss preparations for the 1992 Symposium and to elect executive officers of the IISSC Corporation for the period March 1991 to March 1992. Elected officials are:

President and Meeting Chairman	Andrew Jarabak
Vice President	Charles Anderson
Secretary	Eric Gregory
Treasurer	Thomas Kirk
Administrative	Phyllis Hale

On behalf of the IISSC Board of Directors, we thank all who attended, participated in, and contributed to the 1991 Symposium and we cordially invite all readers to attend the 4th annual IISSC meeting in New Orleans, March 4-March 6, 1992. It promises to be another exciting, informative, and entertaining program.

Paul H. Gilbert
Chairman
IISSC '91

9. Markets for SSC Technology

MEDICAL APPLICATIONS OF ACCELERATORS

Maurice M. Sabado

Science Applications International Corporation
4161 Campus Point Court
San Diego, CA 92121

ABSTRACT

Accelerator technology has been significantly developed and enhanced as a result of the need for better microscopes in physics research. Applications of accelerator technology have also resulted in its enhancement.

Commercial applications of accelerators are expanding. Accelerator technology is used in: neutron, gamma, electron, and proton, cancer radiation treatment; isotope production; explosive detection systems; food irradiation; gem stone production; sterilization; etc. The direct spinoff technology is estimated to be millions of dollars. The indirect spinoff of this technology is likely in the billions of dollars.

The medical market for accelerator technology is reviewed. In particular, the cancer therapy market is characterized. We will review accelerators applications in: radiation treatment for cancer therapy, isotope production and imaging. Significant progress has been made in the application of accelerator technology to radiation oncology for cancer treatment. Status of particle therapy and improvements in technology is briefed. The new emerging clinical applications of heavy ions and protons is discussed. Some new and innovative applications ideas are also discussed.

INTRODUCTION

Challenges and opportunities for the International Industrial Community in the 1990's are varied. The U.S. is faced with these issues: THE CHANGING GLOBAL MARKETPLACE; AMERICA'S QUEST TO MAINTAIN LEADERSHIP IN THE COMPETITIVE WORLD ECONOMY; THE ENVIRONMENT AND ENERGY; AND THE AGING OF AMERICA (medical implications of cancer radiation therapy based on accelerator technology).

APPLICATION TO CANCER TREATMENT

Since the largest commercial application of accelerators and their technology is in cancer treatment and medical diagnostics, the scope of this problem is worthy of some discussion. The United States population is aging as shown in Figure 1. The number of retired will grow from roughly 28 million to 51 million, nearly double, in approximately thirty years. The Statistical Epidemiology and End Results (SEER) Group of the National Cancer Institute (NCI) reports the annual incidence for all newly diagnosed cancers for males aged 65 years and older is 28.095 per 1,000 population; and for females aged 65 years and older, this rate is 14.985. The AVERAGE cancer incidence in the United States

RETIRED POPULATION

YEAR	U.S. POPULATION	NEW YORK
1985	28,608,000 (12%)	2,319,799 (13.1%)
2000	34,921,000 (13%)	2,700,742 (14.6%)
2010	39,196,000 (14%)	2,942,623 (15.5%)
2020	51,422,000 (17%)	NOT AVAILABLE

THE AVERAGE CANCER INCIDENCE IN THE UNITED STATES IS 3.9 PER 1,000 POPULATION

STATISTICAL EPIDEMIOLOGY AND END RESULTS (SEER) GROUP OF NCI, THE ANNUAL INCIDENCE FOR ALL NEWLY DIAGNOSED CANCERS FOR MALES AGED 65 YEARS AND OLDER IS 28,095 PER 1,000 POPULATION; AND FOR FEMALES AGED 65 YEARS AND OLDER, THIS RATE IS 14,985

SOURCE: U.S. CENSUS BUREAU AND NY OFFICE FOR THE AGING

Figure 1. Projected Graying of America, Size of the Market

is 3.9 per 1,000 population. According to the American Cancer Society (ACS), in 1986 the number of new cancer cases was 930,000 and 472,000 deaths, of which 100,000 were from localized disease, where perhaps accelerator based radiation treatment with localization control may have contributed to a reduction. In 1988 the new cancer cases had grown 985,000 and total deaths to 494,000, with the deaths from local disease remaining the same. Clearly, the cancer incidence rate increases with population and age of the population. Figure 2 is a survey of cancer type and some typical treatments. Radiation has become a part of most cancer treatments. According to a National Center for Health Statistics (NCHS) study, cancer accounts for 10 percent of the total cost of disease in the United States, and according to congressional estimates, by the year 2025, programs for the elderly might consume two thirds of the federal budget compared to one quarter now.[1]

Cancer management has changed significantly during this era. Radiation therapy has played an important role, and the evolutionary maturing of technologies has been a direct result of research projects providing spin-off technology. The emergence of radiation therapy can be thought of in four eras as shown in Figure 3: Discovery Era (Kilovoltage), beginning around 1890, and maturing around 1920; Orthovoltage Era, 1920 - 1940; Megavoltage Era, 1940 -1980; and the Era of Precision. Advances were made in physics/engineering and medical scientific fields which followed the development of technologies developed in the laboratory, particularly accelerators and their related technologies.[2]

ACCELERATOR TECHNOLOGY DEVELOPMENT

In the early part of the twentieth century the nuclear model was becoming clearer and needs for better experiments to explore it were identified and constructed. By the second half of the twentieth century physicists knew more about electrons and protons and particles were now Strange and had CHARM and FLAVORS!

To explore matter, physicists build huge and powerful atom smashing machines. They accelerate particles as a type of microscope to explore the nuclear and sub-nuclear structure and unfold the building blocks. These machines split up not only atoms, but also the particles making up the atoms.[3]

The first accelerators built were to investigate the fundamental particles of matter. Each new accelerator project brought new and confirming information of the Unifying Theory, and new puzzles to be investigated and explained. The study of elementary particles through the use of high-energy accelerators became a separate and distinct aspect of physical research shortly after the end of the Second World War. Accelerator technology has been significantly developed and enhanced as a result of the need for better

SURVEY OF CANCER TYPE AND TREATMENTS

CANCER TYPE	LUNG	COLORECTAL	BREAST	PROSTATE	URINARY TRACT (incl. bladder/kidney)	UTERINE (incl. cervical)
INCIDENCE:						
OF ALL CANCERS	16%	15%	14%	10%	7%	4.90%
CASES	150,000	145,000	130,900	96,000	67,300	47,800
MALE	66%	52%	99%		70%	
FEMALE	34%	48%	1%		30%	
INCREASE FROM 1975 TO 1984	18%	4%	8.60%	15.30%	17.3% (kidney)	
DECREASE FROM 1975 TO 1984					3.8% (bladder)	27.1% (cervical) 26.1% (uterine)
TREATMENT	Surgery. If cancer has spread then surgery combined with chemotherapy & radiation.	Surgery, sometimes combined with radiation.	Surgery, combined with chemotherapy, radiation and/or hormone treatment.	Surgery and/or radiation. Hormone treatment and anti-cancer drugs may also be used.	Surgery followed by chemotherapy or radiation.	Surgery and/or radiation.

CANCER TYPE	SKIN (malignant melanoma)	STOMACH	OVARIAN	ORAL	LEUKEMIA	PANCREATIC
INCIDENCE:						
OF ALL CANCERS	2.70%	2.50%	2%	3%	2.70%	2.70%
CASES	25,800	24,600	19,000	29,800	26,400	26,200
MALE	53%	61%		68%	56%	50%
FEMALE	47%	39%		32%	44%	50%
INCREASE FROM 1975 TO 1984	29.70%			Same		0.40%
DECREASE FROM 1975 TO 1984		10.50%	1.10%	Same	11%	
TREATMENT	Surgery; cryosurgery, electrocoagulation, radiation.	Surgery, combined with chemotherapy, radiation and/or hormone treatment.	Surgery; chemotherapy; radiation.	Surgery and radiation.	Chemotherapy; radiation and bone-marrow transplant in some cases.	Surgery, chemotherapy, radiation.

Figure 2. Survey of Cancer Type and Treatments

microscopes in physics research.[4] Applications of accelerator technology have also resulted in its enhancement. For example, the Strategic Defense Initiative brought resources to bear on Ion sources, High Brightness Accelerators, particularly Linac's in areas as: Improved Radio Frequency Quadruple (RFQ) accelerator performance; high current Linacs, etc.

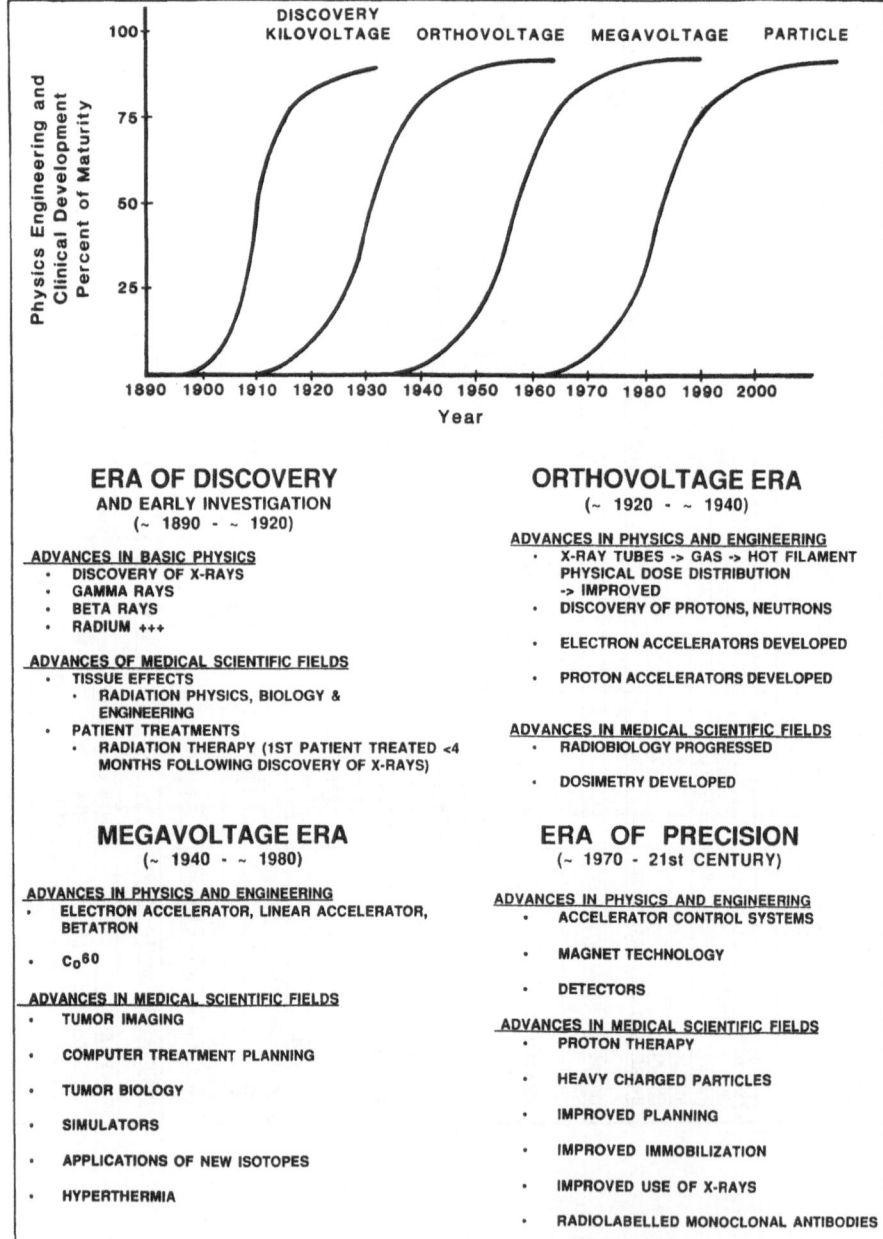

Figure 3. Cancer Management by Radiation Era

COMMERCIAL USES OF ACCELERATOR TECHNOLOGY

Commercial applications of accelerators are expanding. Accelerator technology is used in: neutron, gamma, electron, and proton, cancer radiation treatment; isotope production; explosive detection systems; food irradiation; gem stone production; sterilization; etc. The direct spinoff technology is estimated to be millions of dollars. The indirect spinoff of this technology is likely in the billions of dollars.

The application of accelerator technology to the medical field has been fairly rich, particularly radiation cancer therapy. Radiation treatment modalities now include x-ray, pion, gamma ray, neutron, proton, heavy ion, gamma knife, using linear accelerators, cyclotrons, synchrotron, with superconducting and normal magnet technologies. Dramatic improvements have been made in medical diagnostics through the production of isotopes from accelerator systems for Positron Emission Tomography (PET), and Single Photon Emission Computed Tomography (SPECT). Digital Subtraction Angiography (DSA) and Computed Tomography (CT) benefitted from improvements and advances in x-ray tube technology, whose genesis was the research from our laboratories. Also, Magnetic Resonance Imaging (MRI) benefitted from work done at the laboratories in superconducting magnets. New ground is being broken with insitu Spectroscopy with MRI due to developments in high field linear superconducting magnets and squid technology. New and exciting work to enhance imaging of tumors will likely emerge from the use of monoclonal antibodies tagging with isotopes and perhaps someday a treatment modality. [5]

Improvements in the imaging technology brought benefits to radiation therapy by providing greatly increased tumor definition from these diagnostics. Coupled with improvements in computer three dimensional graphics, tumors could be spatially defined within a human and characterized with respect to normal tissue and organs. With these improvements, precision delivery of a radiation dose to a tumor site with little or no morbidity was at hand. With these improvements, it was now proper to exploit the Bragg Peak as suggested by Robert Wilson. Figure 4 shows a comparison of depth dose distribution for x, gamma-ray and electron beams with distribution of a proton beam. The control of the proton is so superb that normal structures can largely escape the high dose region, therefore, therapy can continue until the tumor is destroyed. Also lower side effects and complications (morbidity) result from lower organ radiation therapy.[6]

Maturing accelerator technology has provided potential clinics with choices in technologies available for treatments. For example, the Proton Therapy Cooperative Group (PTCOG) workshop on accelerators concluded that an accelerator with energy of 250 MeV, and variable was desirable. Three devices developed in the laboratories, a cyclotron, linear accelerator (LINAC), and synchrotron are all possible candidates. With the Cyclotron one has moderately high intensity, moderate size (20 Ft. diameter), competitive cost with a synchrotron, but cannot vary energy. The linac has extremely high intensity, and would be extremely large in size (~ 500 Ft. long) and fairly high in cost. Synchrotrons have adequate intensity, are competitive with cyclotron in size and cost, and can EASILY VARY ENERGY! Clearly, trade studies produced at Loma Linda University Medical Center (LLUMC) of the various accelerator technologies and treatments led to their selection of a hospital based proton synchrotron.

There have been roughly 12,479 charged particle treatments world wide, of which 9,223 were proton treatments. Until 1990, Proton therapy has been limited to a few locations and provided at a accelerator laboratory environment as shown in Figure 5.[7] With the opening of the LLUMC Proton Beam Therapy System (PBTS) this new and exciting technology is now available in a clinical setting. LLUMC has treated its first patient on the eye beam line and recently completed one on the fixed beam line Treatments with the isocentric gantries are scheduled for late April to early May.

Other clinical applications of accelerators are for neutron therapy. Four clinical cyclotrons are used for neutron therapy, in addition the parasitic line at FERMILAB operated by Rush Presbyterian - St. Luke Hospital. The University of Washington utilizes a cyclotron for isotope production used in research and PET.

Some novel and/or innovative ideas and applications of accelerator technology to the medical market are also emerging. A hospital-based 70 MeV proton linear accelerator for particle and radioisotope production has been proposed for Rush- Presbyterian St. Luke Hospital in Chicago. This proposal uses the Linac for Fast Neutron Therapy, Boron Neutron Capture Therapy, and Isotope production. Their design allows for four linac energies, the lowest at 21.4 Mev, for common PET isotopes, and the highest at 70.4 MeV for fast neutron therapy and proton therapy for ocular melanomas, with the current significantly reduced. The LINAC would be used as the injector for a synchrotron to accelerate the beam up to 250 MeV for proton Therapy. An interesting multi-mission research facility.[8]

Compact accelerator applications using Radio Frequency Quadrupole (RFQ) technology for PET are available or being developed. ACCSYS Technology Inc has developed a commercial product for the production of isotopes based on this technology.[9] SAIC is developing a 8 MeV Helium 3 doubly charged RFQ also for PET isotope production. The particle species was selected to enhance operational availability and reduce maintenance requirements.[10]

Boron Neutron Capture Therapy (BNCT) is a new radiation treatment modality in which a tumor is loaded with a boron-doped compound and then irradiated by epithermal neutrons.

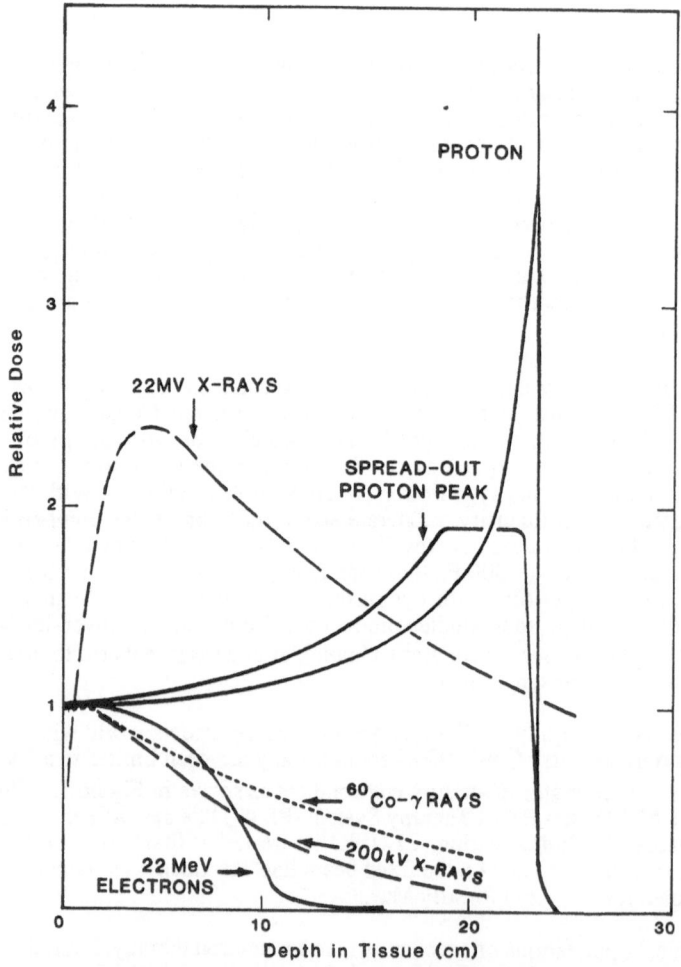

Figure 4. Comparison of Depth Dose Distribution for X,γ-ray & Electron Beams with Distribution of a Proton Beam

RADIATION ONCOLOGY TREATMENT MODALITY IS DOMINATED BY THE ELECTRON LINEAR (E.G., VARIAN LINAC). OTHER PARTICLES HAVE BEEN TRIED WITH THE BEST SUCCESS IN PROTONS.

PRESENT SUMMARY: PROTONS, LIGHT AND HEAVY IONS
WORLD WIDE CHARGED PARTICLE PATIENT TOTALS
January 1, 1991

WHO	WHERE	WHAT	DATE FIRST RX	RECENT PATIENT TOTAL	DATE OF TOTAL
Berkeley 184	CA USA	p	1955-1957	30	
Berkeley 184	CA USA	it	1957-1987	899	
Uppsala	Sweden	p	1957-1976	73	
Harvard	MA USA	p	1961	5120	Dec-90
Dubna	USSR	p		84	
Moscow	USSR	p	1964-1974	1945 *	May-90
Los Alamos	NM USA	π-	1969	230	
Leningrad	USSR	p	1974-1982	685	Sep-90
Berkeley Bev.	CA USA	heavy	1975	1422	Oct-89
Chiba	Japan	p	1975	65	Jul-90
TRIUMF	Canada	π-	1979	227	
PSI (SIN)	Switzerland	π-	1980	478	Dec-89
Tsukuba	Japan	p	1983	178	Apr-90
PSI (SIN)	Switzerland	p	1984	913	Nov-90
Dubna	USSR	p	1987	6	Sep-90
Uppsala	Sweden	p	1988	13	May-90
Clatterbridge	England	p	1989	114	Nov-90
Loma Linda	CA USA	p	1990	3	Dec-90

935	pion beams	
2321	ion beams	
9223	proton beams	
12479	all particle beams	

TOTAL.

*Revised number; re-treatments excluded.

UNITED STATES CYCLOTRONS EXCLUSIVELY USED FOR NEUTRON THERAPY

INSTITUTION	SUPPLIER
M.D. ANDERSON HOSPITAL HOUSTON, TEXAS	CYCLOTRON CORPORATION
UNIVERSITY OF WASHINGTON MEDICAL SCHOOL SEATTLE, WASHINGTON	SCANDITRONIX
UNIVERSITY CALIFORNIA LOS ANGELES MEDICAL SCHOOL LOS ANGELES	CYCLOTRON CORPORATION
HARPER GRACE HOSPITAL (UNDER CONSTRUCTION) DETROIT, MICHIGAN	MEDCYC

ACCELERATORS USED FOR PROTON THERAPY OUTSIDE U.S.A.

Institution	Accelerator	Energy	Yr. of 1st Operation	Number of Patients
Donner Lab. Berkeley	Synchrocyclotron	340 MeV	1954	26
Werner Inst. Uppsala, Sweden	Synchrocyclotron	85 MeV	1957	69
Harvard Cyclotron Laboratory	Synchrocyclotron	160 MeV	1958	3000
Institute Theoretical & Experimental Physics, Moscow, USSR	Synchrotron	70-200 MeV	1969	1250
Joint Inst. for Nuclear Research, Dubna, USSR	Synchrocyclotron	90-200 MeV	1967	85
LIJF, Gatchina USSR	Synchrocyclotron	70-1000 MeV	1973	381
Nat'l Inst. Nuclear Phys., Chiba, Japan	Synchrocyclotron	70 MeV	1979	29
PARMS Tsukuba, Japan	Synchrotron	250 MeV	1983	22
Swiss Inst. Nucl. Research	Synchrocyclotron	70-590 MeV	1985	121

IN FOREIGN COUNTRIES, ACCELERATOR BASED NEUTRON TREATMENTS ARE DONE ON EXISTING LABORATORY CYCLOTRONS USING PARASITIC BEAMLINES

FINAL WITH RUSH-PRESBYTERIAN HOSPITAL REQUIRES NEUTRON THERAPY

Figure 5. Limited Capability to Treat Patients

Epithermal neutron beams appropriate for clinical applications have been or are being installed in reactors at Brookhaven National Laboratory and M.I.T. A 10mA beam from a 2.5 MeV proton RFQ onto a molten lithium target has been proposed in the U.S., while a 100 microamp beam from a 30 MeV proton cyclotron has been proposed in Europe.[11]

High intensity x-ray beams from synchrotron radiation is being utilized for imaging of coronary arteries studies. Visualization of coronary aortas with injection of a contrast agent through the veins and exposure to high intensity x-ray beams. This diagnostic would reduce the risk associated with the conventional method of inserting a catheter into a artery and threaded up to the heart so that a contrast agent can be injected directly into the artery being imaged.[12]

TECHNOLOGY TRANSFER

Figure 6 is similar to the typical "Livingston Plot" which graphically shows that accelerator energy (and therefore technology) plotted on a log scale versus a linear scale of time produces a linear graph. But if one also plots the industrialization of the accelerator a "lag time" of roughly 40-50 years appears. Clearly, the infusion of this technology into applications is slower than need be. The limiting factor seems to be availability of an industrial base to provide it and the large capital cost. Perhaps with the increasing presence of industrial contractors in the laboratories and real technology transfer programs this lag time can be shortened.

In the United States, to address this issue, Congress passed the Stevenson-Wydler Technology Innovation Act of 1980 (Public Law 96-480)and the National Technology Transfer Act (NCTTA) of 1989, (Public Law 101-189), amendment were written so as to benefit the competitiveness of United States Industry. It provided for technology transfer programs between the laboratories and industry. The Department of Energy, in its recent report to Congress on the status for their technology programs profiled the Loma Linda University Medical Centers, Proton Therapy System accelerator technology transfer from FERMILAB as a biomedical application.[13]

SSC TECHNOLOGY TRANSFER

Accelerator based research projects have historically developed technologies that have benefitted medical cancer treatment and diagnostic fields. Accelerator technologies are being transferred into industry from the laboratories. Research projects like the Superconducting Super Collider (SSC) are likely to provide a wealth of spin-off technology that industry can commercialize. Some new technology examples might be:

TECHNOLOGY	POTENTIAL APPLICATIONS
Large Scale Superconducting Magnets	Magnetically Levitated Transportation (MAGLEV), Superconducting Magnetic Energy Storage (SMES), Compact MRI, Compact Accelerators and Gantries
Computer Control System	Improved Factory Automation, Diagnostics
Radiation Hardened Electronics	Improved Nuclear Power Diagnostics
Remote Handling of Targets	Fusion Power Development

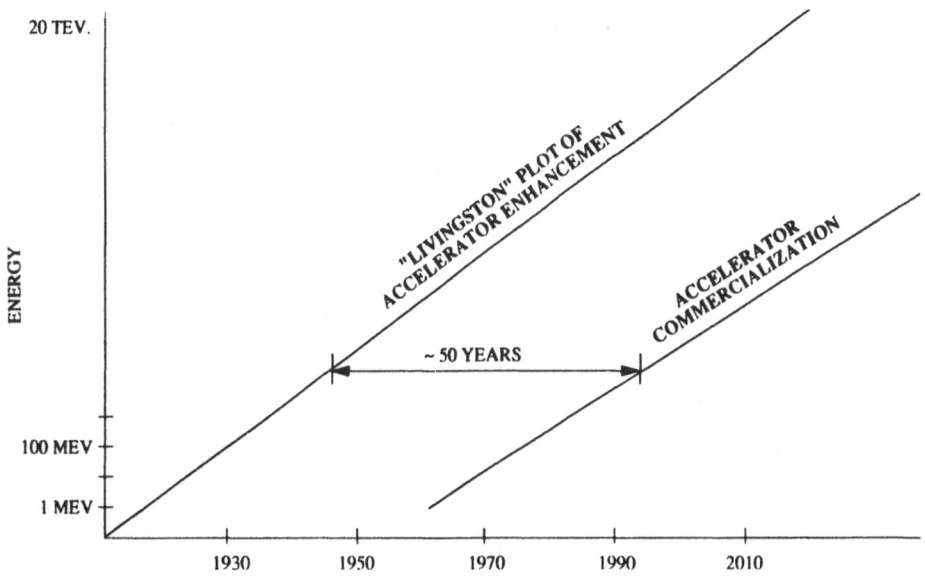

Figure 6.   Lag Time of Accelerator Technology Development To Commercialization

CONCLUSIONS

The medical accelerator market is roughly 160 million dollars per year in the U.S. and one quarter of a billion dollars per year world wide and expanding. A continuing development of accelerator and related technologies on laboratory projects like the SSC, will provide ample technology for industry to exploit. New and innovative ways to shorten the period from laboratory development of a technology and its transfer to industry are needed. The investment we, through the Government, have made in laboratory projects like the SSC, is likely to return many fold, in the exploiting of commercial applications of technology to address many of the needs of society.

REFERENCES

1.   Susan N. Nathanson, and Lan Lerman, "Outpatient Cancer Centers Implementation and Management", American Hospital Publishing, Inc. (1988).
2.   Dr. James Slater, Loma Linda University Medical Center, Proton Beam Therapy Cancer Center, private discussion.
3.   Milton A. Rothman, The Laws of Physics, Fawcett World Library, (1963).
4.   Waldeman Scharf, English Version Edited by Francis T. Cole, Particle Accelerators and Their Uses, Harwood Academic Publishers, (1986).
5.   National Geographic, Volume 171, No. 1, Medicine's New Vision, (January 1987).
6.   ibid 2.
7.   Janet Sisterson, Editor, "Particles Newsletter", Sponsored by the Proton therapy Cooperative Group.
8.   Arlene J. Lennox, "Hospital-Based Proton Linear Accelerator for Particle Therapy and Radioisotope Production", Fermilab PUB-90/217.
9.   Robert W. Hamm, "Commercial Applications of Linacs", Proc. 1990 LINAC Conference.
10.   William K. Hagan, et al., "A Helium 3 Doubly Charged Radio-Frequency Quadrupole Accelerator for Positron Emission Tomography", ANS Nuclear Technology, Vol. 92, (Oct. 1990).
11.   Rolf F. Barth, et al., "Boron Neutron Capture Therapy For Cancer", Scientific American, (October 1990). Also REF. 10.

12.   A.C. Thompson, et al., "Imaging of Coronary Arteries Using Synchrotron Radiation", <u>Proc. Nuclear Instruments And Methods In Physics Research</u>, B40/41 (1989).

13.   U.S. Department of Energy, Report to Congress, Federal Laboratory Technology Transfer Activities, "Accomplishments and Plans", As required by the National Competitiveness Technology Transfer Act of 1989, (February 1991).

SMES - A POWERFUL APPROACH

TO ENERGY CONSERVATION

B. James Lowe
and Robert Scully

EBASCO Services Incorporated
Two World Trade Center
New York, NY

INTRODUCTION

In 1987, the Strategic Defense Initiative Office (SDIO) initiated competitive Phase I contracts for the conceptual definition of a Superconducting Magnetic Energy Storage - Engineering Test Model (SMES-ETM) program. These contracts were conducted through the contract auspices of the Defense Nuclear Agency (DNA). The funding for the 2-1/2 year Phase I effort was approximately $17 million for two contractor teams. The Electrical Power Research Institute (EPRI) co-sponsored approximately $1 million to each of the contractor teams. The announced intention of the government was to choose one of the competitive contractor teams to accomplish a design-build of an Engineering Test Model (ETM) in a Phase II effort that would commence in 1991, which would extend for approximately four years to be "on-line" for testing in 1995.

SMES, as an energy storage technology for the electrical utility industry, has been under extensive development at the University of Wisconsin for almost two decades. Under funding by the Wisconsin Utilities Electric Research Foundation, Roger W. Boom, G. E. McIntosh, Harold A. Peterson and W. C. Young in 1973 [1] first investigated and proposed the potential of direct storage electricity in a dc-solenoid which could be alternately charged and discharged through a three-phase Graetz bridge. Additional studies were conducted by the United States Department of Energy (DOE) and EPRI in the early 1980's. A small scale SMES unit, developed under DOE was operated in the State of Washington under the auspices of the Bonneville Power Administration (BPA) [2] in the early 1980's. This unit was used to demonstrate SMES systems' potential for use in stabilization against power oscillations in the Pacific HVAC intertie system. While the BPA tests provided significant test data, the unit operated only a brief time due to cryogenic/refrigeration system problems.

Starting in 1985, at the University of Wisconsin - Madison campus - another SMES system has operated on-line with the Madison Gas and Electric utility system for a period of over six years. This system operation which has a maximum stored energy capacity of 250 kilojoules, has been directed by Dr. James Skiles [3], and it has been singularly successful in studying, analyzing and developing power convertor technology.

SDIO's awareness of these programs made SMES of potential interest as a reliable pulse power source for directed energy weapon systems. As indicated by Verga [4], a senior executive of the SDIO Directorate, the features of SMES which prompted considerable interest by the SDIO, included the projected capability of system rapid response, high conversion efficiency, high reliability and projected life cycle costs.

As any usage of the terrestrial SMES system for defense system pulse power applications also required that it be used in conjunction with a major electric power generating unit, the DOD recognized that any conceptual design should also embrace both defense system applications and the electric utility industry needs. The government recognized early the benefits of these dual needs. The DNA Contract Statement of Work for Phase I stated explicitly that the conceptual design study should embrace the requirements for both SDI defense applications as well as commercial electrical power needs. As stated in the DNA RFP, the mission was two-fold:

"1.   To develop a power source to drive SDI ground based laser systems (GBFEL*)

2.   To assess the feasibility of using the same power source to perform a load-leveling function for commercial electrical utilities"

One of the Phase I competitive design studies was contracted to EBASCO Services Incorporated and its subcontract team members. They included:

o   Chicago Bridge & Iron

o   Teledyne - Wah Chang Albany

o   Westinghouse Electric Corporation

o   University of Wisconsin - Madison Campus - Applied Superconductivity Laboratory

o   Central Wisconsin Development Corporation

o   El Paso Electric

It is the purpose of this paper to provide a review and synopsis of the Phase I conceptual study that was developed by the EBASCO team under DNA contract. It is also intended that the critical design factors that led to the design be discussed and that the critical experiments that qualify the design selection be reviewed.

CONTRACT TECHNICAL REQUIREMENTS

The contract technical requirements for the SMES-ETM stated:

"...the contractor shall review the requirements for a full scale SMES, for use by both SDI and commercial electrical power utilities, establish technical criteria and test objectives...develop a conceptual design for an ETM which will economically demonstrate the efficacy of...a full scale SMES system. The output parameters must fall within the following range of values;

* Ground Based Free Electron Laser

a. SDI ground based laser power

    1. Less than 2 second rise time

    2. 0.4 - 1.0 gegawatt peak output maintained for 100 sec burst

b. Utility applications for power applications

    1. 10 - 25 megawatts power output

    2. 2 - 3 hours of storage capacity"

Another key aspect of the conceptual design study was that the contractor should address as necesary in a component development and testing program all those materials and components which would assure that there were no "show stoppers" that would preclude successful demonstration of the ETM.

SMES SYSTEM GENERAL DESCRIPTION/DISCUSSION

A SMES system operates on the basis that the energy is stored in a magnetic field that is produced by the introduction of a dc current into a superconducting dc coil or solenoid. For effective use in its intended mission applications, the dc current introduced into the coil is charged by a utility ac energy source which is converted to dc through a power conversion unit, which is an integral part of the SMES total system. On discharge of the SMES coil, the dc current from the coil is converted, through the power conversion, to ac for its specific SDI mission need or for utility application.

While the technology on a laboratory scale (both the 30 MJ unit in the Western Power Grid and the University of Wisconsin utility interface unit [2, 3]) can and has been demonstrated, the demonstration of SMES on the size and scale that can be technically significant and commercially valuable is a daunting engineering problem. The problems relate to the solution of the containment of the radially directed Lorentz forces, the delivery of the requisite power requirements and the effective integration of all other SMES system components and elements within the stress, temperature and material criteria required of a cost-effective and reliable large scale SMES systems. On a large scale SMES unit, the various criteria and their design integration is uniquely different than those normally involved with most superconducting magnets.

The EBASCO/SMES design team proposed a superconducting coil that would be made of a low temperature niobium-titanium superconductor contained in a superfluid helium pool bath. It proposed that the coil design be rippled in order to reduce conductor stabilizer stress and that epoxy-fiberglass struts would be used to transmit the Lorentz induced forces from the helium containment (the cold side) to the vacuum vessel (the warm side) structure. While this design was similar in nature to that proposed by R. W. Boom, et al. [1], the conceptual design proposed by the EBASCO team went beyond that early study in maximizing the effectiveness and economy of the design by using a multiple-layer lower-current coil design and by using a coil design that employed a constant tension, low-bending ripple shape. Major design development was achieved in designing an efficient support system to transfer the loads from the superconducting coil through the helium vessel to the vacuum vessel and then to the surrounding physical geographic site.

This concept design development and analysis extended to the following critical SMES sub-systems:

- o coil/conductor

- o superconducting bus and leads

- o containment and civil support structure

- o refrigeration systems

- o vacuum system

- o power conditioning system

- o monitoring and control.

With respect to these systems, it was the firm objective of the EBASCO/SMES team that the Phase I design concept was to be presented to the government for evaluation as a "closed design" and that all sub-systems of that concept be "internally consistent". By these terms, it is meant to be understood that the design had been established through iterative processes to meet functional and mission objectives; to meet acceptable design and material standards; that all high risk or critical components had been validated by analysis and test; and that all design interfaces between SMES sub-systems had been reviewed and accepted for interface compatibility.

OVERALL DESIGN APPROACH/CRITICAL PARAMETERS

A number of preliminary design studies and analyses indicated that total system issues could be addressed as a group of sub-system problem sets/topical areas. They included:

- o coil/conductor design

- o SMES geological foundation design

- o intermediate sub-system functional/structural design elements.

The coil is the center of the ETM and it is the focal point of major system design decisions on design current, conductor configuration, electromagnetic forces, support structure, coil winding scheme and voltage considerations.

The ETM coil (or solenoid), to meet specification requirements, was designed to store 21 MWh. A residual energy of 1 MWh is retained in the coil at all times since it is impractical to discharge the coil to zero energy level.

The energy of the ETM solenoids is related to the current and the inductance through the equation $E = 1/2 \, LI^2$. The inductance is a function of the coil geometry and involves a number of parameters including the number of radial coil layers, number of turns per unit height, the coil diameter, the Beta ratio (i.e. - height to diameter of the coil). The operating current of the coil is also based on a number of considerations including conductor design, critical current densities, operating voltages and conductor fabrication and installation limitations.

Trade-off studies and cost evaluations were made for a variety of parameters. These included: single layer, two layer and four layer coils; various vertical turn-to-turn separations of the conductors in a layer; for a variety of Beta ratios and coil diameters; and for different radial pressures resulting from the Lorentz forces.

The extensive trade-off studies converged on a four layer design with a Beta ratio of 0.03 with a diameter of 134 meters and a height of 4.1 meters. The inductance with this configuration is 61.2 henrys which, with a normal operating current of 50 KA, gives a stored energy slightly greater than 21 MWh.

The conceptual design is a four layer coil with 104 turns per layer. Each radial layer is separated into a top section of 52 turns and a bottom section of 52 turns. This provides for a modular construction which, as explained by J. R. Logan and D. T. Hackworth [5], could be scaled to higher energy levels as well as permitting the option for isolation of coil modules for reduced power operations.

Figure 1 shows a cross section of the coil/coil structure. In a fully charged coil the axial Lorentz forces are directed downwards in the top section and upwards in the bottom section; the radial Lorentz forces are directed outwards in the two inner layers and inward on the outer two coils. The coils are installed in an aluminum plate which is part of a total coil aluminum support structure.

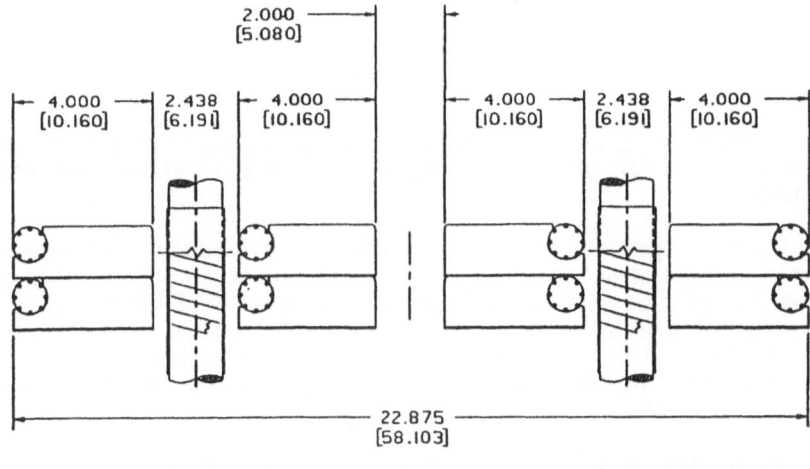

Dimensions in inches
[cm]

Figure 1. Coil Cross Section

The conductor consists of eight superconducting strands helically wound around a high purity aluminum (HPAC) stabilizer. The strands are soldered into the stabilizer slots. A cross section of the conductor is shown in Figure 2. The superconducting strands are composed of niobium titanium (NbTi) alloy filaments in a copper matrix. The round conductor is capable of manufacture. Its selected one inch diameter size also facilitated the ability for quality manufacturing of the conductor off-site and being able to spool the conductor for installation and winding in the field. This permitted de-coupling of the conductor manufacturing from the conductor on-site installation requirements.

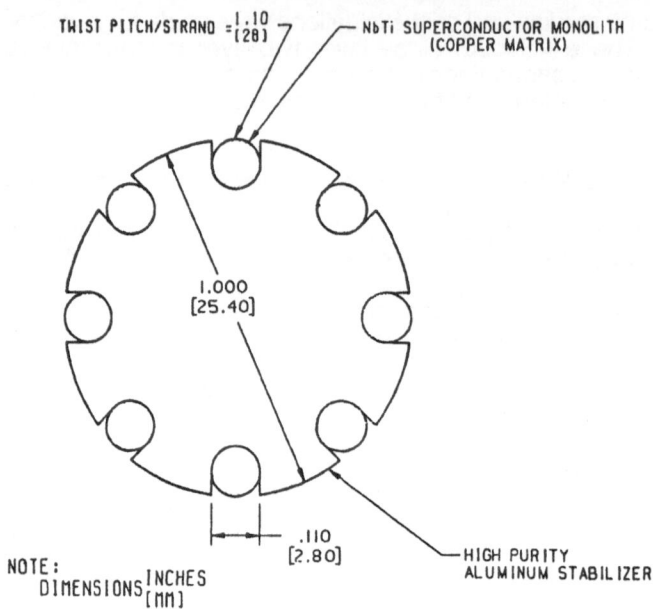

TWIST PITCH/STRAND $=\frac{1.10}{(28)}$

NbTi SUPERCONDUCTOR MONOLITH
(COPPER MATRIX)

1.000
[25.40]

.110
[2.80]

HIGH PURITY
ALUMINUM STABILIZER

NOTE:
DIMENSIONS $\frac{INCHES}{[MM]}$

Figure 2. Conductor Cross Section

In this coil configuration, the maximum magnetic mid-plane field is 3.3 tesla and the maximum ETM magnetic field occurring at the top and bottom inner coils is 4.8 tesla.

Figure 3 depicts the performance criteria of the conductor strand current versus the maximum magnetic field. The total conductor current is the sum of the currents in eight superconducting strands that comprise the ETM conductor. As Figure 3 indicates, a strand current of 6250 KA (total conductor current of 50 KA) is 50% of the critical current at 4.8 tesla, or 66% of the 100% load line. For considerations of dynamic stability of the conductor, the base concept design was limited to 50 KA.

The selection of the conductor was a most critical problem. While manufacturability and the facility of installation were necessary, the performance of the conductor in meeting stability conditions for reliable SMES performance was essential.

Two types of stability must be examined: transient stability and dynamic stability. As explained by I. Eyssa, X. Huang and M. Hilal [6,7,8] in a series of papers, transient stability refers to the recovery of the superconducting state following the generation of a normal conducting zone by some external source of heat. Failure of the conductor to recover the superconducting state would initiate a quench. Dynamic stability refers to the generation of a traveling normal zone, i.e. a constant length normal zone propagating the length of the conductor while recovery of the superconductive state occurs behind the zone. This type of instability does not cause a magnet quench, however it would be a constant source of heat input to the helium bath. Dynamic stability considerations impose more stringent requirements on the conductor design than transient stability.

The dynamic stability of the conductor is parametrically dependent upon the RRR (Residual Resistivity Ratio) of the copper in the strand, the RRR of the aluminum stabilizer, conductor cooling, the solder bond between the strand and stabilizer and the current sharing temperature. Various numerical stability analyses[5] and transport current measurements were made to confirm the conductor selection. Implicit in the dynamic stability studies was the design requirement that the conductor RRR (i.e. - strands and the HPAL stabilizer) maintain a minimum RRR under the cyclic loading imposed by the imposition and dissolution of Lorentz forces during SMES charge and discharge.

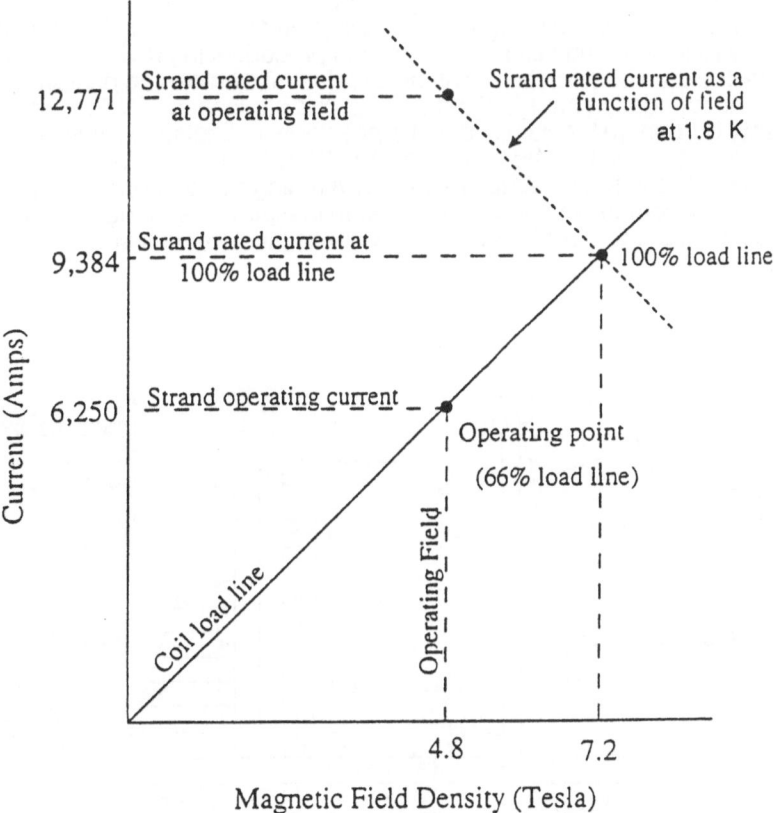

Figure 3. Strand Current vs. ETM Magnetic Field Density

In order to minimize the hoop stresses that would be incurred in a circular coil, which is restrained against radial movements, the coil shape was rippled. The election of the most optimum coil ripple is discussed by X. Huang et al.[9]. This shape was designated a "Constant Tension Ripple" since it minimized overall stress during the conditions of initial vacuum loading of the containment, cool-down to super critical helium, and finally Lorentz loading under full energy storage. The average conductor strain with this design was less than .03%. Strains of this magnitude have negligible impact on RRR degradation.

Further amplification of the coil structure design and the analyses regarding optimum radial strut spacing is presented by J. T. Dederer, et al. [10].

As discussed earlier in the overall design approach, another most critical element of the SMES design was to elect a geological foundation design that would contain the radial Lorentz forces of the coil under full energy storage conditions. The foundation must be capable of meeting both radial load conditions as well as assuring that radial deflections are not in excess of those permissible for minimization of conductor stresses. Moreover, the foundation design should be generic in its ability to meet load conditions in varied soil and/or rock conditions throughout the United States.

The geological foundation design selected was that of a "gravity foundation". It is depicted in Figure 4. As explained by K. Ramachandra, et al. [11], the gravity foundation had established design precedent in previous reservoir/hydroelectric dam construction. The gravity dam shown can be partially above grade. The foundation base is sized to meet civil engineering friction criteria and the gravity burden on the dam averts the possibility of tipping. A gratuitous advantage is found in that the spoils from the SMES entrenchment can be used as a gravity burden for the structure thereby averting the environmental problem of spoils displacement and handling. As seen in Figure 4, an inner wall is required to contain vacuum and cool-down forces prior to the commencement of coil energization.

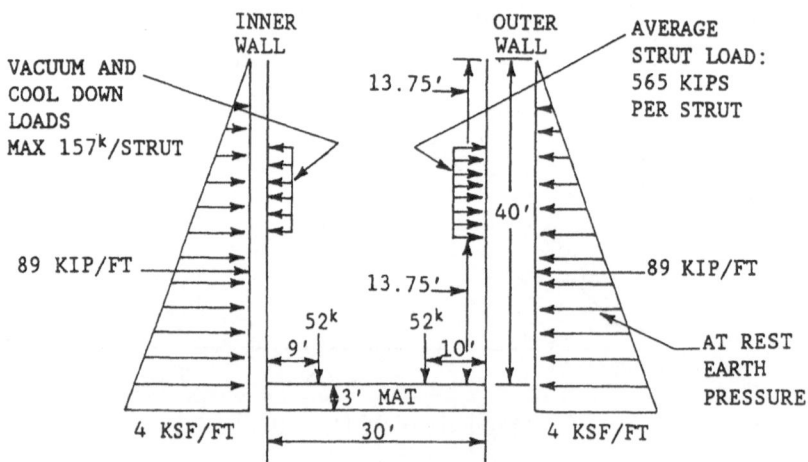

Figure 4. ETM Applied Loads

Parametric considerations of various beta factors and the costs of coil structure and containment versus material criteria for struts and foundations, had led to the evaluation that radial loads into the geological structure would be of the order of 8-12 atmospheres. Gravity type structures and geological siting indicated that the upper limit of 12 atmospheres could be met and still be consistent with the requirement that the remainder of the SMES structure and the gravity foundation would insure that cyclic strains in the conductor not exceed .03 percent.

With the two driving design elements being the conductor/coil structure and the civil support structure, the remaining problem sets related to the aforementioned intermediate sub-system functional and structural design elements. These were not trivial. They included consideration of:

o   helium containment vessel

o   cool-to-warm struts design

o   vacuum vessel design

o   intermediate nitrogen shrouds for thermal intercept

o   design of finger plates between coil and helium vessel to meet both structural criteria imposed by Lorentz forces and standoff distance to prevent voltage breakdown.

All of these conditions were met and extensive testing was accomplished to verify structural strength, dielectric characteristics, cryogenic compatibility and thermal isolation capabilities of the various components.

Figure 5 provides an isometric view of the coil, helium vessel radial support struts and the vacuum vessel final design concept.

Figure 6 depicts a cross-section of the SMES system as it might be installed in a civil engineering structure.

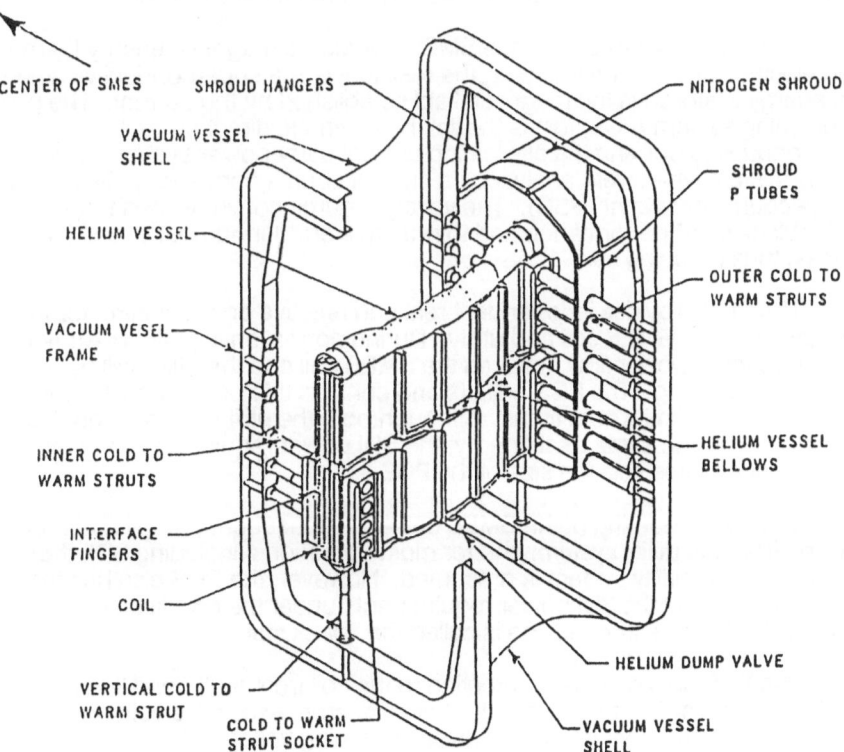

Figure 5. ETM Cross-Section/Isometric Design Concept Recommendation

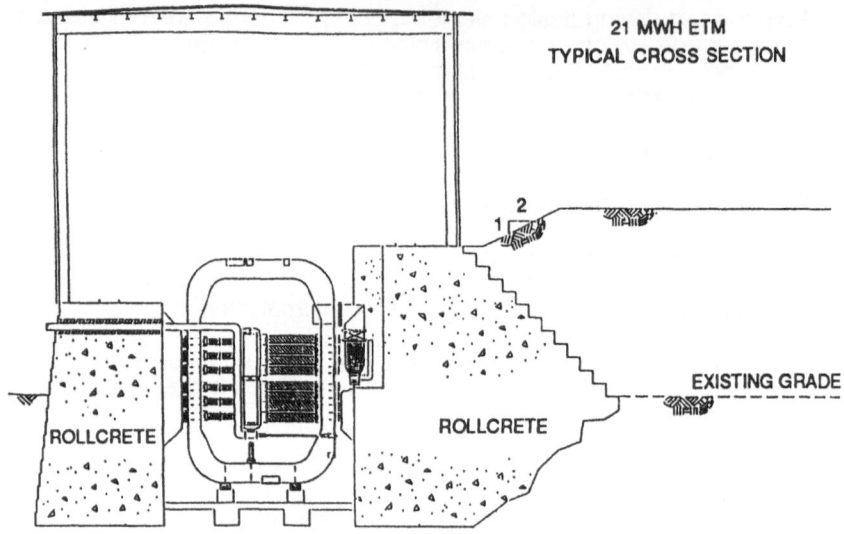

Figure 6. SMES Concept Design in Gravity-Type Foundation

One of the remaining key elements in the system is the Power Conditioning System (PCS). This system is essential to the SMES performance requirements but it does not impact the design characteristics of the SMES device.

As previously discussed, the SMES coil stores magnetic energy by means of an electric current in an inductor, the resistance free superconductive solenoid. This energy is stored in the magnetic field established by the current. The power conditioning system (PCS) forms the link between the direct current superconducting coil and the alternating current utility power system. Two generic system types were analyzed: a current source converter (CSC) and a voltage source converter (VSC). The voltage source converter was found to offer better "black start" performance and to require a smaller number of solid state devices, thus reducing cost.

The PCS provides independent real and reactive power control for the interface between SMES coil and utility. During normal anticipated operation both real and reactive power flow between the SMES coil and the utility will be constantly changing. The PCS adjusts and controls this power transfer within utility determined limits at all times. Although both the utility and second SDI scenario deliver the same total energy, the SDI power requirements are many times larger and this sets the size of the PCS.

The initial charging of the SMES coil and subsequent energy additions will be done from the utility system, and for most operations, including SDI, the availability of the utility system is presumed. However, the PCS also has the capability to supply the SDI power requirements under the condition that there is no utility interface. This condition is called the "black start".

The PCS also provides protection to the coil from fault conditions such as utility line-to-line or line-to-ground short circuits; from excessive voltages due to lighting strikes; and from internal transient switching voltages.

In keeping with the modularized concept for the ETM, the PCS is comprised of two identical 200 MW PCS units in parallel, one for each coil module. Each PCS unit is again subdivided into two 100 MW units connected in series on the dc side, providing the required 400 MW capacity for SDI operations.

In all ETM operations, both coil modules must be maintained at the same current level and charged and discharged at the same rate. For the utility operation requirement each PCS would deliver 5 MW, utilizing one of the 100 MW units of each PCS.

Extensive literature exists on power conditioning systems. As explained by R. Lasseter [12], a comparison of two systems is made with relevance to the SMES system. For reasons related to both control of harmonics and reduced costs, the EBASCO team elected the use of the voltage source inverter approach.

COMPONENT TESTING AND DEVELOPMENT

Approximately 50% of the contract monies received by the EBASCO team were spent on the testing and qualification of SMES components, materials and sub-system elements. This included:

o   conductor manufacturing qualification

o   qualification of radial struts

o   rollcrete qualification for the civil foundation

o   welding processes for the helium containment vessel

o   dielectric/voltage breakdown test of gaseous, liquid and superfluid helium

o   qualification of aluminum structure for coil support

o   residual resistivity ratio testing of strands and conductor at all stages of the manufacturing and installation cycle

o   testing of all coil constraint mechanisms for both dielectric characteristics and structural strength.

Perhaps the most significant test was the Proof of Principle Experiment (POPE) which was conducted at the University of Wisconsin - Applied Superconductivity Laboratory. As explained by J. Pfotenhauer et al. [13], the content of the POPE testing effort was to test a prototype SMES conductor. The test facility exactly duplicated the electric, magnetic and thermal conditions expected for the SMES ETM conductor.

The significant results from the POPE test confirmed the predicted performance under both transient and dynamic stability conditions. The Proof of Principle Experiment thus provided one of the most significant test validations. That is, the selected conductor, manufactured to identical characteristics of the prototype ETM, would be dependable for ETM operating conditions and that normal zone propagation would not occur.

SMES BENEFITS

Besides the obvious usages of pulse power for defense applications, the 2-1/2 year SMES concept definition study provided the opportunity for an extensive analyses of the benefits that could be realized by utilities in the use of SMES. These extended far beyond the RFP Statement of Work which looked primarily at diurnal load leveling. The benefits now recognized include:

o   system/grid stability

o   spinning reserve

o   peak capacity value

o   black start capabilities

o   ramping/load leveling

Studies by S. M. Schoenung [14] et al. have indicated that these savings are significant. For example, it is indicated that the use of a 20MWh capacity SMES with a 400 MW pulse capability could result in a 5 to 10 million dollar annual savings for a 10,000 MW size utility.

The DOE study [15] (Contract DE-AC06-76RLO-1830) indicates that potential energy storage of 10 to 15% of United States generating capacity could be merited and on-line by the year 2010. It is estimated that 4-6% of the total capacity could be provided by SMES type systems. Based on projected total U.S. generating capacity at the turn of the century of 750 giga watts (GW) (750,000,000,000 watts), SMES energy storage savings could approach 30 GW of total capacity. This is approximately twice the U.S. total capacity of existing pumped hydroelectric storage.

Most significantly, SMES units would have almost minimal to zero environmental impact. There are no smoke affluents, acid rain consequences or combustion/radiation by-products.

Its favorable impact on energy costs will be contributory to enhanced U.S. competitiveness on the international scene as well as reducing the need for petroleum imports.

SMES has also received special consideration in the 1991 Defense Appropriations and Authorization Conference as well as recognition in the Nunn/Warner DOD Environmental R&D Initiative of 1990.

CONCLUSIONS

The extensive SMES-ETM concept definition study sponsored by the DOD has resulted in a penetrating analysis of the design and performance characteristics of a recommended prototype design.

Extensive testing of critical components and materials have validated the key elements of a SMES that would be composed of a low-temperature niobium-titanium superconductor contained in a superfluid helium bath. No obstacles to the successful construction and demonstration of a prototype SMES-engineering were defined or anticipated.

References

1.   R. W. Boom, G. E. McIntosh, H. A. Peterson and W. C. Young, "Superconducting Energy Storage" in Advances in Cryogenic Engineering, Vol. 10, Plenum Publishing Corporation (1973), pp 117-126

2.   J. D. Rogers, R. I. Schermer, B. L. Miller and J. F. Hauer, "30 MJ Superconducting Magnetic Energy Storage System for Electric Utility Transmission Stabilization", Proceedings IEEE, Vol. 71, No. 9, pp 1009-1107, September 1983

3.   J. J. Skiles, "Experimental Experience with a Superconducting Magnetic Energy Storage (SMES) Device on Line", Proceedings of the American Power Conference, April 24-26, 1989

4.   R. L. Verga, CEC Conference, UCLA, August, 1989

5.   J. R. Logan and D. T. Hackworth, "Modular Winding Patterns for SMES Coils", Applied Superconductivity Conference, September, 1990

6.   X. Huang and Y. M. Eyssa, "Stability of Large Composite Superconductors", IEEE Pulsed Power Conference - 1991

7.   X. Huang, Y. M. Eyssa and M. A. Hilal, "Dynamic and Transient Stability of He II Cooled Conductors", in Advances in Cryogenic Engineering, Volume 35, Part A

8.   Y. M. Eyssa, "Conductor and Vessel Losses in Bath Cooled SMES Systems", Applied Superconductivity Conference, September 1990

9.   X. Huang, "Low Bending Rippled Structure Design and Frictional Energy Disturbance Analysis for Superconductive Magnetic Energy Storage", Applied Superconductivity Conference, September, 1990

10. J. T. Dederer, R. J. Hillenbrand and D. T. Hackworth, "Structural Considerations and Analysis Results for a Large Superconducting Magnetic Energy Storage Device", IEEE Pulsed Power Conference - 1991

11. K. Ramachandra, J. L. Ehasz, B. J. Lowe, "Ground Support for SMES Systems", Applied Superconductivity Conference, September 1990

12. R. H. Lasseter, S. G. Jalali, "Dynamic Response of Power Conditioning Systems for Superconductive Magnetic Energy Storage", IEEE Pulsed Power Conference - 1991

13. J. M Pfotenhauer, M. K. Abdelsalam, F. Bodker, D. Huttleston, Z. Jiang, O. D. Lokken, D. Scherbarth, B. Tao and D. Yu, "Test Results From the SMES Proof of Principle Experiment", IEEE Pulsed Power Conference - 1991

14. S. M. Schoenung, R. C. Ender, T. E. Walsh, "Utility Benefits of Superconducting Magnetic Energy Storage", Proceedings of the American Power Conference, April 24-26, 1989

15. DOD Study (Contract DE-AC06-76RLO-1830)

10. Superconductors

SUMMARY OF THE PERFORMANCE OF STRAND
PRODUCED FOR THE 1990 SSC DIPOLE PROGRAM

D. Christopherson, D. Capone II, J. Seuntjens, D. Pollock,
and C. R. Hannaford

Magnet Division
Superconducting Super Collider Laboratory*
2550 Beckleymeade Avenue
Dallas, TX 75237

Abstract: In 1990 and at the beginning of 1991, more than 4 million feet of
wire was delivered to support the SSC Dipole Program. This wire was
fabricated to meet specification SSC-MAG-M-4141, and test results and
various statistics are compiled here. Certain strengths and weaknesses in the
performance of the delivered strand are discussed, including analysis of
strand breakage in certain billets. Test results of cable manufactured for 40
mm dipole magnets and 50 mm dipole magnets are reported, and a brief
overview of the 1991-1992 Conductor Program is included.

INTRODUCTION

This work is a compilation of test data and analysis results on superconducting wire
delivered during the past year. The focus is on material that has been fabricated into cable and
used for the SSCL Magnet Program. The billet numbers used here are chosen strictly for the
purpose of vendor neutrality and have noy significance beyond this. Table 1 is a list of the
billets included in this paper and the basic statistics of each. Unless otherwise stated, the
reported values are derived from the raw data from the vendors' measurements. A companion
paper presented by Erdmann et al.[1] addresses critical current (Ic) measurement variation
between different facilities.

BILLET PERFORMANCE DATA—MECHANICAL

As we near the production scale-up period, parameters such as piece length and yield
become very important issues. Historically, manufacturing performance in these areas has
been inconsistent. However, progress has been made over the last year, and we have seen
some promising results from the 13 batches of material covered here. Figure 1 illustrates the
piece-length performance of each batch of material against the SSC-MAG-M-4141
requirement of 90% of the order delivered in lengths greater than 10,000 feet.

*Operated by the Universities Reasearch Association, Inc., for the U.S. Department of Energy under Contract
No. DE-AC02-89R40486.

Table 1. Basic statistics of billets produced during 1990. Type 1.8, 1.5, and 1.3, strand are for nominal Cu/SC ratios of 1.8:1, 1.5:1, and 1.3:1 respectively.

Billet No.	Billet Type	Total Length ft	%L>10K ft	Yield	Mean I_c	Mean Cu/SC	Alloy Source
1	1.8	333976	26	65	316	1.80	TWCA
2	1.8	382489	77	74	325	1.73	TWCA
3	1.8	371245	62	72	323	1.75	TWCA
4	1.5	236831	29	NA	370	1.52	TWCA
5	1.3	117680	95	NA	362	1.27	TWCA
6	1.3	126021	97	NA	365	1.27	TWCA
7	1.5	100773	60	NA	330	1.48	TWCA
8	1.5	166931	66	NA	333	1.46	TWCA
9	1.8	372346	87	72	311	1.80	Cabot
10	1.8	370859	45	72	313	1.79	Cabot
11	1.8	396047	85	77	306	1.79	Cabot
12	1.8	395765	93	77	311	1.79	Cabot
13	1.8	350699	93	68	309	1.79	Cabot

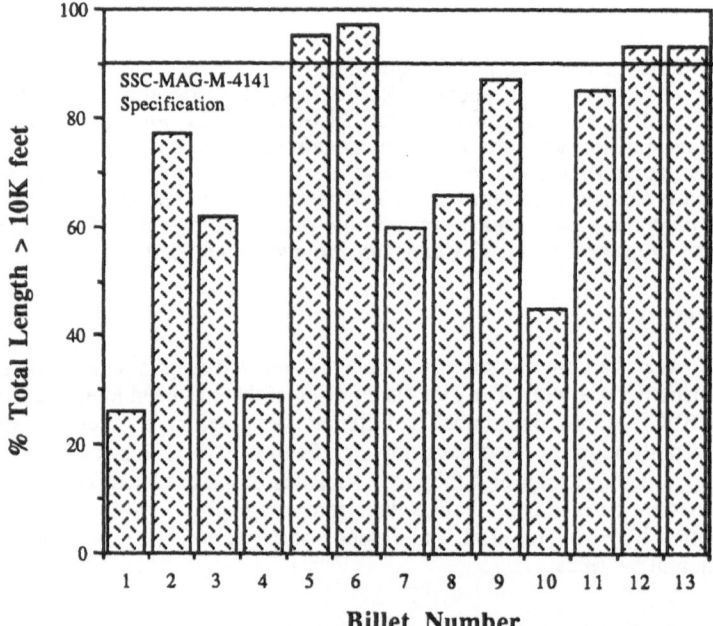

Figure 1. Piece-length performance of billets produced during 1990. SSC-MAG-M-4141 specifies that 90% of total delivery be in pieces greater than 10,000 ft.

Material with poor piece length and yield has been examined; in some cases reasons for this performance have been identified. Batches 7 and 8 were initially designed to be Type 1.3 strand, and they represent the second half of billets 5 and 6. At the request of the SSCL, this material was delivered to meet the Type 1.5 specifications by skipping an important occlusion-removing step in the vendor's normal process. The short lengths of billet 10 were attributed in part to a machine problem during the drawing of this material. The extremely poor piece lengths of billets 1 and 4 were determined to be due to inclusion problems by wire break analysis performed at the SSCL. Billets 2 and 3 were of marginal piece length and have also been proven to contain inclusions that limited their drawability. Discussion of this break analysis work is included later in this paper.

An excessive number of wire breaks can substantially reduce the amount of final strand delivered for an order, showing that piece length is a contributing factor to the yield characteristics of a billet. In many cases, the nature of the R&D contracts governing this work called for extensive sampling and the processing of partial billets, preventing comparable yield statistics. However, SSC Outer billets 1, 2, 3, and 9–13 did not involve such contracts. Multifilament yield discussions will concentrate on these billets.

The multifilament yield was calculated by assuming a 12-in.-diameter billet with a 28 in. (2.33 ft) core length. Using the idea of conservation of volume, we obtain the equation:

$$L_f = (2.33 \text{ ft})\left(\frac{12 \text{ in}}{0.0255 \text{ in}}\right)^2,$$

where L_f is the final length in feet of 100% yield of Outer conductor at the final diameter of 0.0255 in. This equation gives us a 100% theoretical yield of 516,000 ft for an SSC Outer billet. Dividing this number into the total length of delivered wire for a billet gives us the yield illustrated in Figure 2. This is an approximation based on an assumed multifilament billet size and does not account for any pickling or shaving steps. From these calculations, the multifilament billet yields have ranged from 65% to 77% during the last year. Increasing this yield is one of the main goals of development efforts of 1991.

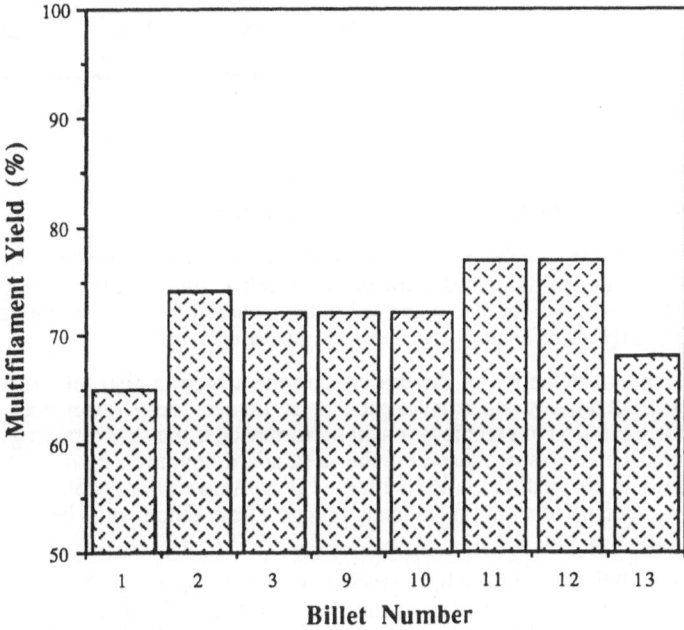

Figure 2.   Multifilament yield statistics. These are approximate values derived from calculations involving an assumed initial multifilament diameter and core length.

Another important mechanical characteristic is the springback of the strand. This tendency of the wire to spring back toward its original shape after deformation is critical to cable fabrication. Strand with a high springback tendency will not lie flat after cabling and will "pop out" along the faces of the cable. This pop-out problem has been observed in cable recently manufactured using R&D strand with springback values greater than the specified limit. Springback is quantified by using a specialized fixture that measures in angular degrees. With a 5 lb. counterweight on one end, the other end of the wire is wound around a spindle for the count of 10 revolutions. The weight is then secured so as not to affect the reading. The spindle is released and allowed to unwind slowly as the wire springs back toward its original shape. Each counter-revolution adds 360 degrees to the value, plus the last partial revolution, for a total springback measurement that cannot exceed 980° for SSC Inner billets and 1090° for SSC Outer billets.

Springback is dependent upon wire processing, heat treatments, and strand geometry, and varies among vendors. All of the material discussed in this work met the specifications. Table 2 shows the springback statistics of the Outer billets.

Table 2. Springback statistics of SSC Outer billets. (SSC-MAG-M-414 requires the spring back to be lesst han 1090°f or SSC Outer material

Billet No.	No. of Tests	Mean°	Min.°	Max.°	Std Dev.°
1	89	825	736	920	31
2	32	820	766	858	23
3	55	805	746	900	28
9	3	797	782	814	16
10	9	835	811	846	11
11	32	840	794	890	25
12	22	820	780	896	26
13	25	806	752	870	28

BILLET PERFORMANCE DATA—ELECTRICAL

Figure 3 shows how the mean Ic of each billet compares with the Ic specification for each type of wire. The mean Ic data show that each billet comfortably meets its respective Ic requirement. However, we are also interested in the standard deviation of the Ic and the slope of the Ic vs. field (H) plot. Table 3 lists these results for each billet.

The standard deviation of the critical current, which is also related to the standard deviation of Cu/SC ratio, is important to cable performance. Large differences in Ic from one strand to another can lead to variations in coil performance if the strand mixing during cabling is not sufficient. For the same reasoning, the Ic standard deviation *between* billets is just as important as the Ic standard deviation within a billet. As we near production stage, we will produce cable made up of strand from many billets. These billets must be nearly identical in all specified variables.

Although billets 9–13 have been processed identically, our database is still too small to be able to draw any firm conclusions about billet to billet variations. However, preliminary indications are promising. The Ic standard deviation between all pieces of billets 9–13 is 5.2 A (1.7% variation from a mean of 311 A), with the Ic ranging from 290 A to 322 A at 5.6 T. Deviations of this magnitude are further narrowed simply by strand mapping the cable by piece length alone, independent of the strand Ic. A strand map of these 5 billets resulted in a cable Ic range of 70 A (0.6%) before degradation.

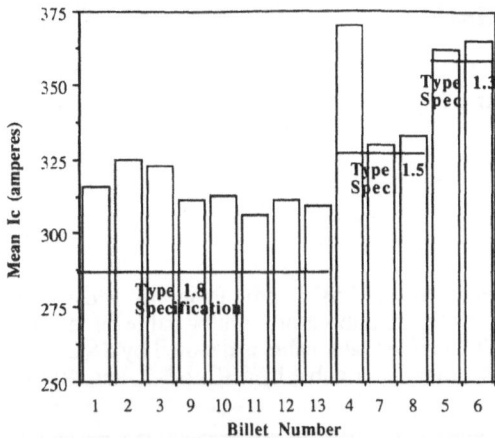

Figure 3. Mean critical current of each billet compared to its respective specification. SSC-MAG-M-4141 requires Type 1.8 material to be 285 A at 5.6 T, Type 1.5 to be 328 A at 7.0 T, and Type 1.3 to be 357 A at 7.0 T.

The slope of the critical current vs. field values reported in Table 3 are calculated from measurements made at Brookhaven National Laboratory (BNL). Ic measurements at 5 T and 6 T on samples of strand determine this slope, which is essentially linear over intermediate fields. The reduction of this slope is important in order to reduce the low field magnetization of the strand at fields near the injection field of 0.6 T. Because magnetization is proportional to Ic, a steeper slope leads to higher magnetization at low fields. For this reason, a flatter Ic vs. H slope, while maintaining acceptable Ic at high fields, is desirable.

Table 3 also lists the mean and standard deviation of the Cu/SC ratio of each billet. All billets meet their respective Cu/SC specifications, and in most cases the standard deviation is small. For reasons of material uniformity, standard deviations of 0.04 and 0.05 seen in a few of these billets are less desirable. With a well-defined and controlled process, Cu/SC ratio standard deviations of 0.02 can be attained with a negligible decrease in material yield. This should become the case as larger lots of material flow though each vendor's production system in the next year.

Table 3. Electrical measurement statistics of billets produced in 1990.

Billet	Mean Ic (A)	STD Dev Ic (A)	Slope Ic vs. H	Mean Cu/SC	STD Dev Cu/SC
1	316	3.5	1.26	1.80	0.02
2	325	6.3	1.25	1.73	0.02
3	323	7.1	1.25	1.75	0.05
4	370	5.7	1.26	1.52	0.03
5	362	7.7	1.24	1.27	0.04
6	365	5.5	1.24	1.27	0.02
7	330	2.5	1.24	1.48	0.04
8	333	2.1	1.24	1.46	0.02
9	311	3.5	1.25	1.80	0.02
10	313	3.4	1.25	1.79	0.02
11	306	4.8	1.25	1.79	0.02
12	311	4.3	1.25	1.79	0.02
13	309	7.4	1.25	1.79	0.04

CABLE PERFORMANCE

Table 4 is a list of the electrical test results from BNL on cable fabricated using strand from the batches mentioned earlier. Descriptions of these test methods and accuracies can be found elsewhere.[2]

Table 4. Cable fabricated for the SSC Dipole Program. Measurements were performed at Brookhaven National Laboratory. Outer cable is tested at 5.6 T, Inner cable is tested at 7.0 T. 50 mm Outer cable is denoted by SSC-4-*-*, 50 mm Inner cable by SSC-3-*-*, 40 mm Outer cable by SSC-2-*-*, and 40 mm Inner cable by SSC-1-*-*.

Cable I.D.	Length (ft)	Ic(A)	Spec. Ic(A)	Degrad. (%)	Billet
SSC-3-I-00021	1060	11024	9990	-4.30	4
SSC-3-I-00022	1120	10827	9990	-2.50	4
SSC-2-I-00023	4031	9294	7860	1.70	2 & 3
SSC-2-I-00024	2310	9470	7860	-0.10	2 & 3
SSC-2-I-00025	1969	11415	10152	-0.50	2 & 3
SSC-4-I-00026	1969	11341	10152	0.10	2 & 3
SSC-4-I-00027	5520	11615	10152	-1.50	2 & 3
SSC-4-I-00028	1320	11075	10152	2.30	2 & 3
SSC-4-I-00029	1160	11075	10152	2.30	2 & 3
SSC-4-I-00030	660	10944	10152	0.10	1 & 3
SSC-4-I-00031	645	10944	10152	0.10	1 & 3
SSC-4-I-00036	1381	10886	10152	0.20	12
SSC-4-I-00037	5568	10339	10152	4.90	9-13
SSC-2-I-00032	4305	8789	7860	4.30	1
SSC-2-I-00033	3428	8953	7860	2.80	1
SSC-2-I-00034	1015	8901	7860	3.20	1
SSC-3-S-00021	2340	10079	9990	-3.70	7 & 8
SSC-3-S-00022	1052	10079	9990	-3.70	7 & 8
SSC-3-S-00023	4200	10079	9990	-3.70	7 & 8
SSC-3-S-00024	1258	10917	NA	-2.20	6

These cables have been manufactured for use in the 40 mm Collider Dipole Program, the 50 mm Model Dipole Program, and the 50 mm Accelerator String Test Program. Each cable meets its respective critical current specification, and in some cases greatly exceeds it. One reason for this is the low cabling degradation values, especially for the 50 mm Inner design cables. This cabling degradation as reported by BNL is calculated by dividing the measured Ic of the cable by the product of the average measured Ic of uncabled strand samples and the number of strands in the cable. The reported degradation is an estimate based on the average strand Ic from a limited number of samples. In addition, the cable measurements are corrected for self field effects while the virgin strand is not. These factors combine to make it possible for the calculated degradation to have a negative value, although realistically it cannot be less than zero. Explanations for the low degradation in the 50 mm cables are being investigated; initial indications are that the width-to-thickness ratio in these cables is improved over that of the 40 mm designs.[3]

DISCUSSIONS

All of this material has been manufactured to meet SSC specification SSC-MAG-M-4141. Material performance comparisons against this specification reveal only one major deficiency: piece length. Historically, billets meeting the piece-length requirement of 90% of delivered pieces longer than 10,000 ft have been the exception rather than the rule.[4] Considering the finished strand diameter and filament size, this is an agressive requirement to meet, but the volume of conductor procurement required to supply the SSC project necessitates long piece lengths. In fact, the specifications to which future material is being ordered, SSC-MAG-M-4145 (Inner strand) and SSC-MAG-M-4146 (Outer strand), are more demanding in terms of piece length. These documents call for the acceptance of only those pieces longer than 1500 m. This can seriously reduce the yield of a billet experiencing drawing difficulties.

Wire Break Analysis

Wire break analysis has been done at the SSCL on several billets experiencing severe piece-length problems. When possible, samples of both ends of the breaks were mounted and polished for metallographic examination. Because of the possibility of losing an important feature of the break during polishing, many repetitions of delicate polishing and microscopic observations were performed on selected samples. Even when exercising great care, there is a danger of polishing away an inclusion or other artifact of the failure. Scanning electron microscopy (SEM) and energy dispersive X-ray analysis (EDX) were performed on samples in which an interesting feature was seen. Approximately 25 samples from various billets were examined for the cause of failure. Only a few are singled out here and discussed at length.

SEM and EDX analysis were performed on two selected breaks of billet 1, one at 0.064 in. diameter and the other at 0.0397 in. diameter. These breaks were similar in appearance, although a cup and cone morphology was more evident in the latter sample. In both specimens, the filament array showed no signs of distortion until very near the failure region. When material was removed in the systematic manner described above, it appeared that the failure originated at a point between the copper jacket and the filament array. In the case of the 0.0397 in. break, several foreign particles were discovered at that location and were analyzed. These inclusions exhibit silicon, manganese, and tin signals. Because our equipment is unable to accurately detect elements having an atomic number less than flourine, we can only conjecture that these particles are oxide or carbide inclusions that were imbedded on the copper can wall during can cleaning or billet packing.

Three of the wire breaks from billet 4 were chosen for similar analysis. The diameters of these breaks were approximately 0.07, 0.064 and 0.058 in. The two smaller diameter samples were very much the same as the breaks described above in billet 1. A failure near the can wall revealed foreign inclusions at the origin. These particles contained Al, Si, and Ca. The 0.07-in.-diameter sample differed; it appeared that the failure originated at the interface between the filament array and the copper core. An inclusion containing silicon was found at the bottom of the cup portion of the fracture.

Wire breaks from the order that includes billets 9–13 show stainless steel inclusions; Figure 4 shows an SEI image and spot X-ray spectrum on one of these inclusions. Although these billets had decent piece-length performance, the presence of these recurring particles suggests that a much better performance is possible. Also included in this order is a series of five Inner billets which are currently being delivered. Preliminary indications show the piece length of these billets to be extremely poor. Wire break analysis results from the vendor and SSCL agree that many stainless steel inclusions, similar to those seen in Figure 4, are present in the billets.

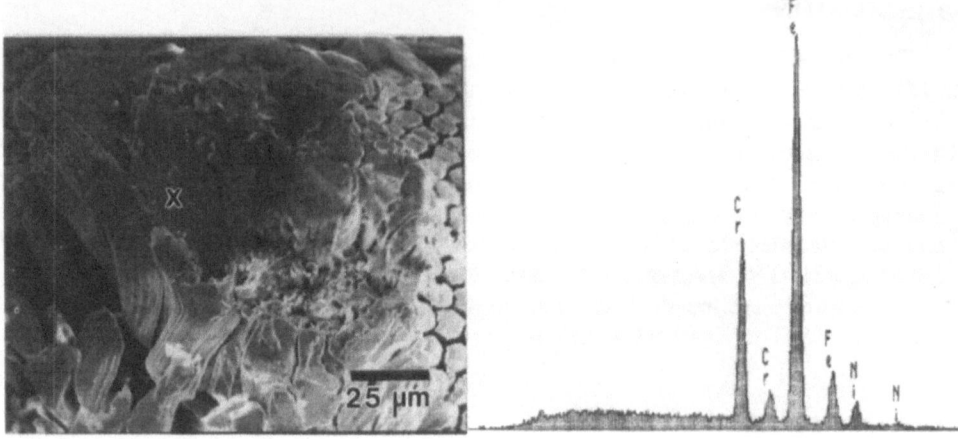

Figure 4. SEM image and EDX spectrum of a stainless steel inclusion. The point of the spot X-ray analysis is marked.

Severe piece-length problems were also experienced in a series of R&D billets separate from those discussed above. Similar analysis techniques were used to determine the cause of these drawing difficulties. Many samples were examined, both at the SSCL and at the vendor's location; all had very similar and peculiar characteristics. Figure 5 shows an SEM image of a break cross-section. The large filament and surrounding distorted array are not the puzzling aspect of this fracture; however the large Nb-rich region covering nearly half of the large filament area led to questions of alloy homogeneity, barrier integrity, and monofilament assembly procedures. EDX showed this region to be essentially pure Nb, which suggests that it may be the diffusion barrier. When analyzed by EDX, the alloy composition of the anomalous filament appeared homogeneous and similar in composition to the surrounding filaments. Deeper polishing revealed that the large filament and barrier actually reduced to normal size and geometry approximately 50 microns away from the fracture area.

Figure 5. SEM image of a strand break cross section. The light colored region is the Nb diffusion barrier.

The vendor's retrospective investigation of the processing of this material revealed a possible set of circumstances which resulted in this material failure. The monofilament anneal was performed at a relatively small diameter in a nitrogen atmosphere. At this size, the copper jacket was thin enough, and the temperature high enough to allow grain boundary diffusion of nitrogen into the niobium barrier, forming locally hard areas. These hard regions caused a filament failure, leading to a strand failure several drawing steps later.

ALLOY SOURCE COMPARISONS

During the past year, drawing difficulties in billets, such as those previously discussed, and observable microstructural differences between alloy from different suppliers have drawn attention to the NbTi alloy stock as a possible concern. So far, we have not found any reason to believe that the alloy is the cause of these difficulties; in fact, we have shown that these piece-length problems most likely originated elsewhere.

Visible "tree rings" are present when some Cabot alloy at a filament size of 200 microns or larger is etched with hydroflouric acid, and line scan EDX analysis on *unetched* surfaces show macro inhomogeneities of ± 1 wt% Ti. This is within the specification window of ± 1.5 wt% Ti. For comparison, line scan EDX of samples from 5.75-in.-diameter ingots of Teledyne material showed *micro* segregation of ± 1wt% Ti. Due to our machine resolution, the reported sizes of these samples are the limit below which this segregation cannot be seen.

A review of billet performance gives no evidence that wire problems or variations can be associated with the product of one specific alloy supplier. Critical current variations between billets 1, 2, and 3, which used Teledyne alloy, and 9–13, which used Cabot alloy, resulted from Cu/SC ratio differences and diameter variations. Furthermore, the performances of billets 9–13 prove that SSC specifications can be achieved with Cabot alloy regardless of the macrosegregration present in the material.

1991–1992 CONDUCTOR DEVELOPMENT PROGRAM

The 1991–1992 Conductor Development Program consists of two phases. Phase I is separated into two parts running in parallel and designated as Phase IA and Phase IB.

Phase IA is a development effort designed by each vendor to improve the manufacturability and production efficiency of his baseline process. Included in the development plan of each vendor is a required investigation of certain parameters dictated by the SSCL. These required parameter studies include piece length, yield, and alternative alloy sources.

Phase IB is a baseline process order for a required amount of finished cable (3400 kg for SSC Inner and 3500 kg for SSC Outer). Each vendor is to use his baseline design to deliver the required amount of conductor to meet SSC Specification SSC-MAG-M-4147 (Inner cable) or SSC-MAG-M-4148 (Outer cable).

Phase II is an order for a required amount of cable (6120 kg for SSC Inner and 6300 kg for SSC Outer), and is to begin when Phases IA and IB are completed. Work on Phase II will not start until the vendor passes a production readiness review. This material is to be manufactured to meet the same specifications, using knowledge gained through Phase I.

At this time there are seven vendors negotiating contracts to participate in this two-year program. This number may decrease depending upon negotiation and performance during the qualification program. We are confident that the SSCL, along with the superconducting cable vendors, will gain the experience and knowledge necessary to supply the Magnet Program with ample high-quality conductor through the duration of the project. In addition, the SSCL is committed to assisting other potential wire vendors who wish to qualify using another funding source.

SUMMARY

The electrical performance of strand produced for the SSC Magnet Program over the past year has been very encouraging. Critical current specifications have been satisfied using alloy from two vendors, and the variations of Ic within a single billet are promising. Likewise, the critical current measurements among 5 billets processed identically had a standard deviation of only 5.2 A at 5.6 T.

However, the piece length and yield statistics of much of this material is a major weakness. The largest factor contributing to these difficulties is cleanliness. The large frequency of inclusions found in the limited number of tested samples gives an idea of the amount of impurities that are present throughout the length of these billets. The need for the improvement of piece length and yield is obvious, and as we move toward supplying conductor for the industry prototype magnets, the 1991–1992 Conductor Development Program will emphasize this requirement.

ACKNOWLEDGEMENTS

The authors would like to thank R. Scanlan, Lawrence Berkeley Laboratory, and M. Garber, Brookhaven National Laboratory, for their input, and D. Harrison and F. Clark for their laboratory assistance.

REFERENCES

1. M. Erdmann, D. Capone II, D. Pollock, L. Goodrich, R. Scanlan, *Quantification of Systematic Error in Ic Testing*, Paper V-F-9, IISSC, Atlanta, GA, March 13-15, 1991.

2. M. Garber and W.B. Sampson, *Quality Control Testing of Cables for Accelerator Magnets*, IISSC, New Orleans, LA, Feb. 8-10, 1989.

3. R. Scanlan, Lawrence Berkeley Laboratory, private communication, February 1991.

4. D. Christopherson, R. Hannaford, R. Remsbottom, M. Garber, R. Scanlan, *Evaluation of SSC Cable Produced for the Model Dipole Program During 1989 and Through February 1990*, IISSC, Miami Beach, FL, March 14-16, 1990.

FILAMENT AND CRITICAL CURRENT DEGRADATIONS

IN EXTRACTED STRANDS OF SSC CABLE

J. M. Seuntjens and D.W. Capone II

Magnet Division
Superconducting Super Collider Laboratory*
2550 Beckleymeade Avenue
Dallas, TX 75237

and

W. H. Warnes

Oregon State University
Corvallis, OR 97331

Abstract: Previous work by Goodrich et al. have shown that virtually all of the voltage in a strand of cable at I_c is concentrated locally at the edges where the damage is severest. We have worked to characterize this local degradation both microstructurally and electromagnetically. Scanning electron microscopy and image analysis have been used to characterize filament defects and uniformity throughout the edge and flat regions of the strand. A summary of this work is given and is correlated to recent electromagnetic results on the distribution of critical currents along the strand.

INTRODUCTION

Rutherford-type cables used for SSC magnets typically suffer a 5 to 10% loss in critical current over the sums of the critical currents of the strands that comprise the cable. The reported amount of degradation is usually quoted between -2 and +5% because a self field correction is used for the cable I_c, but is rarely used for strand measurements. Understanding the nature of the degradation will minimize the degradation and will directly increase the operating margin of accelerator magnets. Previous work by Goodrich and Bray[1] has shown that virtually all of the voltage, at I_c, across an extracted strand from a cable is localized at the narrow and wide edges of the cable. Therefore, it is apparent that the degradation is not uniform throughout the strand. We have attempted to characterize the nonuniformity of

*Operated by the Universities Research Association, Inc., for the U.S. Department of Energy under Contract No. DE-AC02-89ER40486.

filaments along extracted strands of cable as well as the distributions in critical currents. Our early results are presented in this work.

EXPERIMENTAL

A 40 mm outer cable, No. SSC-O-S-00015, was studied in this work. The strand from this cable was made at Supercon, Inc., (billet No. 2071). The reported cabling degradation for this cable was 10.3%. This value is typical for 40 mm cables and compares to 2% for current 50 mm cable production. After the winding, the cable was cut about 1 m from the cabling machine turks-head, the turks-head was opened up, and the cable was backed through the turks-head. The strands were then cut at the spools and labeled. The section of the cable with its virgin strands was removed intact. The cable was dismantled with minimal plastic deformation providing continuous samples of virgin and extracted strand. This procedure allowed for unambiguous identification of each strand by referencing the re-spooling map for the cable.

Metallurgical mounts were made of the virgin strand and of the sections of the extracted strand from the narrow edge, wide edge, and face of the cable. A sharp bend specimen made from a section of virgin strand was also mounted for comparison. Care was taken to ensure that the sections were normal to the wire axis at the section. A sharp bend section of virgin Sample 6 was made to give an extreme case for strand distortion. A section was made normal to the strand axis in the middle of the bend. A JOEL 6100 SEM with a Link Analytical eXL image analysis system was used to determine the relative cross sections of filaments from each section. Specimens were analyzed as polished, with no surface relief of the filaments. Backscatter imaging was found to have insufficient resolution, and the large back-scatter signal from the Nb-Ti filaments made them appear artificially large. All images analyzed were thus created using a 25 KeV primary beam with secondary electron imaging at 750×. Approximately 50 frames were averaged to give the best resolution image. Each filament contained approximately 300–650 pixels to give an accurate filament representation. Approximately 200–300 filaments were analyzed in each image, and 4–5 images were obtained for each sample.

Virgin and extracted strands had their critical current and current distributions measured at Oregon State University. Brass sample holders were used which allowed extended transitions to be measured to several hundred microvolts without excessive heating. The current carried by the barrel was determined from independent tests as a function of field and voltage and was then subtracted from the total current. (It was assumed that the barrel is simply a parallel circuit.) Stable, reproducible, and non-hysteretic currents were measured to the point where voltage was proportional to current (i.e., $V \propto I^n$, where $n = 1$ and no additional super current exists). Measured samples were 30 cm long; in the extracted strand case, this results in the sample having two cable wide edges and two cable narrow edges. Details of the critical current distribution work will be published elsewhere.[2]

RESULTS

Microstructural Uniformity

Two strands of this cable (Nos. 6 and 7) were analyzed in this work. They had reported diameters of 0.0255 in. (0.648 mm) and copper to superconductor ratios (Cu:SC) of 1.71 and 1.67 to 1, respectively. Figure 1 contains schematic cross sections of the virgin, sharp bend, cable broad face, cable wide edge, and cable narrow edge sections of No. 6 traced from photomicrographs. Significant distortion of the filament array can be seen as compared to the bull's-eye-like pattern of the virgin strand. In general, however, the filaments adjust themselves within the array, and little distortion can be seen at the filament level.

A plot of the frequency vs. gray level is made from the stored image during the image analysis. A typical plot is shown in Figure 2, where two distinct peaks are shown, one for the filaments and one for the matrix. A gray level is chosen at the minimum between the two peaks to define the boundary between the filaments (shaded) and the matrix. A binary image is then created with all pixels with gray level below the boundary set to be black to represent the filaments. All remaining pixels are set to white to define the matrix. A representative binary image is shown in Figure 3. The number of filaments in this image is restricted to only about 200 for clear journal reproduction. A variety of statistical analyses are then performed on the filaments, and the pixel statistics are reported here.

Table 1 contains the data from this study, with values from an ideal 6 μm Outer conductor given in the first row. Samples are identified by the following letters: V = virgin strand, VS = virgin sharp-bend, F = cable broad face, W = cable wide edge, and N = cable

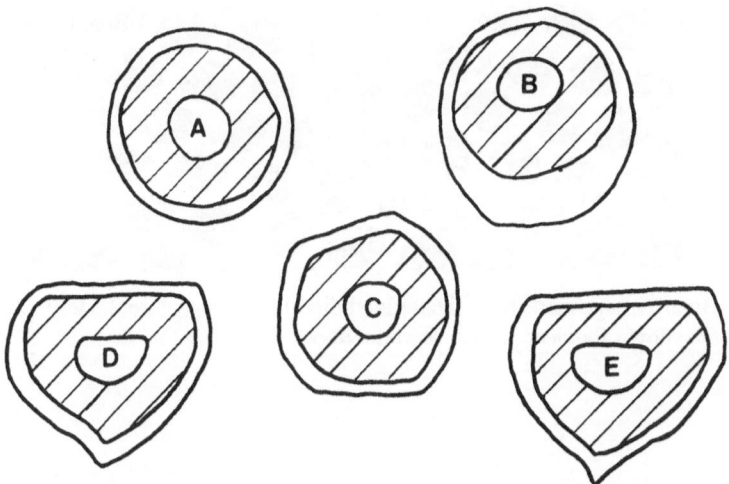

Figure 1. Schematic cross sections of samples in this study showing the strand cross section with filament array shaded. A) virgin strand, B) sharp-bend, C) cable broad face, D) cable wide edge, and E) cable narrow edge.

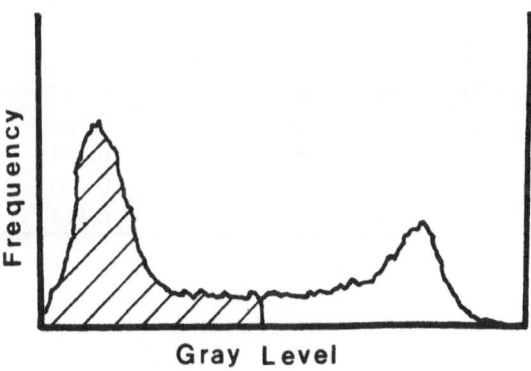

Figure 2. A plot of frequency vs. gray level from a stored image. The minimum in the curve defines the border between the filaments (shaded) and the matrix.

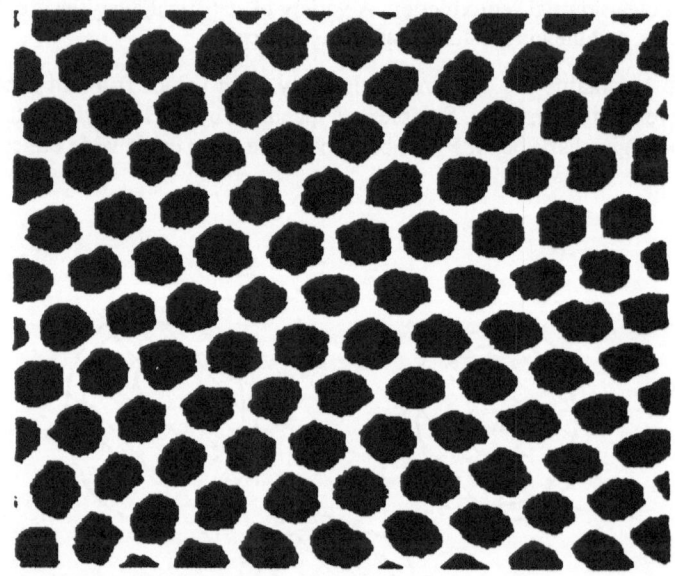

Figure 3. Binary image made from the defined distribution in Figure 2. Only ~200 filaments are shown for clear journal reproduction.

Table 1. Virgin and Cabled Strand Data.

Sample	N	Min	Mean	Max	Std. Dev. (%)	Area (S/C)	Area (Sample)	Cu:SC ratio
(IDEAL)	4173	*	28	*	*	0.118	0.33	1.80
6V	515	20	28	35	9	0.118	0.33	1.80
6VS	1400	13	32	126	57	0.135	0.398	1.95
6F	924	21	29	40	9	0.119	0.329	1.76
6W	1257	21	28	35	8	0.116	0.317	1.73
6N	959	22	27	40	10	0.112	0.309	1.76
7V	585	23	29	38	7	0.122	0.345	1.81
7F	825	22	30	38	8	0.124	0.337	1.72
7W	756	20	29	34	8	0.120	0.337	1.72
7N	944	23	28	37	10	0.118	0.317	1.69

narrow edge. The Min, Mean, and Max columns are the filament cross-section values (in μm^2). Area (SC) is the area of the superconductor (in mm^2) calculated from multiplying the mean filament cross-section by 4173, the number of filaments in an ideal outer conductor. Area (sample) is the measured cross-section of the given section (in mm^2) with the virgin strand calibrated to match the measured 0.0255 in. (0.648 mm) diameter. The Cu:SC ratio is calculated from the Area (SC) and Area (sample) data.

Approximately 12% of the filaments in the virgin strand were surveyed. The mean filament size matches the ideal value of 28 μm^2 for virgin strand No. 6. Thus, the area of superconductor also matches the ideal value, and since the diameter of the strand was measured to be 0.0255, we have calibrated the sample area to be 0.330 mm^2. The Cu:SC ratio is thus 1.80:1, which is slightly higher than the measured value of 1.71:1. In comparison, the sharp bend sample has a larger mean filament size and measured strand cross section. The inside of the bend is at the bottom of "B" in Figure 1, and it can be seen that locally the Cu:SC ratio is increased. The measured value of 1.95:1 is consistent with this observation. The most noteworthy observation is the extreme range of filament cross sections in the sharp-bend. The minimum filament size is almost a factor of 10 smaller than the largest filament size, and the standard deviation of 57% in the sharp-bend is about 6 times higher than the virgin strand standard deviation of 9%.

Approximately 20-30% of the filaments in the extracted strand sections were surveyed. An interesting result in the strands extracted from the cable is that the standard deviation (sigma) in percent increases only slightly from the virgin strand. In fact, if the sigma is expressed in μm^2, rather than in percent, there is no significant difference in sigma between virgin and extracted strands. This result suggests that despite the deformation of the filament array, the filament uniformity is not degraded by the cabling process.

It can be seen from Figure 1 that as one proceeds from the virgin strand to the broad face to the wide edge to the narrow edge, the strand cross section is progressively more distorted. The measured cross section of the strand decreases as the distortion increases. The narrow edge cross section of the the extracted strand is about 6% smaller than the virgin strand. In addition, the mean filament size also decreases accordingly with the decreasing strand cross section. The narrow edge section is again the extreme case where the filament cross section is about 5% smaller than the virgin section. Since the filament and the strand cross section both decrease concurrently, the calculated Cu:SC ratio does not change greatly. However, a weak trend is seen where lower Cu:SC ratios occur for the more distorted, smaller cross-section sections.

Strand No. 7 was measured in a manner similar to No. 6 to test the consistency of this work. In general, the same trends are seen as in strand No. 6. The Cu:SC ratios tend to be lower than strand No. 6, which reflects the lower reported value of 1.67 from the vendor. As in strand No. 6, the mean filament cross section and strand cross sections also decrease as the strand distortion increases. However, all values for strand No. 7 are slightly larger than the strand No. 6 counterparts, suggesting that virgin strand No. 7 is larger than strand No. 6, despite the fact that the measured virgin strand diameters are identical for both strands.

Critical Current Distributions

The second derivative of the V vs. I plot gives the critical current distribution in the strand. Figure 2 contains distributions from virgin and cabled strand No. 10. The I_c is 306 and 296 A, respectively. The distributions are about 10 A wide or about 3% of I_c for both the virgin and extracted strand samples. The peak in the distribution for the extracted strand is about 8 A lower in current than the virgin strand, in agreement with the lower I_c result.

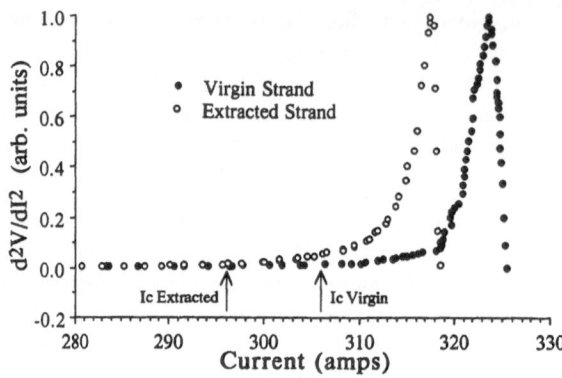

Figure 4. Second derivative of V vs. I data on strand No. 10. Virgin strand (top) and extracted strand (bottom). I_c (10^{-14} Ωm) for the virgin and extracted strands are shown.

DISCUSSION

It is thought that significant filament distortion would induce "sausaging" of the filaments, where the cross section of the filament changes along its length. As described in a model by Warnes,[3] current transfer can take place between sympathetic sausaged filaments. However, this transfer is a resistive link; thus significant sausaging reduces I_c as compared to unsausaged filaments. It is expected that increased sausaging would be apparent as an increase in the filament cross-sectional area standard deviation. The unchanging standard deviation observed suggests that the cable degradation is not strongly related to filament sausaging.

It is known that most of the cabling degradation is located at the wide and narrow regions of the cable.[1] These results show that these regions have a cross section reduced as much as 6% compared to the virgin strand and the face of the cable. This locally reduced cross section will then have a corresponding decrease in I_c due to the smaller superconducting material cross section. I_c may be further decreased by a possible reduction in J_c associated with the additional strain in the edge region; however, even at a strain of 6% ($\varepsilon=0.1$), this is not large enough to reduce J_c significantly.

Since it is difficult to survey all filaments without repetition, operator judgement is used to select representative regions for examination. However, the lack of change in the filament cross section σ suggests that any section chosen is representative. The validity of this image analysis technique requires more verification. Future work will be performed on extracted strands of 50 mm cable to determine whether the amount of cross section reduction is significantly less than with 40 mm cable, as predicted by the greatly reduced cabling degradation in 50 mm cable designs.

The critical current distributions indicate that the filament deformations due to cabling are both small and localized. If the deformations were uniformly distributed along the strand length, the distribution in I_c would broaden on cabling. The similarities between the virgin and extracted strand I_c distributions show that this is not the case. In addition, since the distribution apparently has been shifted to lower currents in the extracted strand, the deformation is occurring locally and is uniformly lowering the I_c. This occurs mostly at the cable edges, where the distortion is greatest. The shift in the I_c distribution is only about 3%, which is in agreement with the measured change in the mean filament size at the narrow edge. The changes in critical current density and filament size are, however, near the measurement limits, so they should be treated with caution. The overall cable degradation was reported to be 10%, significantly higher than the drop in I_c for strand No. 10 described here. Work is underway to examine the remaining extracted strands in this cable using transition shape analysis.

ACKNOWLEDGEMENTS

The authors thank F. Clark and D. Harrison for technical assistance in this work.

REFERENCES

1. L. F. Goodrich and S. L. Bray, "Electromagnetic Characteristics of NbTi Strands Extracted from Rutherford Cables," NISTIR 89-3912, (1988).

2. W. H. Warnes, W. Dai, and J. M. Seuntjens, *ICMC*, Huntsville, AL (1991).

3. W. H. Warnes, *J. Appl. Phys.*, 63, (5), 1651, (1988).

DEVELOPMENT OF SUPERCONDUCTING WIRE AND CABLE

FOR THE SSC PROJECT IN SUMITOMO ELECTRIC INDUSTRIES

T. Sashida, S. Saito, G. Oku, K. Kurimoto
Y. Yamada, M. Yokota, K. Ohmatsu and M. Nagata

Sumitomo Electric Industries, LTD. Osaka Works
1-1-3, Shimaya, Konohanaku, Osaka, 554 JAPAN

INTRODUCTION

As a large production volume of NbTi superconducting wire and cable is required for the SSC project, a production process has been developed at Sumitomo Electric to optimize critical variables of wire properties.

To achieve high electrical properties and a high overall yield of NbTi alloy in the fabrication process, we have employed carefully designed large size multi-filament billets weighing more than 350kg to decrease the number of billets in large production scale.

The collider dipole magnet consists of inner and outer cables, and the cable should be as uniform as possible to ensure the performance of the magnets.

We studied two aspects to obtain such uniformity of superconducting wire; one is the selection of unit weight and the other is the property of critical current density of a strand.

UNIT WEIGHT OF A BILLET

In the case of large production volume, it is very important to use a large unit weight. This means a large extrusion billet should be selected for multi-filamentary NbTi wire. If the number of billets is decreased by employing large unit weight billets, the workload to assemble many billets decreases proportionally with the number of billets. The time and cost of some processes such as extrusion is proportional to the number of billets, not directly depending on the total weight.

The billet length should be as long as the limits of the facility allow in order to obtain the highest yield of NbTi alloy. The metal flow during the extrusion process is not uniform at the beginning and ends of NbTi segments. The non uniform portion volume is almost constant even if the billet diameter is the same but the length is different. A longer billet gives a higher yield. In order to improve the NbTi yield, we need to choose a higher aspect ratio (billet NbTi segment length/billet diameter).

The largest diameter of billet possible also depends on the capacity of the extrusion machine. To increase unit volume and to decrease the number of billets,

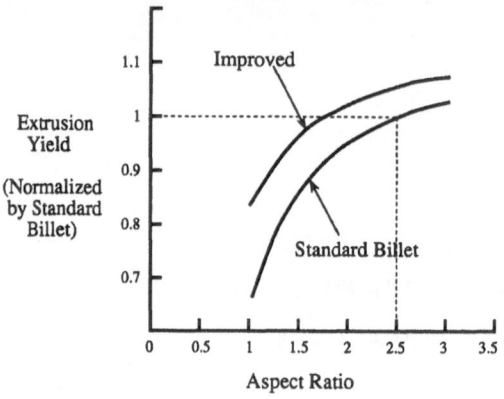

Figure 1. Extrusion Yield vs. Billet Aspect Ratio

we need to choose the largest diameter, but the choice is a trade-off between unit weight and NbTi yield because a larger diameter decreases the aspect ratio.

Thus, we selected a 10-inch diameter monofilament billet and a 12-inch diameter multifilament billet. The monofilament billet has an aspect ratio greater than 2.9, and the aspect ratio of the multifilament billet is 2.7.

For the multifilament billet, we employed a special design to improve the yield and metal flow at the extrusion process. Figure 1 shows the extrusion yield vs. billet aspect ratio. Our special billet design improves the extrusion yield from 6 to 8% compared to the conventional billet design. Thus we expect to obtain a good yield of NbTi, and to produce a 350kg strand from one billet.
The production process for the dipole inner conductor is shown in Figure 2.

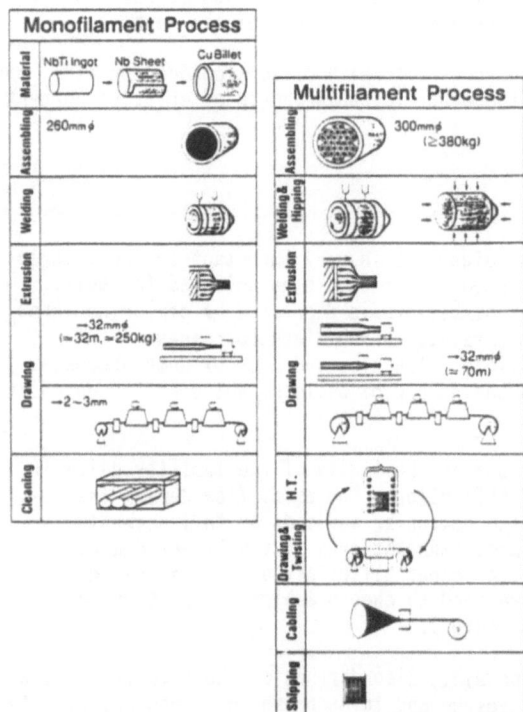

Figure 2. Production process of SSC dipole cable

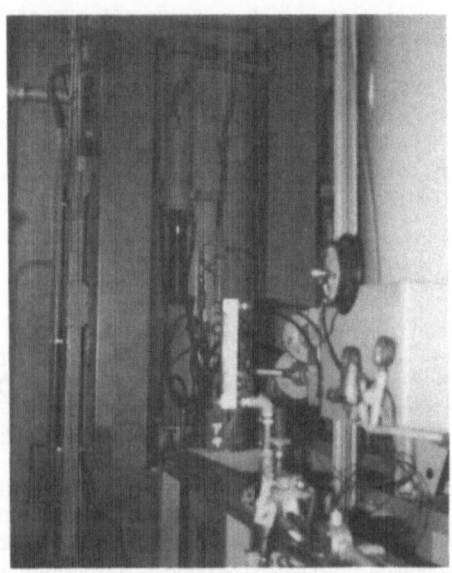

Figure 3.   Hot Isostatic Press

One monofilament billet corresponds to one multifilament billet, thus the data acquisition and data management become simple and easy to trace for quality assurance.

Recently, the HIPing process has been commonly used for improving the packing factor of the multifilament billet before hot extrusion. This process ensures that the metal flow is uniform during the extrusion process. Figure 3 shows our own HIP machine. Figure 4 and 5 show our multifilament billet before and after

Figure 4.   Appearance of a billet before HIPing

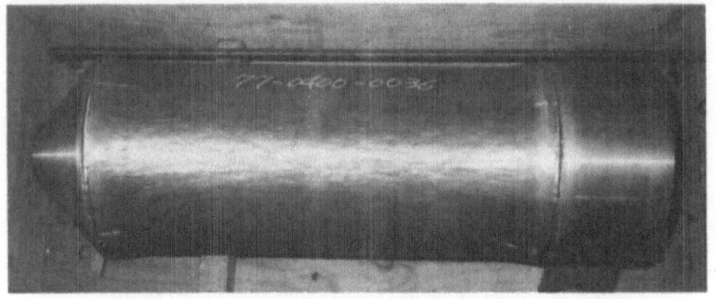

Figure 5.   Appearance of a billet after HIPing

HIPing respectively. Utilizing this process we can expect to maintain the filling factor of every billet at over 99%. This helps to make the products uniform and of high quality.

Moreover we need to control the clearance between the container bore of the extrusion machine and the billet outer diameter to prevent irregular deformation, because the clearances both in the billet and outside of the billet cause the metal to deform in a radial direction which is not desirable.

DEVIATION OF CRITICAL CURRENT DENSITIES

The distribution of critical current densities between strands for large scale production should be minimized to ensure the quality of the dipole magnets. The properties should be averaged as closely as possible. The SSCL's requirement is not only to maintain uniform critical current densities of strands, but to average the critical current of cables by combining high and low Ic strands.

In order to decrease the deviations of critical current densities of strands, raw materials should be uniform and process conditions should be unchanging. The critical current density of Nb Ti is well known to be a function of heat treatment temperature and time, number of heat treatments, reduction ratio/strain and strand size.

We investigated the correlation between the critical current density of the SSC strand and these parameters by utilizing statistical methods. We obtained an empirical equation of the critical current density by multi-variable analysis. The estimated value from this equation and the measured critical current density at 8T are shown in Figure 6. The correlation factor of this equation exceeds 95%, and similar equations can be obtained at 5T or other field values.

For quantity production, we need to choose a heat treatment condition which is not sensitive to the critical current density. From the above mentioned equation, we can determine the least sensitive point by first differential.

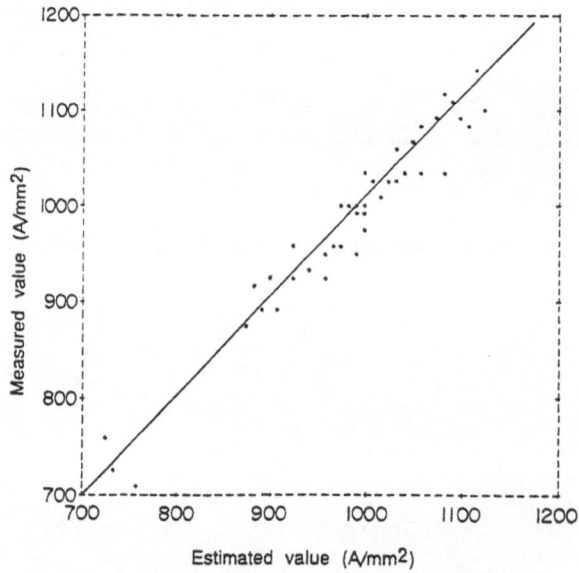

Figure 6.   Measured critical current density vs.
estimated critical current density at 8T.

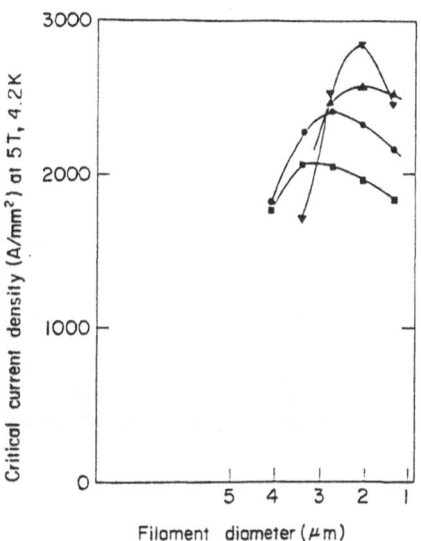

Figure 7.   Optimization of critical current density regarding filament size
around 2μm.

The least sensitive point coincides with the optimum value of the current
density.

Around this optimum point, for example, a temperature deviation of ±5℃ only
changes the critical current density 0.2%. Since the temperature distribution
and fluctuation of our furnace is smaller than ±2℃, it has enough accuracy to
maintain small deviations of the critical current density. Heat treatment time
is controllable within 1% although there is uncertainty of temperature
increasing and decreasing time. The largest uncertainty comes from the extrusion
process, and we are trying to introduce this condition in our empirical equation.

Another parameter which affect to the current density is a reduction ratio
between multiple heat treatments. Heat treatment size and reduction ratio are
controllable by using precise dies. Sumitomo Electric produces drawing dies
inhouse, and we utilize natural diamond, sintered diamond and sintered alloy for
die materials.

Sumitomo Electric has established the technology to make a single stack
billet with 17,300 filaments through the development of cable for a superconducting generator.   The strand is 0.5mm in diameter and the filament size is
2.1μm. The optimized critical current density is 2900A/cm² at 5T and 4.2K.

Figure 7 shows an example of optimization of critical current density
regarding filament size.

True strain for the NbTi filament is related to the reduction ratio between
NbTi ingot size and final filament size. But the critical current density of the
final product is largely affected by the reduction ratio between last heat
treatment size and final strand diameter.

From this figure, it is clear that the optimum condition becomes narrower
regarding filament diameter as the critical current density increases.

We can easily control the wire diameter precisely by maintaining the die
diameter at ±0.5%. The change of wire size by 0.5% gives a 0.1% decrease of the
optimum critical current density in a 2.1μm filament.

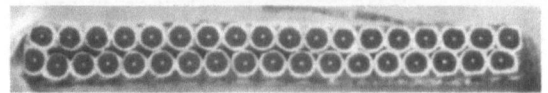

Figure 8. SSC outer cable with 36 strands.

It is necessary to maintain a uniform cross-section through long strand because the filament irregularity easily varies in filament diameter 5%-10%. Uniform deformation of each filament is very important because it affects not only critical current density, but also n-value and filament coupling.

These technologies are fairly useful for development of the HEB 2.5 μm filament strand. Moreover, using our stranding machine with variable back twist, we produced some keystone cable of 36 strands as shown in Figure 8.

CONCLUSION

Corresponding to the requirement to produce a large volume of superconducting cable of high uniformity and high yield, we employed a twelve-inch billet with special design and the HIPing process before extrusion for multifilament extrusion.

We analyzed the critical current density to minimize the deviation in quantity production. We will continue to pursue the uniformity of the properties of SSC strand.

REFERENCES

(1) S. Saito et al., " Recent Activities for NbTi Superconducting Wire at SEI ", Sumitomo Electric Technical Review, No29, pp51-59 January 1990.

(2) S. Saito et al., " Superconducting Properties of Fine Filamentary Supercon- ducting Wires ", Proc. of ICFA, New York U.S.A. May 1986.

(3) S. Saito et al., " Development of NbTi Superconducting Cable for the SSC ", Sumitomo Electric Technical Review, No31, pp124-127 January 1991.

(4) M. Nagata et al., " AC losses of NbTi Superconductor for A Quick Response Superconducting Generator ", MT-11, Tsukuba Japan August 1989.

SUPERCONDUCTORS WITH 2.5 MICRON NbTi FILAMENTS

H. C. Kanithi, P. Valaris and B.A. Zeitlin

IGC Advanced Superconductors Inc.
1875 Thomaston Avenue
Waterbury, CT 06704

ABSTRACT

Multifilamentary superconductors with 2.5 micron sized NbTi filaments are being proposed for the High Energy Booster part of the SSC project. The small filament size and the large number of filaments in the superconductor strands make the fabrication a challenge, especially in view of the high critical current performance required of them. IGC has produced two 309 mm diameter prototype billets and fabricated both the "Inner" and the "Outer" grade strands. The production approach was based on a triple extrusion process incorporating Cu - 0.6 wt.% Mn as the interfilament matrix material. The results of this pilot development effort, in terms of critical current density and mechanical properties, will be presented in this paper. Appropriate comparisons with past performances will be made.

INTRODUCTION

The national and international need for advanced superconductors for accelerator magnets is very clear to the high energy physics community. The superconductors should have small diameter filaments that are uncoupled, adequate ductility, high critical current density (J_c) at high fields while maintaining a small J_c at low fields and should be manufactured reliably for a large quantity production. Such materials are being actively developed in Europe, Japan and the U.S. as their need is being created by projects like the Superconducting Super Collider (SSC) and the Large Hadron Collider (LHC)[1-7].

SSC has decided on a 6 micron filament conductor for its main ring dipole and quadrupole magnets although 2.5 micron filaments might have been more advantageous from the view point of field quality. That decision was made some four years ago at a point when 2.5 micron filament conductor development was at its beginning stages, and the uncertainty of conductor performance and manufacturing ability were too great to be ignored.

Since that time, however, the need for 2.5 micron filaments has been kept alive because of the critical demands of the High Energy Booster (HEB) ring of the SSC[6,8,9]. Many superconductor manufacturers in the U.S. and abroad have been working vigorously to achieve, in 2.5 micron filament conductor, electrical and mechanical performance that is typical of present-day 6 micron filament conductors.

As smaller and smaller filaments are produced they tend to deform non-uniformly leading to filament necking and subsequent degradation of the critical current density. SSC requires high J_c (2420 A/mm^2 at 5.6T and 1600 A/mm^2 at 7T) together with long piece lengths, a combination that has not been demonstrated so far with 2.5 micron filaments.

As interfilament spacing decreases below 0.5 micron and proximity effect coupling becomes apparent, costly correction coils would be necessary in magnet design. Alternate matrix material for the conductor have been proposed such as copper-nickel or manganese doped copper, which will eliminate the requirement for correction coils. Conductor incorporating manganese doped copper is currently being developed[6,9]. This matrix material change has to be implemented with minimum compromise in the electrical stability and thermal conductivity which are being provided by the present high purity copper matrix.

MANUFACTURING METHOD

Simple calculations show that approximately 24,000 and 42,000 filaments are required in the SSC Outer and Inner size strands if the filament diameter is to be 2.5 microns. Single stack method which is now the norm for the 6 micron filament SSC dipole conductor is not suitable for the 2.5 micron filament HEB strand because of the large numbers of very fine monofilament elements that have to be assembled. Therefore, we have employed a bundle and rebundle approach which is shown schematically in Figure 1. It involves three separate extrusion steps and, hence the name, triple extrusion process. The first stage, which is the manufacture of the monofilament happens in the same way as for the present SSC dipole conductor manufacture. A number of monofilament rods are assembled in the second stage billet, extruded and drawn to the next restack size suitable for the third stage billet. The last stage billet is extruded and processed to wire using manufacturing steps common to any multifilamentary superconductors.

In the present development effort, the first stage extrusion can was made by Cu-0.6wt% Mn alloy in order to eliminate proximity effect coupling. The first stage composite was drawn to a hexagonal shape and cut in to straight lengths. 199 monofilament rods were assembled in the second stage copper can. Second stage material was drawn to two hexagonal sizes corresponding to Inner and Outer stacking size. The third stage was assembled with 192 second stage rods to make a short Inner billet and 114 second stage rods to make a short Outer billet. Since the requirement was only for 115 kg of each grade, we designed shorter billets than our production billets. These third stage billets were extruded and drawn to approximately 25 mm dia. at which point sample heat treats were performed and evaluated. Based on the results of the heat treat study, both Inner and Outer billets were released for the balance of the processing.

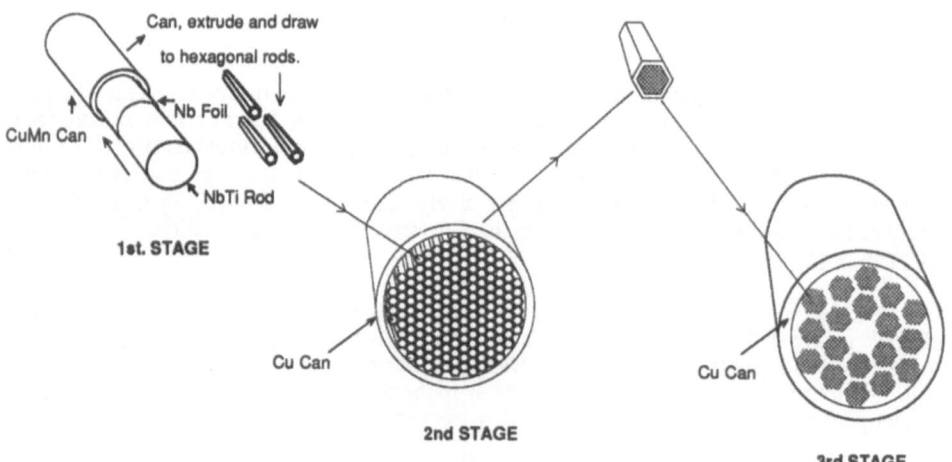

Fig. 1. Flow diagram of the triple extrusion process.

CONDUCTOR PERFORMANCE

Composite sections were examined after each extrusion step and the filament geometry was observed to be regular. Figures 2 and 3 show cross-sections of the Inner and Outer strands at 0.808 mm and 0.648 mm diameter, respectively. A high magnification view of a filament bundle is also shown for each. The filaments appear to be uniform in cross-section and well-separated by the Cu-Mn. It is interesting to note that the last row of NbTi filaments around each bundle has undergone a non-uniform deformation compared to the ones on the inside.

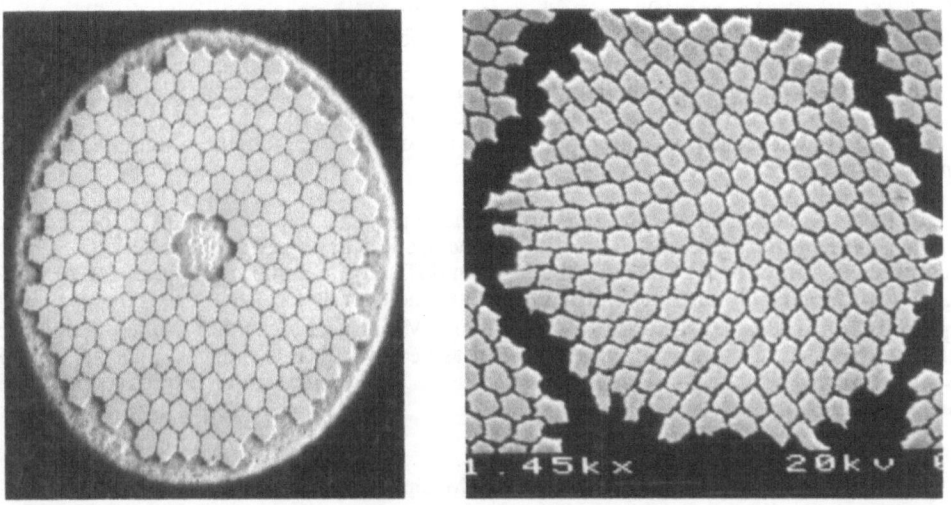

Fig. 2. Inner grade conductor at 0.808 mm dia. containing 38,208 - 2.5 micron filaments.

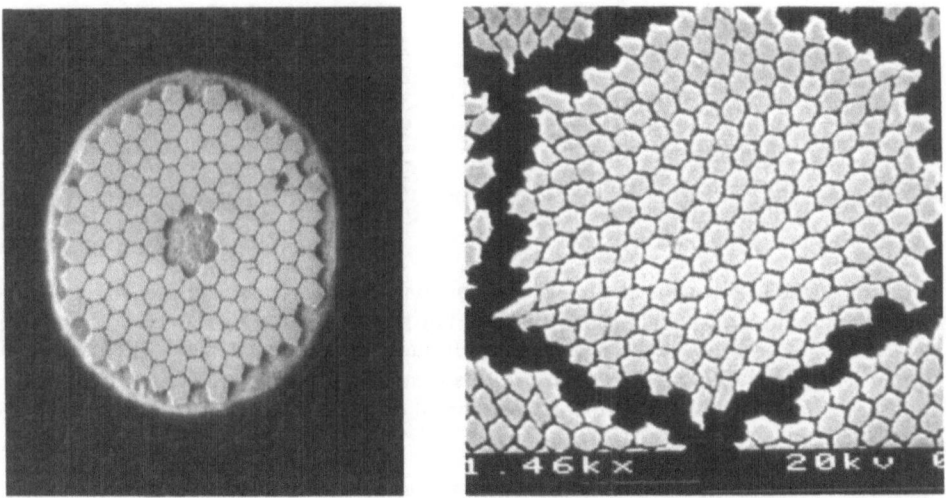

Fig. 3. Outer grade conductor at 0.648 mm dia. containing 22,686 - 2.5 micron filaments.

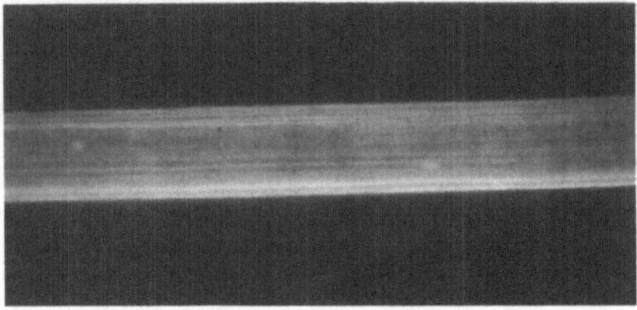

Fig. 4. SEM photograph of typical NbTi filament from the Inner grade strand at final size.

During the hot and cold processing steps of the third stage composite, the filaments on the outer row are flowing into the soft copper webbing between the bundles and becoming rectangular in cross-section. The cross-sectional area of individual filaments, however, appear not to be affected indicating a little or no undesirable filament sausaging. Copper and Cu-Mn were removed by a nitric acid etch and observed randomly in SEM. Figure 4 shows a typical view of the filament surface. They appear to be uniform in size along the length and free from intermetallics.

The purpose of the pilot heat treatment sample processing was not only to optimize the critical current performance but also to evaluate the ductility of the composite materials. A few samples were drawn down to 0.5 mm dia. without difficulty. Samples were tested for a) Cu/SC ratio (Cu to non-Cu) by acid etching and weighing, and b) critical currents using 1m long helical specimens at an overall resistivity of 10^{-14} ohm-m. Mechanical properties, namely spring-back characteristics and sharp-bend behavior, were measured by simple techniques as specified by the SSC Specification Mag-M-4141. Test results from the best samples at the respective final sizes are given in Table 1.

<u>Table 1. Electrical and Mechanical properties</u>

	Inner	Outer
Strand Diameter, mm	0.808	0.648
Cu/SC ratio (Cu to Non-Cu)	1.45	1.72
J_C (7.0T), A/mm^2	1708	1682
J_C (5.6T), A/mm^2	2432	2395
"n" value at 7.0T	29	24
"n" value at 5.6T	32	27
Ultimate Tensile Strength, MPa	978	938
Yield Strength, MPa	584	576
Spring-back Dia.(mm) and Angle	12.8-800^0	12.5-728^0
Filament Damage in Sharp-bend	none	none

The above results indicate that we can obtain in the Inner conductor a J_C(7T) of ~1700 A/mm^2 which means a margin of ~6% above the specification value. In the Outer grade, however, the J_C(5.6T) is at the specification limit. If extrapolated, the J_C(5T) would be approximately 2720 to 2760 A/mm^2. These results were obtained in the as-drawn samples which did not have any final anneal. A final copper anneal of 1 hour at 260^0 C has had virtually no effect on the J_C either at 5.6T or at 7T.

As of this writing, two 150 kg batches are at various points close to the final strand sizes. The wire breakage during multi-die processing has been typical of 6 micron filament dipole conductor manufacture at IGC/ASI.

If we compare the superconducting performance with that of recent 6 micron filament strands[1], we find that the J_C is low by approximately 8%. The "n" value in the 2.5m filament strands is reduced by as much as 40%. The filament uniformity is qualitatively comparable to the 6 micron filaments. The absence of filament sausaging from the micrographic examination seems to suggest that the "n" value should be high around 40. It is possible that other factors such as the closeness of 2.5 micron filaments and the transverse resistivity of Cu-0.6 wt% Mn may be playing a role in the "n" value.

In 1986, IGC- ASI had produced 180 kg each of Inner and Outer grade strand with 2.7 micron filaments[9] using a two-stage extrusion process. The first stage billet consisted of assembling 19 NbTi rods individually wrapped in Nb foils and placed in pure copper tubes. Hexagonal restack rods were made from this 19 filament composite after extrusion and drawing. The final billet had either 1927 or 1086 restack rods corresponding to the Inner or Outer design. Thus the number of filaments was 36,613 in the Inner and 20,634 in the Outer. Relative to the present conductors, the 1986 strands were more brittle and the filament quality was poor. In addition, the filaments were magnetically coupled by proximity effect due to the unalloyed copper between filaments. The $J_c(5T)$ ranged from 2350 to 2500 A/mm^2 with an "n" values in the low 20's. Thus, compared to our earlier attempt, improvements in all aspects of the conductor performance have now been made. Further enhancement of J_c should be the goal of future development.

SUMMARY

A triple extrusion approach was used to produce SSC Inner and Outer strand with 2.5 micron NbTi filaments and Cu-0.5 wt% Mn interfilament matrix. The conductor is being fabricated for use in the prototype magnets for the High Energy Booster ring. In samples tested, a J_c of 1700 A/mm^2 at 7 Tesla was achieved for the Inner wire and 2400 A/mm^2 at 5.6 Tesla was achieved for Outer strand. Spring back and sharp-bend tests of these samples met SSC specification requirements. Filaments are free from intermetallic compounds and are more uniform in cross section than our earlier development program. The latest conductor, when compared to the one manufactured at IGC-ASI in 1986, has shown a marked improvement in J_c, filament quality and in the general mechanical properties. If the critical current requirements remain as high as for the present dipole conductor, further development is necessary to increase the J_c of especially the Outer conductor at 5.6 Tesla.

ACKNOWLEDGEMENT

This work is being performed under a fixed price contract from the Superconducting Super Collider Laboratory.

REFERENCES

1. P. Valaris, H. C. Kanithi and B. A Zeitlin, IEEE Trans., vol. 27, No. 2, Part III, pp. 1752-1754, 1991.

2. H. C. Kanithi, C. G. King, B. A. Zeitlin, IEEE Trans., vol 25, 2, pp. 1922-1925, 1989.

3. T. S. Kreilick, E. Gregory and J. Wong, Adv. in Cryo. Eng., A.F, Clark and R. P. Reed, eds. Plenum Press, New York (1986) vol. 32, pp. 739-745.

4. H. G. Ky, G. Grunblatt and P. Bonnet, IEEE Trans., vol. 27, No. 2, Part III (1991), pp. 1822-1824.

5. S. Saito, T. Sashida, G. Oku, K. Ohmatsu and M. Yokota, Supercollider-2, M. McAshan ed., Plenum Press, New York,pp. 593-599, 1990.

6. G. Iwaki, S. Sakai, Y. Suzuki, H. Moriai and Y. Ishigami, "Current Developments of the Cu/NbTi superconducting cables for SSC in Hiathi Cable Ltd., *ibid*, pp. 621-630.

7. M. Ikeda, H. Ii, S. Meguro and K. Matsumoto, *ibid*, pp. 349-356.

8. P. Valaris, T. S. Kreilick, E. Gregory and E. W. Collings, Supercollider-1, M. McAshan ed., Plenum Press, New York (1989) pp. 449-456.

9. C. King, K. Hemachalam and B. Zeitlin, IEEE Trans., MAG-23, 2, pp. 1351-1354, 1987.

10.T. S. Kreilick, E. Gregory, J. Wong, R. M. Scanlan, A. K. Ghosh, W. B. Sampson, and E. W. Collings, Adv. in Cryo. Eng. Materials, A.F, Clark and R. P. Reed, eds. Plenum Press, New York (1988) vol. 34, pp. 895-900.

ISSUES AND RESULTS IN THE DEVELOPMENT OF

2.5 MICRON FILAMENT STRAND FOR THE HEB

J. M. Seuntjens and D.W. Capone II

Magnet Division
Superconducting Super Collider Laboratory
2550 Beckleymeade Avenue
Dallas, Texas 75237

INTRODUCTION

The High Energy Booster (HEB) ring which injects protons into the Superconducting Super Collider (SSC) will be ramped with a cycle time on the order of 8 minutes. The current magnet design incorporates a strand with 2.5 μm filaments to reduce the AC loss heating component due to the magnetization hysteresis. The reduced magnetization also makes the correction of the sextupole moment easier. Past experience has shown that Nb-Ti filaments in a copper matrix should have a spacing-to-diameter ratio (s/d) of 0.15 for good fabricability.[1] For 2.5 μm filaments, this results in a spacing on the order of 0.4 μm at final wire size. A Cu-Mn alloy has replaced the copper between the filaments in prototype conductors to effectively eliminate proximity coupling between 2.5 μm filaments which would otherwise occur with copper at this spacing. The resistivity of the material between the filaments is limited by the mean free path being equal to the filament spacing; therefore, cryogenic stability has not been compromised.

This paper discusses design and process parameters in the development of 2.5 μm filament conductor production. Data from current 2.5 μm work from several vendors, some of which are on FFP-type contracts sponsored by the SSCL, are included to demonstrate how close industry is to meeting 2.5 μm conductor needs. If the SSCL were unable to bring 2.5 μm conductor into production within about two years' time, a fall-back HEB design with 6 μm filament conductor would be utilized. Therefore, a strong emphasis on manufacturability is reflected in the key issues. The sole purpose of this paper is to establish a common level of understanding of fine filament needs of the SSC; therefore, the data will be purposely presented without vendor affiliation. The authors have paid careful attention to details to avoid the release of proprietary information in this paper.

ULTRA-FINE FILAMENT CONDUCTOR PRODUCTION ISSUES

Production processing of 2.5 μm conductor in a minimum of 3000 m lengths with satisfactory superconducting and mechanical properties will require extending our current technology. In the design and processing of 2.5 μm filament conductor, the following issues must be addressed and are receiving emphasis in the present 2.5 μm conductor development program at the SSCL:

1. Reduction of proximity effect coupling between filaments at final wire size.

2. Reduction or elimination of compound formation during the thermo-mechanical processing operations associated with very-fine filament conductors through suitable barrier construction and heat-treatment schedules.

3. Reduction or elimination of filament distortion during all phases of processing.

4. Understanding of the mechanical properties of the matrix material as filament spacing approaches submicron dimensions.

5. Control of mechanical properties of final strand through composite design or strand processing details.

6. Development of adequate critical current density (i.e., attempt to meet or exceed specifications.) while at the same time ensuring suitable mechanical properties in the conductor, especially piece-length.

7. Demonstration of adequate mechanical performance to ensure satisfactory cabling performance and a minimum of cabling degradation.

SSC specifications for 2.5 μm conductor (SSC-M30-000160 for Inner and SSC-M30-000161 for Outer) are essentially the same as those for 6 μm conductor with several modifications. Obviously, the filament size is specified to be nominally 2.5 μm and the spacing between filaments is approximately 0.4 μm to maintain a similar s/d ratio as in 6 μm filament conductor. The critical current specification for 2.5 μm material, which is the same as for the current 6 μm conductor, incorporates additional conductor margin. While it appears that this current level is obtainable, not enough material has been processed to date to evaluate manufacturability.

The total number of filaments is fixed by the copper-to-superconductor ratio (nominally 42,000 for Inner conductor and 24,000 for Outer conductor). In the case of a double multifilament extrusion design, considerable flexibility exists in the number of filaments in the first stacking and the geometric design of the final multifilament stack. The SSC will require that a single billet design be used for all Inner conductor and an equivalent design for all Outer conductor. Vendors must choose a billet design and diameter which offers the best manufacturability, performance, and cost.

A small amount of manganese (~0.5 weight %) has been added to the copper between adjacent filaments (i.e., the copper in the monofilament) in previous 2.5 μm material. This effectively diminishes proximity effect coupling, which would produce an effective filament diameter larger than the actual diameter. The exact amount of manganese and the spacing between filaments (i.e., design of the filament array) for optimal coupling-free superconducting properties are not known; nor are the mechanical properties for producing high-quality cable. Because of the limited amount of time available for development, 2.5 μm conductor production probably will not deviate far from the 0.5 wt.% addition.

Current 6 μm filament diffusion barrier design must be modified for 2.5 μm filaments. The option of using single wrap or multiple wrap barriers still exists; however, all Nb must be procured to the SSC specification for Nb sheet, SSC-MAG-M-4001. As a guide to billet design, the SSC expects that the nominal barrier thickness for 2.5 μm material should have a thickness at the last heat treatment size equivalent to that of the 6 μm filament material with a 4% barrier, i.e., on the order of 9%. The barrier thickness required is dependent on the uniformity in the thickness of the barrier at heat treatment sizes. The goal here is to establish sufficient process margin by determining an optimal barrier thickness, and to ensure a combination of good piece length and adequate critical current density performance.

The thermo-mechanical optimization of 2.5 μm conductor is likely to be different from 6 μm conductor. It is important to use sufficient material during the thermo-mechanical optimization to obtain some information on piece length and on critical current vs. field properties. It is anticipated that a minimum of 5 kg for each heat treatment is required to accomplish this task. It is important to obtain adequate, but not excessive, I_C from this material. But it is more important that mechanical properties and piece length are not compromised.

The mechanical properties characterized primarily by the sharp bend and spring back tests are as important as the superconducting properties in the production of useful conductor. Dramatic changes in mechanical properties of metal composites can occur when the filament separation is less than 1 μm². Part of final development work on 2.5 μm conductor is to understand any potential for deleterious work hardening. This knowledge should ensure that processing can be developed to avoid excessive filament distortion while maintaining satisfactory mechanical properties for good cabling performance and a minimum of cable degradation.

RESULTS IN ONGOING 2.5 MICRON DEVELOPMENT

We have evaluated prototype conductor from several vendors. All material surveyed used Nb barriers and a triple extrusion process. The billet designs are considered proprietary and are not disclosed here. A brief description of the image analysis is given here. A plot of the frequency vs. gray level is made from the stored image during the image analysis. A typical plot is shown in Figure 1, where two distinct peaks are shown: one for the filaments and one for the matrix. A gray level is chosen at the minimum between the two peaks to define the boundary between the filaments (shaded) and the matrix. A binary image is then created with all pixels with gray level below the boundary set to be black to represent the filaments. All remaining pixels are set to white to define the matrix. A representative binary image is shown in Figure 2. Note that in order to ensure clear journal reproduction, there are fewer filaments in this image than are seen in a typical analysis. A variety of statistical analyses are then performed on the filaments and the pixel statistics are reported here.

Table 1 contains a list of key parameters of some of this material. The Ic, filament uniformity (mean and standard deviation of the filament size measured on SEM images[3], spring back value, and sharp bend results are tabulated to give some perspective on the current level of manufacturability. The I_C values are quoted from the vendor. The reader is referred to the work of Suzuki et al.[4] and Kanithi et al.[5] for additional 2.5 μm results.

DISCUSSION

Gregory et al.[6] have shown that Cu-Mn alloy between filaments effectively eliminates proximity coupling until the filament size is well below 1 μm. Therefore it is expected that the proximity coupling is not a major problem. Recent results from Iwaki et al.[4] demonstrate that no proximity coupling exists at any significant magnetic fields in final size 2.5 μm filament conductor.

The barrier thickness in the conductors that come close to or exceed current density specifications appears to be satisfactory for preventing intermetallic formation. Conductors with greatly reduced I_C have shown some intermetallic formation on filaments. However, it is also clear that intermetallic formation is not the sole source of filament non-uniformity.

Conductor made by a triple extrusion process can have local filament distortions, especially for filaments near the outside edge of the sub-elements. The copper between the sub-elements is typically several filament diameters wide. Therefore, in this region the s/d ratio is much larger than for the bulk of the filaments. Under certain processing conditions, the un-mechanically supported filaments can distort into a ribbon-like shape. Images in Figure 3 demonstrate varying degrees of this problem. This filament distortion does not affect the standard deviation of the filament cross-section (σ) until the distortion is severe. The σ is about 13–15% for the current 2.5 μm conductor. This compares to a typical value of 8% for current production 6 μm conductor. It is possible that a three-extrusion process will inherently produce filaments with broader distributions than a two-extrusion process used for 6 μm conductor. It is not yet clear what parameters in the thermo-mechanical processes of conductor processing most affect filament uniformity, given the present case where intermetallic formation has been effectively eliminated with the proper barrier design.

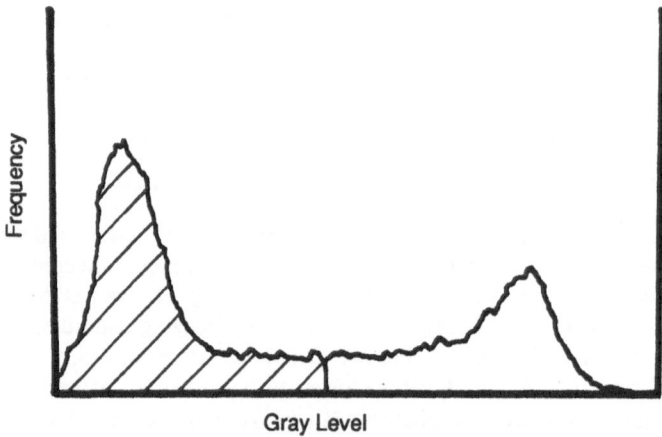

Figure 1.   A Plot of Frequency vs. Gray Level from a Stored Image. The minimum in the curve defines the border between the filaments (shaded) and the matrix.

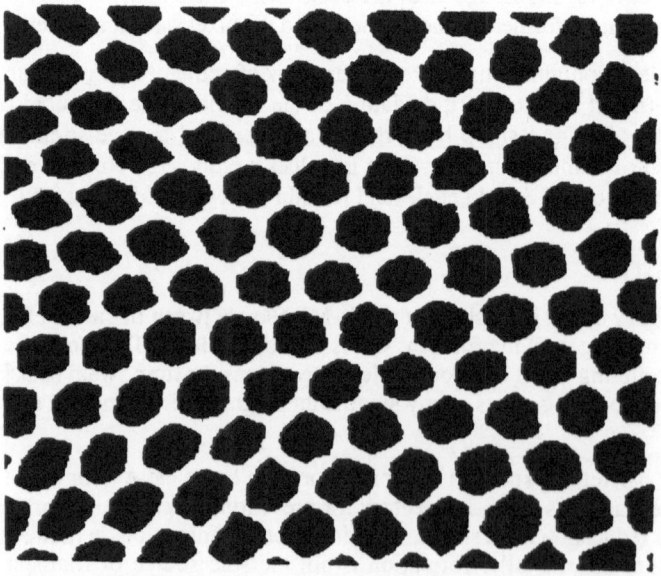

Figure 2.   Binary Image Made from the Defined Distribution in Figure 1. Fewer filaments are shown than the number in a typical analysis.

Table 1. 2.5 μm Conductor Data.

Vendor	Inner/Outer	I_c (amps)	Mean Filament Area (sq. μm)	Filament Std. Dev. (%)	Springback (degrees)	Sharp Bend C or NC/%**
SSCL spec.	Inner	>339	4.9		<980	NC/<1
	Outer	>286	4.9		<1090	NC/<1
Green	Inner	337	5.1	14	840	NC/0
	Outer	287	5.1	15	904	NC/0
Blue	Inner	*	3.6	28	1050	C/70
	Outer	*	#	#	1092	C/70
Red	Inner	328	4.7	15	768	C/70
	Outer	#	#	#	#	#
Black	Inner	345	4.2	13	730	NC/0
	Outer	281	4.8	14	925	NC/1
Orange	Inner	200	4.8	18	813	C/90
	Outer	264	3.9	15	883	C/20

* Only non-heat treated wire could be drawn to final size.
# Data not available.
** C = cracked, NC = not cracked; % = percent of broken filaments.

Figure 3. Filament distortion between sub-elements in 2.5 μm conductor made by triple extrusion processes: (left) little distortion, (middle) some distortion, (right) severe distortion. The mechanical properties of these composites degrade as the filament distortion increases.

ACKNOWLEDGEMENTS

The authors thank D. Harrison and F. Clark for technical assistance in this work. This work was supported by the Director, Office of Energy Research, Office of High Energy and Nuclear Physics, High Energy Physics Division, U.S. Department of Energy, under Contract No. DE-AC02-89ER40486.

REFERENCES

1. E. Gregory, *Adv. in Cry. Engr.*, **34,** 993, (1987).

2. J. Bevk, J. P. Harbison, and J. L. Bell, *J. Appl. Phys.*, **49,** 6031, (1978).

3. J. M. Seuntjens, B. H. Warnes, and D. W. Capone II, *IISSC*, Atlanta, GA (1991) to be published in *Supercollider 3*.

4. G. Iwaki, A. Sakai, and S. Nishiyama, Hitachi Technical Report No. 4ML6-0022, (1991).

5. H. Kanithi, P. Valaris, and B. A. Zeitlin, *IISSC*, Atlanta, GA (1991) to be published in *Supercollider 3*.

6. E. Gregory, T. S. Krelick, J. Wong, E. W. Collings, K. R. Marken, R. M. Scanlan, and C. E. Taylor, *IEEE Trans. Mag.*, **25**, 1926, (1989).

THE QUALITY ASSURANCE OF SUPERCONDUCTING WIRE

AND CABLE FOR SSC MAGNETS

Douglas A. Pollock, Penny Baggett and Donald Capone II

Superconducting Super Collider Laboratory*
2550 Beckleymeade Avenue
Dallas, TX 75237

ABSTRACT

The success of the SSC depends on the consistency and uniformity of the superconducting magnets used in the main collider rings and the high energy booster. To a great extent the success of the magnets depends upon the quality of the superconductor wire and cable used in coil windings. As the SSC project has begun its transition from Research to Development, a new laboratory organization has been established to carry the design requirements from concept to reality. The SSCL Magnet Systems Division Quality Assurance Group has been working on the development of a quality management and analysis system for insuring superconductor uniformity through the understanding and control of manufacturing variation. Key areas of the QA activity include: 1.) the design and development of a computer database and analysis system for the collection and statistical analysis of superconductor materials data (containing: source physical and chemical properties, billet process history, and final product performance data); and 2.) the development of wire and cable product specifications which focus on the control of variation. As a result of this work several new concepts have been developed which will affect the traditional approach to superconductor wire and cable production.

INTRODUCTION

The main purpose of this paper is to describe the present state of planning for the management of superconductor material quality, as viewed by the SSCL Magnet Systems Quality Assurance organization. Since all of the material produced to date for the SSC has been made under R&D (rather than Production) contracts, many of the quality management ideas presented here have yet to be adopted by industry. The initiation this spring of the two year Conductor Qualification Program [1] with industry, containing up to 280 billets in seven contracts, provides a vehicle for the introduction and preliminary verification of these plans. Future technical and management reviews will confirm the soundness of these ideas. The major focus of this paper will be the "means to achieving quality" which are currently within the direct scope of SSCL QA activity; (in particular, portions of the Magnet Components Database along with the Material Specification requirements).

*Operated by the Universities Research Association, Inc., for the U.S. Department of Energy under Contract. No. DE-AC02-89ER40486.

SUPERCONDUCTOR MATERIALS REQUIREMENTS FOR COLLIDER DIPOLES

In order to measure the relative magnitude of the task facing the SSCL and the superconducting materials industry, a summary of the conductor requirements for Collider Dipole cable is presented.

	INNER CABLE		OUTER CABLE	
Strand Number:	30		36	
Strand Diameter:	0.808 mm	(0.0318")	0.648 mm	(0.02551")
Cable Length:	19,973,904 m	(12,408 mi)	13,946,524 m	(8,664 mi)
Cable Mass:	1,256,104 kg	(1,385 Tons)	1,331,746 kg	(1,468 Tons)
Billets:	3,641		3,805	

The cable quantities listed are based on absolute magnet requirements, and have no allowance for yield losses [2]. For representative purposes a "billet" of multifilament conductor is 305 mm diameter prior to extrusion and yields approximately 350 kg of bare cable. Actual billet dimensions may vary according to supplier process requirements.

In the past year less than twenty billets of dipole conductor have been produced for the SSCL by industry. Starting in 1994 the annual production rate will need to be approximately one thousand billets per year each, for Collider Dipole inner and outer conductor.

QUALITY ASSURANCE GOALS FOR COLLIDER CONDUCTOR

The following are the primary goals of the QA program for superconductor materials [3,4].

A. Conductor Defect Prevention.

B. Conductor Defect Detection.

C. Reduction and Control of Manufacturing Process Variability.

D. Overall uniformity of magnet coil performance.

The program being designed to meet these goals is intended to be practical not legalistic. If successful, the result of this effort will be to eliminate much of the mystique surrounding the production of superconductor materials. Simply stated, a major objective of this plan is to reduce or eliminate the possibility of a supplier "losing the superconductor recipe" for unidentifiable reasons, once it has been established.

THE MEANS TO ACHIEVING QA GOALS

Achieving quality is not something that will just happen because someone somewhere has written a specification or invoked a mil standard. As in any worthy endeavor, quality will be the result of careful planning and hard work. We believe the key to meeting the stated goals are "attention to detail" and "communication" at all levels of the program. This means paying the price in effort to predict, plan, and document the inputs necessary for a stable manufacturing process, followed by observation and evaluation of the feedback coming from the process. The following outline describes the fundamental steps or "means" to achieving quality upon which the SSCL QA Plan is based. In terms of the philosophy described above, these steps may be viewed as indicating a relative order of significance.

The Means to Achieving QA Goals:

A. Close SSCL interaction with superconductor suppliers.

B. Supplier establishment of a well characterized "baseline" process containing clearly defined and measurable controls throughout the process.

C. Continuous monitoring and analysis of product performance and process variation by the Supplier and the SSCL.

D. SSCL QA-1: Seller Quality Assurance Requirements, (Quality Management System requirements, appended to product specifications).

E. The Wire and Cable Specifications.

To the extent that producers typically look upon the specifications as an "end" in themselves, the ordered outline above indicates a need for adjustment in such thinking. The formal specifications are a starting point for communicating fundamental requirements. As both the SSCL and Industry learn more about the process, product specification requirements may be more sharply focused to satisfy overall program goals. Successful cable performance in the collider ring is the "end".

THE MAGNET COMPONENTS DATABASE

To implement the monitoring and analysis elements of the plan, a computer database has been designed and installed at the SSCL [5,6]. The database will be used for storing important information regarding all materials used in the collider magnets. The database is known as MAGCOM. The MAGCOM system, a network accessible computer database, resides on a Sun 4 Computer running the UNIX operating system. The actual database is built on SYBASE, a SQL relational database application. For superconductor materials, the database has been organized to contain information about: source material physical and chemical properties, manufacturing process history, manufacturing process variation data, and final product performance. The database is not all inclusive, it contains only the minimum information which the SSCL considers critical to maintaining component uniformity and traceability.

To simplify setting up a link with industry, an IBM PC version of the conductor database has been developed [7,8]. The definition of data requirements for the system has been appended to the wire and cable specifications [9] in the form of manual data sheets, (see Table 1 below). During the start up phases of the Conductor Qualification Program, procedures for computerized data management and transfer will be worked out between the SSCL and each contracted supplier.

Monitoring Conductor Process Stability

From a QA perspective, this database will be essential for establishing confidence in the stability of the suppliers' process. We will use the information collected to monitor the magnitude of manufacturing variation and try to correlate particular process conditions (dependant variables) to conductor performance requirements (response variables) [10]. We will also try to answer the following general questions by evaluating the data.

A. How do supplier controllable process parameters effect final product performance?

B. Are supplier established manufacturing tolerances and SSCL specification requirements sufficient to assure uniformity and consistency in the final cable?

C. Can statistically significant shifts be detected and corrected before performance is adversely affected?

Monitoring Source Material Consistency

Source material data (NbTi, Nb, Cu), will be used to provide traceability from the cable back through the process to the raw material. We will also monitor material variability from lot to lot. As above, we hope to answer some fundamental questions with the data.

A. To what extent does measurable variation of the physical and chemical properties (within the established tolerance limits) relate to final product performance?

B. As with the conductor process, can significant material performance shifts be detected and corrected before they become problems?

C. Are SSCL tolerances sufficient to assure uniformity and consistency in the final cable?

DATABASE CONTENT

The minimum data collection requirements for conductor have been appended to the material specifications for NbTi, Nb, Wire and Cable, as summarized below.

TABLE 1. DATABASE CONTENT

DOCUMENT	DATA SHEET	REQUIRED INFORMATION
M50-000003	COPPER MATERIAL SOURCE DATA	LOT CHEMICAL ANALYSIS
M50-000004	NbTi ALLOY INGOT CHEMICAL ANALYSIS DATA	INGOT / MELT CHEMICAL ANALYSIS
M50-000005	NbTi ALLOY LOT CERTIFICATION DATA	LOT CHEMICAL ANALYSIS
M50-000006	NIOBIUM SHEET CHEMICAL/MECHANICAL DATA	LOT CHEMICAL & MECHANICAL ANALYSIS
M50-000007	MONOFILAMENT BILLET PRODUCTION DATA	BILLET ASSEMBLY DATE SOURCE MATERIAL
M50-000008	MULTIFILAMENT BILLET PRODUCTION DATA	BILLET ASSEMBLY DATE SOURCE MATERIAL
M50-000009	BILLET WELD / COMPACTION DATA	WELD & COMPACTION DATES, WELD INSPECTION & COMPACTION EFFECT PRESSURE / TIME / TEMP
M50-000010	BILLET EXTRUSTION DATA	DATE / VENDOR / SEQUENCE PRE-HEAT INFORMATION EXTRUSION RUN FORCE BILLET TEMPERATURES
M50-000011	MULTIFILAMENT STRAND HEAT TREAT HISTORY.	IDENTITY / QUANTITY FURNACE INFORMATION DATES IN & OUT VISUAL CONDITION TIMES / TEMPERATURES

TABLE 1. (continued)

DOCUMENT	DATA SHEET	REQUIRED INFORMATION
M50-000012	STRAND CHEMICAL / MECHANICAL / .... ELECTRICAL TEST DATA	PRESENCE OF TWIST CU:SC RATIO SPRING BACK SHARP BEND TEST FIELD (TESLA) DIAMETER CRITICAL CURRENT, I_c QUALITY INDEX, "n" RESISTANCE R295 K RESISTANCE R10 K
M50-000013	STRAND PRODUCTION DATA................................	FINAL ANNEAL STATE PIECE LENGTH PIECE WEIGHT DIAMETER STATISTICS
M50-000014	CABLE RUN DATA..	GENERAL INFORMATION MID-THICKNESS WIDTH KEYSTONE ANGLE
M50-000015	CABLE STRAND-MAP DATA....................................	MATERIAL IDENTITY / LOCATION / WELDS / I_c
M50-000016	CABLE EXTRACTED STRAND CRITICAL................. CURRENT TEST DATA	TEST FIELD (TESLA) STRAND I_c QUALITY INDEX "n" CU:SC RATIO STRAND AVERAGE I_c CABLE EST. I_c

CURRENT FUNCTIONAL STATUS OF THE QA DATABASE

The IBM/PC conductor database presently in use in the SSCL QA Group has been developed on Ashton/Tate dBASE III PLUS Version 1.1. The data flow diagram (Table 2) outlines the present functional activities associated with superconductor data management.

PERFORMANCE DATA

Conductor performance data is being manually provided by the suppliers and SSCL Conductor Engineering. The data is currently limited to the following items:

WIRE DATA	CABLE DATA
A. Wire Identity	G. Cable Strand Map
B. Critical Current	H. Cable Measuring Machine Data
C. Cu:SC Ratio	I. Cable Run Report
D. Strand Diameter	J. Cable Test Data (from BNL)
E. Piece Length	
F. Sharp Bend and Spring Back, (periodically)	

SOURCE MATERIAL AND MANUFACTURING PROCESS DATA

No source material or manufacturing process data is being submitted by industry at this time. This is one of the major areas of the QA program in which the SSCL is breaking new ground with industry. It is expected that during the next year this link will be firmly established.

TABLE 2. DATA FLOW DIAGRAM

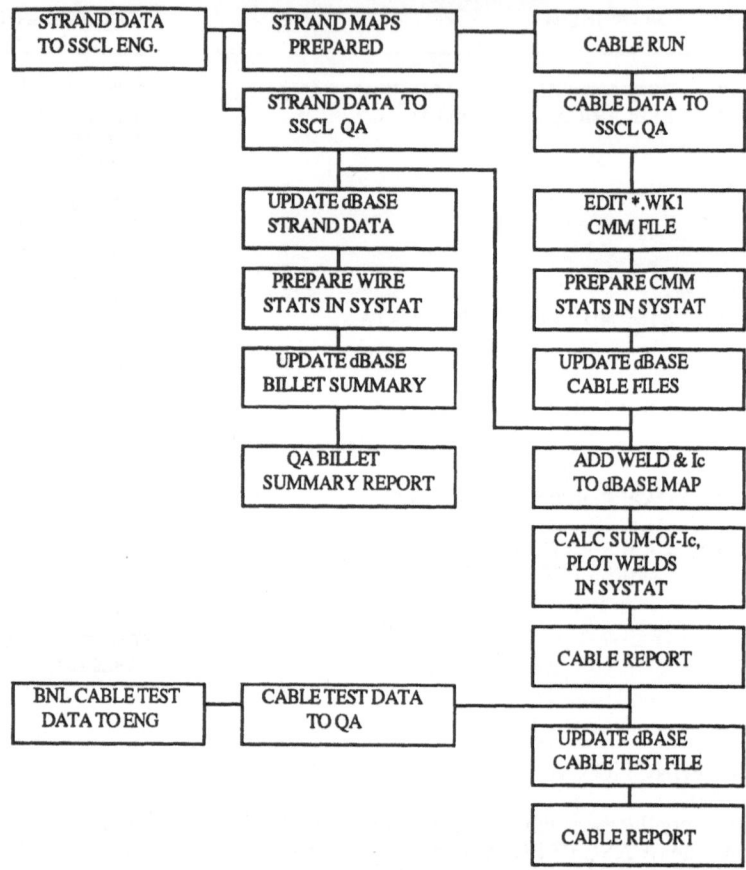

DATA ANALYSIS

Data analysis is performed with SYSTAT version 5.0 [11] a command language statistical analysis application available for IBM/DOS, Apple/Macintosh, DEC/VAX and UNIX operating systems. The goal of the QA Group is to establish "automated" routines to perform primary data reduction tasks. At this time conductor performance data is transferred to SYSTAT format for statistical analysis and graphing through the use of commands as desired. Statistics for cumulative billet pieces are used as input for a billet summary/history dBase file. Periodically, reports of conductor performance are prepared for management using these tools [12]. One of many types of data manipulation available with SYSTAT is shown below, Chart-1: Box Plot: Outer Billet I_c (@ 5.6T). Before considering the graph, a brief explanation of box plots is provided.

Box plots are a tool for exploratory data analysis. They provide a means of represent data distribution values graphically, but without showing each and every point. A box plot provides a "picture" of the middle of a data set, roughly how spread out the middle is, and how the tails (or extreme values) relate to the middle. With one look the viewer can easily form impressions of overall level, amount of spread, and symmetry in the data. What makes them particularly useful is that with SYSTAT an entire data set (for example: I_c by billet) containing hundreds of individual observations can be summarized with one simple command. In other words box plots provide a "quick and dirty" way of getting meaningful information out of a lot of data [13]. The specific elements of a box plot as implemented in SYSTAT are described below [14].

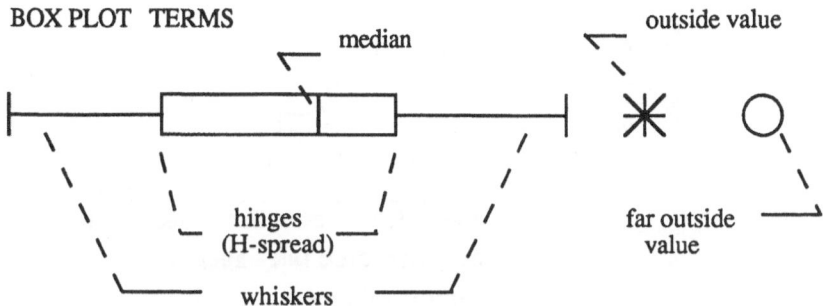

BOX PLOT TERMS

MEDIAN   The middle value of an ordered set of observations. The median splits the data set in half.

HINGE   The middle value of an ordered half from the data set.

EXTREME VALUES   (Outliers)

Outliers are values so low or high they seem to stand apart from the rest of the data set. When calculating simple statistics, such as the mean and standard deviation, outliers can have an unrealistic or exaggerated effect on the results. Extreme values may or may not be "real" however they should be investigated. In box plots two terms are used to denote outliers, "outside" values and "far outside" values.

Outside values are defined by "inner fences", and plotted as asterisks.

Inner Fences: lower hinge - (1.5 x H-spread)   upper hinge + (1.5 x H-spread)

Far Outside values are defined by "outer fences", and marked with empty circles.

Outer Fences: lower hinge - (3 x H-spread)   upper hinge + (3 x H-spread).

BOX   The box is simply a drawing defined by the hinges (hinge spread).

WISKER   A line which approximately denotes the range (min to max) of the data. Specifically, wiskers extend to the outermost data value on each end that is still not beyond the corresponding inner fence. In SYSTAT the whiskers show the range of values which fall within 1.5 H-spreads of the hinges. They do not necessarily extend to the inner fences.

EXAMPLE FROM THE SSCL CONDUCTOR DATABASE

Outer strand I_c for wire delivered in 1990 will be described in Charts 1, 2 and 3. Chart-1: Box Plot displays individual billet I_c distribution for 8 billets (a total of 304 wire sample measurements). Added to the display are ± 3 sigma limits, based on the data set average and sample standard deviation. As a comparison the same data is displayed as a typical Frequency Histogram, in Chart-2. Summary statistics for I_c and Cu:Sc Ratio are provided as well.

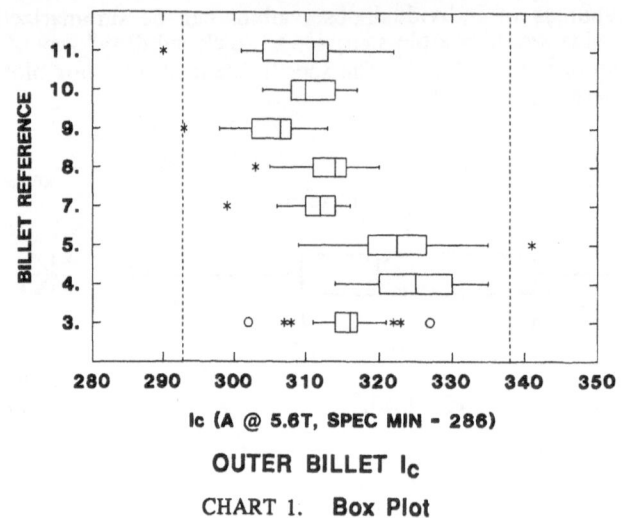

Ic (A @ 5.6T, SPEC MIN = 286)

OUTER BILLET I$_C$

CHART 1. **Box Plot**

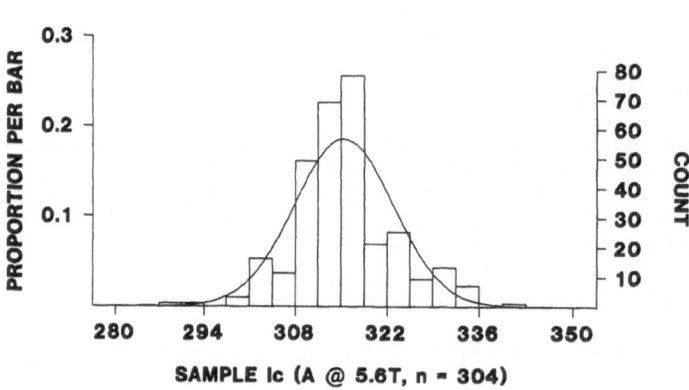

SAMPLE Ic (A @ 5.6T, n = 304)

CHART 2. **Frequency Histogram**

BILLET NUMBER:	≤ 5		≥ 7		COMBINED	
	I$_C$	Cu:Sc	I$_C$	Cu:Sc	I$_C$	Cu:Sc
MEASUREMENTS	160	160	144	144	304	304
MEAN	319.66	1.77	310.63	1.79	315.38	1.78
STANDARD DEV	6.69	0.04	5.21	0.02	7.53	0.04
%C.V.	2.09	2.51	1.68	1.24	2.39	2.06

(%C.V. = [(1 Sigma / Mean) * 100])

Billet Numbers ≤ 5 and ≥ 7 are from different production lots. All were produced by one supplier. With the box plot one can see billets 4 and 5 stand out from the others, both in average I$_C$ and in the magnitude of I$_C$ spread. This difference is closely related to Cu:Sc ratio, as shown in Chart-3. To the degree that strand Cu:Sc variation along the length of a billet is both "real" and "controllable", uniformity of conductor performance will be affected.

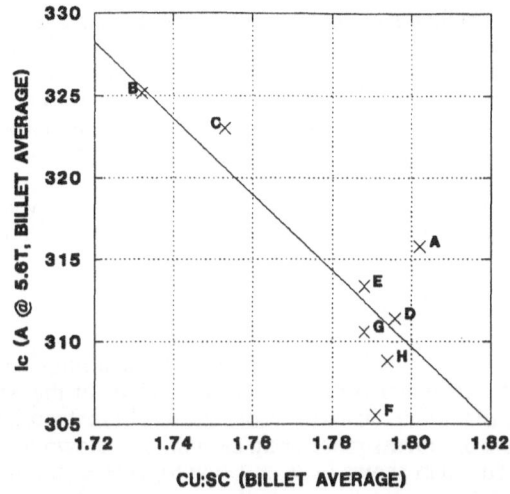

CHART 3. **Outer Billet I$_c$ & Cu:Sc Variation**

Note: The letter labels above, correspond to the billet reference numbers in chart 1, (i.e. A = 3).

CABLE MEASURING MACHINE DATA

A CMM Data disk with LOTUS .WK1 file of required data is being provided by SSCL Engineering after each cable run. The Lotus file is manually edited for use in SYSTAT. Summary reports and graphs are generated from the CMM data, (for individual cables as well as groups of cables, by measured attribute). Much of this work is handled in QA by two programmed menu systems for 50mm and 40mm CDM Cable. Chart-4 is a summary of cable mid-thickness for 8 pieces of 50mm CDM cable totalling 4,458m (14,624 feet). The chart shows the cable mid-thickness average and ± 3 sigma range. Tolerance limits are placed at 0.0457" and 0.0453".

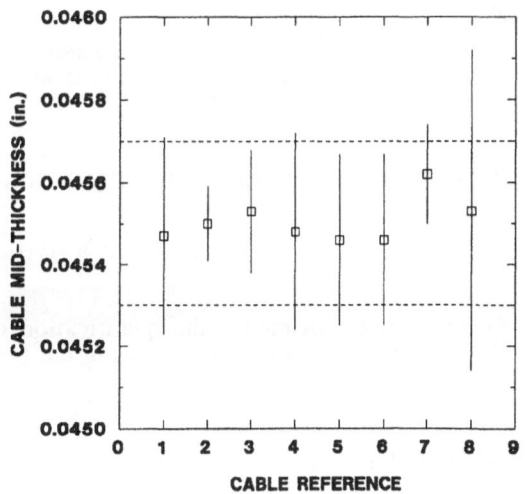

CHART 4. **50mm CDM Outer Cable Mid-Thickness**

Chart-4 above shows cabling is being controlled such that the average thickness is generally near the nominal. However, thickness variation within these cables, as described by the ± 3 sigma range, is generally consuming the allowable specification range. The cable specification is not currently being interpreted "statistically". Presently, cable acceptance is based on CMM results within the tolerance limits. The CMM is programmed to need to see 3 data points out-of-spec before alarming. Analysis of the implications of these difference is needed.

Additional cable run data sheets, including strand maps, are received from Engineering with the CMM disk. This data is still being manually entered into the QA database.

CABLE ELECTRICAL TEST DATA

Cable test data is currently being provide by BNL [15] through Engineering on a periodic basis, (BNL measures "full" cable performance). Once received at the SSCL, this data is being stored in a QA dbase file which relates the cable test results to the original cable strand map I_c data, Another example of process monitoring capability is shown in Chart-5: Strand Map To Cable Degradation. The chart shows I_c degradation for recent 40mm and 50mm CDM cable, as calculated from BNL measurements and also as compared to the Sum-of-Strand I_c from strand map data. The difference between the two is believed to be due to the relative "accuracy" of the measurement sources (BNL and the wire supplier), as discussed in a separate IISSC paper [16].

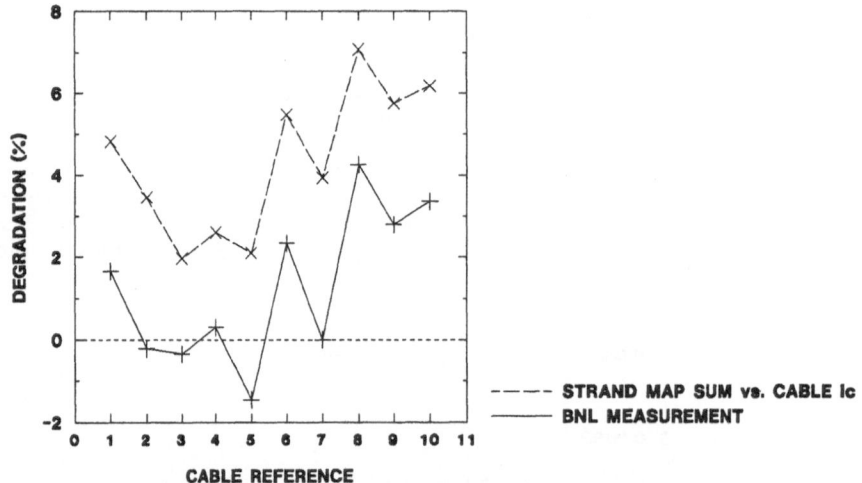

CHART 5. **STRAND MAP TO CABLE I_c DEGRADATION (OUTER)**

NOTE: Cable Reference Numbers above do not match those shown in chart 3.

The current cable specifications provide for the qualification of "extracted strand" measurements for cable verification by the supplier during the Conductor Qualification program.

MAGNET PRODUCTION DATA

Magnet Production Data beyond cabling is not being processed as part of the QA database at this time. Work in this area is being organized by the Magnet Systems Test and Data Management Section and by the several sites for magnet production (FNAL, BNL, LBL, KEK, SSCL MDL, Saclay, and Industry). However, in order for magnet producers and users to gain maximum benefit from the conductor data, work needs to be done to match actual cable usage during coil winding to the source cable strand map. Additionally, a reference table matching magnet, coil, and cable identities needs to be established and maintained.

dBASE TO SYBASE TRANSFER

The current method of transfer is to give the Test and Data Management Group a data disk with the appropriate QA dbase files. The data is converted directly to SYBASE form for use in the publicly available database MAGCOM. This happens about once a month or as requested. Direct electronic transfer is now possible, all we need to do is establish the desired procedure.

MATERIAL SPECIFICATION QA REQUIREMENTS

The current SSCL CDM Conductor Specifications are listed in the references [9]. The following are the "Quality" specific requirements from the wire and cable specifications. Items marked with a "*" in the left margin indicate requirements, topics, or levels of activity which are "new" to the superconductor materials industry. Many of these ideas have been developed in consultation with others in the national laboratory community [17,18].

	A.	Traceability to source materials:
		(a.) Wire: Mono-Billet, NbTi, Nb, Cu.
*		(b.) Cable: Multi-Billet, "Production Unit", Billet "Segment".
*	B.	Process History by billet "Production Unit".
*	C.	SSCL "Change Control" over Manufacturing Plan.
*	D.	SSCL "Change Control" over Manufacturing Machines and Methods.
*	E.	Application of Statistical Process Control, (SPC).
	F.	Control of Non-Conforming Material.
*	G.	Cable "I_C Strand Mapping", for Sum-of-Strand I_C Variation Control.

The following are highlights of the key Quality Management Requirements from SSCL-QA-1:

- A. Establishment of an effective supplier QA program.
- B. Development of written detailed process procedures.
- C. Establishment of cleanliness control procedures.
- D. Use of a supplier program for subcontract control.
- E. Measuring and test equipment calibration.
- F. SSCL source inspection/surveillance of the conductor supplier and his subcontractors as necessary.

Test and Inspection of wire and cable

*	A.	Physical & Chemical Properties of source materials.
	B.	Mechanical & Electrical Properties.
*	C.	Sampling Levels for tests, during Phase I of the Qualification Program.

CONCLUSION

The past year of SSCL operations has provided a unique opportunity to prepare a comprehensive "Quality Plan" as well as data collection and analysis tools for SSC superconductor materials. Of invaluable significance has been the opportunity to integrate these plans and ideas as part of the Engineering Specifications for wire and cable. The establishment of close ties between Engineering, Test & Data Management, and Quality at the early stages of planning for the SSC is key to our ultimate success. In the year ahead, we look forward to completing the building of a strong foundation for the future, with the addition of Industry participants.

Acknowledgement

This work was supported by the Director, Office of Energy Research, Office of High Energy and Nuclear Physics, High Energy Physics Division, U.S. Department of Energy, under Contract No. DE-AC02-89ER40486.

References

1. Request For Proposal SSC-90-A-000573, Development And Production For Superconducting Wire and Cable, SSCL, September 7, 1990.

2. SSCL-SR-1056, Site-Specific Conceptual Design Report, July 1990, Table 4.2.1.1-1, pg. 286.

3. D. Pollock (SSCL). Introduction to the SSCL MSD QA Group. Presentation to US Industry, April 1990.

4. D. Pollock. Collider Dipole Conductor: A review of Current Quality Assurance Issues. Presentation to Japanese Industry, July 1990.

5. T. Ferbel (Chairman). General Remarks on the Organization and Contents of the Proposed Cable Data Base. SSC-N-582, March 23, 1989.

6. M.J. Baggett (BNL), R. Leedy, C. Saltmarsh, J. Tompkins (SSCL). The Magnet Components Database System. IISSC 1990.

7. D. Pollock. Database Development and Quality Analysis: A report on recent efforts to establish a Quality Information System at the SSCL. Magnet Systems Integration Meeting (MSIM) Presentation at LBL, January 17, 1990.

8. D. Pollock. SSC Cable Database System, (dBaseIII Plus application programs and data files).

9. Current SSCL CDM Conductor Specifications:
    NbTi Bars, SSC-Mag-M-4000A; Nb Barrier, SSC-Mag-M-4001;
    NbTi Wire (Inner), SSC-Mag-M-4145; NbTi Wire (Outer), SSC-Mag-M-4146;
    NbTi Cable (Inner), SSC-Mag-M-4147; NbTi Cable (Outer), SSC-Mag-M-4148,
    Seller Quality Assurance Requirements, SSCL QA-1.

10. D. Pollock, Superconductor Critical Current Variation: A review of current work on the study of SSC Conductor Variation, as described by Critical Current and related parameters. MSIM Presentation at SSCL, March 20, 1990.

11. SYSTAT, Version 5.0 for DOS. Copyright @ 1990, SYSTAT, Inc. 1800 Sherman Avenue, Evanston, IL 60201-3793.

12. D. Pollock, Superconductor Wire & Cable Quality Assurance Activities. MSIM Presentation at SSCL, November 13, 1990.

13. P. Velleman and D. Hoaglin. Applications, Basics, and Computing of Exploratory Data Analysis. Boston, MA: Duxbury Press, PWS Publishers, 1981.

14. L. Wilkinson. SYGRAPH: The System for Graphics. Evanston, IL. SYSTAT, Inc., 1990.

15. M. Garber, et.al. (BNL). Superconductor Cable Critical Current Measurements.

16. M. Erdmann (SSCL), et.al. Quantification of Systematic Errors In I_c Testing, IISSC, V-F-9, 1991.

17. A. Greene, (BNL). Discussions on RHIC and SSC conductor specification development, meeting February 8, 1990.

18. D. Capone (SSCL), et.al. Cable Specification Review Meeting, held at LBL, January 17, 1990.

QUANTIFICATION OF SYSTEMATIC ERROR IN I_c TESTING

Mark J. Erdmann, Douglas A. Pollock, and
D.W. Capone II

Magnet Division
Superconducting Super Collider Laboratory*
2550 Beckleymeade Avenue
Dallas, TX 75237

Abstract: Critical current (I_c) testing of the superconducting wire for the Superconducting Super Collider (SSC) is an important Quality Assurance concern due to its significance in magnet variability. Established industrial QA procedures to quantify measurement variability have been adapted to I_c measurements. To implement these ideas, a round robin experiment was developed and six test sites were invited to participate in the evaluation of I_c measurement uncertainty. An SSCL Quality Assurance representative witnessed all measurements performed in this program. The definition of each component of variability, the test plan procedure that quantified each component, preliminary findings, and future plans are reported. Preliminary results show that four of the six round robin participants achieved accuracy within +/- 2% of the National Institute of Standards and Technology critical current Standard Reference Material value. Reproducibility results show that five of six test sites have a 2% or less uncertainty, with one test site as high as 6.5%. In addition, the repeatability results show that only one participant has less than 2% uncertainty; the remaining labs falling within 2–5%. Cumulative uncertainty is determined to be 3–10% depending on the method used to calculate repeatability.

INTRODUCTION

In order to make good management decisions from reported I_c measurement data, gage calibration must be guaranteed by all testing sources. By measuring and controlling test error, the Superconducting Super Collider (SSC) Laboratory hopes to ensure that long-term accurate I_c values are reported from all possible sources during the entire SSC construction program. This is vital to implement the current philosophy of cable strand mapping.[5,6] Strand I_cs from different billets are to be intermixed into cable to minimize variation from magnet to magnet. In the enclosed test plan, the systematic error has been broken down into components

---

*Operated by the Universities Reasearch Association, Inc., for the U.S. Department of Energy under Contract No. DE-AC02-89R40486.

which include accuracy, linearity, reproducibility, and repeatability. These components take into account both the device-specific and operator-induced error experienced in I_c testing. Of particular interest is the component of variability due to operator technique. To our knowledge, this element has never been quantified. This method of error evaluation is based on an established Quality Assurance model for measurement calibration and, to our knowledge, has never been performed for this particular application.[1-4]

DEFINITION/PROCEDURE

Accuracy

Measurement accuracy is defined as the difference between the observed average of measurements and the true average of the same parts using precision instruments.

Linearity

Measurement linearity is defined as the difference in the accuracy over the expected operating range.

To determine this portion of the uncertainty, each participant was given a National Institute of Standards and Technology Standard Reference Material (SRM) 1457. SRM 1457 is a NbTi superconducting wire certified by NIST as an I_c standard.[7,8] The SRMs were purchased consecutively, guaranteeing that each came from the same section of the parent billet.

One certified operator from each site mounted the SRM onto the sample holder and lowered the sample into a liquid helium-cooled superconducting magnet. The magnet was ramped up to 8.0 T and three voltage vs. current curves were generated at this same applied field. The applied magnetic field in the magnet was not changed between the three tests. The critical current was then calculated from each curve using a resistivity criterion that was converted from the SRM certified electric field criterion of 0.2 μV/cm.[7,8] The same operator repeated the same procedure at magnetic fields of 7.0, 6.0, 5.6, and 4.0 T.

The I_c measurement accuracy was determined by taking the difference between each field's SRM value and the observed average of three measurements for the laboratory. The difference was converted to percent uncertainty by dividing by the SRM value. The linearity for the test site was computed using the accuracy values determined above. The field with the lowest percent uncertainty was subtracted from the field with the greatest percent uncertainty.[1,2]

The SRM I_c value was corrected for each test site's liquid helium bath temperature using the I_c temperature correction table in NBS Special Publication 260-91, which was received with the SRM.[8] The liquid helium bath temperature was computed from the absolute atmospheric pressure at the time of the test.[9] Because the SRM is certified only at even fields, the 5.6 and 7.0 T values were determined by interpolation from the even fields.[8] This method is valid because the I_c vs. magnetic field curve is nearly linear at the fields in question. Sample self field corrections on the I_c were included only if the test site typically corrects the measurements that it submits.

Reproducibility

Measurement reproducibility is defined as the range between the average of the I_c measurements made by different operators using the same test facility when measuring identical wire samples. All certified I_c measurement operators were invited to take part in this test. Each operator was given a contiguous SSC-type 1.3:1 inner grade wire sample (from the same source spool) which he/she mounted to the I_c sample holder. Three voltage vs. current curves were produced by each operator at 7.0 T, and the I_c was determined using a resistivity criterion of 10^{-14} Ω-m. The magnetic field of 7.0 T was chosen for data comparison as it is the SSC inner grade wire specified test field. The field was held consistent for all three measurements.

To determine reproducibility, first the average I_c value was calculated separately for each operator. The range in amperes was determined by subtracting the smallest average I_c from the largest average I_c. This value was then converted to percent uncertainty by dividing the average I_c range by the grand average of all the operators' I_c results in the given facility.[1,2]

Repeatability

Measurement repeatability is defined as the variation in measurements obtained when one certified operator uses the same test facility for measuring the I_c of the same source sample. One certified test site operator was given three contiguous 1.3:1 type inner grade SSC wire samples which he/she mounted on an I_c sample holder. Three voltage vs. current curves were produced at 7.0 T for each sample, and the I_cs were determined, again using a resistivity criterion of 10^{-14} Ω-m. A total of nine measurements were made on the same source wire by each test site.

Repeatability was determined using the two methods described below. Both assume that the I_c measurement variation is normally distributed, an assumption that has been verified by the data.

Method I

The I_c range in amperes was determined for each of the three samples, and the average range was computed from the three ranges. The average range was divided by d_2, which is a factor to estimate the standard deviation of the population using the range.[10] The 95% (+/- 1.96 sigma) spread was determined by multiplying the estimated standard deviation by two times the respective Z factor. Finally, the spread was converted to percent uncertainty by dividing the +/- 1.96 sigma spread by the average I_c of the test site's nine measurements, then multiplying by 100.[2-4]

Method II

The calculation in Method II is similar to Method I except that the standard deviation was determined from all nine measurements directly. To compare results with Method I, the 95% spread was calculated (as in Method I) and converted to % uncertainty by dividing the +/- 1.96 sigma spread by the average of the nine measurements and multiplying by 100.[10]

The reason for calculating the percent uncertainty using Method II is that currently the I_c data is taken using the "1-test-per-1-spool-of-wire" technique. Although the Method I test plan was performed by taking three I_c tests per sample, Method II looks at the data as if there were nine separate measurements, any one of which could have been reported had this been a typical industry verification. Method II uncertainty can be compared to that determined using Method I where the range of three tests was used to estimate the population standard deviation.

Miscellaneous Data

In addition to the critical current results, other pertinent data were collected such as the strand diameter and the length between voltage taps on the sample holder. The copper to non-copper ratio was calculated using both the chemical method and the electrical resistance method on one sample of each type of wire. This data will not be presented in this paper, but each method will eventually be statistically evaluated.

Note: The test sites listed in the following results are identified by an arbitrary number (1-6). This number has no bearing on the outcome or rating of the test sites. Another paper[12] presented at the 1991 IISSC uses colors instead of numbers to identify certain test sites. For comparison, the number with its corresponding color is given: 2 = brown, 3 = blue, 5 = black, 6 = purple.

RESULTS

Table 1 lists all the participants accuracy data for each tested field.

Figure 1 graphically compares the percent uncertainty for each laboratory at all fields. Currently, a limit has not been determined for an acceptable accuracy uncertainty, but if a +/- 2% limit is arbitrarily picked, two of the six laboratories would fail. Test Site 2 was the only participant that corrected its measurements for self field and, as seen by the data, produced the lowest percent uncertainty due to accuracy. The SRM does not assume the application of self field corrections, however.

The accuracy results also show that the temperature dependence of I_c (SRM I_c variation) is significant, and that a temperature correction should be included when I_c data is reported. This confirms previously reported results.[9]

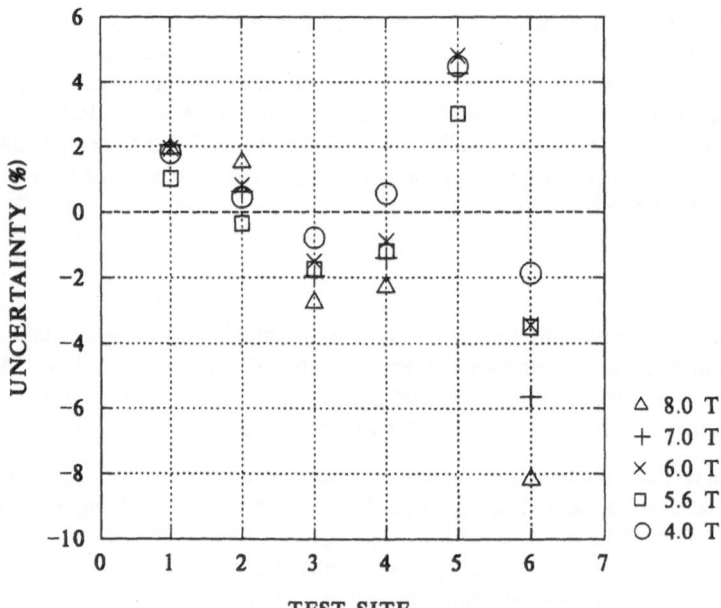

Figure 1. Accuracy percent uncertainty for each test site.

Table 1. Accuracy.

4.0 T LAB	I_c AVERAGE (Amps)	SRM I_c (Amps)	% DIFFERENCE
1	191.27	187.88	1.803
2	186.87	186.03	0.450
3	185.00	186.48	-0.794
4	187.87	186.82	0.563
5	195.00	186.63	4.485
6	182.67	186.16	-1.874
5.6 T LAB			
1	139.13	137.72	1.026
2	135.50	135.98	-0.353
3	134.00	136.41	-1.767
4	135.07	136.72	-1.209
5	140.67	136.55	3.015
6	131.33	136.10	-3.502
6.0 T LAB			
1	127.60	125.18	1.933
2	124.47	123.47	0.806
3	122.00	123.89	-1.526
4	123.10	124.20	-0.886
5	130.00	124.03	4.813
6	119.33	123.59	-3.444
7.0 T LAB			
1	99.67	97.68	2.034
2	96.57	95.98	0.611
3	94.50	96.39	-1.961
4	95.33	96.71	-1.418
5	100.67	96.54	4.280
6	90.67	96.10	-5.652
8.0 T LAB			
1	71.53	70.17	1.943
2	69.53	68.50	1.516
3	67.00	68.90	-2.758
4	67.63	69.21	-2.278
5	N/A	N/A	N/A
6	63.00	68.61	-8.177

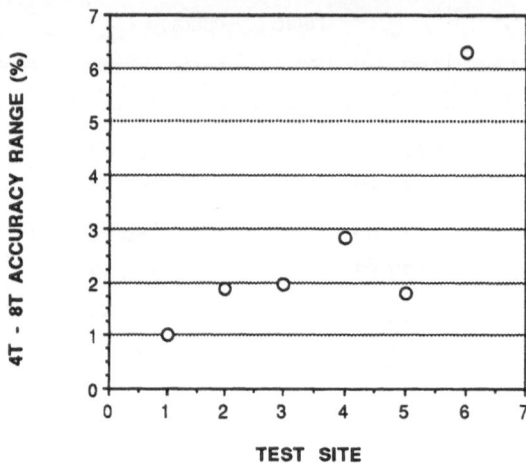

Figure 2. Linearity for each test site over the magnetic fields of 4–8 T.

Figure 2 lists each test site's linearity using the above accuracy data.

If a test site exhibited a 0% range in accuracy over the fields measured, this would mean that the error (or lack of error) was consistent at all fields, i.e., all the I_c values deviated by the same amount. Applying an arbitrary 2% limit, again two labs failed, with three others near the limit. This suggests that the variation in I_c with field is not consistent for most of the test sites, and that causes of field-dependent error should be found and corrected.

Table 2 lists the reproducibility data for each participant.   These results can be seen in Figure 3.

Table 2. Reproducibility Data.

LAB/OPERATOR	I_c AVG. (Amps)	RANGE (Amps)	% UNCERTAINTY
1/A	392.27		
1/B	386.87	5.40	1.39%
2/A	364.97		
2/B	367.20		
2/C	390.10	25.13	6.72%
3/A	375.33		
3/B	373.00	2.33	0.62%
4/A	375.97		
4/B	382.20	6.23	1.64%
5/A	396.67		
5/B	398.00	1.33	0.34%
6/A	377.33		
6/B	374.00	3.33	0.89%

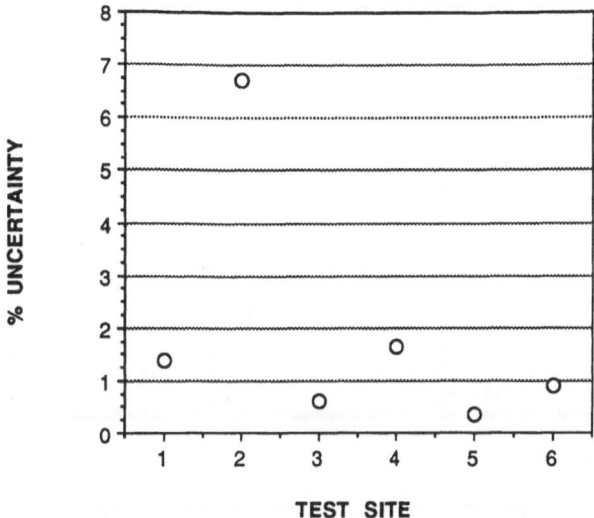

Figure 3. Percent uncertainty due to reproducibility for each test site.

From Figure 3, it appears that Lab B has a very large percent uncertainty due to reproducibility. This demonstrates that Lab B needs to carefully evaluate operator technique. All of the other remaining labs have roughly the same percent uncertainty ($< 2\%$). An even better 1% limit appears to be attainable by five of the six test sites through further process control.

Table 3 lists the repeatability data in tabular form.

Table 3. Repeatability.

Method I (95% spread = (I_C Avg Range/d_2) * 1.96_z * 2)

LAB/SAMPLE	I_C AVG. (Amps)	95% SPREAD (Amps)	% UNCERTAINTY
1/A	392.27		
1/B	384.67		
1/C	390.20	4.13	1.06%
2/A	364.97		
2/B	360.93		
2/C	364.67	13.51	3.72%
3/A	375.33		
3/B	373.00		
3/C	373.00	0.74	0.20%
4/A	375.97		
4/B	381.33		
4/C	376.40	3.10	0.82%
5/A	396.67		
5/B	395.67		
5/C	397.67	11.07	2.79%
6/A	377.33		
6/B	376.67		
6/C	377.33	8.86	2.35%

Table 3. Repeatability (Continued).

Method II (95% spread = sample $\sigma * 1.96_z * 2$)

LAB	I_c AVG. (Amps)	95% SPREAD (Amps)	% UNCERTAINTY
1	388.76	13.84	3.56%
2	363.71	17.64	4.85%
3	373.81	4.71	1.26%
4	378.26	10.44	2.76%
5	397.15	12.55	3.16%
6	376.78	7.95	2.11%

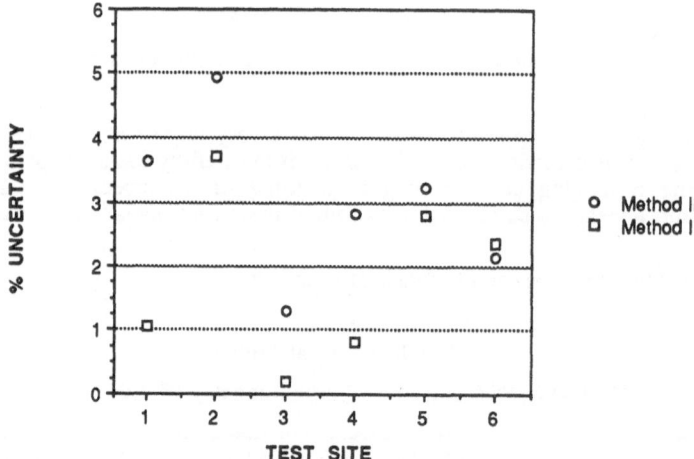

Figure 4. Percent uncertainty due to repeatability for each test site.

The results of Table 3 can be seen in Figure 4.

Figure 4 shows that if an arbitrary 2% limit is applied to the data, five of six participants fail when calculating the repeatability using Method II and three of six fail using Method I. These results suggest that one way to decrease uncertainty would be to increase the number of measurements per sample and to report the average.

Figure 5 shows the sum of the uncertainty measured for each lab. The mid-point for each test site is the relative accuracy determined with the SRM. The spread around accuracy is based on the repeatability plus the reproducibility. The repeatability method depicted uses the standard deviation of all nine measurements (Method II) to simulate the error of one I_c test per spool.

The sum of the absolute uncertainty is given in Figure 6. The ASTM B714-82 procedure for I_c measurements states that using the procedure should give an I_c accurate to within 5%.[11] It is unclear, however, how ASTM is defining "accuracy." If the 5% benchmark is used on the total uncertainty, four of six labs would fail.

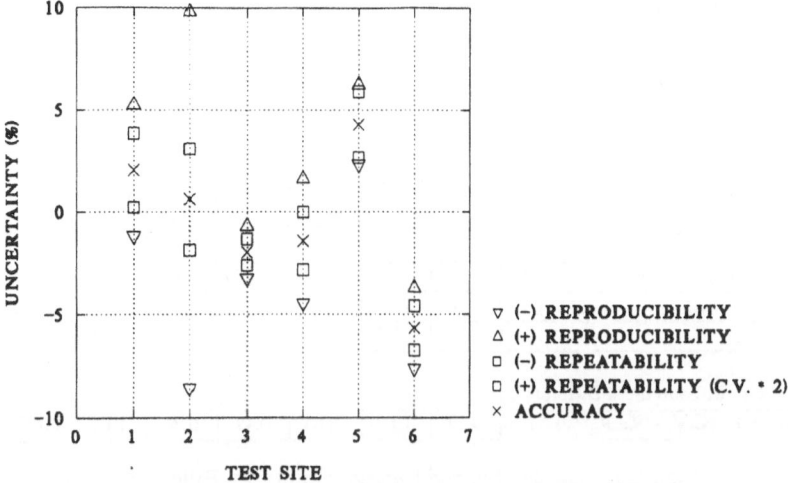

Figure 5. The sum of uncertainty for each test site.

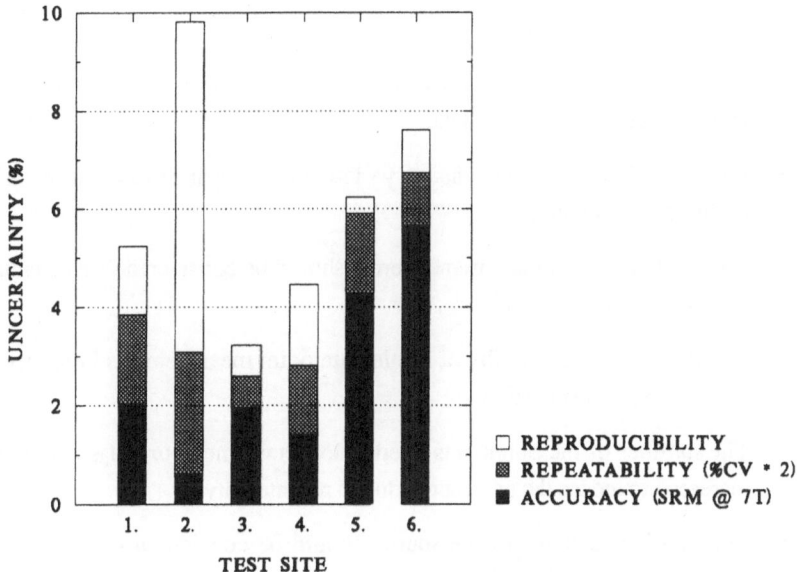

Figure 6. The sum of absolute uncertainty for each test site.

Impact on Cable Mapping

Figure 7 is a plot of the one sigma variation of several inner and outer grade SSC superconducting billets produced by one of the test sites. It also lists the one σ uncertainty for that particular test site due to reproducibility and repeatability. Upon comparison, it appears that uncertainty in the measurement is nearly equal to the variation about the billet mean for the samples measured. This condition suggests that it may be difficult to determine whether the variation was due to local differences in the wire or to measurement error.

TEST UNCERTAINTY (LAB 5)	%	
REPEATABILITY	0.807	(1 * Sigma / Mean) * 100, n = 9
REPRODUCABILITY	0.335	(Avg. Tech Range / Avg. Ic)
R&R	1.141	(Repeatability + Reproducability)

Billet Ic Variation								
Wire Type	IN	OUT	OUT	OUT	OUT	OUT	OUT	OUT
Number of Samples	55	29	55	28	17	83	29	48
Mean	369.6	311.4	313.4	305.5	310.6	315.8	325.1	323.0
Standard Deviation	5.68	3.51	3.42	4.76	4.35	3.49	6.28	7.09
Coefficient of Variation (C.V.)	0.015	0.011	0.011	0.016	0.014	0.011	0.019	0.022
% C.V.	1.54	1.13	1.09	1.56	1.40	1.10	1.93	2.19

Materail - Measurement Difference								
BILLET %C.V. - R&R (%)	0.39	-0.02	-0.05	0.42	0.26	-0.04	0.79	1.05

Figure 7. Repeatability and Reproducibility vs. Billet Variation.

IMPLICATIONS OF THE RESULTS

- Temperature dependence on I_c measurement is significant; a correction should be included when reporting I_c data.

- It is difficult to determine whether variation in I_cs is due to local wire variation or to measurement error.

- The implications of measurement error should be considered when evaluating I_cs near the specification limit.

- The significance or validity of single data point measurements in I_c performance studies may be in question.

- The number of measurements needed to determine source I_cs may have to be increased to reduce the uncertainty due to repeatability.

- The relative accuracy of the source should be considered when comparing or judging the results from different sources.

FUTURE PLANS

The strategy to be used in evaluating the test sites will be to first determine an agreeable uncertainty for both the components and the total variability, and compare it to each participant's error. The results will then be reviewed with each participant. Causes of unacceptable uncertainty will be identified and corrective action taken. Finally, participants with unacceptable uncertainty will be asked to repeat certain portions of the test plan after corrective action has been taken.

SUMMARY

Two of six labs have uncertainty due to accuracy of less than 2%.

Two of six labs have large variation over their applied field, suggesting that their error is not consistent at all fields.

The reproducibility of five of six labs was less than 2%, with the other lab exceeding 6% uncertainty for this component.

The 95% spread for repeatability using the range to estimate the standard deviation (Method I) was between 0.2% and 4%, while the 95% spread using the sample standard deviation of the nine measurements (Method II) was between 1% and 5%.

The total percent uncertainty for each lab fell within the range of 3% to 10% (Method II repeatability). Four of six labs have greater than a 5 % uncertainty.

ACKNOWLEDGMENTS

The authors would like to thank Dr. Loren Goodrich of the National Institute of Standards and Technology and Dr. Ron Scanlan of Lawrence Berkeley Laboratory for their help in developing and evaluating the test plan.

REFERENCES

1. "Measurement System Evaluation," General Motors Statistical Process Control Manual.

2. H.C. Charbonneau and G.L. Webster, "Evaluating Gage Capabilities," *Industrial Quality Control*, Prentice-Hall, Inc., New York, 1978.

3. *Statistical Quality Control Handbook*, Western Electric Co., Indianapolis, 1956.

4. R. Travor, "Measurement Equipment Repeatability—The Rubber Ruler," *ASQC Convention Transactions* (1962).

5. SSCL Spec No. SSC-Mag-M-4148, "NbTi Superconducting Cable for SSC Dipole Magnets (Outer)," 1990.

6. SSCL Spec No. SSC-Mag-M-4147, "NbTi Superconducting Cable for SSC Dipole Magnets (Inner)," 1990.

7. S.D. Rasberry, *National Institute of Standards Certificate for Standard Reference Material 1457, Superconducting Critical Current—NbTi Wire*, Gaithersburg, MD, 1984.

8. L.F. Goodrich, D.F. Vecchia, E.S. Pittman, J.W. Ekin, and A.F. Clark, NBS Special Publication 260-91, *Standard Reference Materials: Critical Current Measurements on an NbTi Superconducting Wire Standard Reference Material*, U.S. Government Printing Office, Washington, D.C., 1984.

9. L.F. Goodrich and F.R. Fickett, "Critical Current Measurements: A Compendium of Experimental Results," *Cryogenics*, May 1982.

10. J.M. Juran and F.M. Gryna, *Juran's Quality Control Handbook*, Fourth Edition, McGraw-Hill, 1988.

11. "ASTM Standard Test Method for D-C Critical Current of Composite Superconductors," Designation B714-82.

12. J.M. Seuntjens and D.W. Capone II, "Issues and Results in the Development of 2.5 Micron Filament Strand for the HEB," IISSC Atlanta, GA, 1991.

11. Using the SSC

HOW DETECTOR COLLABORATIONS EVOLVED;

A PERSONAL RECOLLECTION

John Peoples, Jr.

Fermi National Accelerator Laboratory*
P.O. Box 500
Batavia, IL 60510
USA

Abstract: The scientific goals of the SSC mean that its energy and luminosity will be greater than any previous accelerator-- and consequently that experimenters will be building bigger and more complex detectors in larger collaborations than ever before. I trace the evolution of detectors and experimental groups, and their relationship to the laboratories at which the experiments are carried out, over the thirty years that I have spent in particle physics, to illuminate the lessons of the past for shaping comfortable, practical, manageable, and effective experimental collaborations at the SSC.

INTRODUCTION

The experimental physicist's perspective of using the SSC, the theme of this session, will depend on how the SSC scientists and SSC Laboratory will organize themselves to build and use their detectors. These are active issues for the SSCL management. I thought that it would be useful to recall how detector collaborations have evolved into their current size and organization. When I was a Ph.D. student nearly thirty years ago, an experiment consisted of five to six collaborators from one institution. The success of those experiments has led us to ask questions whose answers often lie in less easily accessible realms. As detectors have had to increase in size and complexity, collaborations and laboratories have had to create more formal and elaborate organizations. The prospect of research groups of a thousand people causes many to long for the days when a handful of people decided on Friday to do an experiment and started assembling equipment Saturday. Those days are long gone, but with the help of

---

* Operated by the Universities Research Association, Inc. for the U.S. Department of Energy under Contract No. DE-AC02-76CH03000.

industry and the potential to adapt we have demonstrated in the past, we will figure out how to accommodate ourselves.

THE CURRENT SITUATION

The Letter of Intent submitted by the Solenoid Detector Collaboration (SDC) lists a huge number of collaborators: 631 physicists from 95 institutions. Why does the collaboration have to be so immense? Why does it need so many institutions? This detector is a significant step beyond existing detectors because the SSC performance specifications are very dramatic steps forward in energy and luminosity. The Tevatron, today's forefront hadron collider, is by any measure ahead of its competition, and yet the SSC energy is twenty times greater than Fermilab's Tevatron and its luminosity is two hundred times greater than the Tevatron will achieve in 1992. An SSC detector must be sufficiently large and sufficiently complex to complement the SSC collider parameters. Each detector by itself is a major research instrument, exceeding the accelerators in complexity and the degree to which the limits of technology are pushed. It will take all 631 people eight or nine years to design it, build it, commission it and use it.

It will also require the help of industry. The technology to discover and process the information at the SSC simply does not exist in the market place at this time. A vigorous R&D effort will be required, and scientists cannot achieve this alone. Industry will have to play a role.

A SIMPLER TIME (1962 - 1969)

Fifty years ago, the nucleus of the atom was the high-energy frontier. Smart, ambitious physicists like Fermi, Lawrence, and Rabi formed teams of five or six people hoping to be first to reach an understanding of the properties of the nucleus. They used their special talents to divide the experiments into a few well-defined pieces and then integrated them. The quality of the discoveries they and their proteges made attest to the success of their approach.

I don't want to begin my story quite that early, but rather in 1962, when I started my thesis research at Nevis Laboratories. Nevis had a 400 MeV synchrocyclotron (a very good accelerator for its day), a good large machine shop, three mechanical engineers, two electronics engineers (one of whom was Walter LeCroy), some draftsmen, perhaps fifteen electronics technicians, and a primitive computer.

At that time, a typical experimental team consisted of a senior professor, a junior faculty member, a postdoc, and two or more graduate students. When I was a student there were five such groups at Nevis. The apparatus was built and assembled by the students and the postdocs under the loose direction of the faculty. Components were built at Nevis by the technical staff. Commercial equipment was used when it was available. With the support that Nevis Laboratory provided one could do just about anything. As a rule, the lifetime of an experiment was too short to be able to rely on the development of equipment by industry.

An idea for a detector could be evaluated in a matter of a month or two, barely longer than the time needed to negotiate a contract with a company. Advanced planning was relatively minimal. Nevis experiments did not even require a written proposal. A presentation of an idea to the senior faculty was enough to launch an experiment. Spontaneity and ingenuity were amply rewarded, because it was possible to recover from a mistake in a matter of months. It made sense to be bold, to try an idea even if you weren't sure it would work.

My thesis experiment took two and a half years from the start of construction until the presentation of the first result. It was another year before I finished my thesis. The experiment was longer than most because I spent most of the first year developing and perfecting the basic detector elements. When the experiment was conceived, the only thing that was clear was that my thesis advisor, Alan Sachs, and two staff scientists, wanted to make an exact measurement of the positron momentum spectrum from the decay of positive muons. For the measurement to meet the desired precision they realized that they needed to measure the positron momentum from a hundred million muon decays. They had established earlier that it was feasible to do the experiment if positron momenta could be measured at a rate of 25 hertz to within a few tenths of a percent.

A relatively new device at the time, the spark chamber, offered the promise of making accurate position measurements of a particle track in space. A spark chamber consisted of two flat plates separated by 5 mm immersed in a neon atmosphere. When a charged particle traversed the gap between the plates of the chamber it left a trail of ionization that would persist for a few microseconds. This time was more than enough to detect the presence of the particle with scintillation counters and then apply a 5 kv pulse to the plates of the chamber. A spark discharge formed whenever a charged particle passed through the gap in the previous several microseconds. The spark was visible and even audible. While the spark made a very fine image on film, 35 mm cameras couldn't advance film much faster than 10 frames/second. To make matters worse, after the film had been developed, each view of the event had to be measured. Roughly 20 measurements were required, and a human measurer could not do this much faster than one per minute. The time to measure 50×10^6 pictures was a million hours or 160 measurer-years. This was too slow.

Fortunately, about the time that I asked if I could join the experiment, a colleague reported hearing of a European physicist who had detected the noise rather that the light of the spark. It occurred to my colleagues that if a microphone was placed at each corner of the chamber and if the time it took the sound wave to travel from the spark to each microphone was measured, then the position of the spark in the chamber could be determined. Four microphones would make it possible to determine if there was more than one spark and hence more than one track. Since the velocity of sound in neon is about 0.3 mm/μsec and the dimensions of the plates of the chambers were roughly 30 cm by 30 cm, one could then design a detection system measuring the four times in less than 1000 µsec with a resolving time of 0.2 µsec or 0.1 mm. Since three spark chambers were needed to define the momentum, by using four chambers the set of measurements would allow a consistency verification. My initial responsibility was to build a sonic chamber, the high-voltage pulsing system, and make it work. Already at this time the front-end for the microphones electronics were too difficult to be put in the hands of a student; Nevis had two superb engineers who took care of these matters.

In addition, there were five scintillation counters in the experiment. The pulse area of one of these counters was digitized and recorded. An event consisted of 17 four-digit, binary coded decimal numbers. Treating each four-digit number as a word, the peak data rate was less than 1000 words/second. By today's data analysis standards this was no big deal, but then it required a computer. We used an IBM 1401, a business machine, to allow us to write to the new 556 Bpi magnetic tapes. From the standpoint of the technology of the day the experiment was quite advanced, and some people viewed it as a big unwieldy enterprise.

The experiment took a little more than a year and a half to build once the development work was complete. Everything in the experiment- except for the magnet, its power supply, and the 1401- was built at Nevis. Everyone involved with the experiment worked at Nevis. Communication was informal. My thesis advisor visited me and the other graduate student in the evening several times a week to find out how we were doing. If we were stuck he made suggestions. The staff scientists saw us more often and helped with the labor of data acquisition and analysis. Initially they guided the work of the engineers and programmers until we, the students, could do the work. Over the course of two years the whole group, all five of us, met several times. We each knew what the others were doing. There was no need for transparencies or a group directory. Yet, in spite of its informal organization, this thesis experiment produced the most precise measurement of the Michel parameter, the principal parameter that defines the momentum spectrum of positrons from the decay of unpolarized muons. Twenty-six years later it is still the most precise number.

My next adventure in high energy physics was to do an experiment at Brookhaven with six colleagues from Nevis, by this time I was a junior faculty member. The highest energy accelerator of the day was the Brookhaven AGS at 30 GeV. Competition to do an experiment at the AGS by a user group from Nevis was stiff, so a little formality was needed. Brookhaven experiments required a proposal although one-page was sufficient. The idea had to be read by their High Energy Discussion Group (HEDG). While this committee wasn't even called an advisory committee, it made recommendations to Brookhaven management that had the force of approval or rejection. Brookhaven provided the beam of particles, hydrogen targets, magnets and a limited amount of general purpose electronics. The Nevis group provided everything else. Beam time was awarded in units of a few hundred hours. When a Brookhaven experiment was completed the students and the postdocs disassembled the apparatus and returned to their home institutions with the data. Teams were formed to do one experiment, and generally it could be done in two years, the duration of a postdoctoral appointment. Less than a year of this time was spent at Brookhaven.

CHANGING TIMES (1970 - 1979)

By 1970 I was teaching at Cornell. With colleagues from Columbia, Hawaii, and Illinois, we wrote a proposal to look for the photoproduction of high-mass dimuon pairs at the National Accelerator Laboratory (NAL), now called Fermilab. The period 1970-1975, witnessed a further step in the evolution of the size of collaborations. Our collaboration consisting of four diverse, as well as geographically dispersed, institutions-- Columbia, Illinois, Fermilab and Hawaii -- was formed because we needed the special talents of each person. (By 1973 I had left Cornell and joined Fermilab.) Moreover, there was too much work for five or six people to do in a reasonable length of time. There were perhaps ten collaborators-- all teaching faculty-- who submitted a five-page proposal in 1970. By 1975, when we published our first paper on the first observation of the photoproduction of the J/ψ, there were nearly twenty authors. The increase was due to the arrival of graduate students and postdocs in 1972. This "large" group consisted of smaller subgroups, organized according to areas of interest in physics topics and detector responsibilities rather than by institutions. This pattern of small subgroups persists to this day. Group meetings were held several times a month, a supplement to informal meetings of a few people every day. Decisions were made and promulgated at group meetings.

Construction of the detector started in the fall of 1971 at Columbia, Illinois, and SLAC (the Hawaii piece). Since I had a Sloan Fellowship I took a leave of absence from Cornell to get things started. My first job was to work on a new invention of Jim Sanford and Bob Wilson: the Agreement between Fermilab (NAL) and the collaboration. The Agreement defined what the Laboratory would provide and what each collaborating institution would provide. Prior to Fermilab, groups just showed up at a laboratory with a truckload of apparatus at a prearranged date.

Installation in Fermilab's Proton area began in the summer of 1973. The first test of the detector occurred in the spring of 1974, and the first significant data run occurred in November and December of 1974. The initial group of students finished their Ph.D.'s in the fall of 1975, roughly four years after they had started. (During this time a new group of graduate students and postdocs had joined the enterprise.)

There were a number of differences from my thesis experiment started ten years earlier or even the Brookhaven experiment:

The proposal was a very serious matter and approvals were tightly contested. In the past it had been relatively rare that two groups wanted to do the same thing. Fermilab opened up a new energy domain and everybody wanted to be the first to use it.

These experiments were more complex and more sophisticated. A typical event might have ten charged particles in the final state. The event rates were approaching several hundred kilohertz or even a megahertz. Everything had to be ten or a hundred times faster than ten years earlier. There were ten times as many analog channels that had to be digitized than there were ten years earlier. As a result the detectors were more expensive and more specialized skills were required to design and build them. Experiments were beginning to require several hundred thousand dollars worth of equipment. Since the grants of experimenters did not allow this degree of support, the Laboratory was asked to provide much of the equipment.

Given the magnitude of the required resources a signed agreement with Fermilab specifying what the experimenters would provide and what Fermilab would provide was essential. Initially, Fermilab was expected to provide general purpose or reusable equipment, but later, anything that was beyond the resources of the collaboration's university research grants came to be expected.

The experiments took a much longer time. Four years was not unusual. As a result postdocs and students took up residence at the Laboratory. Some postdocs never saw their universities. My student's experience was not unusual; he lived at Fermilab from 1972 until he finished his thesis in 1975, and then he joined the Laboratory as a postdoc on another experiment for several more years. The Laboratory became the place of activity for the group.

When the experiment was over, the "core" of this same group had either another approved proposal to do an experiment in an improved version of the same apparatus or it had a pending proposal. The detector built for E-87 was not dismantled until 1984, twelve years after it was first installed. In the intervening time many changes were made and many experiments were done by essentially the same group of people using upgraded versions of their original equipment. Successful experiments all had this continuity.

Although the scientific interest of the group remained basically the same, there were changes in the scientific leadership. By 1984 none of the people from the original 1970 proposal remained on the experiment. The leadership was in the hands of people who had been postdocs and students in 1977.

More of the equipment used in experiments was built commercially. The time to develop the skills to design and build electronics demanded a new approach. Specialized electronics companies sprang up to produce the general-purpose electronics for the experiment.

The capital value of an experiment had reached several million dollars by the time the detector was dismantled. The value had accrued incrementally, a few hundred thousand every year, through replacements and improvements.

Software had became as important as hardware. The software needed to process twenty million complex events had to be efficient and reliable. The computers of the day were not friendly, but they were fast, and cleverness with algorithms paid off. Some of the basic software remained intact for ten years although it was rewritten many times to accommodate a variety of ever-more powerful computers. Only laboratories had the big number-crunchers like the CDC 7600.   As the computers became more powerful, the experiments took more data. This was one of the reasons why the the postdocs and students lived at Fermilab.

THE RISE OF LABORATORIES WITHIN THE LABORATORY (1981-1990)

When Fermilab began to prepare for the Tevatron, the Program Advisory Committee still reviewed each experiment and on the basis of its advice the Director approved or rejected it. There was no guarantee of approval and there were no official long-term commitments to a collaboration. The Laboratory provided a beam, magnets, general-purpose electronics, and a few other things. The job of building and running an experiment was left entirely to the university-based experimentalists. Fermilab experimentalists brought minimal support to an experiment. The notion that Fermilab would make a major contribution to the building and operating of a detector was still in the future. The large fifteen-foot bubble chamber was an exception to this pattern, although it was not at the center of physics.

Fixed-target experiments involving 30 people from five or six institutions were common by 1980, and much larger groups had been formed to attempt the bold new colliding-beam experiments conducted at frontier energy ranges. The fixed-target experiments at Fermilab did not have dedicated support groups to help with the design, construction and maintenance of a detector. The engineering and technical support came from the university groups. Fermilab physicists who worked on experiments did so as a part-time activity, typically an activity of love subordinated to running an accelerator or an accounting department. Under these circumstances it was not surprising that there were no strong in-house groups dedicated to designing, building and operating detectors. Nevertheless it was clear that the successful experiments were facilities of long-standing duration. The organization model of the sixties no longer worked.

At other laboratories in-house groups with substantial resources of technicians and engineers were the cores around which experiments formed. While the early colliding-beam experiment teams at SLAC and DESY were not much larger than the fixed-target experiment teams at Fermilab, there were some important differences in organization and management. Early collider detectors at SLAC and DESY typically had a dedicated staff of laboratory employees; the physics group leader was the intellectual head as well as the top manager. Typically there was a project engineer and a number of dedicated technicians. There was also, in stunning contrast to contemporary fixed-target experiments at Fermilab, a a budget. Graduate students were not the main source of the engineering, but they did help with the construction.

Proposals (1976-77) to build colliding-beam detectors at the new electron-positron collider were not taken lightly: cost had become as important as scientific merit. Each detector required at least several million dollars in capital funds. The colliding-beam experiments at Fermilab and CERN required even larger and more complex detectors. By 1982, one of the collider detectors at CERN was being fed collisions at the rate of 4 kilohertz by the SppS. By 1989 the rate had climbed to almost 100 kilohertz. Each event typically produced thirty or forty particles.

In 1977, when the two groups of physicists at CERN, UA-1 and UA-2, embarked on the search for the W and Z bosons, they knew that they would have to sift through a billion events to find a handful of W's and Z's. They also knew that to build a detector would take more than five years and at least a hundred physicists. The management style had to adapt once again to the demands of the search. More time was spent in meetings, and transparency presentations were de rigueur at group meetings. Collaborations needed phone books and administrators, not just a coffee pot. Fund raising by the collaborations had become a full-time activity: the UA-1, UA-2, and CDF detectors all cost more than 30 million dollars. With these big collider detectors the laboratories for the first time approved a program for a detector knowing that the detector would take five years or more to build, plus at least five more years to take data. Since this time was so long, the physics goals would almost certainly shift once the initial goals were achieved. Program Advisory Committees could only recommend that such an enterprise be launched. While the Committee could monitor progress, the scientific priorities were now set by the collaboration and monitored by the Laboratory management.

It was also clear that the loose style of management that characterized earlier experiments would not be adequate for a collider experiment. This lesson was learned later at Fermilab than elsewhere. In January 1977, Fermilab approved a physics program for colliding beams and at the same time created an administrative department in the Laboratory responsible for the design, construction, and operation of the detector: the Colliding Beams Department. Since money and human resources were very scarce it remained a very small group. Nevertheless it served as an administrative focus for a design effort. Very quickly 30 or 40 physicists from universities joined Laboratory scientists to form a Collaboration. Collaboration meetings and subgroup meetings were held weekly. The Collaboration began to document its work in notes, as a means of communicating with those who were absent and to provide a means for retrieving what had been learned. In time these internal department notes assumed the same level of professionalism as preprints.

After three years, participatory democracy was able to produce a very good detector design, but there was no realistic cost estimate or schedule. Finally Fermilab was prepared to commit people and resources to building the detector. A full-time Project Manager was chosen by Fermilab to develop a cost estimate and schedule and subsequently manage the construction. The first step was to prepare a conceptual design report that established the project goals, cost, and schedule. When the dust settled, the cost estimate in 1982 dollars was around 40 million dollars, and it was clear that it was going to take at least six years to complete the Collider Detector at Fermilab, known by its acronym; CDF.

Meanwhile, the collaboration organized itself to govern its activities, under the guidance of the cospokespersons, Alvin Tollestrup and Roy Schwitters, who was also project manager. Many activities needed to be regulated, the most important of which was the integration of the efforts and resources of what became one hundred collaborators from outside institutions. But there were many other tasks such as determining the names attributed with the publications and the procedures for admitting new collaborators. I suspect that the need for collaboration governance was first recognized with these collider collaborations at CERN and Fermilab. It was also recognized that this part of the scientific management was distinct from the project organization. They had somewhat separate, but related, tasks. The cospokespersons had to keep the two overlapping organizations on track. The Collaboration executive board took the initiative in establishing the scientific program through meetings, internal workshops, and the CDF notes. It was clear that over the six-year construction period the scientific program was going to evolve in response to the science that emerged around the world during that time. In the middle of this six year period the UA-1 and UA-2 collaborations discovered the W and Z. The goals of CDF shifted to what lay beyond the W and Z. It was the responsibility of the cospokespersons to shape the transition of the evolving program, balancing scientific desirability with administrative discipline. The Fermilab Physics Advisory Committee which had been in the business of approving or rejecting specific experiments, or choosing among different paths to a particular goal, found itself acting as an oversight committee.

The responsibilities of the experimenters had evolved as well. In essence, the Laboratory Director expected the CDF Collaboration and the CDF Department to build the detector at a fixed cost to the Laboratory. (The CDF Department was the administrative unit of the Laboratory with the project responsibility of building the detector.) This was not a fixed cost in the usual sense of the term, since the CDF Collaboration had grown to include Japanese and Italian participation, which ultimately contributed 25% of the total funds. Nearly all of the rest came from the Laboratory, which is to say DOE. Even with this infusion, some ambitions needed to be stilled to keep the project within the resources that the Laboratory was prepared to provide. In addition some descoping occurred because the detector R&D could not be completed in time. It is important to note that the detector technology of 1981 was not sufficient to achieve the performance goals stated in the 1981 conceptual design report. Although a vigorous R&D program was initiated in 1981, the detector design had to be fixed in 1983, and certain parts of the original plan were dropped. The construction of a silicon microvertex detector was an example of a detector subsystem that had to be postponed for a decade.

Another necessity, and a break with Fermilab tradition, was to build-up the personnel in the CDF Department so that it could provide the engineering and technical support to build the detector. In 1980 there were less than 10 people in the CDF Department; by 1985 the number had grown to 100. During the peak period of construction the number of full-time equivalent laboratory staff on CDF reached almost 200.

In 1983 another collaboration was approved to build a second detector, D0, which will be put on beam for the first time in late 1991. Today the CDF and D0 collaborations are essentially laboratories within the Laboratory, because they set the details of their scientific programs. The CDF and D0 detectors and collaborations are likely to remain in their current form, with considerable evolution in technical performance, until 1998. One can hardly call these organisms--with changing memberships, hardware, software, and scientific goals and lifespans measured in decades-- "experiments" in the traditional sense of the word. Nevertheless, they do experiments.

It is very important to realize that the science is still done in relatively small groups of five or six people just as it had been done when I was a student. By "doing science" I mean the work that is most essential to the core of an experiment: developing the instrumentation that does not yet exist and analyzing the data. But, to get to the point where the science can be done, several hundred people have to work together for many years. It is also important to realize that scientists in the CDF and D0 collaborations designed the detector, managed its construction, and made certain it would work. A detector is not a set of specifications that could have been subcontracted to an industrial concern.

Today both the CDF and D0 Collaborations contain roughly 300 scientists. The capital cost of each detector is about $60M, exclusive of perhaps $30M of labor that was provided by the collaborations. Each detector has of order 10^5 channels of analog information data -- a stunning increase from the 17 channels for my 1965 thesis experiment. Each of the major analysis programs contains of order two million lines of source code. Contributions to that code were made and are being made by a large fraction of the collaboration. Each detector has a weight of order 2,000 tons to 3,000 tons. Enormous as they may seem, in terms of gross weight and the number of electronic channels they are a factor of 10 to 100 smaller respectively than proposed SSC detectors.

USING THE SSC

Is this factor too big a hurdle to leap? I don't think so. CDF and D0 have ten-thousand times the channels of information of my thesis experiment. They will take data at the rate of ten megabytes/secs, whereas thirty years ago my experiment's rate was a thousand bytes/sec. I believe that most of the challenges faced by experimenters at the SSC will be solved, but there is a caveat: clearly no critical subsystem that has to be fixed by 1993 can depend on technology that will not exist until 1998.

The construction of SSC detectors is going to require strong and dedicated management because the initial cost estimates for these detectors appear to exceed $500M. The detectors will have to be descoped until they fit within the available funds. I am sure that this can be done because each collaboration will want to be ready with a basic detector when the SSC turns on.

I suspect that the SSC detectors will have to have far greater industry participation than was the case with CDF and D0. The factors of ten in complexity means that there are millions of pieces to be designed, procured, evaluated, and then assembled into a subsystem. Many of these tasks can be done more effectively in industry. Industry will have a role in systems integration, but not an exclusive role. The scientists must stay with the detector. Nevertheless, there is a lot that they can learn by including an industrial systems' integration team in the design and construction team. It is reassuring that a number of industries are already participating in the early stages of SSC design and R&D.

CONCLUSION

There is one underlying message throughout this story about the way science has been successfully practiced by teams of ever-expanding size. First there has to be a set of scientific goals that the team embraces. Then there has to be an agreed-upon approach that distinguishes one team from the other. As a rule the features of a detector that are adopted by a team will lead to a differing emphasis on the scientific goals among different teams. The team's effort has to be subdivided by expertise to design and build the detector. Regardless of the size of the project, each subgroup has to rely on the others if the project is to succeed. But as the size increases, the organization providing for this productive involvement of many must have the following elements: (1) internal communication is necessary for progress, and to allow for this there must be (2) scientific leadership to set the goals and satisfy the aspirations of the bright junior participants, who need creative freedom from the (3) management leadership, which is more experienced and dedicated to effectively guiding the project to completion within a budget and on a schedule established by a (4) governing body that is receptive, and not resistant, to the whole. Science does not flourish under tyranny; spokespersons serve at the pleasure of the collaboration.

Let me conclude by making some observations about these collaborations. As the components of a complicated piece of machinery are all interdependent, so is any high-energy physics experiment. The scientists planning to use the SSC are united in their mission to design their detectors and organize their collaborations to complete their experiment. There will be innovations throughout the decade of construction and the decade of use. An organization of a thousand people will need some rather creative rules for governance if it is to be a vital and comfortable place for people to work for 20 years. I suspect that ultimately the science will be done, as it has always been done, in small teams, although these teams will be much more tightly coupled than before. The physicists and engineers building the detectors will force the development of some extraordinary technology that will touch the lives of our children and perhaps even our own lives if we are hardy enough to be here in the 21st century. I am confident that the scientific discoveries and technical innovations that come from the SSC will show that the effort and expense was well worth the support of the nation.

ACKNOWLEDGEMENTS

This manuscript is based on a talk I gave to the Third International Industrialization Symposium of the Superconducting Super Collider, IISSC. Adrienne Kolb and Kate Metropolis made numerous suggestions that clarified the original turbid text. Kate Metropolis did a masterful job of editing when the press of duties made it difficult for me to meet the often missed deadlines.

STATUS WITH DEVELOPMENT AND PRODUCTION

OF EQUIPMENT FOR THE UNK (THE ROLE OF INDUSTRY)

V.A. Yarba

Institute for High Energy Physics
142284 Protvino
Moscow Region, USSR

INTRODUCTION

The IHEP Accelerating and Storage Complex project (UNK)[1-5] envisions the possibility of accelerating protons up to 3 TeV with the beam extracted onto the fixed target and with the beam in a colliding mode. The first stage, incorporating the 3-TeV machine with the system extracting the beam onto the fixed target, is presently under construction.

The UNK is placed in the 5.1-m-diameter underground ring tunnel, which has a circumference of 20.77 km. Figure 1 shows the ring cross-section with the equipment layout. The existing 70-GeV proton synchrotron, the intensity of which is planned to be upgraded up to 5×10^{13} ppp, will be the injector into the UNK. The first stage of the UNK (UNK-1, i.e., the 400-GeV conventional machine) is the booster for the second, superconducting stage (UNK-2), but it can also run as the storage ring. The orbits of the first and second stage actually coincide and intersect at four points in the centers of Matched Straight Sections (MSS) 2, 3, 5, and 6, where the detectors for colliding beams may be placed. In addition to these MSSs, each 490-m long, the project also envisions another two 800-m technological sections, MSS1 and MSS4. MSS1 will house the injection, loss localization, and beam abort systems as well as the accelerating stations for all the stages. MSS4 is intended for the system of extraction and beam transfer from UNK-1 into UNK-2. A part of the MSS6 space is to be occupied by the reverse beam injection from U-70 into UNK-1 in the pp colliding mode.

OPERATION OF THE COMPLEX

After rebunching at a frequency of 200 MHz, the U-70 beam is injected into UNK-1. Here an intensity of 6×10^{14} ppp is stacked within 72 s by multiple injection, up to 12 times. After 20-s acceleration in UNK-1 (to 400 GeV), the beam is transferred into UNK-2 by single-turn injection where it is accelerated further to 3 TeV. The cycle of the superconducting stage, UNK-2, is as follows: 40-s field rise, 40-s flattop, and 40-s drop.

The basic parameters of the fixed-target mode are presented in Table 1.

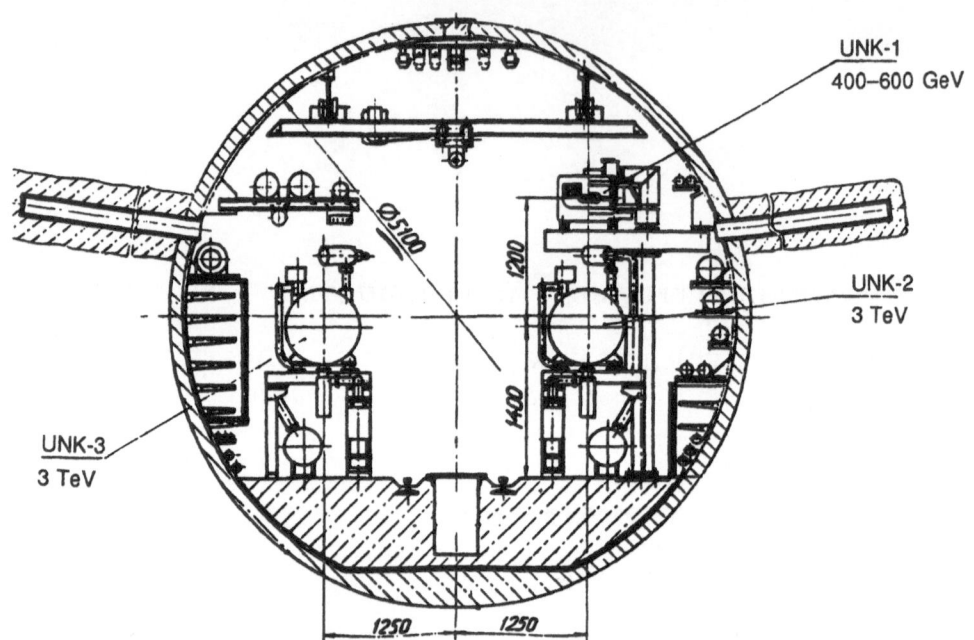

UNK-1
400–600 GeV

UNK-2
3 TeV

UNK-3
3 TeV

Figure 1. The Ring Cross Section with the Equipment Layout.

Table 1. Basic Parameters of Fixed-Target Mode.

Parameter	UNK-1	UNK-2
Maximum energy, GeV	400	3000
Injection energy, GeV	65	400
Circumference, m	20771.9	20771.9
Maximum field, T	0.67	5
Injection field, T	0.108	0.67
Acceleration time, s	20	40
Harmonic number of accelerating field	13860	13860
Total amplitude of accelerating field, MV	7	12
Maximal energy gain per turn, MeV	2.1	4.5
Transition energy, GeV	42	42
Frequency of betatron oscillations		
(without straight sections)	36.7	36.7
Maximal β-function in regular lattice, m	152	152
Maximal β-function in the extraction area, m	400	784
Maximal ψ-function in regular lattice, m	2.43	2.43
RMS transverse emittance, mm·mrad, not more than	150	200
RMS longitudinal emittance, MeV/c·m, not more than	100	120

In addition to beam extraction onto the external targets of the experimental area, the first stage of the UNK envisions the experiments at the internal jet hydrogen target to be performed in MSS3 in the circulating beam of both UNK-1 and UNK-2.

The further upgrading of the UNK is related to its operation in the collider mode. Three different schemes of colliding beams have been investigated: $0.4 \times 3\,\mathrm{TeV}$ pp beams from UNK-1 and UNK-2;[2] $3 \times 3\,\mathrm{TeV}$ p$\bar{\mathrm{p}}$ beams in the UNK-2 ring;[4-6] and $3 \times 3\,\mathrm{TeV}$ pp beams from the superconducting rings of UNK-2 and UNK-3.[3-5]

In the pp collider mode, particles are stacked following the same scheme as in the acceleration mode. In this mode, the luminosity attainable in MSS6 can be as high as $1 \times 10^{33}\,\mathrm{cm}^{-2}\mathrm{s}^{-1}$, where the underground experimental hall for the physics setup MMS (MultiMuon Spectrometer) is planned to be constructed. This experiment is aimed primarily at the search for the t-quark at a pp-collision energy of $0.4 \times 3\,\mathrm{TeV}$ (2.2 TeV in the c.m.s.).

The basic parameters of the $0.4 \times 3\,\mathrm{TeV}$ pp collider are presented in Table 2.

The requirements imposed on the collider mode are more stringent than those imposed on the fixed-target mode. Therefore the UNK project has rather heavy demands on the value of the chamber coupling impedance, the level of noise in the accelerating system, injection and beam transfer mismatches.

STATUS OF THE UNK CONSTRUCTION

Figure 2 shows the status of tunneling. By February 1, 1991, 20.2 km had been bored. Construction of the southern part of the ring tunnel and of the injection beam line is complete; preparation for equipment assembly is underway. The construction of the surface technological buildings designed to house the power supply, cryogenics, and control systems is going on. The equipment for UNK-1 is presently in serial production. The whole electromagnetic and vacuum equipment for the 2.7-km-long injection beam line has been manufactured. More than

Table 2. Basic Parameters of $0.4 \times 3\,\mathrm{TeV}$ pp Collider.

Parameter	UNK-1	UNK-2
Maximal energy, TeV	0.4	3
Injection energy, TeV	0.065	0.4
Number of bunches	348	348
Number of particles per bunch	3×10^{11}	3×10^{11}
Total number of particles per ring	1×10^{14}	1×10^{14}
Bunch-to-bunch distance, m	49.5	49.5
Harmonic number of accelerating field	13860	13860
Maximal amplitude of accelerating field, MV	10	20
Beam acceleration time, s	100	100
RMS invariant beam emittance, mm·mrad	7.5	7.5
Longitudinal bunch emittance, MeV·m/c	120	150
Minimal amplitude function at the collision point, m	0.2	1.5
RMS beam diameter at collision point, mm	0.14	0.14
Bunch length at maximal energy, cm	65	36
Beam-beam tune shift	5×10^{-3}	5×10^{-3}
Luminosity, $\mathrm{cm}^{-2}\mathrm{s}^{-1}$		1×10^{33}
Luminosity lifetime, h		≥ 10
Total beam stacking and acceleration time, h		≤ 0.3
Average number of events per collision		30
Free area for detector, m		± 8

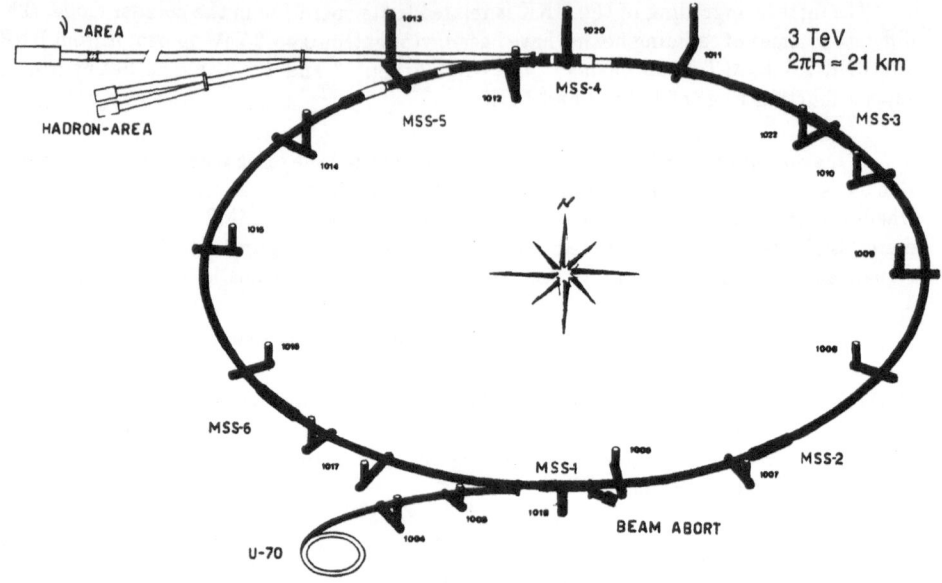

Figure 2. The Layout of the UNK System and the Tunnelling Status.

1000 warm magnets and 11,000 m of the vacuum chamber have been supplied. The equipment for the accelerating system and for the power supply system of the ring electromagnet is being manufactured.

FIRST STAGE OF THE UNK

The first stage of the UNK is the 70–600 GeV proton synchrotron, which is the booster for the superconducting second stage, accelerating the proton beam from 400 GeV to 3000 GeV. The first stage incorporates the following major systems:

- magnet system

- power supply systems

- vacuum system

- RF accelerating system

- beam diagnostics system.

Magnet System

The magnet system of the first stage consists of 160 normal periods, each 91.8 m long and broken into eight identical sections and eight special ones: six matched straight sections and two for orbit length matching.

Each normal period has the FODO structure and consists of 12 bending dipole magnets and 2 quadrupole lenses. Each quadrupole lens has next to it three correction magnets: a quadrupole to correct the betatron oscillation frequencies, Qr, z; a sextupole to correct chromaticity and excitation of the resonance harmonic of sextupole field nonlinearity for arranging slow extraction; and a dipole to correct the closed orbit distortions in the horizontal and vertical directions.

The correctors are designed in such a way that the maximum current of each should be 10 A. This allows one to make 1500 power supplies unified.

The system correcting the closed orbit horizontally includes 228 dipole correctors, each having an aperture of 50 mm; the system correcting the closed orbit vertically has 241 dipoles correctors, each 75 mm in aperture. The maximum strength of each corrector is $B * 1 = 800$ gauss·m; this allows one to correct the orbit distortion up to 30 mm.

The correction system for betatron oscillation frequencies includes 320 quadrupole correctors—each 80 mm diameter and having a strength of 100 gauss$*$m/cm^2—that will make it possible to shift the frequencies by $Qr, z = 0.7$ within the whole energy range.

The chromaticity correction system contains 160 sextupole correctors, each 75 mm in aperture and possessing a strength of 92 gauss$*$m/cm^2; and 160 correctors, each 90 mm in aperture and having a strength of 54 gauss$*$m/cm^2. They will make it possible to correct chromaticity in the $r, z = 120$ range up to energies of 600 GeV.

The correction system for octupole field nonlinearity contains 32 octupole correctors, each 96 mm in diameter and 50 gauss$*$m/cm^2 in strength. This system allows one to correct the effect of the systematic decapole nonlinearity of the basic dipoles, $B_3/B_0 = 1 \times 10^{-4}$. (The value of the field addition due to decapole nonlinearity is taken at the edge of the working region of the accelerator aperture, $r = 35$ mm.)

The existing proton synchrotron, U-70, is the injector of the first stage of the UNK. The U-70 beam is transported to MSS-1 containing the system that injects the beam in the forward direction, and the MSS6 containing the one injecting it in the opposite direction, through the injection beam line (IBL). The dipoles of the first stage will be used as the bending magnets for the IBL. Pulsed quadrupole lenses, each 100 mm in aperture, 1 m in length, and possessing a maximum gradient of 13 T/m, are used to focus the transported beam.

The Efremov Institute in Leningrad is the designer of the technical project and of all design plans and specifications for the whole conventional magnet and optic equipment.

Responsibility for the production of the equipment is shared between the following manufacturers:

Efremov Institute. The major part of dipoles and quads, sextupole correctors 75 mm in aperture; bending magnets for the injection beam line; bump magnets for the beam loss localization, extraction and beam abort systems.

Pskov plant of heavy electric and welding equipment. Dipole, quadrupole and sextupole $d = 90$ mm correctors.

Leningrad pilot-production plant of "Electrophysics," Inc. Quadrupole lenses for the IBL.

IHEP Workshop. Correction magnets for the IBL; pulsed magnets for the injection, extraction. transfer, and beam abort systems.

By 1989 the R&D work on the main magnet and optical equipment of the first stage of the UNK was finished, the preseries manufactured, the technical specifications corrected, and mass production underway. Presently, the conventional equipment for the second phase of the first stage, required for the colliding mode, and that for the second stage are being developed.

More than 1200 dipoles and about 450 quadrupoles have been manufactured, and 1000 dipoles and 360 quadrupoles are already at IHEP. All equipment for the IBL has already been supplied, and 1040 correctors have been manufactured and supplied. The final supplies planned for coming years are as follows:

1990—1000 dipoles and 360 quadrupoles.

1991—600 dipoles and 160 bump magnets.

1992—480 dipoles and 160 bump magnets.

1993—beam loss localization system of the first stage.

1994—beam abort system for the second stage.

1995—extraction system of the second stage.

The equipment is tested and prepared for assembly at the more than 20 IHEP test facilities. By October 1, 1990 all equipment of the IBL had been tested and assembly had begun. The operation of the IBL is to start at the end of this year. The assembling of the equipment for the ring electromagnet is to start in the second half of this year.

Power Supply System

The power supply system for the first stage of the UNK incorporates three major subsystems: the power supply system for the ring electromagnet (REM), one for the correction magnets, and one for magnets to commute beams.

The power supply system for the REM consists of 24 powerful thyristor converters, turned on in series into the electromagnet breaks, which are placed equidistantly along the accelerator circumference in 12 technological buildings.

The initial requirements for the power supply system were formulated by IHEP and Efremov Institute. The detail design and production forms and records were made by the Research Institute of Kharkov Electric and Mechanical Plant.

The manufacturer of the equipment is Kharkov Electric and Mechanical Plant ("KEMP," Inc.). The transformer and choke equipment is supplied by "Uralelectrotyazhmash," Inc. (Sverdlovsk). IHEP received the main prototype of the power supply in 1987; it underwent successful comprehensive tests at the dedicated test facility.

In 1990–1991 IHEP had five serial sets supplied. Further delivery schedule is the following: 1991, 9 sets; 1992, 6 sets; and 1993, 4 sets.

The preassembly tests of these five sets of power supplies will start this year in test facility building N371. Plans call for equipment to be assembled directly in the technological buildings, when these are commissioned, at a rate of six sets per year. The comprehensive running-in of the power supply system for REM1 is planned for 1993. If required, it can be done in 1992 with a 50% set of power supplies, which are sufficient to attain an energy of 400 GeV.

The power supply system for the correction magnets and coils consists of four identical power supplies combined into sets, four supplies in each, in one crate of "Ryabina" type. Each supply is a 2 kW (10 A, 200 V) transistor switch current controller.

The constant-voltage source is placed in the rack and is shared by four power supplies. All in all the power supply system incorporates 1500 power supplies (375 racks). The power supply racks are placed in 14 technological buildings equidistantly along the machine circumference.

The detail design of the power supply for the correction magnets was done at IHEP; the prototypes that underwent comprehensive tests in 1988 were also manufactured here. The development of the production forms and records for them were transferred to the "Quartz," Inc. (Kaliningrad), which is to start mass production this year. The approximate production schedule is as follows: 1991, 100 pieces; 1992, 600 pieces; and 1993, 800 pieces.

The comprehensive tests and running-in will be done by the plant. The racks for the power supplies will be assembled in 1992–1993 in accordance with the commissioning schedule of the technological buildings.

The power supplies of the beam-matching system consist primarily of pulsed power generators for the kicker magnets and septum magnets of the beam injection and extraction systems, as well as of those for emergency beam dump onto the absorber. In addition, the power supply systems include standard power supplies of the "IST" type for the bump magnets. Beam injection into the first stage, emergency beam dump in the two directions, and beam transfer into the second stage require 26 modules of kicker magnets and 26 modules of pulsed septum magnets. To energize these magnets it is necessary to install 52 pulsed generators, 40 pieces in building N1018 and 12 pieces in building N1020. The pulsed generators consist of such storage elements as capacitor banks for septum magnets, interruption switches, thyristors or thyratrons, and charging voltage sources. The charging voltage of the generators for septum magnets is up to 10 kV, and the output current is up to 13 kA; charging voltage of generators for kicker magnets is up to 49 kV, and the output current is up to 5 kA.

The pulsed generators are developed and manufactured at IHEP. The prototypes of the equipment have already been tested. The kicker magnets were developed and are manufactured at IHEP. The septum magnets were developed at Efremov Institute and are manufactured at IHEP.

The main prototypes of the equipment will be made and tested in 1991. The serial prototypes will be manufactured in 1991–1992. The assembly of the equipment in the tunnel and in buildings N1018 and N1020 can start in 1992 and be completed in 1993.

Vacuum System

The initial requirements imposed on the vacuum system are as follows: the working vacuum in the colliding beam mode should be at least 2×10^{-9} torr in the nitrogen equivalent from nuclear scattering without warming up the vacuum chamber after it is assembled in the tunnel. The smoothness of the chamber means the absence of resonance cavities when the beam circulates in it; the chamber is controlled remotely and automatically. These requirements have been formulated by the vacuum laboratory of the UNK Division at IHEP.

The detail design was done by "Electrophysics," Inc. of the Ministry of Atomic Energy and Industry. The production forms and records were made by "Vacuummashporibor," Inc. of the Ministry of Electron Industry.

The technical requirements of the materials used for the production of this equipment are as follows: the magnetization of the stainless steel should be 1.01; materials should be radiation-resistant, and have no admixtures of organic materials, etc.

The major manufacturers of this equipment are the plant of technological equipment (city of Bryansk) and the "Quartz," Inc. from Kaliningrad, which belong to the Ministry of Electronic Industry.

The specific features of the technology are intended to obtain different elliptic cross sections of 6.5-m-long sections of the vacuum chamber, high-temperature baking in vacuum furnaces, electric and chemical polishing, high tightness of the welding seams and connections. The main prototype of one period of the regular lattice, 91 m long, has been assembled and tested. The vacuum obtained without warming-up the chamber was 5×10^{-10} torr. Preliminary pumping-down to 10^{-6} torr is done by the pumping station on the basis of turbo-molecular pump OIAB-450 and to 10^{-9} torr with the help of getter-ion pumps with built-in batch-operated Ti evaporators.

By now the vacuum equipment for 100 out of 160 periods of the regular part has been manufactured, including pumps, slide gates, power supply system, control system, and vacuum gauges. As planned, the production of the remaining equipment for the regular part and for MSS1 will be finished in 1991–1992, and that for MSS 2, 3, 4, 5, and 6 will be finished in 1993.

Assembly and run-in of the vacuum equipment for the system extracting the beam from U-70 and for the head part of the UNK IBL are finished. The preassembly tests of all the equipment of the IBL were completed in 1990, making it possible to assemble it in the IBL tunnel this year. Assembly of the units of the vacuum equipment, together with the electromagnet equipment in the regular part of the first stage, and vacuum tests will be done at test facilities during 1991–1994.

The vacuum equipment will be assembled in the IBL tunnel and surface buildings in 1991 and in the main ring tunnel in 1991–1994.

RF Accelerating System

The RF system of each stage of the UNK includes the accelerating and feeder systems, for RF power and low-power level RF electronics. The system operates at a frequency of 200 MHz. During acceleration in the first stage the frequency varies by 20 kHz; in the second, it varies by 20 kHz.

The basic requirements imposed on the RF system are presented in Table 3.

The accelerating systems (AS) for the first and second stages have a modular structure. The first stage will use eight modules of the AS, and the second 16 modules. The feeder system is a set of RF 40-to-60-m-long RF feeders, with eight feeders in the first stage and 16 feeders in the second.

All bends (75 pieces) required for the first stage and the first batch (100 out of the total number of 250) of linear sections are manufactured. The production of all 16 cavities for the first stage has begun; four are to be commissioned in the first quarter of this year, and the rest in the course of the year. The complete set of a number of other units of the accelerating system are manufactured. These include 30 measuring bridges, 20 units of fixed-frequency tuning, 24 rejection and 40 radial dampers, and 24 units of control loops.

Table 3. Basic Requirements of RF System.

Parameter, Unit	Fixed target		pp collisions, 04 × 3 TeV	
	stage 1	stage 2	stage 1	stage 2
1. Injection energy, GeV	65	400	65	400
2. Maximum particle energy, GeV	400	3000	400	3000
3. Mean beam current in bunch train, A	1.6	1.6	0.32	0.32
4. Maximum energy gain per turn, MeV	2.11	4.5	0.385	2.89
5. Maximum amplitude of RF voltage, MV	7	12	10.6	20
6. Maximum power transferred to beam, MW	3.38	7.19	0.123	0.924

This equipment is designed and manufactured by the Scientific and Industrial Corporation named after Comintern (Leningrad).

Last year 40% (in terms of the cost) of the equipment of the RF power supply system for the first stage of the UNK was manufactured. The rest of the equipment will be manufactured and supplied by mid-1992.

A mock-up klystron prototype for the RF power supply system of the second-stage will be manufactured and tested in the pulsed mode this year; its test in the continuous mode will be done in the first half of 1992. As planned, in 1994 six prototypes of klystrons and associated RF oscillators will be manufactured; the remaining 10 klystrons and RF oscillators will be manufactured and supplied in 1995.

Beam Diagnostics System

The following beam parameters must be measured in the course of the run-in and operation of the accelerating complex during its whole cycle:

- intensity

- position of the beam center of gravity

- transverse density distribution (profile)

- longitudinal density distribution (azimuthal structure)

- characteristics of betatron oscillations

- beam losses.

By now, work on the development of all systems for the IBL, for the first stage of the UNK, and on their modelling is over; the production forms and records for them are made and the serial production has begun.

The beam position monitors (550 pieces) and the crates (600 pieces) for the primary electronics, which will be placed in special recessed of the tunnel, are manufactured at the instrumental plant of Tomsk under a contract between IHEP and the Institute of Nuclear

Physics of the Polytechnic Institute of Tomsk. This type of equipment will be manufactured at IHEP.

At present a device manufactured at the IHEP Workshop and designed to measure the profile of the circulating beam has been assembled and tested at U-70. The operational principle of the device is crossing the beam by a carbon filament with the help of an electromagnetic drive. A prototype of a similar device based on a spacious rotor motor has been developed and manufactured. The parameters of these devices (the velocity of crossing the beam with filaments is about 10 m/s) satisfies completely the demands imposed on measuring the beam profile in the UNK.

A profile meter based on the effect of secondary emission and designed to measure the profile of the beam injected from U-70 into UNK has also been developed and manufactured. It can operate in strong magnet fields.

Jointly with the Institute of Pulsed Technology (Moscow), we have developed and manufactured an ADC-based prototype of the system for measuring the beam azimuthal structure. The bandwidth of this system is 5 GHz. These devices make it possible to record single pulses of the pico and nano range and to present the data in digital form. The computational means of this system can be made compatible with IBM PCs.

SECOND STAGE OF THE UNK

Superconducting Magnets of the UNK

The ring electromagnet of the UNK consists of superconducting (SC) dipoles, quadrupoles, correctors, and special-purpose magnets. The major element of the SC dipole and quadrupole is a two-layer shell-type cold-iron coil. Tables 4 and 5 present the basic parameters of SC dipoles and quadrupoles.

The initial specifications for the SC magnets were developed at the UNK Division of IHEP. The conceptual design of the major components of the system was made by IHEP (dipoles, quadrupoles), and by IHEP jointly with Efremov Institute (correctors). The design plans and specifications were drawn by the same institutions for the same components.

Technical requirements where imposed on the materials used for the production of SC magnets: SC wire, SC cable, insulation for SC cable, stainless steel for beam pipes and collars, electrical-sheet steel for cold iron, and superinsulation.

SC Wire

The technical specifications for the SC wire were developed by IHEP, and the production technology for it was developed by the Institute of Inorganic Materials (Moscow). The manufacturer of the wire is Ul'binsk Metallurgical Plant (town of Ust' Kamenogorsk). The basic parameters of the wire are presented in Table 6.

SC Cable

The technological specifications and production technology were developed by IHEP. The cable is produced at IHEP. Table 7 shows the basic parameters of the SC cable.

Table 4. Basic Characteristics of the Coil Block of a SC Dipole of the UNK.

Parameter	Inner shell	Outer shell
Number of turns	2 × 34	2 × 21
Turn width, mm	8.7	8.7
Mean thickness of turn, mm	1.63	1.63
Angular dimension of shell, deg	2 × 75.96	2 × 44.34
Angular dimension of spacer, deg	4.25	7.24
Angular dimension of median spacer, deg	2 × 0.13	2 × 0.53
Inner radius of shell, mm	40	49.2
Length of rectilinear part of coil, mm	5563	5590
Total length of coil, mm	5770	5770
Ratio between the maximum field in shell and aperture field	1.096	0.806

Table 5. Basic Parameters of a SC Quadrupole of the UNK.

Parameter	Inner shell	Outer shell
Number of turns	4 × 12	4 × 19
Number of strands in turn	19	19
Turn width, mm	8.7	8.7
Mean thickness of turn, mm	1.63	1.63
Inner radius of shell, mm	40.0	49.2
Outer radius of shell, mm	48.7	57.9
Inner angle of shell, deg	0.39	0.86
Outer angle of shell, deg	35.06	31.6
Maximum coil field, T	4.5	4.53
Critical temperature of shell for maximum operating current, K	5.8	5.8
Coil length, mm	3000	3000

Table 6. Basic Characteristics of SC Wires for SC Magnets of the UNK.

SC alloy	Nb + 50% Ti (NT-50)
Diameter of SC wire, mm	0.85 + 0.03
Number of SC filaments	8910
Mean diameter of SC filament, μm	6
Twist pitch, mm	8–10
Matrix material	Cu
Filling factor	0.42
Ratio of matrix resistances at ambient and helium temperatures	> 70
Critical current at 5 T and 4.2 K, A	> 550
Critical current density at 5 T and 4.2 K, A/cm^2	> 2.2 10^5

Insulation for SC Cable

The insulation should withstand temperatures of 4–300°K in the helium environment, high mechanical stresses (up to 100 MPa), and radiation up to 1000 Mrad.

Since the materials hould be radiation-resistant, the SC cable is insulated with a 20-μm-thick polyamide film (kapton). The tolerance for the film thickness is 2 μm. An epoxy-

Table 7. Basic Parameters of SC Cable for SC Dipole of the UNK.

Number of strands	19
Type	zebra
Strand coating:	
9 strands	Sn + 50% Ag
10 strands	bare
Cross section:	
inner shell, mm^2	$(1.30–1.62) \times 8.5$
outer shell, mm^2	$(1.33–1.59) \times 8.5$
Transposition pitch, mm	62
Critical current at 5 T and 4.2 K, A	> 9500
Effective specific transverse interstrand resistance in cable, Ohm × m	$(1–2) \times 10^{-7}$
Ratio of cable resistance at ambient and helium temperatures	> 60

impregnated fiberglass film 100 μm thick and 10 mm wide is wrapped on the polyamide film with a gap of 1 mm. This film ensures cementing of the turns of the coil.

The technical specifications for the film were developed at IHEP. The film is manufactured by the Plant of Synthetic Materials (town of Novocherkassk). The technology of impregnating the fiberglass with epoxy compound was developed at IHEP. The technology of coating the polyamide film with epoxy compound was developed by the Riga branch of the All-Union Institute of Materials for Electric Insulation.

Stainless Steels

The collars of the SC magnets are manufactured from 2-mm-thick austenitic sheet steel of quality 05X20H15AΓ6. Its chemical composition is presented in Table 8.

Table 8. Chemical Composition of Steel of Quality 05X20H15AΓ6.

C	Cr	Ni	Mn	N	Si	Al
< 0.05	8–20	14–14.5	5.5–8	0.1–0.18	< 0.4	< 0.04

The basic demands imposed on this steel are the constancy and small value of its magnetic susceptibility at $T = 4.2$ K and in field up to 5 T.

The technical requirements were developed by IHEP.

The technical specifications for the steel were developed jointly by the Institute of High-Quality Steels of the All-Union Institute of Ferrous Metallurgy (Moscow), Chelyabinsk Integrated Iron-and-Steel Works, Institute of Electric Welding named after Paton (Kiev), and IHEP. The manufacturer of the steel is Integrated Iron-and-Steel Works (Chelyabinsk).

The beam pipe is manufactured from austenitic stainless steel of quality 03X20H16AΓ6. Its chemical composition is given in Table 9. The requirements imposed on this steel are similar to those of steel of quality 05X20H15AΓ6.

Table 9. Chemical Composition of Steel of Quality 03X20HI6AΓ6.

C	Cr	Ni	Mn	N	Si	Al
0.03	20–22	14.5–16.5	5–7.5	0.2–0.3	0.6	0.04

The technical specifications for the steel were developed by IHEP.

The production technology for the steel was developed jointly by the Institute of High-Quality Steels of the All-Union Institute of Ferrous Metallurgy (Moscow), Southern-Pipe Plant (town of Nicopol'), and the Institute of Metallurgic Technology (Dnepropetrovsk).

The round billet is manufactured in Chelyabinsk, and the pipe is produced in Nicopol'.

Electric steel is of quality 2081. The chemical composition of the steel is given in Table 10, and the basic characteristics are presented in Table 11.

Table 10. Chemical Composition of Steel of Quality 2081.

C	Si	Mn	S	P	Al	Cr
< 0.01	< 0.02	< 0.1	< 0.01	< 0.01	< 0.15	< 0.01

Table 11. Basic Properties of Electric Steel of Quality 2081.

Sheet thickness, mm	3
Coercive force at $T = 293$ K, A/m	< 160
Induction at $H = 2500$, $T = 293$ K, T	1.65
Saturation magnetization at $T = 4.2$ K and in 5 T field, T	> 2.15

The technical requirements for the steel were developed at IHEP.

The production technology was developed jointly by the Institute of Precision Steels of the All-Union Institute of Ferrous Metallurgy, Novolipetsk Integrated Iron-and-Steel Works (town of Lipetsk), and IHEP.

The steel is produced in Lipetsk.

The manufacturers of the electromagnets are: IHEP, "Electrosila" plant (Leningrad)—dipole magnets; and Efremov Institute—quadrupole electromagnets and correction magnets.

The design of dipole and quadrupole magnets has been repeatedly described and discussed.[3,5] Its basic part is a two-layer shell-type cold iron coil. Figures 3a and 3b show the cross sections of the dipole and quadrupole magnet for the UNK, respectively. The magnets are produced from the same superconducting cable of the "zebra" type, consisting of 19 0.85-mm-diameter strands, each containing 8910 6-μ-thick Nb-Ti filaments embedded into a copper matrix. The critical current density at 5 T and 4.2 K is at least 2.3×10^5 A/cm^2.

DEVELOPMENT OF NONSTANDARD EQUIPMENT FOR MASS PRODUCTION OF SC MAGNETS FOR THE UNK

To facilitate the mass production of SC magnets, IHEP is developing a complex of equipment including a stamping line, a line for assembling collar blocks and collars, a line for SC transposed wires, a line for manufacturing half-coils and assembling coil blocks (collared), and a line for the general assembly of magnets and assembly with crystals.

The stamping line is equipped with a press tool (produced by the Swiss company Feints), a billet complex Safan, a Soviet washing complex AKSOD, and setups for cassetting, assembling and welding the collar modules and collars (developed at IHEP).

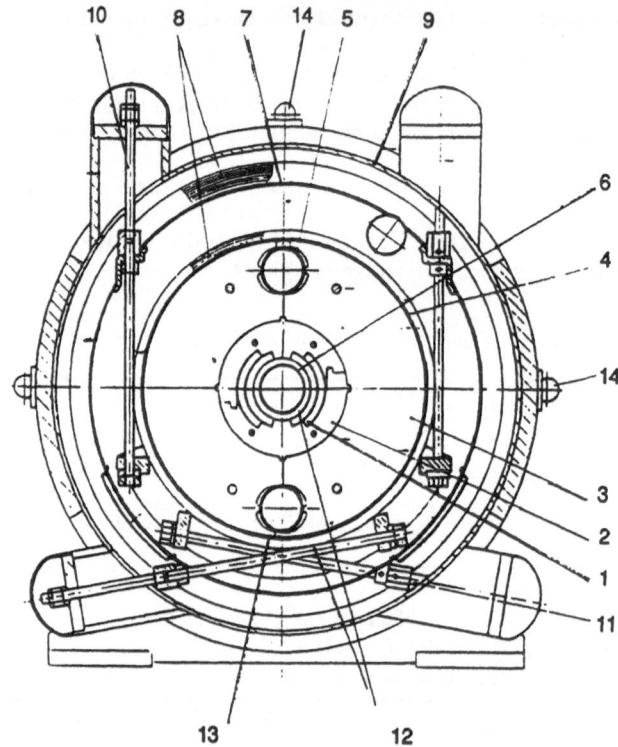

Figure 3a. Cross Sectional View of a SC Dipole Assembled in a Force-Circulating Cryostat: 1-Coil, 2-Collars, 3-Magnetic Shield, 4-Helium Vessel, 5-Two-Phase Helium Pipe, 6-Ion Pipe, 7-Thermal Shield, 8-Superinsulation, 9-Vacuum Vessel, 10-Suspension, 11-Extension Rod, 12-Single-Phase Helium, 13-Two-Phase Helium, 14-Geodetic Marks.

The line for manufacturing half-coils and assembling coil blocks includes facilities for winding coils, hydraulic presses for curing half-coils and collaring coil blocks, devices for checking the half-coil dimensions under positive pressure, a stand for electric measurements of coil blocks, and a test facility for warm measurements of coil blocks of SC magnets.

The line for assembling the helium blocks and magnets will be equipped with facilities for assembling half-yokes, for vacuum and pneumatic tests of helium blocks, for high-voltage tests, and for optic and mechanical tests to adjust the reference marks with respect to the magnet axis and other equipment.

The realization of the UNK project will require about 3,500 km of superconducting transposed wire (SCTW), which will involve treating 72,600 km of superconducting wire (SCW). The production processes for SCTW can conditionally be broken into two stages: obtaining twisted SCW and preparing it for transposition, and obtaining transposed wires to be coated with insulation materials.

The first stage is realized primarily at standard equipment, i.e., presses, heat-treating furnaces, vacuum and welding equipment, chain and other types of drawbenches, etc. The final operations—twisting and finishing drawing—and operations on preparing the SCW for transposition—control. cutting, and tinning SCW—are performed at special technological equipment developed at IHEP.

Before the SCW is twisted, the superconductor arrives at the line producing the SCW and is checked for the twist pitch, for the tolerances of the accuracy of the diameter in the SCW, for defects in the matrix, etc. A special control and re-reeling tool manufactured following the scheme of kinematic strength-locking, was developed for the control re-reeling combined with the cutting. This kinematic solution decreased power requirements by 5–6 times compared with the classical two-motor scheme.

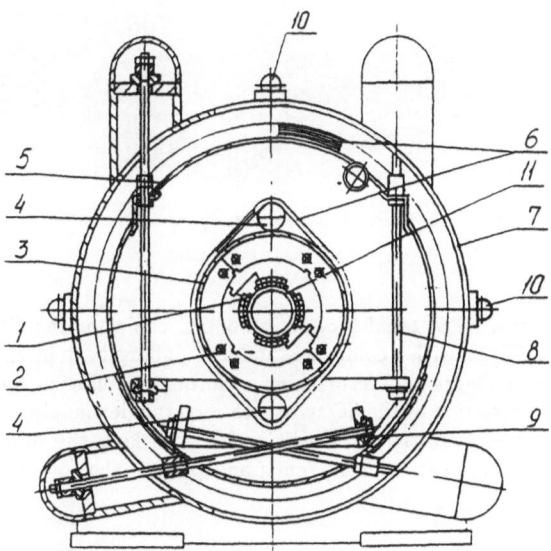

Figure 3b. Cross Sectional View of a SC Quad Assembled in a Force-Circulating Cryostat: 1-Coil, 2-Magnet Shield, 3-Helium Vessel, 4-Two-Phase Helium Pipe, 5-Thermal Shield, 6-Superinsulation, 7-Vacuum Shell, 8-Suspension, 9-Extension Rod, 10-Geodetic Marks, 11-Ion Pipe.

Following this development IHEP has manufactured and put into operation five control and re-reeling tools.

Nine of 19 strands of SC wire are coated with Sn+5%Ag. This coating should ensure the required and reproducible level of the transverse resistances between SC strands of the wire. This in turn determines the maximum permissible level of cooperative energy losses in the wire with simultaneous normal current redistribution between SC strands.

Figure 4 shows a special setup for coating the SC cable. In its trial operation, the results of measuring the transverse resistances and the instability of the measured values of ac loss from magnet to magnet point to certain outstanding problems regarding the required values of the transverse resistances.

Figure 4. Facility for Coating the SC with Sn+5% Ag Layer.

To ensure a high rate of serial production of SC cable for the UNK magnets, a small cabling machine, MKT-28, with a compact stand and pulling-back mechanism, was developed and manufactured (see Figure 5). The machine performs the following operations: transposing and shaping the cross-section of the SC wire into a rectangle; shaping and precision gauging the trapezoid-shaped cross-section of the SC cable; checking the dimensions of this cross-section at an indicated pressure of $200\,N/cm^2$; and placing the rows of the ready-made wire into the drum.

Two MKT-28 cabling machines were manufactured and put into operation at IHEP. Last year they produced 42.5 km of SC cable. The inaccuracy in the median of the trapezoid was less than $5\,\mu m$; the current density deteriorated by about 5.5% at 5 T.

The transposed cable, before being insulated, undergoes mechanical cleaning, ultrasonic washing, and high-temperature drying. The SC cable is mechanically cleaned of matrix dust and other foreign particles with rotating brushes that are cleaned ultrasonically in a bath. The final washing is done in another bath with distilled water.

A 20-μm-thick polyamide film and 100-μm-thick fiberglass impregnated with epoxy compound EC are applied as insulating materials. According to the technical specifications for the SC cable, the fiberglass should contain at least 18±1% of the epoxy compound. To meet such stringent requirements, special impregnation tools were developed. The underlying impregnation technique consists of the hydraulic weighing of resin onto the continuously moving tape.

Two types of insulation machines equipped with the central-type winders were developed. One is based on the cassette-type receiving and pulling devices; the other incorporates the drum-type devices. The tension of insulation tapes is stabilized with the help of powder

Figure 5. Cabling Machine MKT-28.

electromagnetic clutches in the 3–5 N tension range for a polyamide film and in the 20–30 N tension range for fiberglass.

Taking into account the fact that the production of SC quads and correctors and 50% of SC will be transferred to industry, the complex of nonstandard equipment under development should ensure the production of SC magnets at a rate of four per 24 hours. IHEP is to ensure the production of all collars and SC transposed cables.

As planned, the output of the line producing the transposed SC cable will be 10 km per 24 hours with two-shift operation.

CRYOGENICS

The cryogenic system of the UNK is designed for cooling, cool-down, and warm-up of the SC magnets.

The initial requirements imposed on the cryogenic system were worked out at IHEP and handed over to "Cryogenmash," Inc. (Balashikha, near Moscow).

The cryogenic complex should ensure circulation cooling at two temperature levels, 4.4 K and 80 K. The total heat load on the cryogenic system for the second stage of the UNK is 33 kW at a temperature level of 4.4 K. The cooling capacity at a level of 4.4 K is 60 kW. Cooling the main and correction current leads will require 200 kg/h of liquid helium. The total heat load on the second stage of the UNK at a temperature level of 80 K is 110 kW.

The operation of the UNK will require additionally 4 t/h of liquid nitrogen for the vacuum system, experimental setups, etc. The machine will run 5000 hours a year. The total life of the cryogenic equipment will be at least 20 years. The helium liquefaction station, nitrogen plant, system for helium cooling, purification and storage are located in the surface buildings. The refrigeration stations are placed in 12 surface buildings positioned equidistantly along the ring.

Proceeding from these requirements, the "Cryogenmash," Inc. has worked out the "Detail Design of the Cryogenic Complex for the UNK" and "Additions to the Detail Design of the Cryogenic Complex for the UNK."

The State Union Designing Institute, as the chief designer of the whole UNK complex, has entrusted to the State Institute for Designing the Plants of Oxygen Industry (SIDOI) the task of designing the cryogenic system for the UNK. This institute used the "Detail Design of the Cryogenic Complex for the UNK" and "Additions to the Detail Design" worked out by Balashikha for the initial data. The technological, electrical and technical aspects, and the instrumentation of the project were developed by this institute. As assigned by "SIDOI," the State Union Designing Institute has undertaken the construction phase of the project, the provision of plumbing fixtures, and the survey of the buildings with the general layout.

CONTROLS

The computerized control system for the UNK is developed according to the technical standards and the requests for proposal formulated by IHEP.

The projects for the controls of the injection beam line—the first and second stages of the UNK—have already been developed. The development of the project of the upper-level controls for the UNK is at its completion stage, i.e., the specifications and cost of the computational facilities and computer networks are particularized. Negotiations with companies on concluding contracts for the supply of equipment and computational facilities are underway. This phase of activities is planned to be completed by the end of this year.

The electronics for the control of the technological level of the IBL for the first stage includes the following parts:

- controls for power supplies

- timing and beam abort system

- heat setting control system

- electronics for diagnostics system

- master oscillator for the RF system

- beam transfer timing system

- magnetic field timing system

- controls for the accelerating stations.

These systems are manufactured mainly at IHEP. Some blocks, e.g., power supplies, controller modules and others, as well as constructives, are manufactured at the "Quartz," Inc. The total amount of electronics required is as follows: MULTIBUS crates—938 pieces; beam diagnostics crates—167 pieces; built-in controllers—1600 pieces; and instrumental crates—535 pieces.

The State Union Designing Institute will finish the design of the main control room by mid-1991 and start construction so that commissioning could occur in 1993.

To control the technological systems of the second stage of the UNK, the following subsystems have been developed:

- control system for the cryogenic complex

- quench protection system

- field correction control system

- beam diagnostics system

- timing systems

- beam abort systems

- radiation monitoring system

- control system for the accelerating stations

- master oscillator of the RF system

- magnetic field timing system.

In accordance with the project, the low-level electronics is based on three major constructives. Its major part is the interface electronics, performing A/D, D/A, and logic conversions. It is implemented in the I-41M (BUS-1) Standard, i.e., it is based on the electric protocol MULTIBUS and Euromechancis constructive.

Groups of crates are interfaced with the upper level of the data transfer system via a multiplexer channel in the modified Standard MIL-1553.

As planned, the construction of the 3-TeV machine with the minimum required equipment is to be completed in 1994–1995, and the 0.4×3 TeV collider is to be constructed in 1996–1997.

References

1. Balbekov V.I. et al., *Proc. of 12th Internat. Conf. on High Energy Accel.*, Batavia, 1983, p. 40.

2. Ageev A.I. et al., *Proc. of XIII Intern. Conf. on High Energy Accel.*, Novosibirsk, 1987, v. 2, p. 332.

3. Ageev A.I. et al., *Proc. of European Part. Accel. Conf.*, Rome, 1988, v. 1, p. 223.

4. Myznikov K.P., *Proc. of Workshop Physics at UNK*, Protvino, 1989, p. 5–15.

5. Yarba V.A., *Proc. of XIV Internat. Conf. on High Energy Accel.*, Tsukuba, 1989.

6. Vsevolzhskaya T.A. et al., *Proc. of VII All-Union Conf. on Charged Part. Accel.*, Dubna, 1981, v. 1, p. 229.

12. Accelerator Systems and Controls

AN OVERVIEW OF THE SSC SYNCHROTRON RF SYSTEMS

J. D. Rogers

Accelerator Division
Superconducting Super Collider Laboratory*
2550 Beckleymeade Avenue
Dallas, Texas 75237

Abstract: The Superconducting Super Collider (SSC) has five synchrotron systems: three booster synchrotrons plus two main synchrotron rings which make up the collider. The three booster synchrotron RF systems will utilize a multiplicity of high-power VHF, 150 KW tetrode amplifiers, whereas the collider RF systems will utilize four 360 MHz, 1-MW cw klystrons. The RF systems include DC power supplies, local control and monitoring systems, and accelerating cavities as well as RF amplifier systems. The system requirements and the conceptual designs for the RF systems for the five SSC synchrotrons are presented. The status of the Low Energy Booster synchrotron accelerating cavity and tuner assembly development programs are specifically addressed. Opportunities for industry to participate in the design and manufacture of RF subsystems and/or major components are also identified and briefly discussed.

INTRODUCTION

RF systems for the five Superconducting Super Collider (SSC) synchrotrons are addressed in this paper. Synchrotron RF system requirements, initial design concepts, status, and schedules are presented. SSC LINAC RF systems are discussed in a separate paper by J.H. Ferrell.

The five synchrotrons are:

1. One Low Energy Booster (LEB) synchrotron,

2. One Medium Energy Booster (MEB) synchrotron,

3. One High Energy Booster (HEB) synchrotron, and

4. Two main ring SSC Collider synchrotrons.

---
*Operated by the Universities Research Association, Inc., for the U.S. Department of Energy under Contract No. DE-AC02-89ER40486.

The LEB, MEB and HEB RF systems are very similar. Each of these synchrotrons utilizes several RF amplifier systems that operate in the 47.5–60 MHz frequency range. The primary difference is that the tuning range and cavity accelerating gap voltage requirement varies for the three synchrotrons. The final amplifiers for these systems are gridded vacuum tubes delivering approximately 160 kw each.

The main ring RF systems operate at 360 MHz and utilize one megawatt (1 MW) klystrons for the final amplifiers.

LOW ENERGY BOOSTER RF SYSTEMS

The total accelerating voltage will be provided by six or more ferrite tuned RF cavities, each driven by a gridded vacuum tube amplifier. The vacuum tube amplifier attaches directly to the accelerating cavity, which serves as the resonant tank circuit. The design goal is to achieve the specified ring voltage with eight RF cavities in the normal operating mode, and with six cavities in the extended mode, i.e., when a pair of cavities is not operational. For a ring voltage of 725 kV, operation with only six cavities would require 121 kV per cavity. No operational system to date has cavities that tune 47.5–60 MHz and also produce 121 kV. TRIUMF is in the process of developing a cavity with a slightly larger tuning range and a maximum operating peak voltage of 62.5 kV. The TRIUMF cavity and tuner is shown in Figure 1. Because the desired voltage with such a wide tuning range is much greater than the present state of the art, room is being left in the LEB accelerator lattice for additional cavities.

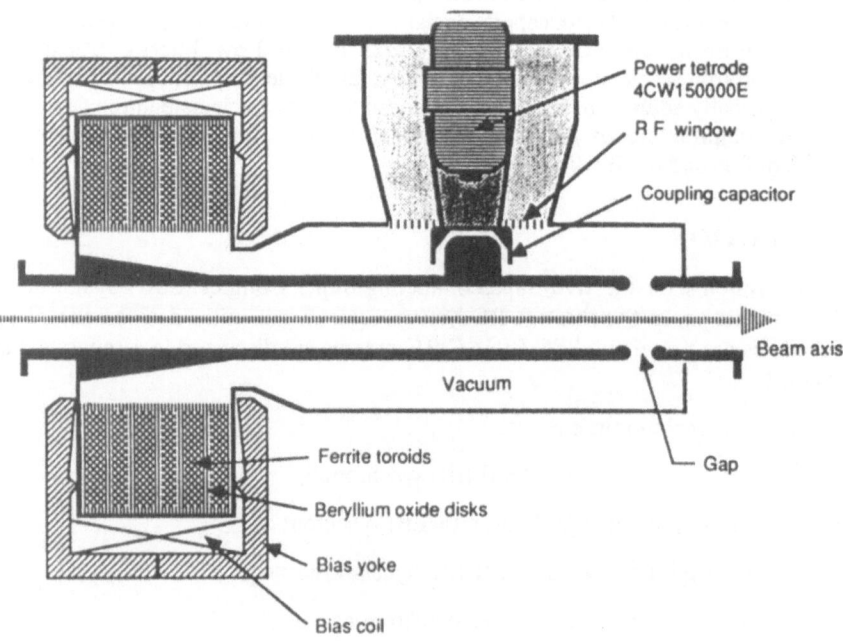

Figure 1. TRIUMF Cavity and Tuner.

The LEB RF System requirements are summarized in Table 1.

Table 1. LEB RF System Requirements.

Frequency Tuning Range	47.5-59.8 MHz
Ramp/Reset Time	50 ms
Maximum Tuning Rate	850 MHz/s
Minimum Acceptable Gradient	40 kV/m
Amplitude/Phase Stability	±0.5%/±1degree
Total Peak Circumferential Voltage	725 kV

The RF system associated with each LEB cavity is shown in Figure 2. The final amplifier is driven by a solid state RF amplifier. The cavity is tuned by varying the bias current to the ferrite tuner. Depending upon the final tuner design, the bias supply could require voltages in the range of 25–200 volts and currents up to 2000 amperes. It will be required to vary the cavity resonant frequency at rates as high as 850 MHz per second. The ferrite bias current will be controlled by a preprogrammed wave shape and feedback control loops. Likewise, the RF phase and amplitude will be controlled by fast feedback circuits.

Each LEB RF amplifier system will have a local control system that may be utilized for checkout and test. There will also be a local RF control system from which one may operate and troubleshoot all of the LEB RF systems. The control systems will be computer based and will facilitate operation, control, monitoring, and troubleshooting from the SSC main control room. Similar control and monitor systems are used for all the synchrotron RF systems.

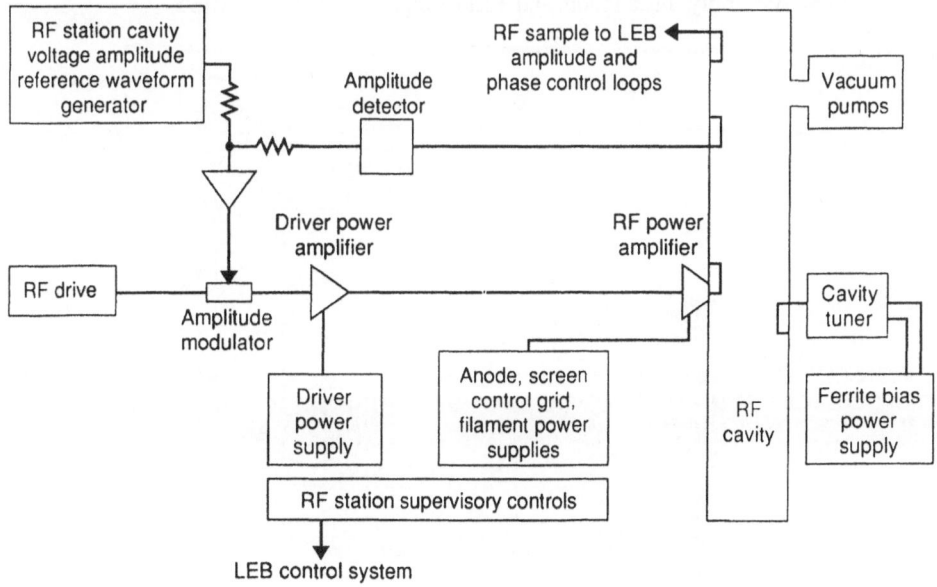

Figure 2. LEB RF System.

STATUS OF LEB CAVITY AND TUNER DEVELOPMENT PROGRAM

Numerous cavity/tuner design concepts were investigated by SSC personnel and by consultants and personnel from other laboratories. An LEB cavity/tuner design workshop was held at the SSC February 5-8, 1991. Six different designs in varying stages of development were presented for review. The review team, comprised of 10 engineers and physicists from outside the SSC, recommended that highest priority be placed on a design presented by C. C. Friedrichs of LANL and B. M. Campbell. Their design is very similar to the TRIUMF cavity and tuner; the primary difference is that the ferrites are cooled by flowing liquid directly over them instead of sandwiching them between beryllium oxide disks, as is done in the TRIUMF tuner. The liquid design should be capable of higher voltages than the TRIUMF cavity because corona in the tuner is avoided and the cooling system is more efficient. Also under investigation is a design backup approach based on a modification of a FERMI Lab cavity and tuner.

SSC has an RF test stand that will be used for testing the LEB cavity and tuner. At the present time we have a cavity that was developed by LANL which is driven by an EIMAC 4CW 150,000E tetrode. The cavity and tuner is shown in Figure 3. We plan to use the cavity to test tuner concepts.

The LEB schedule milestones are presented in Table 2.

Table 2. LEB Schedule Milestones.

Prototype Development	12/91-7/93
RFP's Issued for Production	1/93-8/93
Issue Purchase Orders	10/93
All Subsystems and Components Delivered	10/94
Assembly, Installation, and Test Complete	10/95

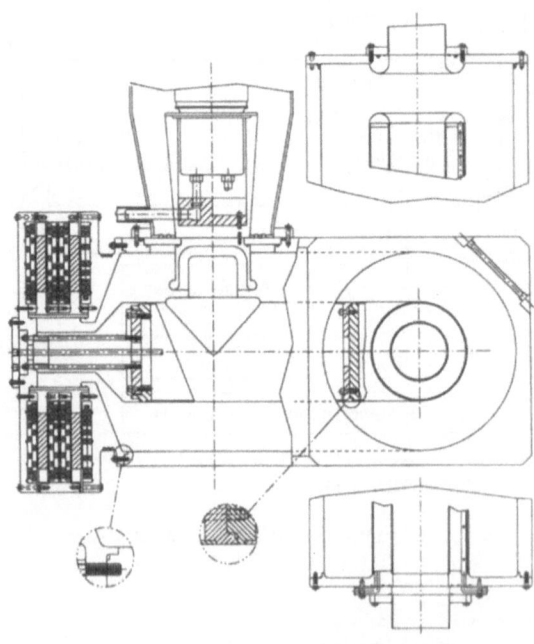

Figure 3. SSC RF Test Stand Cavity and Tuner.

MEDIUM ENERGY BOOSTER RF SYSTEMS

The MEB RF System requirements are summarized in Table 3.

Table 3. MEB RF System Requirements.

Frequency Tuning Range	59.77-59.96 MHz
Pulse Length	2 sec
Duty	67%
Total Peak Circumferential Voltage	2300 kV

The total accelerating voltage will be provided by twelve 160 kw RF systems. Each RF system will be very similar to an LEB RF system except for the RF accelerating cavity, tuner and tuner bias supply. The MEB cavities must provide 2300 kV total or 192 kV per cavity when 12 systems are operational. However, the systems will be designed and tested for 230 kV per cavity so that all MEB requirements will be met with only 10 systems operational.

Unlike the LEB, detailed design has not yet begun for the MEB and HEB systems. Preliminary design concepts for the MEB cavity and tuner are based on the Fermilab main ring cavity, shown in Figure 4.

SSC plans to build another test stand, very similar to the LEB test stand, to test the MEB RF system. MEB schedule milestones are presented in Table 4.

Table 4. MEB RF Schedule Milestones.

Prototype Development	2/92-9/93
RFPs Issued for Production	3/93-10/93
Issue Purchase Orders	1/94
All Subsystems and Components Delivered	1/95
Assembly, Install and Test Complete	6/96

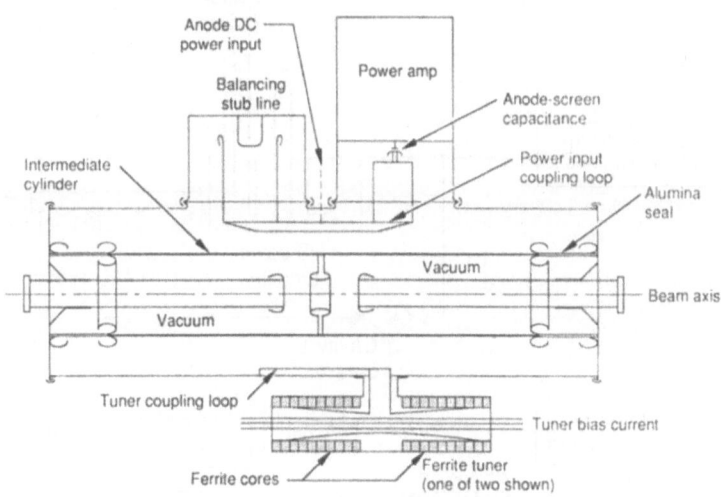

Figure 4. Fermilab Main Ring Cavity.

HIGH ENERGY BOOSTER RF SYSTEM

The HEB RF system requirements are summarized in Table 5.

Table 5. HEB RF System Requirements.

Frequency Tuning Range	59.95–60 MHz
Pulse Length	237 sec
Duty	50%
Total Peak Circumferential Voltage	1600 kV

The total accelerating ring voltage of 1600 kV will be provided by seven 160 kw RF systems very similar to the MEB systems, except that the final RF amplifiers will not be located directly on top of the accelerating cavities. As shown in Figure 5, the amplifiers will be located an integral number of half-wavelengths away, in the RF gallery. This arrangement is possible because the tuning range is almost negligible. Placing the amplifiers in the RF gallery is desirable because they are shielded from radiation and are easily accessible for maintenance.

In normal operation each of the seven systems will be delivering a peak accelerating gap voltage of 229 kV. The systems will be designed to deliver full voltage with any one system inoperable. That is, the systems will be designed and tested to deliver a peak voltage of 267 kV per cavity.

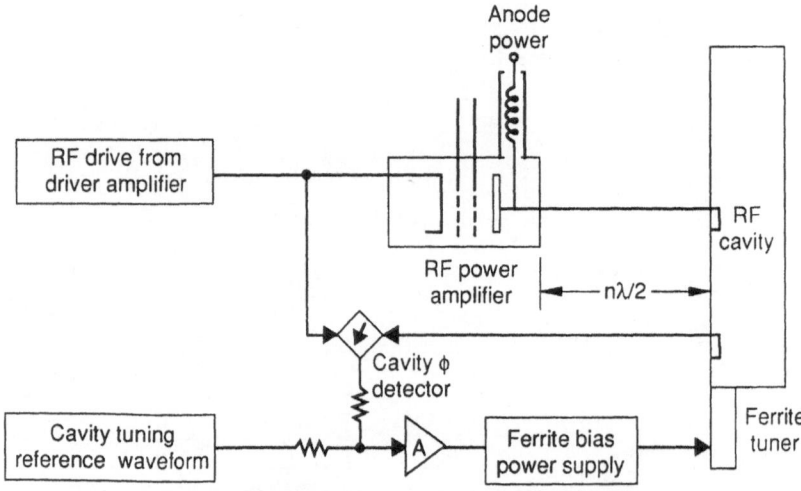

Figure 5. HEB RF Cavity Resonance Control.

The HEB schedule milestones are shown in Table 6.

Table 6. HEB RF Schedule Milestones.

Prototype Development	5/94-12/95
RFPs Issued for Production	6/95-1/96
Issue Purchase Orders	4/96
All Subsystems and Components Delivered	4/97
Assembly, Install and Test Complete	9/98

COLLIDER RF SYSTEMS

The Collider consists of two synchrotrons with beams traveling in opposite directions. The RF systems for the two Collider synchrotrons are essentially identical. The RF requirements for each of the two synchrotrons are given in Table 7.

Table 7. Collider RF System Requirements.

Frequency	360 MHz
Duty	100%
Peak Circumferential Voltage Per Ring	20 MV
Accelerating Voltage Per Turn	5.3 MV

The total RF power will be provided by four one megawatt cw 360 MHz klystrons; two per synchrotron. The power from each klystron will be divided by magic-Tees so as to deliver equal power to each of four cavities; eight cavities per synchrotron, sixteen total. The RF system for each synchrotron ring is shown in simplified block diagram form in Figure 6.

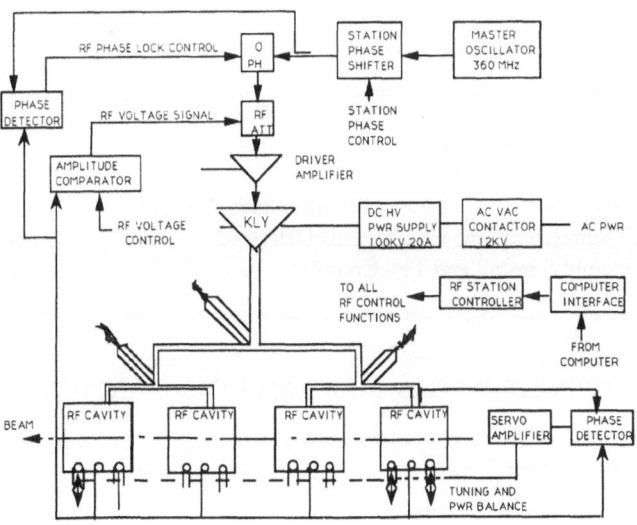

Figure 6. Collider RF System.

The collider cavities have five cells and are tuned with two mechanical slugs. A conceptual diagram is shown in Figure 7.

The klystrons are approximately 16 feet long and will be located in an RF gallery approximately 200 feet under ground along with most of the other RF components. It has not been decided whether to locate the 100 kV 20 ampere HV power supply transformer-rectifier assemblies in the RF gallery or at ground level.

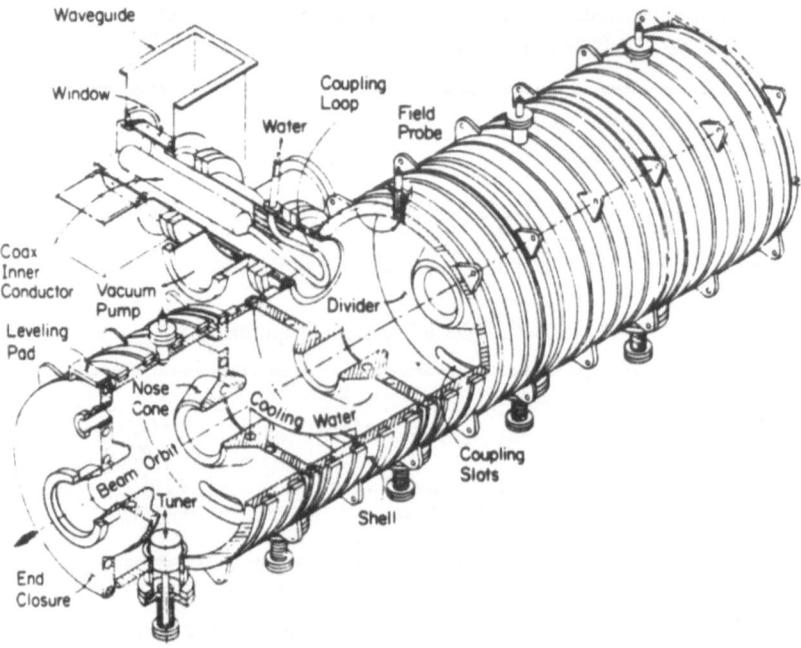

Figure 7. Collider Cavity.

The Collider RF schedule milestones are shown in Table 8.

Table 8. Collider RF Schedule Milestones.

RFPs Issued for Production	2/95
Issue Purchase Orders	10/95
First System Components and Subsystems Delivered	4/97
All Subsystems and Components Delivered	4/98
Assembly, Install and Test Complete	4/99

POTENTIAL OPPORTUNITIES FOR INDUSTRY PARTICIPATION

Detailed procurement plans have not yet been prepared for the synchrotron RF systems. However, the following provides insight into our preliminary thoughts on procurement.

We plan to design and develop each of the three different booster RF cavity and tuner designs as well as their associated RF amplifiers. The RF amplifiers will then be competitively procured on a fixed-price contract from a design and manufacturing firm with a proven ability to design and manufacture high-power VHF amplifiers. The cavities and tuners will likewise be procured from firms with a proven ability to design and manufacture such equipment. The performance requirements and critical dimensions will be provided, but the manufacturers will have some freedom in how the components are actually produced.

Another critical high-power component is the ferrite bias supplies, which must deliver and control very high, variable DC current to inductive loads. We will develop such a bias supply for our test program, and we plan to utilize what we learn to procure production units from manufacturers who have proven capability to design, manufacture and test such power supplies.

Most of the other high power components will be procured from performance specifications. In some cases, identical components will be required for all three booster systems. In these cases multiyear procurements, or procurements with option quantities, will probably be utilized.

We plan to design and develop the RF supervisory control systems and the high speed low level RF systems. Our procurement plan for the Collider RF system is very similar to that of the booster systems. We plan to design the RF system and procure all components from specifications. We plan to carry out the RF system integration and to be responsible for all RF system tests. We plan to use temporary or contract labor.

High reliability and availability are of prime importance to the SSC. The highest engineering standards will be imposed on all components and subsystems.

ACKNOWLEDGEMENT

The author wishes to acknowledge the assistance of J. H. Ferrell in obtaining schedule requirements.

LINAC RF SYSTEMS FOR THE SSC

J. H. Ferrell

Superconducting Super Collider Laboratory*
2550 Beckleymeade Avenue
Dallas, Texas 75237

Abstract: An overview of the RF system design for the SSC Linear Accelerator is presented. The RF systems for the Radio Frequency Quadrupole, Drift Tube Linac, Coupled Cell Linac, and bunchers are described. Performance requirements for each system are presented. The Linac low-level RF amplitude and phase control concept is described, as is the Linac RF control system design concept. Present RF system design status and plans are also presented.

LINAC RF SYSTEM GENERAL DESCRIPTION

The SSC Linear Accelerator (Linac) provides the RF energy for accelerating the 0.035 MeV beam from the ion source to the 600 MeV level for injection into the Low Energy Booster (LEB). Figure 1 is a block diagram of the Linac.

For reliability purposes, two ion sources are planned, each with its associated Radio Frequency Quadrupole (RFQ) and buncher. The RFQ accelerates the beam to 2.5 MeV. The buncher following each RFQ and the two bunchers preceding the Drift Tube Linac (DTL) match the output longitudinal emittance of the RFQ to the DTL.

The DTL accelerates the beam to 70 MeV and is followed by two bunchers for matching to the coupled cell Linac (CCL). The CCL accelerates the beam to 600 MeV. The CCL is followed by an energy compressor located in the transport line between the Linac and the LEB. Table 1 describes the general requirements for the Linac RF system.

The RF system operates at the ninth harmonic of the LEB injection frequency through the DTL. The bunchers following the DTL, the CCL, and the CCL/LEB compressor operate at the third harmonic of 427.617 MHz, namely 1282.851 MHz. For simplicity, the frequencies are often referred to as 428 MHz and 1284 MHz. Initially the RF frequency will be generated within the Linac RF system. At some future date, the Linac system may be locked to the LEB injection frequency.

---

*Operated by the Universities Research Association, Inc., for the U.S. Department of Energy under Contract No. DE-AC02-89ER40486.

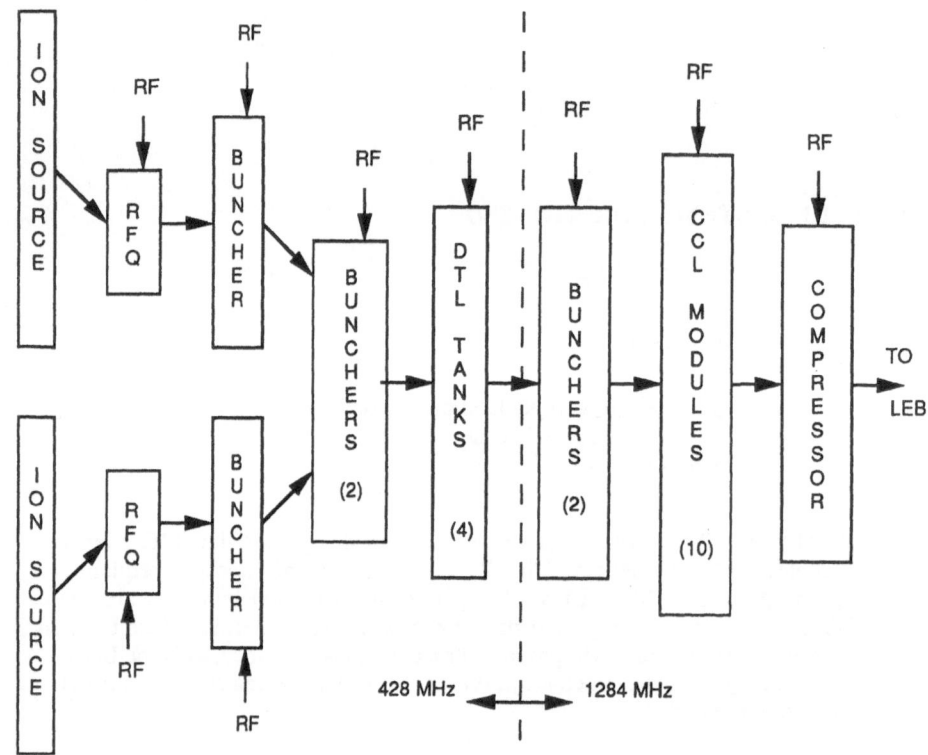

Figure 1. Linac Block Diagram.

Table 1. General Requirements—Linac RF System.

Input Energy	0.035 MeV (from source)
Output Energy	600 MeV
Beam Current	
Input	30 mA
Output	25 mA
RF Frequency	427.617 (428) MHz
	1282.851 (1284) MHz
RF Frequency Stability	Less than 1 part in 10^8
	or locked to LEB
Pulse Repetition Rate	10 Hz (1-10 Hz capability)
Beam On Time	7 μsec for Collider
	mode, 35 μsec for
	test mode
RF Voltage Stability	
Amplitude	$\leq 0.5\%$
Phase	≤ 0.5 degrees

As indicated in Table 1, the Linac RF system is a pulse RF system with a low duty factor; therefore, the average power requirements are quite low compared to the peak power required.

The RF voltage amplitude and phase stability requirements are quite stringent and represent one of the major influences on the RF system design. All of the amplifiers and modulators must be designed with the stability requirements in mind.

Figure 2 is a block diagram appropriate to either the RFQ or the bunchers following. Each RFQ and buncher amplifier will have its own amplitude and phase control system. Local control and monitoring will be provided at each amplifier station. A transfer switch and dummy load will be provided at each amplifier for test purposes. This will be the case for all Linac amplifier stations.

Table 2 describes the RFQ RF requirements. The peak power required is within the capability of gridded tubes or klystrons.

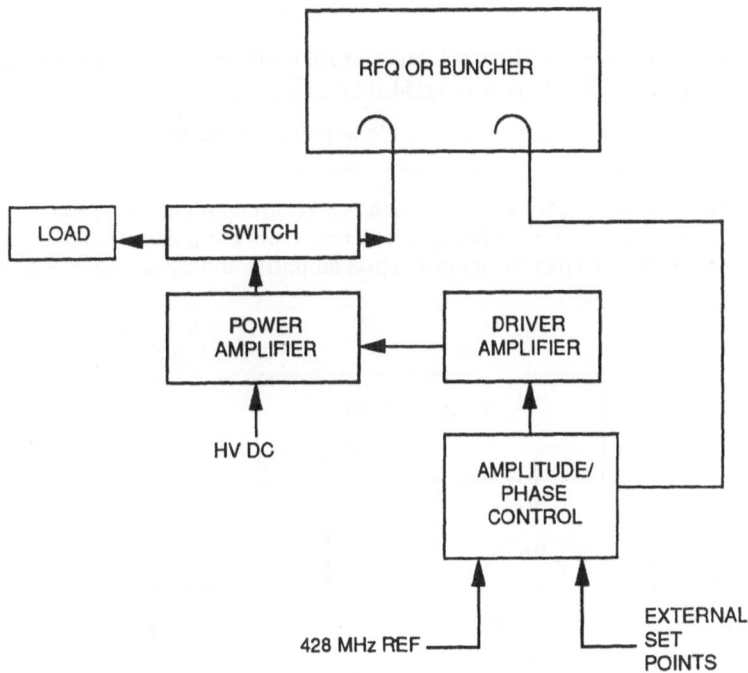

Figure 2. Block Diagram—RFQ or Buncher RF System.

Table 2. RFQ Requirements.

Number of RFQs	2
RF Frequency	427.617 MHz
Tank Q	7500
Peak RF Power Required	419 kW
Amplifier Power Capability	600 kW
Amplifier Bandwidth	500 kHz minimum
Amplifier Type	Gridded tube or klystron
RF Pulse Length	50 μsec

Table 3. RFQ/DTL Buncher Requirements.

Number of Bunchers	4
Tank Q	10,000
RF Frequency	427.617 MHz
Peak RF Power Required	22.5 kW, 22.5 kW, 13.3 kW, 0.8 kW
Amplifier Power Capability	30 kW, 30 kW, 20 kW, 2 kW
Amplifier Bandwidth	1 MHz minimum
Amplifier Type	Solid state or gridded tube
RF Pulse Length	50 μsec

Table 3 describes the RFQ/DTL buncher RF requirements. The peak power required is within the capability of solid state or gridded tube amplifiers.

Figure 3 is a block diagram applicable to either a DTL or CCL RF station. The first CCL station differs somewhat, as will be described later.

There are four DTL RF stations. Each is a klystron amplifier capable of providing 4 MW peak power. Due to the higher Q of the tanks, the RF pulse length is longer for the DTL. Each DTL station consists of the klystron amplifier and associated HV power supply,

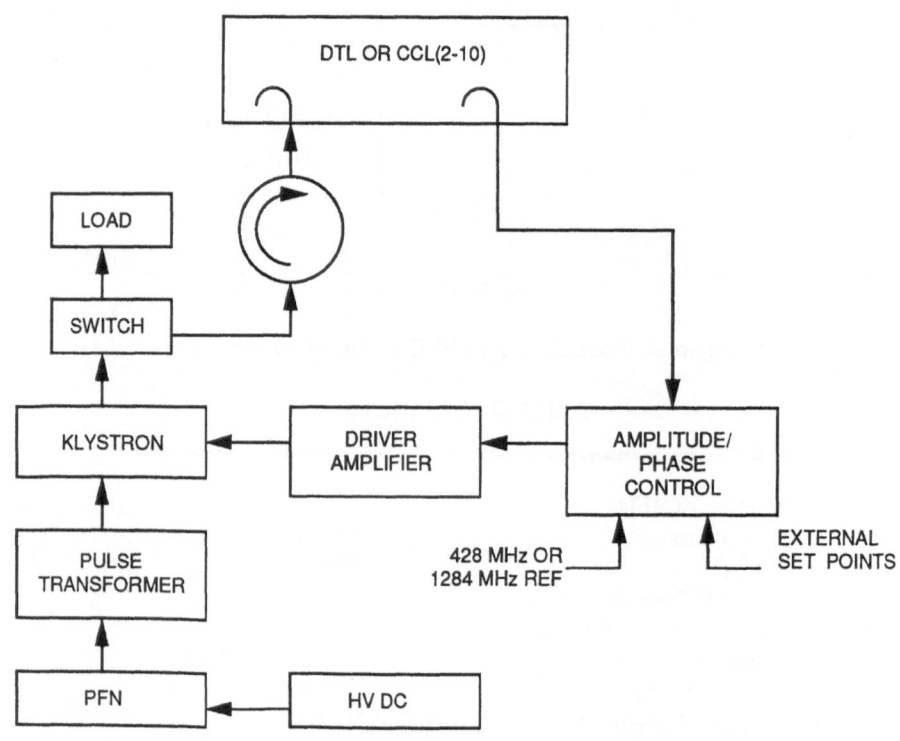

Figure 3. Block Diagram—DTL or CCL (2-10).

modulator, solid state driver amplifier, and amplitude/phase control loops. An isolator is provided between the klystron and the cavity at each station. The klystrons will be cathode modulated by a line type modulator. Present plans include a DTL test RF station for the purpose of maintaining a "hot" spare klystron. Table 4 describes the DTL RF system requirements.

Table 4. DTL Requirements.

Number of Tanks	4
RF Frequency	427.617 MHz
Tank Q	35,000
Peak RF Power Required	3 MW
Amplifier Power Capability	4 MW
Amplifier Bandwidth	500 kHz minimum
Amplifier Type	Klystron
RF Pulse Length	100 μsec maximum

CCL RF stations 2 through 10 are similar in concept to the DTL RF stations. The klystrons are cathode modulated by line type modulators. A test CCL station is also planned. Table 5 describes the CCL RF system requirements.

Table 5. CCL Requirements.

Number of Modules	10
Tank Q	15,000
RF Frequency	1282.851 MHz
Peak RF Power Required	15 MW
Amplifier Power Capability	20 MW
Amplifier Bandwidth	1 MHz minimum
Amplifier Type	Klystron
RF Pulse Length	50 μsec

As mentioned previously, the first CCL station is configured differently from stations 2-10. Figure 4 shows the first CCL station configuration. The klystron output will be divided to provide RF power to the two bunchers between the DTL and CCL as well as to the first CCL.

A motorized phase adjustment will be provided in the feed to each buncher. Analysis is in progress to determine if the amplitude and phase stability requirements can be met using this configuration. If not, a separate amplifier will be required for each buncher. Table 6 describes the buncher requirements.

The single CCL/LEB compressor RF station is located on the transport line between the CCL and the LEB. This system is a lower power system, presently planned for 100 kW peak power output. Studies are underway to determine if any economic advantage is achieved in reducing the power requirement by increasing the number of cells in the accelerating structure. Table 7 describes the compressor RF requirements as presently envisioned.

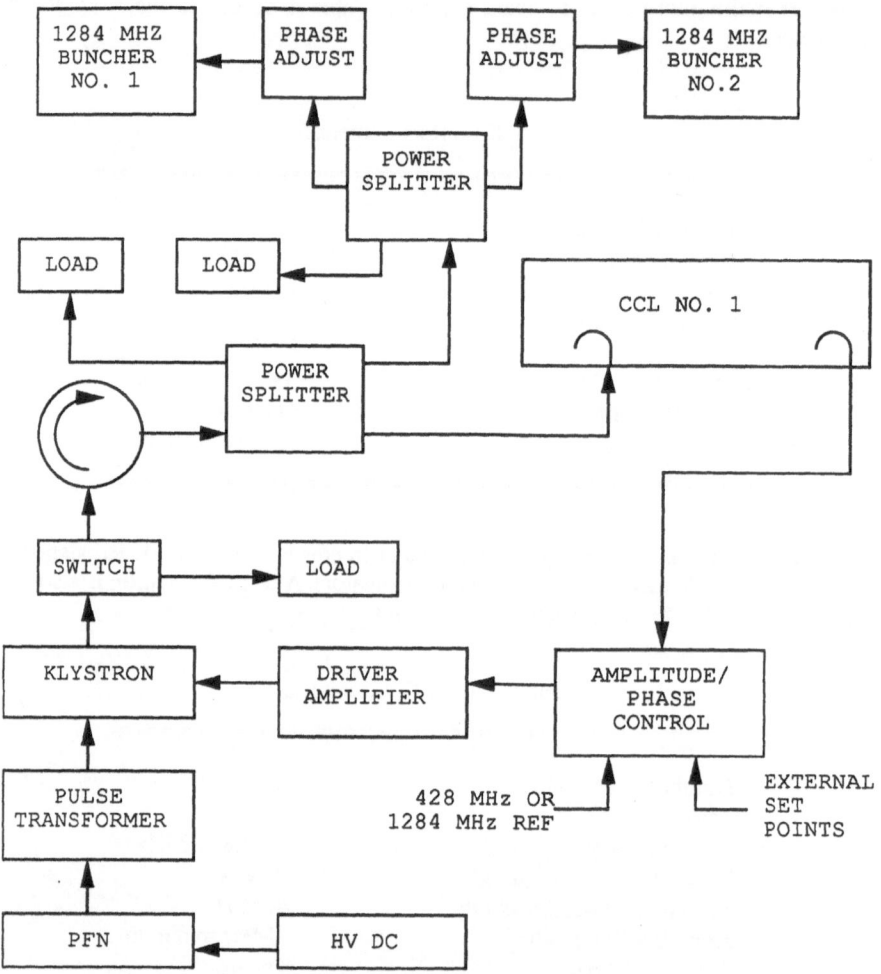

Figure 4. Block Diagram—1284 MHz Bunchers and CCL No. 1.

Table 6. DTL/CCL Buncher Requirements.

Number of Bunchers	2
Tank Q	15,000
RF Frequency	1282.851 MHz
Peak RF Power Required	250 kW, 125 kW
Amplifier Type	Power derived from first CCL

Table 7. CCL/LEB Compressor Requirements.

Tank Q	15,000
RF Frequency	1282.851 MHz
Peak Power Required	50 kW
Amplifier Power Capability	100 kW
Amplifier Bandwidth	1 MHz minimum
Amplifier Type	Gridded tube or klystron
RF Pulse Length	50 μsec

The low level portion of the LINAC RF system includes the amplitude and phase control loops as well as the RF frequency distribution system. Figure 5 is a generic block diagram applicable to each RF amplifier station.

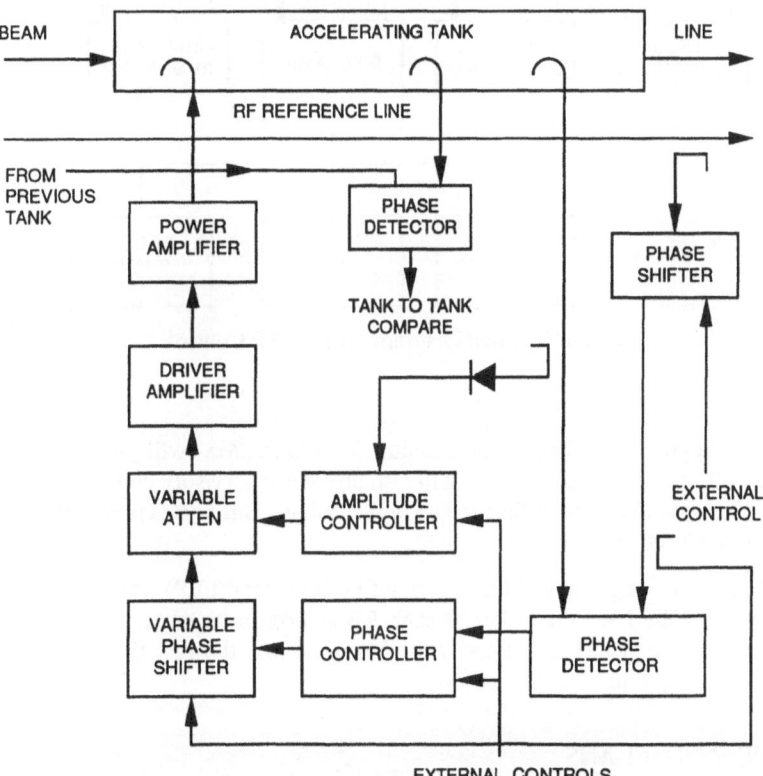

Figure 5. Block Diagram—Generic Low Level RF.

Pressurized and temperature-controlled rigid coax lines, located in the beam tunnel, will be used for frequency distribution. Since the RF voltage stability is directly dependent on the stability of the reference frequency, maintaining tight temperature control of these two lines (one for each frequency) is critical. The phase and amplitude detectors will also be located in the tunnel and will be temperature-controlled.

The Linac RF system will have its own control system. Figure 6 is a block diagram of the envisoned control system.

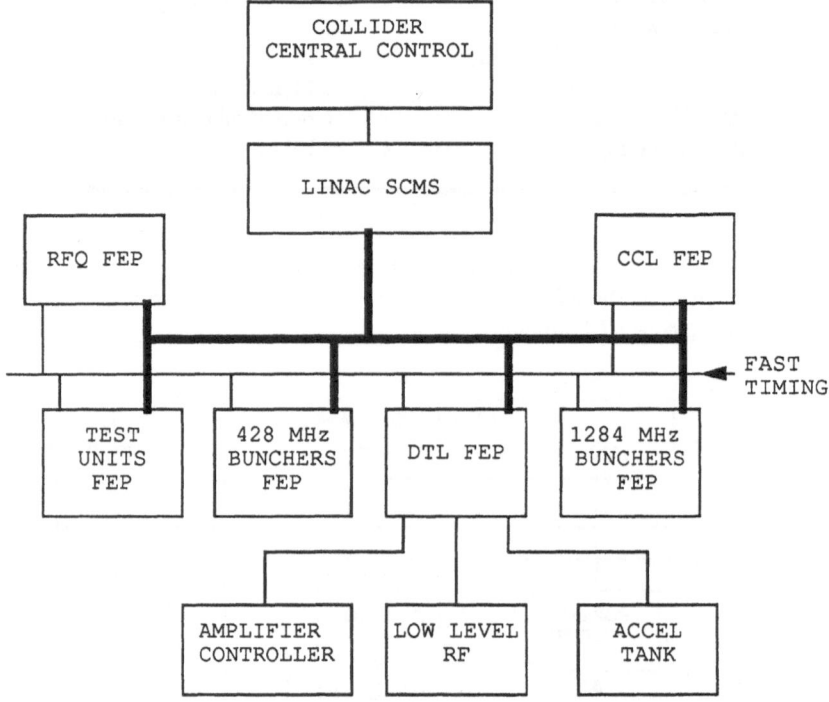

Figure 6. Block Diagram—Linac RF Control.

The Supervisory Control and Monitor System (SCMS) will provide communication with the Collider Central Control and will also provide supervisory control and monitoring of the Linac RF system during testing and commissioning. This unit is presently planned to be a Unix type workstation.

Each Linac segment will have a Front End Processor (FEP) which communicates with the SCMS and the local controller at each RF station in its segment. The FEPs will be configured as VME/VXI crates and will also distribute the fast timing signals from the Collider timing system to the RF stations.

STATUS AND PLANS

The SSC plans to procure the amplifier and modulator systems from industry. Procurement specifications are being prepared for each system as the design progresses. The high-power klystron specification should be released to industry in March or April 1991. The modulator specifications for these klystrons are nearing completion and will be finalized and released as soon as klystron vendors are selected. High power RF components will be specified once the final spatial relationship between the beam tunnel and RF gallery is known.

A computer model and prototype of the amplitude and phase control loops is in progress. Present plans call for design of the low level RF system by SSC personnel with build-to-print procurement following successful prototype tests. A detailed analysis of the supervisory control requirements is also in progress. The Linac control system will be designed by SSC personnel utilizing standard industry boards whenever practical.

ACKNOWLEDGEMENT

The author gratefully acknowledges the contributions of D. Dohan, J. Grippe, E. Marsden, R. Rodriguez and J. Wallace in providing information for this paper.

EARLY STAGES IN THE DEVELOPMENT OF THE GLOBAL RF

FEEDBACK FOR THE SSC LOW ENERGY BOOSTER

L.K. Mestha, J. Mangino, J. Santana and R. Webber

Accelerator Division
Superconducting Super Collider Laboratory*
2550 Beckleymeade Avenue
Dallas, Texas 75237

Abstract: A new method to generate the controlled master frequency has been designed to meet the requirements for accelerating the beam in the Low Energy Booster of the Superconducting Super Collider. This method is also designed to achieve rf synchronization for beam transfer to the Medium Energy Booster. The sweeping of the rf during acceleration complicates the programmed synchronization. The architecture discussed in this paper can be tailored to include all types of synchronization methods. Global synchronization timing for collider injection is discussed briefly. With the available hardware the frequency is updated at intervals of less than a microsecond using a TMS320C30, floating-point Digital Signal Processor. We are investigating the feasibility of generating the frequency with a Direct Digital Synthesizer as the frequency source for smooth acceleration within the LEB frequency range. Some of the problems associated with the products are discussed with a view to improve the design.

INTRODUCTION

The basic Low-Level rf system comprises several feedback control loops. Well-controlled master frequency is necessary to (1) capture the Linac beam, (2) accelerate the beam while correcting for the radial excursions due to field errors, and (3) transfer the beam to the next machine. Several feedback loops are involved to accomplish these functions. A precise control of the master frequency helps in filling the collider rings with less emittance dilution. Since the filling process is quite involved, we have attempted to set the goals by describing the timing sequence for filling cycles. There are several novel schemes in which the precise timing sequence can be generated to extract the beam. In developing the controlled master frequency, emphasis is given to the flexibility needed to incorporate any of the novel transfer schemes under investigation.

It is normal practice to use a Voltage Controlled Oscillator (VCO), or a phase-locked-loop frequency synthesizer with a VCO as the source of frequency.[1-3] An alternative way to produce frequency is by programming a Direct Digital Synthesizer (DDS), a device that

---

*Operated by the Universities Research Association, Inc., for the U.S. Department of Energy under Contract No. DE-AC02-89ER40486.

generates frequencies with digital techniques. For the Low Energy Booster (LEB) at the from 47 MHz to 59 MHz are considered. Various feedback loops discussed in this paper to control the master frequency are centered around the DDS. Most of the digital synthesizers available in the market have serious phase discontinuity problems, with some exceptions. We highlight a test set up to observe these defects.

GLOBAL RF FEEDBACK SYSTEM

A conceptual diagram of all the feedback loops associated with the master frequency generator is shown in Figure 1. A Digital Signal Processor (DSP1) is loaded with the required frequency table to sweep the rf signal frequency between 47.518 MHz and 59.776 MHz in 50 ms. The maximum rate of change of frequency, on the order of 1036.8 MHz/sec, occurs at about 5.7 ms into the LEB cycle. The DSP1 processor is able to read a single frequency value from the onboard static memory and to update the DDS frequency at a specified interval. The DSP and DDS combination can give an update interval of as short as 180 ns. The DDS we have evaluated is capable of changing the frequency in about 25 nanoseconds with 1 GHz clock frequency. Since the technology is rapidly advancing, it would not seem unreasonable to consider a complete digital approach to close all the associated feedback loops, as in Figure 1.

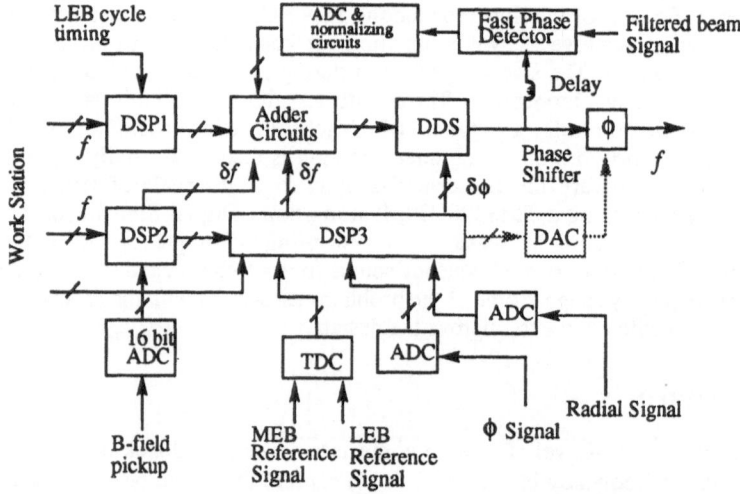

Figure 1. Schematic of the active feedback loops involved with the master frequency oscillator. (Interfacing circuits are not shown for simplicity).

Field Loop

The frequency table stored in DSP1 is corrected in real time every 2 s by using the measured B-field signal with an off-the-shelf 16-bit Analogue-to-Digital Converter (ADC) with a 2 s sample period. The processor in the B-field loop (DSP2) has a frequency table similar to that in DSP1 but with appropriate resolution which will correspond to each bit combination of the ADC. Thus these values will form a frequency look-up table with respect to the B-field. With this we can generate a frequency shift, δf, which is equal to the difference between the ideal frequency and the frequency corresponding to the ADC value (error in the B-field). In this way we can allow the DDS to sweep the frequency even if we disable the B-field loop.

Radial Loop

In some accelerators, such as the Fermilab Booster and DESY III proton machine, a B-field loop of the type described above is not included.[1] However, in the Fermilab Booster the correction for the field error is done by measuring the radial offset of the beam and varying the phase of the rf signal driving the high-power rf system. A similar loop can be implemented here for the SSC LEB. It is desirable to have the radial beam signal sampled once every Medium Energy Booster (MEB) turn. We then process it inside a DSP by using the advanced feedback controllers (under investigation), so that a limited phase shift is applied to the rf signal. A necessary phase shift can be implemented digitally in the DDS or by means of a wideband analogue phase shifter. Since the radial position can also be corrected by modulating the ramping frequency at the input end of the DDS, the digital implementation will help to achieve optimal trade-off between the frequency shift and the phase shift. The radial feedback loop may not be needed for collider operation since the beam phase loop (explained below) and the B-field loop may be sufficient to handle the unwanted beam excursions. However, for machine studies, on some occasions we may need to steer the the beam radially. To do this a δR program may have to be loaded into DSP3 from the workstation without disturbing the real time operation. We are evaluating the hardware for such architecture. Also, when we consider the LEB-MEB synchronization using the phase-control scheme, the radial position information may be useful to implement the multi-variable feedback controller.[4] Should we encounter unexpected problems during commissioning of the accelerator, an analogue radial feedback loop could be added without major changes with the digital system described above.

φ Loop

This loop may be necessary for machine studies period, or while commissioning the accelerator. It includes the accurate measurement of the phase of the center of the bunch with respect to the rf wave, φ. A φ program can be loaded into the DSP. Measured values are compared with theoretical values to obtain the phase error. The phase error is then applied to the DDS or the phase shifter, thus forcing the beam to follow the new program. The B-field and the radial feedback loops are used appropriately, if necessary, to the input end of the DDS. If we need to use this loop to damp the synchrotron oscillations, then the loop bandwidth must be designed well above the highest synchrotron frequency.

Synchronization Loop

This feedback loop is configured depending on the approach one can adopt to achieve synchronization. In Figure 1, a Time-to-Digital Converter (TDC) is shown in the synchronization loop. It is used to measure the position of the LEB reference signal for each MEB revolution. This will enable us to record the arrival time of the LEB reference wave. Using this time-of-flight information, the phase of the LEB reference wave is calculated in terms of the path length for each MEB reference wave. To achieve this goal, we may need information on the average radial position offset during the arrival time measurement. This information is available from the output of the radial loop ADC. The phase error is calculated by subtracting the measured LEB phase values from those stored in the processor. (The phase values stored in the processor are for the ideal case.) This phase error is then used to compute the frequency shift required to keep the phase error zero. By coupling the radial loop with the synchronization loop, we are in a position to compute the optimal frequency shift and the phase shift, if necessary. The frequency shift from the synchronization loop is fed to the adder circuits to control the master frequency. This feedback loop can be turned on at any time during the acceleration cycle.

If we use a phase-locking approach to synchronize the LEB with the MEB, then, as in Fermilab Booster,[1] we need to close the synchronization loop a few milliseconds before the flat B-field region. A phase detector will be used instead of the TDC to detect the phase information between the down-converted frequency from the LEB and the MEB reference waves. The phase values are compared with a table to follow a smooth program back to the extraction frequency. If the DDS is stable enough, we may not need an alternative frequency source to bring the phase difference to zero. However, an accurate matching of the frequency, phase, and rate of change of frequency is needed to accomplish the phase-locking scheme. The hardware architecture seems to be able to provide the necessary flexibility. Also, if necessary, a feedback loop to the beam can be included to smear out the noise on the frequency shift while phase-locking the two rf waves.

Beam Feedback Loop

A conceptual diagram of this loop is shown in Figure 1. The implementation may include more electronics. The loop can be closed during ramping so that the DDS can be phase-locked to the beam. The beam pickup signal may need pre-filtering through a band pass filter before it is sent to the fast phase detector,[1] depending on the hardware.[5] The fast phase detector output is digitized and normalized to integrate with the DDS. A small delay may be necessary between the DDS output and the fast phase detector as in the Fermilab Booster system. By dc coupling the beam phase loop using an additional circuitry after the ADC, we can use the loop merely to add a small frequency shift to the DDS adder circuits while ramping.[5] However, if the dc coupling is not done, the beam frequency is generated from the beam phase loop by disconnecting the ideal frequency table from DSP1. Under such a condition, the B-field loop may have to be disabled since the beam frequency would act as the reference to the DDS. Also, if we use this loop to damp synchrotron oscillations, the bandwidth must be well above the highest synchrotron frequency in the LEB. We have included this to work with a worst-case scenario.

Linac rf Signal

A few microseconds before the Linac-to-LEB injection, the LEB master frequency hardware may be required to generate the correct injection frequency to drive the Linac rf system. This is because the Linac rf system will be running at the 9th harmonic of the LEB injection frequency. Linac frequency must remain unchanged while the LEB frequency is allowed to ramp during acceleration. We have two schemes, shown in Figure 2, in which this could be achieved. In Figure 2(a), we show the design using a separate DDS for the Linac reference source. The data lines to the Linac DDS are hard-wired, and the output signal is multiplied to give nine times the LEB injection frequency. Reset lines are used to accurately start the two DDSs at zero phase. A minimum of interfacing between the two DDSs would allow significant flexibility in controlling the Linac frequency relative to the LEB injection frequency. It is not clear whether the phase discontinuity created by the DDS reset signal would have problems to the high-power rf system. Alternatively, in Figure 2(b), the Linac reference frequency is generated using a Voltage Controlled Oscillator (VCO). Here the gated track and hold circuits will be used to phase-lock the VCO to the LEB DDS at the appropriate time before and during injection. The hold circuits maintain the required frequency to the Linac rf system.

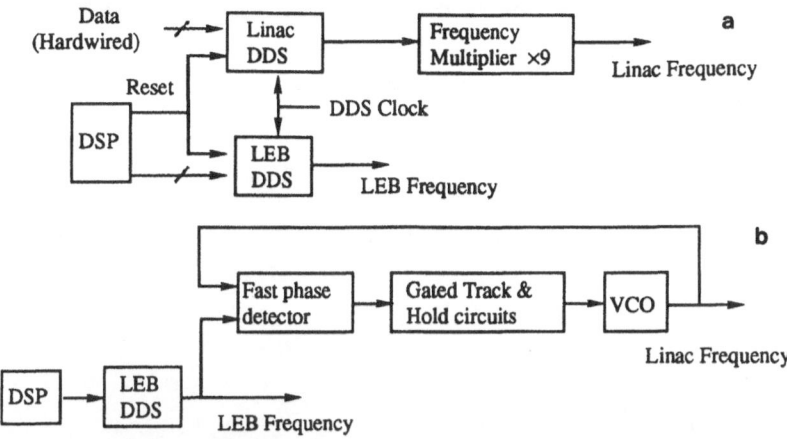

Figure 2. (a) Schematic diagram of the setup to generate Linac frequency using DDS.
(b) Schematic diagram of the alternate setup to generate Linac frequency use
VCO.

GLOBAL SYNCHRONIZATION TIMING

Global timing sequence for synchronization depends on the filling sequence required for the collider operation, injection, and extraction kicker rise and fall times and the cycle time of each machine. In Figure 3, the collider filling scenario is given with the number of beam bunches and gaps. Refer to the SCDR[6] for the injection and extraction kicker parameters. In Figure 4, the timing associated with each machine cycle is shown in detail. The injection time required for the Linac to the LEB is 6.819 µs due to three-turn injection. At this time we disable all the feedback loops mentioned above and hold the LEB frequency to the injection frequency. The numbers 1 to 2916 in Figure 4(a) represent the number of Linac beam bunches stacked in the LEB. Figure 4(b) shows the timing between the beam bunches just before acceleration. After acceleration to the top LEB energy (Figure 4(c)) beam pulses are separated by 16.68 ns and are extracted into the MEB. There will be two gaps in the MEB due to injection and extraction kicker rise and fall times. Figure 4(d) shows the MEB filling sequence from the LEB for the first two MEB-to-High Energy Booster (HEB) transfers. At least 24 gaps are needed in the MEB to accommodate the MEB extraction kicker and the HEB injection kicker rise times. In this cycle there are seven LEB transfers; in the third MEB-to-HEB transfer there are only four full LEB transfers and one partial transfer (Figure 4(e)). Thus we would need to include a counter module to the LEB-MEB synchronization loop shown in Figure 1. The partial transfer of only 70 bunches is made by aborting the unwanted LEB bunches at the top energy into the beam dump located somewhere in the transfer line. Furthermore, there will be eight transfers from the HEB to the top collider ring with the filled bunches and the gaps as shown in Figures 4(f) and 4(g). Effectively it takes 1752 sec to fill the top collider ring (Figure 4(g)). A similar filling sequence is followed to the bottom collider ring (Figure 4(h)) by reversing the polarity in the HEB. From Figure 4 it is clear that the complete collider filling time is 3517 seconds with the current design.

From the synchronization chart it must be noted that while the MEB is accelerating, there will be no beam in the LEB until all the beam from the MEB is extracted to the HEB. This means the second set of injections to the MEB takes place only after completing one full MEB cycle. Thus there will be a period in which there will be no beam in many of the LEB cycles. This means the radial feedback loop and the synchronization loops in Figure 1 may need to be opened when the LEB is free-running without the beam. If needed, the timing circuits will be developed to incorporate these considerations.

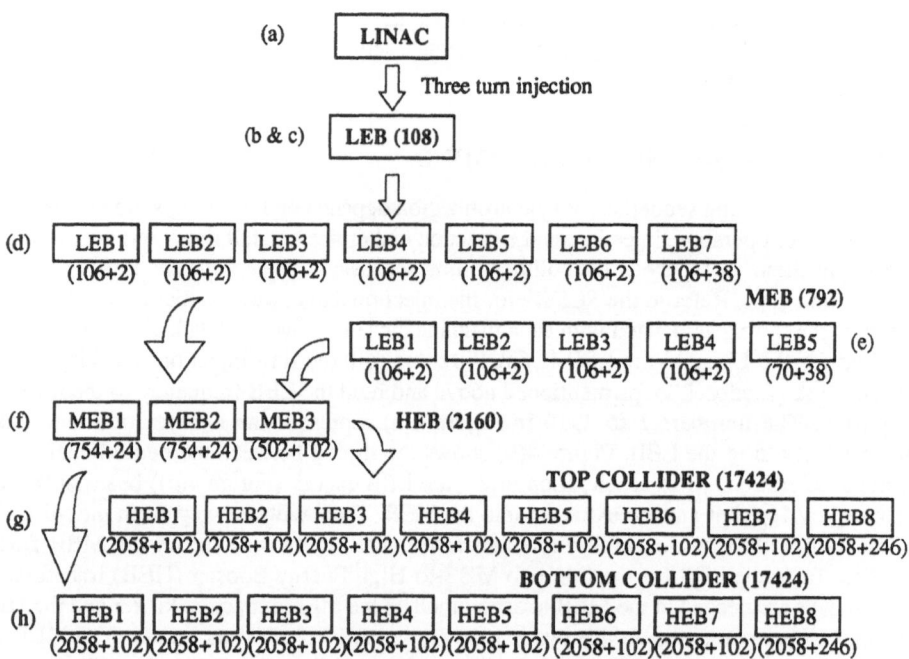

Figure 3. Filling sequence for collider operation.

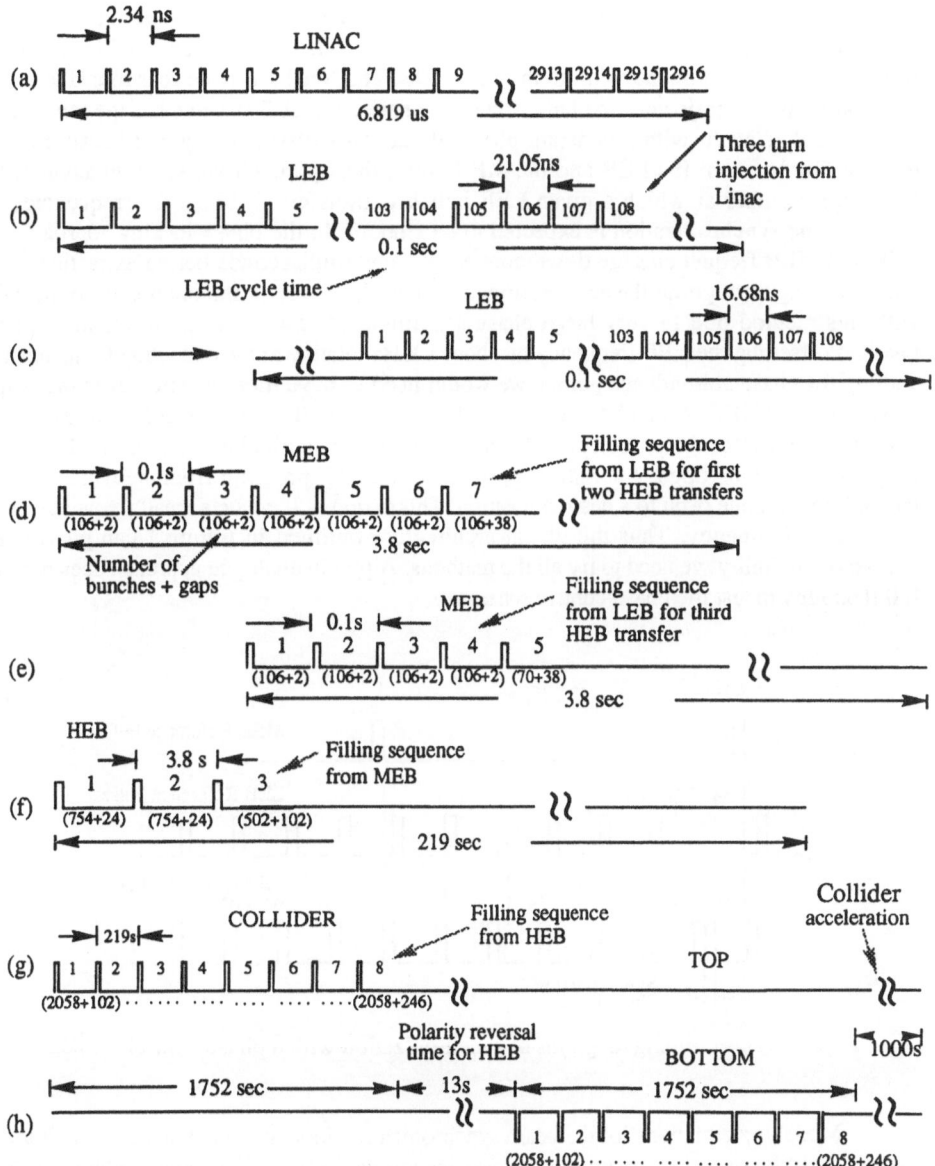

Figure 4. Filling sequence for collider operation with timing operation.

SYNCHRONIZATION SCHEMES UNDER INVESTIGATION

There are several ways to implement beam transfer synchronization. In previous proceedings of this conference we reported three schemes:[7] phase-control, phase-locking, and phase slippage. The phase-control and phase-locking schemes are very similar in principle, but they differ in their implementation. In simple terms, the phase-control approach involves measuring the phase information between the LEB and MEB reference rf waves (see Figure 5) and then estimating the error by subtracting the ideal values stored in the computer memory on a real time basis at every MEB revolution. The error is corrected by adding or subtracting a small frequency shift to the LEB DDS. Thus the LEB rf wave is forced to follow a pre-programmed trajectory. At the instant of transfer, the LEB rf wave has the same phase as the MEB rf wave with a constant phase offset. This offset is required because of the transfer line between the LEB and the MEB. With this approach we know in advance the MEB turn number at which a given LEB bunch is transferred. Since the frequencies are matched, the synchronization is expected to be smooth. In the phase-locking approach, the LEB and MEB frequencies are down-converted a few milliseconds before extraction.[1] The phase-locking throughout the acceleration, or at least half-way in the middle, is not possible with this method due to very large phase accumulation. Since the third transfer scheme involves offsetting the LEB frequency by about 1 KHz relative to the extraction frequency and waiting for phase coincidence points, we would need to have more electronics at the output end of the LEB DDS. A good phase detector between the LEB and the MEB rf signals can be considered to detect the phase coincidence points. In this particular case, we would need to keep the radial loop open. Alternatively, we can load a δR program from the workstation to the DSP to force the orbit to a desired position. This action will create a small frequency offset in the LEB frequency. Thus the global architecture outlined in Figure 1 can provide the necessary flexibility we need to try all the methods. A purely analogue approach does not give full flexibility to test the phase-control scheme.

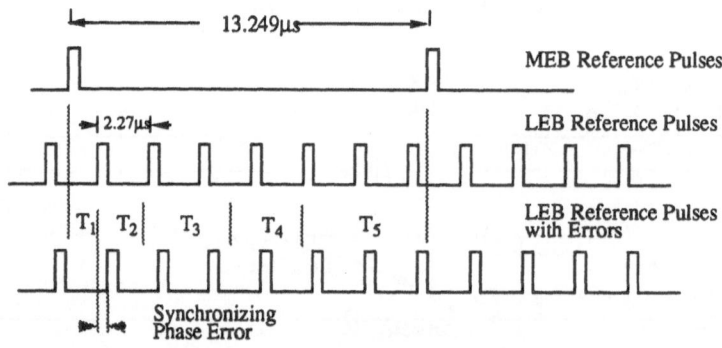

Figure 5. Basic timing diagrams of synchronization with a phase-control method.

Modelling equations for the beam synchronization loop and the radial position loop are derived in Reference 4 by assuming the cavities are well-tuned. The cavity detuning effects due to beam loading and the error due to phasing multiple cavities can be considered secondary to the multi-variable feedback loop model. Using the model, we expect to design a robust controller which would maintain the orbit shift to minimum value and also synchronize with the MEB. We expect to bench test the hardware with different feedback controllers and to simulate field errors. Recent hardware experience with such attempts is discussed below.

HARDWARE EXPERIENCE

The hardware configuration shown in Figure 6 is used to bench test the synchronization loop as a demonstration of the "proof of principle" of the phase-control approach. Later, the hardware will be extended to bench test the phase-locking and phase-slippage schemes with additional essential circuitry. In Figure 7 we show the schematics of the VME/CAMAC data acquisition system used to communicate with the DSP and other electronic modules. A SUN 4/330-based workstation is used as the mainframe computer to develop the assembly-level software for the DSP. Two 32-bit floating-point DSPs, TMS320C30, from Texas Instruments were used as the processing units. The C30 DSP board is made by SKY Computers. It has 2 MB of Dynamic RAM and the usual control registers to interface with the VME bus. All the communication to the UNIX workstation is done through the VME interface bus.

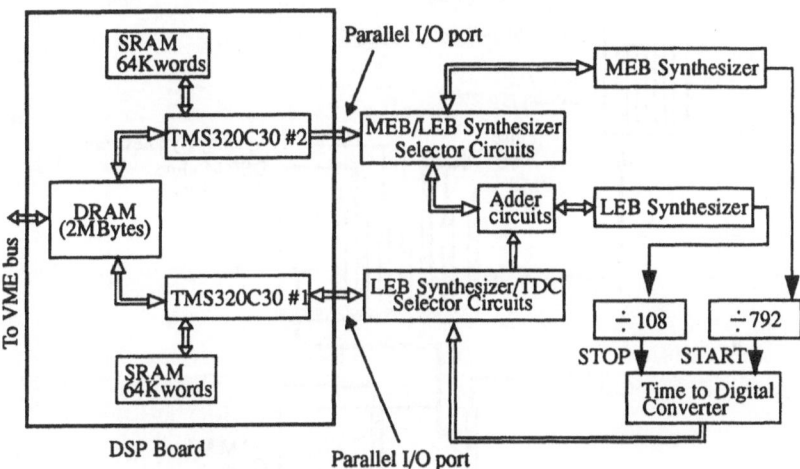

Figure 6. Schematic representation of the system diagram used to bench test the synchronization feedback loop.

The Static RAM used in the SKY board is only 64 K words long. Using C30 #2 on the SKY board, the MEB synthesizer is loaded with an appropriate number to generate the rf at the LEB extraction frequency. Then the same processor is used to generate the master frequency by reading the frequency data stored in the Static RAM and then launching it to the LEB frequency synthesizer at a specified time interval in real time. Since we have only 64 K words of static memory as a storage place for the frequency table, we set the time interval equal to 960 ns to cover 50 ms ramping time for the LEB. Even though the DSP board has 2 MB of global memory, we could not use this due to a memory refresh cycle invoked by the global RAM arbiter on a priority basis. We would like to see a DSP product which has as much as 2 MB or more Static RAM with zero wait states to further reduce the frequency updating interval to as low as 180 ns. It will be useful to ramp the frequency at small intervals so that the frequency jump required from the previous interval is small.

The counters, dividing by 108 and 792, represent the number of rf waves in one revolution of the LEB and MEB, respectively. The TDC measures the arrival time of the first LEB reference pulse after every MEB reference pulse. The data is then read into the C30 #1 and processed using the trip-plan and the feedback controllers explained in Reference 4 to generate the frequency shift, $\delta f(t)$. The frequency shift is fed to the adder circuits on the interface board to correct the master frequency. The operation of the feedback loop is repeated every MEB turn, which is about 13.249 μs long. We expect to include a small phase error at the beginning between the LEB and the MEB reference waves. After we close the feedback loop, the loop will bring the phase error to zero. If the phase error is zero, then at a designated MEB turn we must have the phase synchronization for transfer (see Reference 4). Thus the result of this experiment is expected to demonstrate the ability to phase-lock the fixed frequency oscillator of the MEB during injection, to a variable frequency oscillator of the LEB from any given MEB turn.

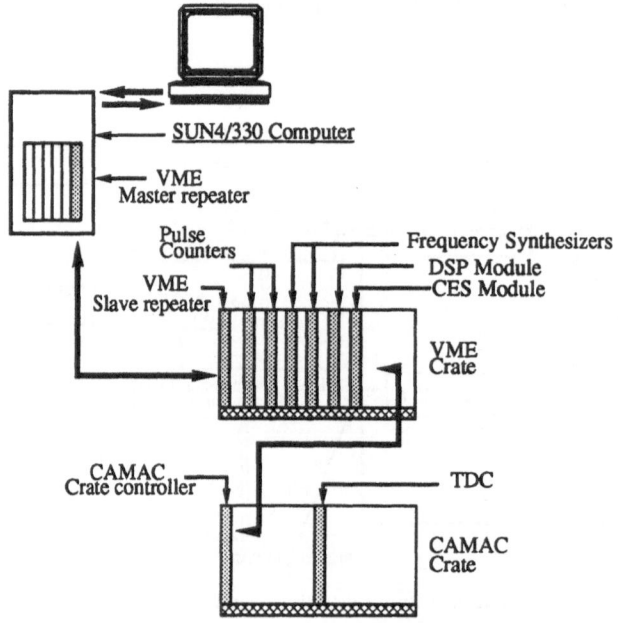

Figure 7. The hardware architecture with VME and CAMAC read-out system.

The TDC we have been using is made by Lecroy Corporation's Model 4204. In this module, the decoded time interval data does not appear instantaneously on the parallel port to read into the DSP. Also, the resolution varies from 0.156 ns to 2 ns in absolute terms. Hence there is inconsistency in the time measurement. Ideally, we would like a TDC to have a resolution of better than 30 ps with no time lapse for the conversion. The maximum time we expect to measure would not exceed 2.2 μs. The DDSs have advantages over the phase-locked loop VCOs due to the fine frequency resolution (1 part in 2^{32}). They have fast switching speed between output frequencies, and good spurious and phase noise performance. The phase discontinuity we have observed on some DDSs is the most undesirable behavior for our application. The discontinuities in the rf signal are normally due to lack of onboard synchronization of the clock and the frequency strobe signal. The best tool to test for phase discontinuity is the accelerator itself. However, we have carried out a simple test (Figure 8) in which we use two identical synthesizers. In the first approach, two DDSs are driven in parallel by the data such that the most significant bit is toggled to offset the frequency by 0.93 Hz. This test will enable us to look for the discontinuity when all the usable data lines on the DDS are fully loaded. In the second approach, they are driven in parallel by a counter. The second approach is used to test when the frequency is swept at a certain rate. In our case we used the counter circuitry to sweep at a 1 MHz rate, although we could go much higher than this. The step size in frequency was again made to be 0.93 Hz by selecting one of the bits on DDS #2 to ground and the same bit on the other DDS high. The output of the two DDSs are then passed through a mixer. The mixer output is filtered using a low pass filter so that the original frequencies of the two DDSs, their harmonics, and their sum are removed. Only the difference frequency is recorded in Figure 9. When the two DDSs are running without the phase-discontinuity, the filtered output of the mixer must be a pure sinusoid with a frequency equal to the difference in frequency of the two DDSs. When there is phase jump in one DDS due to poor circuit layout or no synchronization, the mixer output shows discontinuities, as in Figure 9. This is undesirable for our application since the sudden phase change will lead into coherent oscillations, emittance dilution, and possible beam loss. Hence we are proposing to the DDS manufacturers to include this in their standard test procedures.

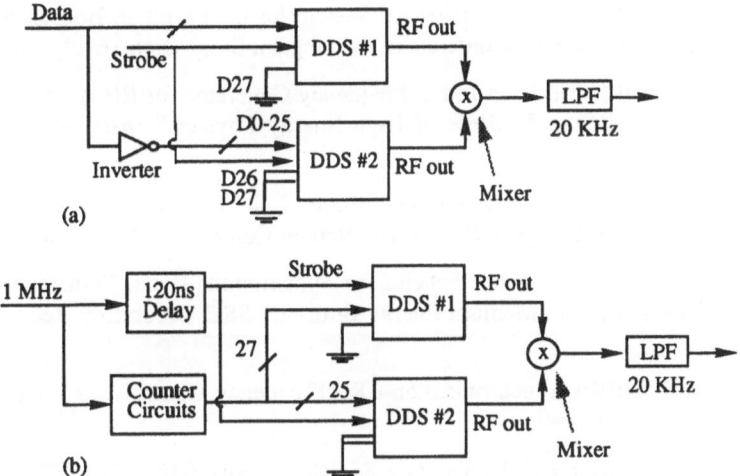

Figure 8. Test setup to observe phase discontinuities in the DDS.

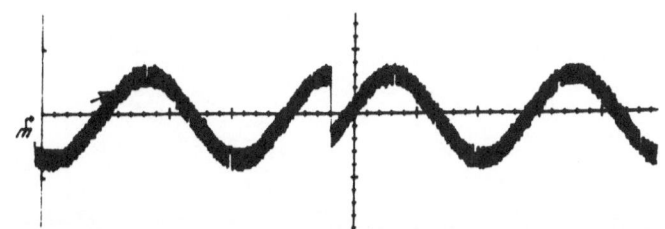

Figure 9. Output signal of the mixer (only difference frequency is retained).

CONCLUSIONS

We have shown in this paper the hardware architecture involved in producing master frequency to the SSC Low Energy Booster which is controlled during capture, ramping, and extraction. The hardware architecture we have proposed allows an environment to test all the transfer schemes under consideration. The radial feedback loop hardware is considered in the system to control the beam orbit due to field errors. The B-field loop to the master frequency oscillator is included to bench test the synchronization loop by imposing random errors in the field signal. The heart of the system consists of Digital Signal Processors and Direct Digital Synthesizers. We would need a DDS to have a spur suppression of up to 70 dB with the possibility of phase modulation included to the rf signal. Because of its speed, the DSP executes feedback control algorithms well. With our existing DSP hardware, we would have difficulties in accessing the processor without interrupting the real time operation. By using the radial position signal and the synchronizing phase signal in a single processor, we can implement a global feedback controller which would produce an optimum frequency and the phase shift to control both parameters. Since this approach is still theoretical, we are proposing to simulate the feedback scheme under various practical limitations. However, should there be unexpected problems, the configuration can be easily tailored to include more conventional frequency and phase-control schemes used in other accelerators.

REFERENCES

1. C. Kerns, et al., "New Low-level RF System for the Fermilab Booster Synthrotron," IEEE Particle Accelerator Conference, Washington, DC, March 16-19, 1987.

2. V.L. Bruk et al., "Precision Initial-Frequency Generator for RF Accelerating Voltage Shaper of Booster of Institute of High-Energy Physics," *Instrum. Exp. Tech.*, **29**, 1986.

3. P. Barratt et al., "RF System and Beam Loading Compensation on the ISIS Synchrotron," 2nd European Particle Accelerator Conference, Nice, June 12-16, 1990.

4. L.K. Mestha, "Phase-Control Scheme for Synchronous Beam Transfer from the Low Energy Booster to the Medium Energy Booster," SSC Laboratory Report SSCL-340, 1990.

5. W. Kriens, "HERA Synchronization—RF Frequency Control and Transfer Timing," Notes, September 1990.

6. *Site-Specific Conceptual Design of the SSC*, SSCL-SR-1056, July 1990.

7. D.J. Martin et al., "Early Instrumentation Projects at the SSC," 2nd International Industrial Symposium on the Super Collider, Miami, March 14-16, 1990.

KICKER MAGNET SYSTEMS FOR THE INJECTOR SYNCHROTRONS

D.E. Anderson, G.C. Pappas, L.X. Schneider, and J.M. Wilson

Superconducting Super Collider Laboratory*
2550 Beckleymeade Avenue
Dallas, Texas 75237

L.L. Reginato

Lawrence Livermore National Laboratory
P.O. Box 808 L-627
Livermore, California 94550

Abstract: A kicker magnet is a device capable of producing a pulsed magnetic field for the purpose of deflecting charged particle beams out of their defined orbit. The kicker magnets at the Superconducting Super Collider (SSC) Laboratory are designed to transfer beam bunches from a series of three booster stages into a counter-rotating 20 TeV collider ring. The required kicker magnetic field is the result of factors such as bunch spacing, maximum allowable lost bunches, required angular deflection, and other characteristics of synchrotron operation.

An overview of the injection synchrotron kicker magnet systems to be used in the SSC is presented. System design includes magnet and modulator design for each of the kicker systems addressed. Design issues for each subsystem are presented, including magnet geometry, magnetic material selection, feed topologies, switch selection, and other key considerations. An overview of system requirements, as well as a brief statement of design status, is also presented.

---

*Operated by the Universities Reasearch Association, Inc., for the U.S. Department of Energy under Contract No. DE-AC02-89R40486

INTRODUCTION

A block diagram of a typical kicker system is shown in Figure 1. The system is made up of a control system, a high voltage charging supply, a trigger generator, a pulse-forming line and switch, the magnet assembly, and a matching load. System charging and firing commands, generated in the main accelerator control room, synchronize all of the kicker magnets to achieve the sequential filling of the injection synchrotrons and main collider rings. Filling of the Medium Energy Booster (MEB) from the Low Energy Booster (LEB) occurs in seven cycles at 10 Hz; the High Energy Booster (HEB) is filled from the MEB in three cycles at 0.3 Hz; each collider ring is filled from the HEB in eight cycles at 0.01 Hz.

Two geometries, depicted in Figure 2, are under consideration for the SSC kicker magnets. The C-magnet (Figure 2(a)) uses a toroid or a C-core with a large air gap to accommodate the beam tube. Current driven through a single-turn busbar surrounding the magnetic

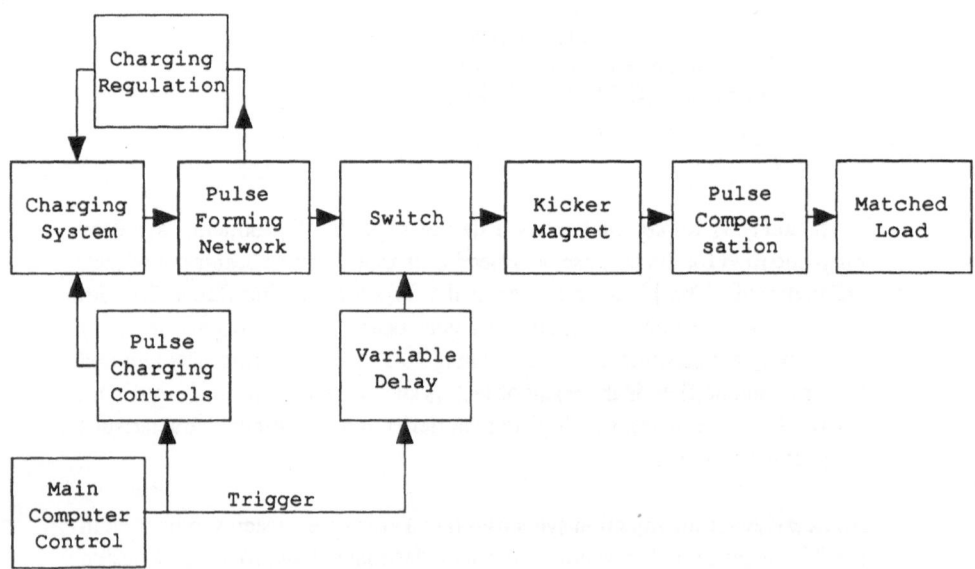

Figure 1. Block Diagram of Typical Kicker Magnet Systems.

material produces vertical flux in the air gap, resulting in horizontal deflection of the proton beam. The H-magnet (Figures 2(b) or 2(c)) employs either an H-core or a toroid with pole pieces to form a closed magnetic path with a large air gap in the middle. The H-magnet uses two equal driving currents of opposite direction on either side of the air gap to produce flux in the air gap. Since the current in each H-magnet bus is half that required in a C-magnet bus, the H-magnet is in many ways analogous to two parallel C-magnets.

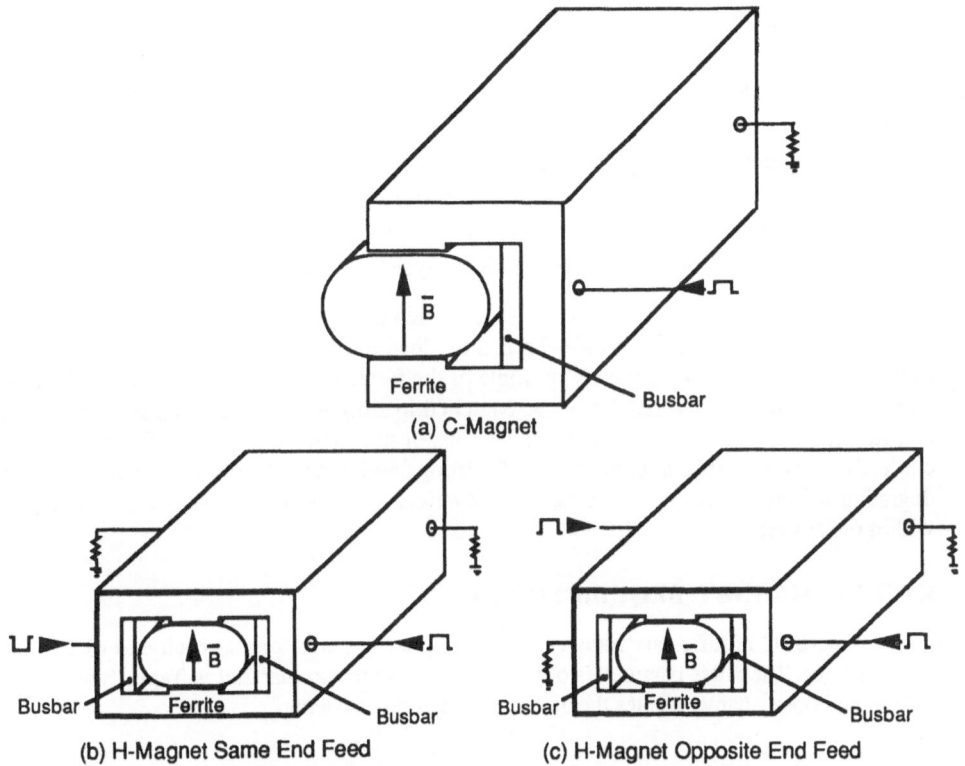

Figure 2. Comparison of H- and C-magnets.

Magnetic material selection depends primarily on the frequency of the drive pulse and the required magnetic field intensity. Generally, ferrite cores can be operated into the 100 MHz range with manageable core losses, but at the expense of lower saturation flux densities and small initial permeabilities. Laminated steel (or Metglas) cores offer much higher saturation flux densities and initial permeabilities, but with significant core loss above approximately 1 MHz (higher for Metglas).

Modulator choices depend on pulsewidth, risetime, and falltime requirements. Typically, for pulse widths of less than 2 µs, a charged-cable pulse-forming line is used. Risetimes, which are dominated by switch turn-on time, switch inductance, and stray capacitance, can be made lower than a few tens of nanoseconds, and flattop ripple is essentially zero. Droop due to cable losses can be compensated by passive correction circuits. Lumped LC pulse-forming networks (PFNs) are traditionally used for longer pulses. Both flattop ripple and rise/falltimes are functions of the number of elements in the network. Most of the SSC kicker systems will employ lumped-element PFNs.

Pulse parameters such as peak current, charge transfer, voltage, and jitter also determine the types of modulator switches. For many of the SSC kicker modulators, hydrogen thyratrons will be used. They are reliable, low-loss, low-jitter devices which, with careful design, can produce the required switching speeds and perform with long lifetimes. For the fastest kicker systems, alternative switches include the back-lighted thyratron (BLT) and spark gap. The BLT is advertised as being faster than a traditional thyratron, but since

BLTs are still in the development stage, very little lifetime data is available on any single device.[1] Triggered spark gaps, which are more compact and generally less inductive than thyratrons, are generally discounted in such high-reliability applications because of limited lifetime and/or excessive longterm jitter due to electrode erosion. Silicon-controlled rectifiers (SCRs) are under consideration for some of the slower kicker modulators, where high currents, long pulse widths, and less-demanding switching speeds are required. Series/parallel arrays of SCRs will be required because single devices capable of operating reliably at the voltages and currents of the kicker systems do not currently exist.

There are many other modulator issues which must be addressed in the design of individual systems. These include pulse-sharpening, load matching, and compensation circuitry. Pulse-sharpening may be necessary on some of the faster kickers if a switch cannot be found which meets all the pulse requirements. Matched loading is required when stringent flattop and falltime requirements demand no reflections or transients in the magnetic field. Finally, cable dispersion, cable attenuation, PFN droop, and other effects can result in flattop degradation outside acceptable limits. In these cases, compensation networks will be required within the system.

KICKER MAGNET REQUIREMENTS

The SSC requires six transfer kicker magnet systems, one at each end of the three transfer lines. The pulse parameters of the six systems are summarized in Table 1, where E or I after the booster name denotes the ejection or injection system.

Table 1. Kicker Magnet Parameters.

System	B_{max}	Aperture	Rise-time[a]	Pulse length	Jitter (rms)	Flat-top	Number magnet
LEBE	0.015T	5 × 9 cm	47ns	1.8µs	1ns	≤0.5%	4
MEBI	0.010T	5 × 10 cm	47ns	1.8µs	1ns	≤1%	4
MEBE	0.065T	5 × 10 cm	400ns	12.7µs	5ns	≤1%	2 × 6
HEBI	0.112T	4 × 4 cm	400ns	12.7µs	10ns	≤1%	2 × 6
HEBE	0.125T	4 × 4 cm	1.7µs	36.3µs	10ns	≤0.5%	2 × 6
ColI	0.060T	6 × 6 cm	1.7µs	36.3µs	10ns	≤1%	2 × 10

[a]0-99% risetime

MAGNET DESIGN ISSUES

Magnet design follows a logical approach based on three parameters. These parameters, set by the beam physics and material limitations, are peak magnetic field, aperture size, and peak operating voltage.

The approximate current required to establish a given magnetic field in the kicker magnet can be determined from Ampere's Law:

$$I \approx (B/\mu_0)[g+(l_m/\mu_r)] \, ,$$

where B is the desired magnetic field intensity, g is the air gap spacing (aperture height) across which the field is to be established, l_m is the mean length of magnetic material, and μ_r is the relative permeability of the magnetic material. In all the kicker magnets discussed in this paper, $g>>(l_m/\mu_r)$, giving rise to the following simplification:

$$I \approx Bg/\mu_0 .$$

Material considerations limit the peak and charge voltages at which the kicker magnet system can operate. These limitations exist primarily on the modulator cables and switching devices, but may also exist on the pulse compensator, pulse-forming network capacitors (where required), and elsewhere. This voltage limit, when divided by the current required to establish the field, sets an upper limit on the operating impedance of the kicker magnet. For a distributed-element transmission line system, the operating impedance is given as

$$Z_0 = (L/C)^{1/2} ,$$

where L and C are the inductance and capacitance per unit length of the transmission line system, respectively.

The distributed inductance for a kicker magnet system can be calculated based primarily on the aperture size. From Faraday's law, we obtain

$$L = \Phi/I = [\mathcal{R}_a + \mathcal{R}_m]^{-1} ,$$

where $\mathcal{R}_a$ and $\mathcal{R}_m$ are the air and magnetic material reluctances, respectively. Substituting the geometrical representation of the reluctances on a per unit length basis leads to

$$L = [l_m/(\mu_r\mu_0 t) + g/(\mu_0 W)]^{-1} ,$$

where t is the thickness of the magnetic material and W is the width of the aperture. If the thickness of the magnetic material is comparable to the width of the aperture, the following approximation can be made:

$$L \approx \mu_0 W/[g + (l_m/\mu_r)] .$$

Furthermore, $g \gg l_m/\mu_r$, and, therefore,

$$L \approx \mu_0 W/g .$$

With the above expressions for the inductances per unit length and the maximum value of the magnet impedance, the minimum required capacitance per unit length of the kicker magnet can be calculated to be

$$C = \mu_0 W/(Z_0^2 g) .$$

The per unit length propagation time (τ) for the kicker magnet is just the square root of the product of the inductance and capacitance per unit length. Substituting the above expressions for L and C gives

$$\tau = \mu_0 W/(Z_0 g) .$$

The magnetic field risetime at any point in the magnet depends on three effects. The first effect, the risetime of the applied pulse, depends on the speed of the switch as well as the stray modulator inductance and capacitance. The second effect is due to the finite electrical length of the kicker magnet and is described by the propagation time given above. Finally, second-order effects include 1) B-field penetration time through the resistively-coated beampipe, 2) variations in propagation speeds due to non-linear effects in the magnetic material, 3) hysteresis losses, 4) skin effect losses, and 5) eddy current losses in the beampipe coating. The net effect is a smoothing of the B-field risetime.

LEB-TO-MEB TRANSFER KICKER DESIGN STATUS

These kicker magnets must deflect the 9-12 GeV beam into and out of the LEB-to-MEB transfer line. Both C- and H-magnet geometries have been considered for this magnet set, and the H-magnet has been adopted as the baseline design.

Non-electrical advantages of the C-magnet include the ability to be designed to slide over the accelerator beampipe without breaking the vacuum and the fact that there are fewer parts to the C-magnet, thus making high tolerance machining and assembly easier.In addition, there is considerable experience with kicker magnets of the C-geometry within the high-energy physics community.[2,3] Disadvantages include the requirement for higher current in a single busbar and longer propagation times through the magnet for a given magnet impedance.

The H-magnet is more difficult to construct to tight tolerances, and replacement of the magnet requires either partial disassembly of the magnet or extraction of a section of accelerator beampipe with the magnet. However, for a given impedance, the H-magnet exhibits one-half the fill time of its C-magnet counterpart.

In the H-magnet, there are two feed configurations that can be used to produce currents of opposite directions in the two busbars. These are shown in Figures 2(b) and 2(c). In Figure 2(b), currents of opposite direction are generated by driving the two busbars from the same end with opposite-polarity voltage pulses. Current travels through the magnet to matching resistive loads. Figure 2(c) depicts another possible feed configuration, with the two busbars fed from opposite ends of the magnet with voltage pulses of the same polarity. Determining which feed system results in optimum performance of the magnet system is a major goal of future experiments at the SSCL. Other experiments, in either the design or procurement phase, include thyratron development, BLT characterization, and cable testing.

MEB-TO-HEB TRANSFER KICKER DESIGN STATUS

This set of kicker magnets transfers the beam from the MEB to the HEB at an energy of 200 GeV. The total current required to drive the necessary magnetic field is approximately 3500 A for the MEBE and 5000 A for the HEBI. Half of these current values are required for each busbar in a dual-bus H-magnet configuration. For the HEBI kicker, the inductance per unit length for a rectangular H-magnet surrounding the specified beampipe can be calculated as approximately 1.2 μH/m for each busbar. For the higher current magnet (HEBI), staying below a pulsed voltage limit of 50 kV requires that magnet impedance be less than 20 Ω. Selecting 16.7 Ω as the design impedance implies a busbar-to-ground capacitance of $L/Z^2 =$ 4nF/m. This high capacitance cannot be achieved by relying strictly on the natural geometry of the magnet. Discrete capacitors, consisting of "pads" of copper-coated Kapton sheets, can be added along the magnet's length to realize an effective magnet impedance of the required 16.7 Ω.

Due to the longer pulse width and slower risetime for this set of magnets, a lumped-element pulse-forming network can be used instead of excessively long pieces of charged cable. If it is assumed that only the risetime is important in the case of ejection kicker magnets (since the MEB will be empty after the pulse), and both the rise and falltimes are important for the injection kicker magnets, then the HEBI again represents the worst case. The PFN must be designed so that the rise and falltimes are faster than 3% of the pulse width. For a PFN tuned to be flat to within ≤1%, this requires a minimum of 46 sections.[4] Switching in this set of kicker magnets will employ standard hydrogen thyratrons, since the pulse requirements are well within the operating range of existing devices.

HEB-TO-COLLIDER TRANSFER KICKER DESIGN STATUS

The final set of transfer kicker magnets is required to move the 2 TeV beam from the HEB to the Collider rings. Due to the high currents involved and long pulse widths necessary to empty the large HEB, a low-impedance C-magnet fed by a SCR-switched PFN is being considered. Again, the inclusion of discrete capacitors within the magnet will be required to achieve low magnet impedance.

The required total current for the HEBE kicker magnet is approximately 5500 A. To maintain pulse voltages below 50 kV requires impedances of less than 9 Ω. An impedance of 8.33 Ω equates to six 50 Ω cables in parallel and is certainly possible in a C-magnet configuration. If discrete capacitor "pads" are added, this value is achievable. If an H-magnet is chosen, the impedance per busbar can increase by a factor of 2, resulting in a possible operating impedance of 16.7 Ω. This will still require the addition of discrete capacitors.

The relatively slow risetime allows for investigation of solid-state switch arrays. SCRs have demonstrated extremely long lifetimes when operated conservatively, and they are capable of conducting several kiloamperes per device. Their voltage holdoff capability, however, is limited to approximately 3500 V. Therefore, series arrays would have to be used and properly biased to allow for operation at tens of kilovolts.

CONCLUSION

A wide variety of pulsed transfer kicker magnets and associated modulators will be developed for use at the SSC. Due to stringent risetime and flattop requirements for the LEB-to-MEB transfer kickers, several magnet and switching options are being investigated. The H-magnet geometry looks very attractive for this particular application. Essentially two tightly coupled C-magnets in parallel, the H-magnet can be designed for shorter fill times when compared against a C-magnet of identical impedance. Though thyratrons are the traditional switch of choice, investigations into BLTs and spark gaps are also underway.

The other kicker systems require a different design approach. The longer pulse widths and risetimes, combined with the large coulomb transfer, allow for the use of pulse-forming networks and slower thyratrons or SCR switches. Matching of magnet to modulator at the low values of impedance necessary, design and tuning of fast, ripple-free pulse-forming networks, and creation of large arrays of SCR switches are important design issues which must be addressed.

REFERENCES

1. K. Frank, E. Boggasch, J. Christiansen, A. Goertler, W. Hartman, C. Kozlik, G. Kirkman, C. Braun, V. Dominic, M. A. Gundersen, and H. Riege, "High Power Hollow Electrode Thyratron-Type Switches," *Proceedings of the 6th Pulsed Power Conference*, Arlington, p. 213 (1987).

2. P.E. Faugeras, E. Frick, C.G. Harrisobn, H. Kuhn, V. Rodell, G.H. Schroder, and J.P. Zanasco, "The SPS Fast Pulsed Magnet Systems," *Proceedings of the 12th Modulator Symposium*, New York, p. 147 (1976).

3. F. Bulos, R.L. Cassel, A.R. Donaldson, L.F. Genova, J.A. Grant, A.M. Mihalka, B.A. Sukiennicki, W.T. Tomlin, F.T. Veldhuizen, D.R. Walz, J.N. Weaver, and D.S. Williams, "Some Fast Kicker Magnet Systems at SLAC," *Proceedings of the 1987 Particle Accelerator Conference*, Washington, p. 1884 (1987).

4. D.E. Anderson, "PFN Considerations for Kicker Magnet Systems," SSCL-PPDN3, unpublished (1991).

A HIGH PERFORMANCE ARCHITECTURE

FOR ACCELERATOR CONTROLS

M. Allen, S.M. Hunt, H. Lue, and C.G. Saltmarsh

Accelerator Division
Superconducting Super Collider Laboratory*
2550 Beckleymeade Avenue
Dallas, TX 75237

C.R.C.B. Parker

SL Division
CERN
CH1211 Geneva 23
Switzerland

Abstract: The demands placed on the Superconducting Super Collider (SSC) control system due to large distances, high bandwidth and fast response time required for operation will require a fresh approach to the data communications architecture of the accelerator. The prototype design effort aims at providing deterministic communication across the accelerator complex with a response time of <100 ms and total bandwidth of 2 Gbits/sec. It will offer a consistent interface for a large number of equipment types, from vacuum pumps to beam position monitors, providing appropriate communications performance for each equipment type.

It will consist of highly parallel links to all equipments: those with computing resources, non-intelligent direct control interfaces, and data concentrators. This system will give each piece of equipment a dedicated link of fixed bandwidth to the control system. Application programs will have access to all accelerator devices which will be memory mapped into a global virtual addressing scheme.

Links to devices in the same geographical area will be multiplexed using commercial Time Division Multiplexing equipment. Low-level access will use reflective memory techniques, eliminating processing overhead and complexity of traditional data communication protocols. The use of commercial standards and equipment will enable a high performance system to be built at low cost.

---

*Operated by the Universities Research Association, Inc., for the U.S. Department of Energy under Contract No. DE-AC02-89ER40486.

REQUIREMENTS OF THE SSC LAB CONTROL SYSTEM

The control system of the Superconducting Super Collider Lab (SSCL) will need to operate at high speed over large distances and offer a fast response time. The data acquisition and control requirements have been categorized,[1,2] and a maximum response time of less than 100 ms is required. Data rates will vary from 2 bytes/sec for vacuum equipment to as much as 50 kbytes/sec for some beam monitoring equipment. More than 150,000 control points must be accessed, including those for the ramp generators, superconducting dipoles, superconducting quadrupoles, vacuum pumps, RF systems, cryogenics, and cooling water. Total data acquisition and control bandwidth requirement approaches 2 Gbits/sec.

Existing large accelerator control systems often use local area networks for equipment access. These were designed primarily for peer-to-peer communication, rather than the hierarchal communication that characterizes much accelerator control. In an accelerator there is generally little data exchange between local processors.

Contention-based CSMA/CD local area networks such as Ethernet cannot guarantee any network station access to the network at any fixed or even minimum rate. This type of network is said to be non-deterministic. CSMA/CD networks also exhibit rapid degradation of performance under heavy load, and suffer from a limited geographical coverage because the propagation delay of the network must be shorter than the time of transmission of a minimum-size packet, leading to inefficient transmission of short messages.

Token-passing networks, while providing deterministic communication, have not adequately met the needs of large accelerator control systems due to reduced response time as the number of stations and size of the network increases. Furthermore, many layers of protocol running on non-real time operating systems have led to unpredictable performance as seen by application programs.

The control system requirement to provide 100 ms response time to all equipment—at whatever data transfer rate is required for that equipment—precludes the use of token passing or CSMA/CD local area networks.

TIME DIVISION MULTIPLEXING

The requirement for high-bandwidth, deterministic communication suggests a highly parallel system of data communication operating in the time or frequency domain. Frequency division multiplexing interfaces are not widely available for general purpose computers and micro-controllers. Time division multiplexing equipment, however, is widely available and extensively used in the telecommunication industry.

Time Division Multiplexing (TDM) is used by telephone companies for the efficient transmission of digitized voice and data traffic. It can also be used as a means of sharing a physical medium (fiber or copper) between different forms of data,[3] such as digitized voice, packet switched data, Ethernet, Token passing networks, digitized video, and others.

Standards have long existed for low-speed TDM interfaces:[4] T1 (1.544 Mbits/sec) is the equivalent of 24 voice channels (each 64 kbits/sec); DS3 (45 Mbits/sec) can carry 24 T1 links. Standardization has now been achieved at higher data rates by the Synchronous Optical NETwork (SONET) standards. SONET is a U.S. standard equivalent to the Synchronous Digital Hierarchy (SDH), defined by the CCITT for international (non-U.S. and Japan) equipment. This standard defines a way of interfacing transmission equipment from different vendors—not only the format of the data on the links, but also management protocols for maintaining the network.

SONET multiplexors are now available up to speeds of 2 Gbits/sec, enabling large amounts of data to be economically transferred over single-mode fiber optic cable. This use of TDM over fiber optic cable will significantly cut down on the amount of copper cable that must be installed in the accelerators.

IMPLEMENTATION OF CONTROL LOOPS

Control loops can be located within the front end interface crates using dedicated microprocessors, regional computers located near the front end electronics, or computers in the central control room. Each of these approaches offers advantages and disadvantages.

Local control loops provide the fastest response time, as no data communication is necessary. This would be the preferred solution where safety issues preclude the integrity of the system from relying on data communications equipment.

Regional control loops would be used where a group of devices must be controlled as a whole but without central intervention. Such cases might be for local commissioning of equipment before the central control room is operational.

Global control loops would be provided by centrally situated computers partitioned by function (rf computer, vacuum computer, access control computer, beam monitoring computer). This approach would simplify maintenance, as most complexity could be grouped at a central location. Application programs would operate faster and be simpler to write as they would be operating in their own program space. This approach would be especially efficient where an application required data from more than one location.

DATA TRANSPORT PROTOCOL

The use of high speed TDM links within the accelerator does not by itself ensure a high-performance control system. If many layers of protocol are used between the application and the data transport medium, the control system is slowed both by the delays in implementing that protocol and by the software overhead of processing the vast number of communications links that are needed. What is needed is a way of reducing both the number of protocol layers and the processing overhead of the data communication.

REFLECTIVE MEMORY

Reflective memory is the transferring of data between processors in a way transparent to the application software using that data. Typically a process will act on (read or write to) local memory, the contents of which are copied to the memory of another, physically separate system where other processes can act on it as if it were local to them. Using this technique, processes are unaware of data sharing and do not have to be involved in the complexities of network transmission and protocol. The size of the data unit transferred, speed of data transfer, and update rate can, of course, be varied to suit the applications.

The reflective memory buffer can be shared between two or more processors and can vary in size. The buffer can be sent at a fixed time interval, when requested by a process (typically when an operation on the data is complete), or when any changes have been made to the data. It is also possible to send only the changes that have occurred.

If the whole buffer is sent, then the update rate is low for larger buffers, but the process writing into the buffer can write at full speed. If the process writes to the same location twice before the buffer has been transferred, then only the second value will be sent. However, if only the data written are transferred the size of the buffer does not affect the update rate. But a mechanism must be provided to prevent the process writing data from accessing the buffer at an aggregate rate greater thanthe data transfer rate.

The physical medium for data transfer can be a serial or parallel bus. A parallel bus typically consists of twisted pairs between the crates: 8, 16, or 32 bits wide. The transfer rate is limited by distance. A serial bus has a lower transfer as a parallel bus for the same distance and medium, but can make use of media such as fiber optic cable that can sustain a high data transfer rate over long distances. The use of standard TDM techniques for serial transfer of reflective memory buffers makes the system non-distance limited, able to operate at potentially high data transfer rates and able to share the physical media (often single-mode fiber cable) with other data transfers.

FORMAT OF DATA

Having a large available bandwidth for data communication means that it is no longer necessary to buffer data at the front end acquisition equipment, transferring only a portion of the data that might be of interest. It becomes possible to transfer and store in the Main Control Room all raw data acquired by sensors. If this data is kept in shared memory, an application program requiring readings from a number of sensors does not have to request data from the sensors, await its return, store it in an appropriate format, and then process it. Instead it can simply access the data in its shared memory space, already formatted.

One possible scenario for a data acquisition and control system would be non-intelligent sensors continually acquiring data, perhaps triggered by the message broadcast or precision timing system. The data would be built into a parcel of, for instance, three 16-bit integer values and sent over a 64 kbit TDM link to a central control computer. The values arriving would be placed into shared memory for access by application programs. Depending on the application, new values could be placed in a circular buffer or they could simply overwrite previous values. In some cases, the data might have to be time tagged.

With data arriving in the shared memory space of centralized functional computers from many sensors, from data acquisition crates, and from regional computers, application programs would have access to the memory-mapped control system. That is, an application would behave as if it were directly connected to the hardware.

A mechanism is needed to access and describe the data. One such mechanism being developed by the SSC controls group is the Self-Describing Data Set (SDS) toolset. This is a set of library routines for the creation, access, and manipulation of data sets within file, network, or shared memory segments of a computer system. Names are assigned to collections of data that can then be accessed and manipulated as a whole or individually. A similar concept, Multiple Object Partitioned Structure (MOPS)[5] is employed at CERN for the control of the LEP accelerator. The approach is analogous to the way in which a computer mass storage file system allows access to collections of data by file names, and allows the grouping of collections of files in directories. As in a disk file system, the description of data does not have to be in the same location as the data itself. Furthermore, the data can be updated (perhaps new values arriving from a sensor) without affecting the description of that data. This can be considered as a way of saving the information describing the structure of data that the compiler throws away.

PROGRAM ACCESS TO DATA

In cases where central computers need to share data, either a pre-assigned or on-demand system of networked memory access could be employed. A pre-assigned networked memory access system might use reflective memory to share data. For the short distances involved in the central control room, a commercial parallel transfer reflective memory system might be used.

An on-demand system would allow any computer to have access to the shared memory region of any of its peers. This global networked shared memory could be implemented using standard local area networks and protocols based in much the same way as files are commonly shared across the network now.

IMPLEMENTATION OF THE CONTROL SYSTEM

The elements of the control system (Figure 1) that have been described in this paper are already largely available from industry. These include single mode fiber, SONET multiplexors, and TDM (T1) interfaces for VME computers. TDM interfaces have been built by the control group for the STD bus computers that will act as data concentrators for low-speed control interfaces. SDS libraries are available for describing and accessing data within the control system.

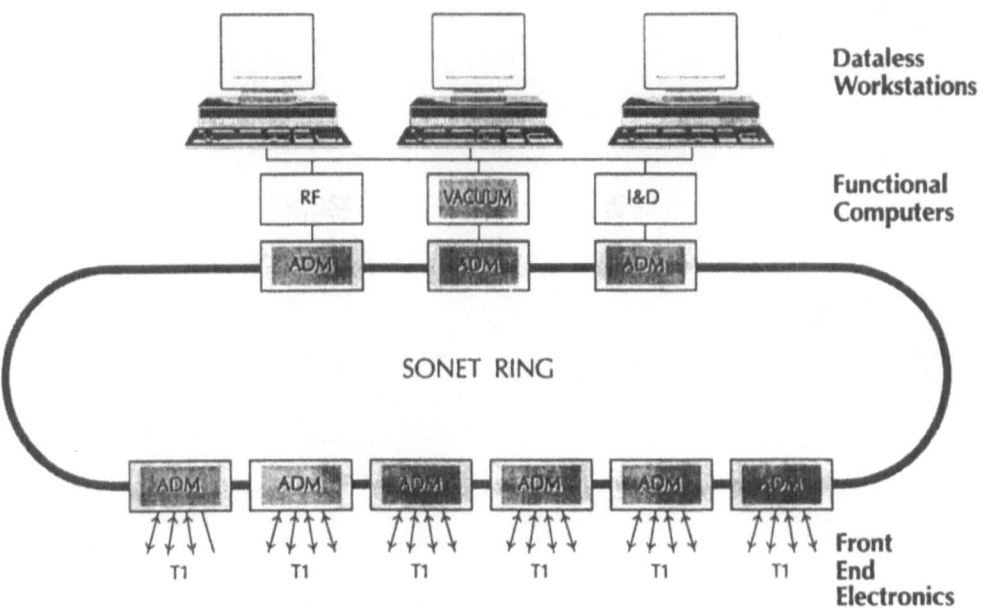

Figure 1. Accelerator Control System Scheme.

Higher rate TDM interfaces, perhaps at SONET rates, will be needed. These should be developed by industry. Standard protocols for program downloading and debugging of real time kernels used in embedded front end applications will have to be ported to run on TDM links. Appropriate error detection and correction protocols will have to be applied to the TDM links.

REFERENCES

1. Site Specific Conceptual Design, SSCL-SR-1056.

2. M.E. Allen, C.G. Saltmarsh, S.G. Peggs, D.A. Dohan, and V.E. Paxson, "Demands Placed on the SSC Laboratory Control System by the Engineering and Operational Requirements of the Accelerators," *International Conference on Accelerator and Large Experimental Physics Controls Systems*, 1989, Vancouver, Canada.

3. M.E. Allen, H.C. Lue, C.R.C.B. Parker, and C.G. Saltmarsh, "A New Data Communications Strategy for Accelerator Control Systems," *International Conference on Accelerator and Large Experimental Physics Controls Systems*, 1989, Vancouver, Canada.

4. ANSI T1.102, "DS1, DS1C, DS2 and DS3 Levels of Digital Hierarchy."

5. W. Herr, "Multiple Object Partitioned Structure," CERN, Geneva CH1211.

13. Cryogenics

REFRIGERATION PLANTS FOR THE SSCL

Mike McAshan, Venkatarao Ganni, Roberto Than, and Ted Niehaus

Accelerator Division
Superconducting Super Collider Laboratory*
2550 Beckleymeade Ave.
Dallas, Texas 75237

Abstract: The basic requirements and operating features of the collider cryogenic system have already been described in other publications.[1,2,3] The general arrangement of the refrigeration plant and its subsystems is presented, and the issue of how to provide redundancy in the cryogenic system is addressed, and some of the basic features of the refrigeration plants are described. The collider cryogenic system design is not final yet, and this report only reflects the direction and current status of the cryogenic system design.

INTRODUCTION

The high on-line operating time requirements of the collider system demands an availability of at least 98% on the SSCL cryogenic refrigeration system.[1] To meet this requirement, some redundancy has to be built into the system. There are many different ways to design redundancy into the SSCL cryogenic refrigeration system. To select the configuration that meets redundancy and operational characteristics requires the consideration of many issues: performance flexibility (load adjustment), high system availability, capability to provide and control single phase flow, inventory management, quench handling, contamination, sector cooldown and warm-up, and capability of upgrading to higher refrigeration capacity or lower temperature operation (3 K). Redundancy is provided by designing plants with larger capacities and shifting load from neighboring plants during emergencies. Design pipe sizes for the cryogenic flows in the cryostat make it difficult to shift 4 K refrigeration and 20 K refrigeration to neighboring plants. In contrast, shifting 4 K liquefaction load can be accomplished relatively easily. The goal is to have a cost-effective design that meets the collider operational and redundancy requirements.

BASIC FEATURES OF THE HELIUM REFRIGERATION PLANT

Ten helium refrigeration plants 8.6 km apart are located around the ring. Each plant has to provide cooling for the magnet system, which requires 4 K refrigeration, lead cooling, and 20 K shield cooling. The overall plant arrangement and its subsystems configurations

* Operated by the Universities Research Association, Inc. for the U.S. Department of Energy under Contract No. DE-AC02-89ER40486.

play an important role in meeting the operational requirements of the collider helium refrigeration system. Development of the plant arrangement is an important activity at the present time.

The interconnections of the major subsystems of the typical refrigeration system at a site are shown in Figure 1. Redundancy and other requirements will determine the final configuration and interconnections in the plant, and only the major operating points, essential features, and components of the plant are indicated on this figure. The coldboxes operate on a common compressor system with the appropriate interconnecting lines for the 4 K and 20 K loop (13,14,15,16). There is capacity to store 18700 standard m^3 (660,000 SCF) of helium gas inventory in ten tanks and 228 m^3 (60,000 gallons) of liquid helium inventory in two dewars at each site. This will be used for inventory management, emergency situations, and cooldown. A 76 m^3 (20,000 gallons) liquid nitrogen storage dewar, to provide emergency cooling for the shields and refrigerator and enough inventory to keep the collider system cool during shutdown periods of the refrigeration plants, is also necessary. The refrigerator coldboxes are located at ground level for ease of installation and maintenance.

One or more coldboxes in the tunnel house a subcooler with circulating pumps supplying the conditioned 4 K, 4 bar helium to the magnets, a cold compressor, and a heat exchanger in the 20 K loop that will be used during quench conditions to manage refrigeration. Another coldbox circulates and recools the nitrogen flow for the 80 K shield. Nitrogen vapor (12) from this recooler is used for precooling in the helium refrigerator. Liquid helium (1) subcooled to 4 K and subcooled liquid nitrogen (6) supplied from their respective coldboxes go to the valve box where the flows are divided up, supplying the four strings with cooling. Liquid helium is returned to the subcooler via a separate line (2). The saturated vapor from the recoolers is also returned in another line (3) to the subcooler, where the cold compressor maintains the appropriate saturation pressure, depending on the required operating temperature of the magnets. Flow exiting the cold compressor (9) is returned to the refrigerator. The warmest magnet is allowed to operate at 4.35 K, which in turn requires the warmest recooler to be at 4.1 K, and pressure drop in the vapor return line requires an even lower (vapor) pressure (approximately 0.8 bar) at the return to the subcooler. A cold compressor is chosen as part of the system, since this allows the collider system to be operated at the design operating temperature and provide the capability to upgrade to low-temperature operation, while simultaneously providing the head pressure and allowing the coldbox the flexibility to operate at above atmospheric pressure. A vacuum-jacketed transfer line containing the low-temperature lines connects the refrigerator to coldboxes in the tunnel. Helium from the power lead cooling is returned through a warm gas header that runs through the collider ring and is fed back to the refrigerator through a separate line (11). Cooling to the 20 K shield is provided by a separate supply and return line (4 and 5) to the valve box in the tunnel. Quench handling is also done through the 20 K system. In order to supply liquid (10) from the dewars at the desired pressure, a make-up pump is required between the dewars and the subcooler. Table 1 lists the nominal sector loads of the collider cryogenic system. Table 2 lists the basic operating points for a refrigeration plant.

REDUNDANCY IN THE SSCL CRYOGENIC SYSTEM

The high operating on-line time requirements of the collider system demand an availability of at least 98% on the SSCL cryogenic refrigeration system. To meet these requirements, redundancy has to be built into the system. There are many different ways to design redundancy into the SSCL cryogenic refrigeration system. To select a configuration that best meets the operating and redundancy requirements, several major issues should be considered: performance flexibility (load adjustment); high system availability; capability to provide and control single phase super critical helium flow; inventory management; quench

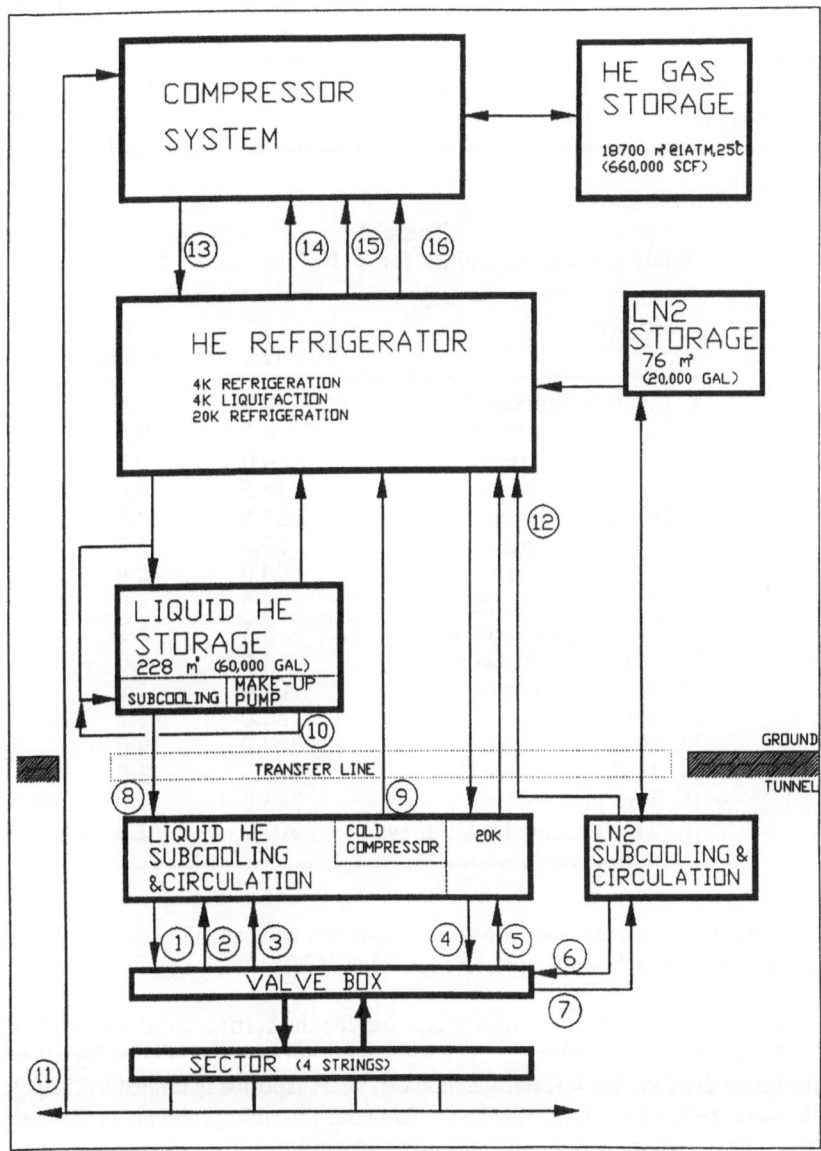

Figure 1. Basic configuration for the SSC refrigeration plant.

Table 1
Nominal Sector Loads of the SSC Cryogenic System

	T (supply) (K)	P (supply) (atm)	T (return) (K)	P (return) (atm)	Nominal Load
4 K Refrigeration	4.1	4.0	4.0	0.8	5200 W
4 K Liquefaction	4.1	4.0	303.0	1.1	28 g/s
20 K Refrigeration	16.0	3.0	25.0	2.0	9600 W
80 K Refrigeration	80.0	6.0	90.0	4.6	42,500 W

Table 2
Basic Operating Points for a Refrigeration Plant

Description	T (K)	P (Atm)
1. Liquid helium supply	4.1	4.0
2. Liquid helium return	4.3	3.5
3. Helium vapor return	4.0	0.8
4. 20 K shield supply	16.5	3.0
5. 20 K shield return	25.0	2.0
6. LN shield supply	80.0	6.0
7. LN shield return	90.0	4.6
8. LHe supply from refrig.	4.5	4.0
9. Cold compressor return	5.3	1.2
10. Liquid helium makeup	4.5	4.0
11. Warm helium gas return	300.0	1.2
12. Nitrogen Gas	80.0	1.2
13. High pressure He	308.0	19.0
14. Medium pressure He	303.0	3.0
15. Low pressure He	303.0	1.0
16. Low pressure He (20 K sys.)	303.0	2.0

handling; contamination; sector cooldown and warm-up; and capability of upgrading to higher refrigeration capacity or lower temperature operation (3 K).

A major issue in the redundancy scheme for the refrigerators is whether or not to retain 4 K refrigeration capability at a site when a failure occurs in the refrigeration system. The maximum distance that 4 K refrigeration can be transported is limited by the pipe size of the 4 K vapor return line in the magnets. To avoid possible problems in maintaining the desired magnet temperature, it is preferable to transport 4 K refrigeration only over the normal operating length of one string (4.3 km). 4 K liquefaction load can be transported over greater distances by moving liquid from one sector to the other. There is only one line for the 20 K refrigeration in the cryostat: the 20 K loop runs through one string and back to the refrigerator through the string in the other ring. The normal flow length of the 20 K loop is therefore 8.6 km. If no local redundancy is provided for the 20 K refrigeration, in failure mode it will be required to design the 20 K loop with flexibility for higher pressure drop and temperature rise to allow the 20 K flow to travel a distance equivalent to a sector and back (17.2 km). A 20 K shield operating at a higher temperature means that the 4 K load will also increase. Concern about this and the large range of operating conditions that the 20 K loop will impose on the operating plant makes it preferable to have local redundancy for the 20 K load. Hence, local redundancy for 4 K and 20 K refrigeration with inter-sector redundancy for 4 K liquid load is a preferred redundancy scheme. This capability can be accomplished through different plant configurations and backup of individual components in the plant.

With local backup of 4 K and 20 K refrigeration as a choice and 4 K liquefaction to be tapped from the neighboring plants, the issue is how to provide the local redundancy necessary to meet collider availability requirements. One way of accomplishing this is to operate the refrigerator unbalanced -- that is, with the low pressure return flows of the 20 K and 4 K at their normal rates and the high pressure supply at a reduced rate. The excess inventory, after being compressed, is returned to the warm helium return header; from there the neighboring plants make up the extra flow that they are providing to the failed sector. Hence, the sector that has a failed coldbox (in the case of a two coldbox configuration) or has lost capacity (in the case of a single coldbox configuration) will still handle the 4 K and 20 K refrigeration and the neighboring plants will provide the additional liquid to makeup for the reduced capacity. The excess inventory at the failed sector is returned through the warm helium return header to the neighboring plants.

SSCL REFRIGERATION PLANT CONFIGURATIONS

As described earlier, there are many issues to consider in the design of the SSCL refrigeration system. One of the major issues is how to design redundancy into the cryogenic system. A previous plant design concept consists of three different coldboxes,[3] which gives greater flexibility in the control of the collider cryogenic system but faces the same redundancy issues, and it may not be cost effective. Plants with fewer coldboxes should be examined in order to design to cost.

In the study by Powell and Quack,[4] various coldbox schemes were examined with regard to cost, availability, and redundancy. One of the solutions provided in their study was a plant with two coldboxes, each providing 67% capacity of the nominal sector 4 K load. However, the 20 K refrigeration load was not considered in any detail. The 67% 4 K coldboxes would meet the redundancy requirements, but the 20 K coldbox must be sized for 150% if no local backup is present for the 20 K load, and refrigeration has to be provided by the neighboring plants.

A separate 4 K and 20 K configuration requires three coldboxes at each site. To provide local redundancy for the 20 K refrigeration, it will be more cost effective and efficient to have an integrated design. With the two coldbox concept the most advantageous configuration would be the one consisting of identical coldboxes. Here the load is evenly split during normal operation, and during failure mode the operating coldbox provides the sector with the 4 K and 20 K refrigeration. For a single coldbox configuration, redundancy is provided by designing the system so that the redundancy requirements are met by backing up the various individual components. In either case the operating coldbox at the site of the failure will run in unbalanced mode.

A helpful tool in studying the redundancy scheme and in approximating the size of the plants is to examine the Carnot work requirements of the system. This will help to examine how the loads can be readjusted in the remaining integrated 20 K / 4 K coldbox. Although it is not possible to exactly interchange the loads (4 K refrigeration, 4 K liquefaction, and 20 K refrigeration) for a given Carnot work, it can be used as a first approximation for sizing the coldboxes and compressor system, recognizing that it will be harder to match fixed-nozzle expanders to follow the different modes of operation.

The SSCL refrigeration system is required to deliver 4 K refrigeration at 4 bar pressure with the vapor returned at 0.8 bar. The proper Carnot work requirements should be calculated from these supply and return conditions. By this method, the losses due to J-T expansion in the collider ring recoolers are essentially being considered as part of the load. Table 3 summarizes the supply-and-return design points of the refrigerator for each load.

Table 3
Refrigerator Supply-and-Return Design Points

	T (supply) (K)	P (supply) (atm)	T (return) (K)	P (return) (atm)	Sector Load	Min. Sector Refrigerator Capacity
4 K Refrig.	4.05	4.0 (Liq.)	4.0	0.80 (vap)	5200 W	6500 W
4 K Liquefact..	4.05	4.0 (Liq.)	303.0	1.05 (gas)	28 g/s	35 g/s
20 K Refrigeration	16.50	3.0	24.5	2.00	9600 W	14400 W

The plant sizing can be better understood by studying the Carnot work requirements for the following operational modes:

1. Nominal Load
2. Peak Load (For adjacent plants providing assistance to the failed plant)
3. Failed Plant (4 K and 20 K refrigeration load by the remaining coldbox)

The sector loads and refrigerator design capacities are tabulated below. The 150% of the 20 K load was initially specified as the margin to provide adequate cooldown capacity and for redundancy. If the loads are integrated into the same coldbox, the margin on the 20 K load also needs to be revisited.

Table 4
Nominal Load and Sector Carnot Work Requirements

	Sector Load	Carnot Work (kW)	Refrigerator Capacity	Carnot Work(kW)
4 K Refrigeration:	5200 W	442	6500 W	552
4 K Liquefaction:	28 g/s	196	35 g/s	245
20 K Refrigeration:	9600 W	215	4400 W	324
		-----		------
	Total	853		1121

During failure mode (plant loses half its capacity) the extra Carnot work requirements for the neighboring plants is determined by the peak capacity of the plant. The missing capacity (half plant) in the failed sector has to be made up from neighboring plants. For a 133% plant the Carnot work requirements are:

If the remaining coldbox provides 576 kW of Carnot work capacity, the balance (657-576 = 81 kW) would theoretically require a 12 g/s unbalance operation (excess mass flow return) to meet the local redundancy requirements of the plant. Two identical coldboxes -- each with a 67% capacity of the total sector 4 K refrigeration, liquefaction, and 20 K load, would provide a total of 1152 kW of Carnot work (Table 7). If each coldbox is sized for 600 kW of Carnot work, it should be capable of providing 100% of the 4 K and 20K refrigeration

Table 5
Peak Load Carnot Work for 133% plant

	Load	Carnot Work(kW)
4 K Refrigeration:	7000 W	595
4 K Liquefaction:	38 g/s	266
20 K Refrigeration:	13000 W	292

	Total	1152

The operating part of the failed plant shall provide the 4 K and 20 K refrigeration for that sector, as follows:

Table 6
4 K and 20 K Refrigeration Mode Carnot Work

	Load	Carnot Work (kW)
4 K Refrigeration:	5200 W	442
4 K Liquefaction:	0 g/s	0
20 K Refrigeration:	9600 W	215

	Total	657

sector load with some unbalance flow when operating in failure mode. The Carnot work requirements of the neighboring plants providing the additional 4 K liquefaction load are easily met since they are sized for 133%. Sizing the plant with this method is only approximate, and final design capacity depends on the plant's operational characteristics and requirements.

Table 7
67% 4 K load Coldbox Carnot Work

	Each coldbox	Carnot Work (kW) Each Coldbox	Carnot Work (kW) Sector Total
4 K Refrigeration:	3500 W	298	595
4 K Liquefaction:	19 g/s	133	266
20 K Refrigeration:	6500 W	145	291
		-----	------
	Total	576	1152

ADDITIONAL FEATURES

Changes were made to the original design concept during the study of the two-coldbox configuration that would allow the system to remain simple but still incorporate the features necessary for collider operations. Some of the changes and additional features incorporated into the system are listed below.

Compressor system. The low and medium pressure flows of the 4 K load from both coldboxes return to a common compressor system, which has its own backup compressors. The use of a common system with the appropriate interconnections allows for greater flexibility and more efficient use of the operating and backup compressors. The coldbox has a separate compressor suction return line for the 20 K system. This makes the 20 K system more flexible by allowing the suction pressure to vary somewhat to follow the 20 K load requirements. This in turn will accommodate the transient operating modes, in particular quenches, and minimize the disturbance to the 4 K refrigeration.

Refrigeration supply. It will be advantageous to supply the make-up helium to the subcooler at 4 bar and 4.5 K directly from the refrigerator and to use the make-up pump only as a backup.

Liquid dewar pressure. The dewar pressure will be allowed to increase to 2 bar, without affecting the refrigerator operation, thus increasing the flexibility for quench handling and inventory management.

Use of quench return line. The quench return line to the dewar can be used for the following applications:

1. Liquid helium inventory can be transported back to the dewar if the pressure in the 4 K, 4 bar loop exceeds maximum operating conditions due to quenches.

2. Will also serve as pump by-pass for both make-up and circulating pumps.

3. During cooldown from 20 K to 4 K and during the ring filling sequence, large quantities of make-up helium will be required. The quench return line can be used to transport liquid helium from the dewar to the magnets. Since the magnet string will be at 1 bar nominal pressure, the liquid helium can be transported from the dewar and the return vapor can be recovered by the refrigerator.

SELECTION OF PLANT CONFIGURATION

With two coldboxes and the 20 K system integrated into both coldboxes it is possible to design a system that meets the redundancy requirements of the collider. The single coldbox may be a cost-effective option if redundancy requirements can be satisfied with the proper design configuration. Further study needs to be done to explore the single coldbox option. Selection of the final configuration also requires further examination of other issues such as cooldown/warm-up, contamination, quench and inventory handling, upgradability, load adjustment, and cost.

Process Cycle Design

A process cycle with the combined 4 K and 20 K refrigerator has been modeled and an exergy analysis has been carried out on the model to aid in optimization. Current studies indicate that nominal coldbox Carnot efficiencies of 50-55% are achievable with a 4-turbine configuration using liquid nitrogen pre-cooling, assuming the load is located at ground level. Since tunnel depth varies from site to site, ground level was used as the baseline design depth. With compressors of 55% isothermal efficiency, an overall plant efficiency of 25-30% could be achieved. A more detailed explanation will follow on how the coldbox and plant efficiency are defined here. Figure 2 gives a flow diagram of a refrigeration plant. It should be emphasized that a coldbox design was used here for illustrative purposes and to obtain performance and capacity results. The process cycle configuration could be different and will be left open for design by the manufacturer of the refrigerator. In addition, the cycle design needs to be further examined and modified to handle the upgrade to 3 K refrigeration. For operation at 3 K, a case study is made with the assumption that approximately the same amount of Carnot work is required as that of the 4 K load. The nominal 3 K refrigeration load is therefore approximately 3400 W. This is accomplished by intercepting the synchrotron radiation with a special 20 K liner in the beam tube, and results in an increase of the 20 K load. Table 8 lists the case study loads for the 3 K operation of the collider ring.

Table 8
Load Estimates of A Collider Sector at 3 K

	Load	Carnot Work
3 K Refrigeration:	3400 W	391 kW
3 K Liquefaction:	28 g/s	202 kW
20 K Refrigeration :	11600 W	260 kW

	Total	853 kW

Definition of Coldbox and Plant Efficiency

As described earlier, the required Carnot work is computed based on the supply and return conditions to the magnet string and not on the saturated conditions in the recoolers, where liquid is vaporized. Coldbox efficiency is defined here as the Carnot work required for the load over the total isothermal compressor work plus the Carnot work required for the nitrogen pre-cooling. The compressor work required for the nitrogen pre-cooling is based on a nitrogen plant Carnot efficiency of 35% in calculating the overall plant efficiency. Following are results from a process simulation of a coldbox design with a capacity of approximately 67% of each nominal load to illustrate the definition of the efficiencies.

Table 9
Results of Process Simulation of a 67% Nominal Load Coldbox

	T (supply) (K)	P (supply) (atm)	T (return) (K)	P (return) (atm)	Load	Carnot Work (kW)
4 K Refrig.	4.05	4.0	4.00	0.80	3500 W	297
4 K Liquefact.	4.05	4.0	303.0	1.05	20 g/s	141
20 K Refrig.	16.4	3.0	25.2	2.00	6500 W	127

					Total	565

	Isothermal Compressor Work *	Actual Compressor Work
First-Stage He Compressors:	212 kW	386 kW (55% isotherm. eff.)
Second-Stage He Compressors	606 kW	1102 kW (55% isotherm. eff.)
20 K System He Compressors:	211 kW	423 kW (50% isotherm. eff.)
Nitrogen Carnot Work:	65 kW**	184 kW (35% Carnot eff.)
	------	-------
Total	1095 kW	2096 kW

Coldbox efficiency: 565/1095=51.7% (of Carnot)
Plant efficiency: 565/2096=27.0% (of Carnot)

* 100% isothermal efficiency
** Carnot work for nitrogen pre-cooling (100% Carnot efficiency)

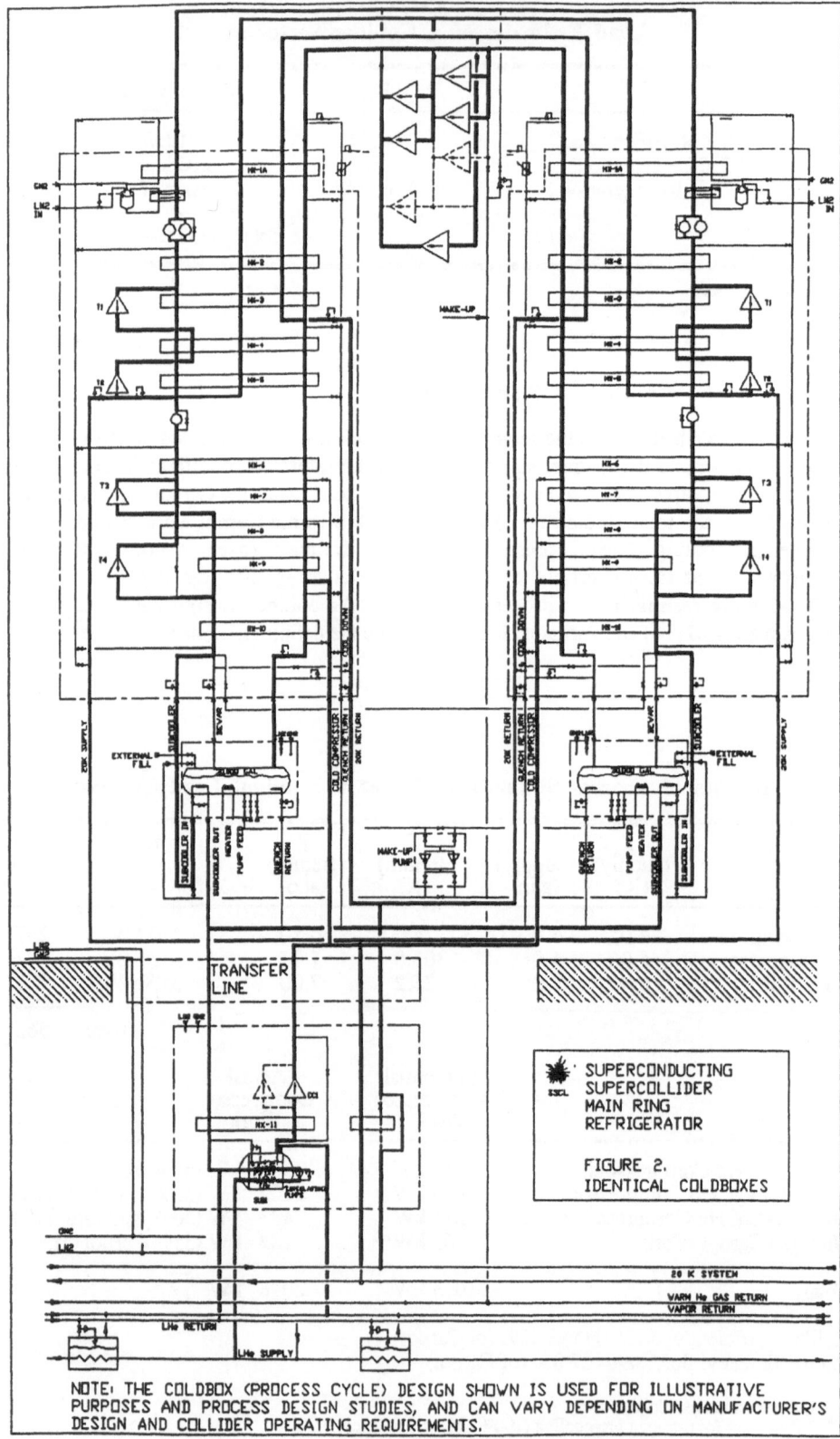

NOTE: THE COLDBOX (PROCESS CYCLE) DESIGN SHOWN IS USED FOR ILLUSTRATIVE
PURPOSES AND PROCESS DESIGN STUDIES, AND CAN VARY DEPENDING ON MANUFACTURER'S
DESIGN AND COLLIDER OPERATING REQUIREMENTS.

Figure 2. Flow diagram of a refrigeration plant.

CONCLUSION

Since there are many different process cycles possible, the arrangement in the current model might not be as desirable as others with regard to actual components, failure mode operation, upgradability, and other transient modes, and further changes are likely when a better understanding is obtained of the related issues. Detailed process design studies need to be carried out on how to handle quenches, liquid and gas inventory, and cooldown and warm-up operations. This may require results from dynamic simulations of a sector string. Work is currently being done in this area by others and should give us a better understanding in designing the refrigeration system.

This has been an overview of the major issues of the SSCL cryogenic system. Plant configurations that incorporate the use of one and two coldboxes are being considered, as a further attempt to reduce the cost and complexities of the refrigeration plants. Current preliminary process simulation results indicate that plants with two coldboxes can be designed that are efficient and able to meet the collider's load, redundancy, and flexibility requirements. The ideas and issues presented here reflect the current status and direction of the cryogenic system design. However, further studies and modeling of the transient operating modes and upgradability of the collider cryogenic system are needed before a suitable and final design can be selected.

Procurement of Plants

The goal of the SSCL is to issue a first RFI in the fall of '91 and the first RFP during the summer of '92. Ordering of the refrigeration plants will start in '93 and will continue through to '99 with the first plant being commissioned sometime in '95. An outline of the Collider Cryogenic System Milestones follows.

Collider Cryogenic System Outline of Procurement Milestones

Issuance of First RFI	9/91
Remarks from Vendors	12/91
Issuance of First RFP	6/92
Issuance of Last RFP	3/93
E2 Refrigeration Plant Complete Commissioning	9/95

References

1. M. S. McAshan, "Refrigeration Plants for the SSC," SSC Central Design Group Report No. SSC-129, May 1987.

2. *Site-Specific Conceptual Design*, SSC Report No. SSCL-SR-1056, July 1990, SSC Cryogenics System by M. S. McAshan.

3. *Conceptual Design of the Superconducting Super Collider*, SSC Report No. SSC-SR-2020, March 1986.

4. R. Powell and H. Quack, "A Refrigeration Plant Concept for the SSC," Proceedings of the International Industrial Symposium on the Superconducting Super Collider, Plenum Press, New York, (1990).

DESIGN OF A LARGE 2.0 K REFRIGERATOR FOR CEBAF

H. Parish (1), G. Gistau (2), C. Hood (1),
K. Kreinbrink (1), and W. Appleton (1)

(1) CVI Inc.   (2) L'Air Liquide
     Columbus, Ohio   Sassenage, France

ABSTRACT

The CEBAF refrigerator is designed to produce 4620 watts of refrigeration at 2.0 K, another 180 watts at 2.14 K plus 4.5 K liquid and higher temperature cooling for shields. Several design features are described, including a modular concept for the 2 K and 4.5 K refrigerator sections and the use of multistage centrifugal cryogenic compressors to achieve the 2 K. The development program for the centrifugal cryogenic compressors is discussed. Some startup information is also provided.

INTRODUCTION

The use of superconducting cavities in the Continuous Electron Beam Accelerator Facility (CEBAF) leads to some very demanding requirements, pushing the state of the art in helium refrigeration technology. The optimum temperature for operating the superconducting accelerator cavities was determined by CEBAF to be 2.0 K. The CEBAF cavities with a 2.0 K operating temperature constitute the primary loads for the CEBAF refrigerator. The machine is also required to serve as a liquefier to provide the liquid inventory prior to the initial operation of the accelerator and to provide shield cooling and a small amount of liquefaction to the end stations during normal operation. These requirements led to the largest refrigerator built to date operating in the 2.0 K range.

The CEBAF refrigerator utilizes state-of-the-art centrifugal cryogenic compressors (CCC) to produce the very low pressure (3.13 kPa) required in the cavities to achieve the 2.0 K temperature. The compressors are installed in a series configuration and operate with independently variable speeds as conditions demand. Successful testing of these compressors in France is expected to be indicative of the performance of these units in normal operation.

One of the challenging aspects in the design of a 2.0 K refrigerator is the contradictory nature of the requirements for physically large components such as valves, heat exchangers and associated piping (due to the large volumetric flow rates and the need for very low pressure drops) and the importance of maintaining a low overall heat leak. One of the results of these requirements was CVI's development of a special, low heat leak 25.4 cm cryogenic valve to isolate the CCC from the loads. The valve and its operator are enclosed in the vacuum insulation space and utilize GHe to drive the pneumatic actuator.

CEBAF Description

The Continuous Electron Beam Accelerator Facility is designed and constructed to be a world-class electron accelerator to serve the nuclear physics community, providing a continuous beam of electrons at any energy between 0.5 and 4.0 GeV[1]. The key component of this linac is the superconducting accelerating cavity which allows for the continuous acceleration of a beam with minimal power loss in the cavity wall. The CEBAF linac will use a 1500-MHz superconducting cavity designed, developed, and tested at Cornell University. Current superconducting RF technology permits the construction of the high efficiency superconducting linac.

Design Bases

Equipment Requirements. The CEBAF Refrigerator was specified to be comprised of separate major interconnected component systems as follows:

1. A vacuum insulated 4.5 K cold box assembly containing expanders, heat exchangers, control valves, purifiers, and other items required to produce 4.5 K refrigeration and/or liquefaction.

2. A vacuum insulated subatmospheric cold box containing CCC, a subatmospheric subcooler and other items required to produce 2.0 K refrigeration.

3. A skid mounted screw compressor system. All intercoolers, aftercoolers, oil removal, and oil injection systems were to be mounted on the skid associated with the respective compressors. This compressor system also included a gas management system, oil coalescer skid located outdoors and an adsorber unit filled with activated carbon.

4. In order to integrate the refrigerator into the overall CEBAF control system, the central controls for the refrigerator were supplied by CEBAF, while skid instrumentation, sensors, local manual controls and personnel and equipment safety interlocks were supplied and installed by CVI. Local manual control stations for the warm compressor system were required to enable local compressor operation for maintenance and start-up use. Local interlock protection was furnished by CVI. Therefore, the equipment will be protected at all times whether or not the remote computer control system is operating. The physical interface for the control system is at terminals in the CEBAF furnished I/O panels.

Modes of Operation. The CEBAF Refrigerator is required to provide the cryogenic refrigeration for CEBAF at use points which are designed to operate below liquid nitrogen temperatures. Three distinct modes of operation are as follows.

1. The Principal Mode of operation of the CEBAF Refrigerator supports the normal operation of the accelerator. During the Principal Mode of operation, the CEBAF Refrigerator is operating with all systems on line to produce a refrigeration temperature of 2.0 K in the primary loads. It is required that the refrigerator/liquefier be capable of carrying the specified loads while operating with any two of the three first stage compressors.

2. The Liquefier Mode is utilized when the production of liquid helium from ambient temperature gas is desired. This would normally be used to establish a sufficient reserve of LHe in the storage dewar for normal operation and to store the large inventory of helium at the site during periods of accelerator warm-up and shutdown.

The gaseous helium storage system consists of six 113.3 m^3 vessels supplied and installed by CEBAF and the liquid helium storage system consists of a 113,600 liter dewar and a 10,000 liter dewar, both designed and fabricated by CVI.

3.   The Turndown Mode requires that 85% of the refrigeration capacity (Principal Mode) be produced when one of the three second stage compressors is off-line.

   Load Summaries. The loads are summarized in Table 1. The primary load for the Alternate Refrigeration Mode and the liquefaction rate are estimated values for the equipment provided, not specification requirements.

   The primary loads remain at 2.0 K utilizing only the latent heat of vaporization to absorb the heat load. The secondary load superheats the 2.0 K vapor to 2.14 K.

Table 1. Performance Summary

Refrigerator (Principal) Mode:

Primary Load	4,620 Watts @ 2.0 K
Secondary Load	180 Watts between 2.0 K and 2.14 K
Shield Load	12,000 Watts @ 38 K Supply and 52 K Return
End Station Load	10 Grams per Second of 4.5 K Liquid Helium
Carnot Efficiency	18.5% Minimum
Duty Cycle	6 Months Continuous Operation

Alternate Refrigeration Mode:

Primary Load	7550 Watts @ 4.5 K
Shield Load	12,000 Watts @ 38 K Supply and 52 K Return
End Station Load	10 Grams per Second of 4.5 K Liquid Helium

Liquefier Mode:

| Liquefaction Rate | 4,569 Liters per Hour |

DESCRIPTION OF REFRIGERATOR

Process Description

   The system was specified, designed and built as a 4.5 K conventional refrigerator/liquefier module and a 2.0 K subatmospheric module. This configuration has the advantage of effectively de-coupling the 4.5 K and the 2.0 K systems, permitting operation of the 4.5 K system without the 2.0 K system, for example during commissioning of the system or when, for any reason, 2.0 K is not required.

   The cycle consists of a reverse Brayton cycle for the shield, coupled with a Claude cycle using a supercritical cold expander and cold compression to achieve the 2 K temperature. The following process description is for operation in the Principal Mode. A simplified flow schematic and TS diagram are included as Figures 1 and 2, respectively.

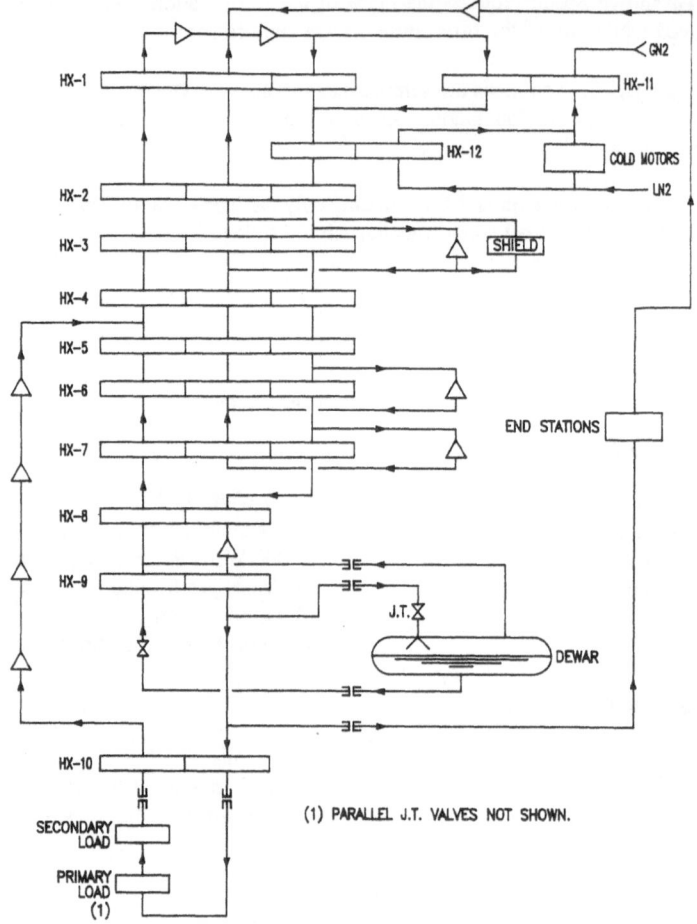

Fig. 1 Simplified Flow Schematic

4.5 K Module. The cycle features LN_2 cooling and four (4) turbo-expanders. This configuration and number of expanders was considered to provide optimum reliability and robustness of operation while still exceeding the required thermodynamic efficiency for the system. The first level of refrigeration is LN_2 cooling at approximately 78 K. It should be noted that the warm heat exchanger is split into two sections operating in parallel. One exchanger is purely a helium exchanger, while the second exchanger involves heat exchange between GN_2 and high pressure (warm) helium only. This was specified in order to preclude freezing of N_2 in the event that the low pressure (cold) return streams would be below the freezing point of N_2 under some conditions of operation.

The next three stages of refrigeration (first, second and third stage expanders) have inlet temperatures of approximately 54, 28 and 17.3 K. These expanders are in a parallel configuration, with inlet pressures of approximately 2040 kPa and discharging at approximately 400 kPa for the first expander and less than 350 kPa for the second and third expanders. The first expander must discharge at a somewhat higher pressure in order to provide pressure drop for flow through the shields. The last stage of refrigeration is the fourth stage expander that is in series with the load. It discharges at approximately 284 kPa, thus utilizing most of the available expansion yet ensuring that the exhaust is supercritical.

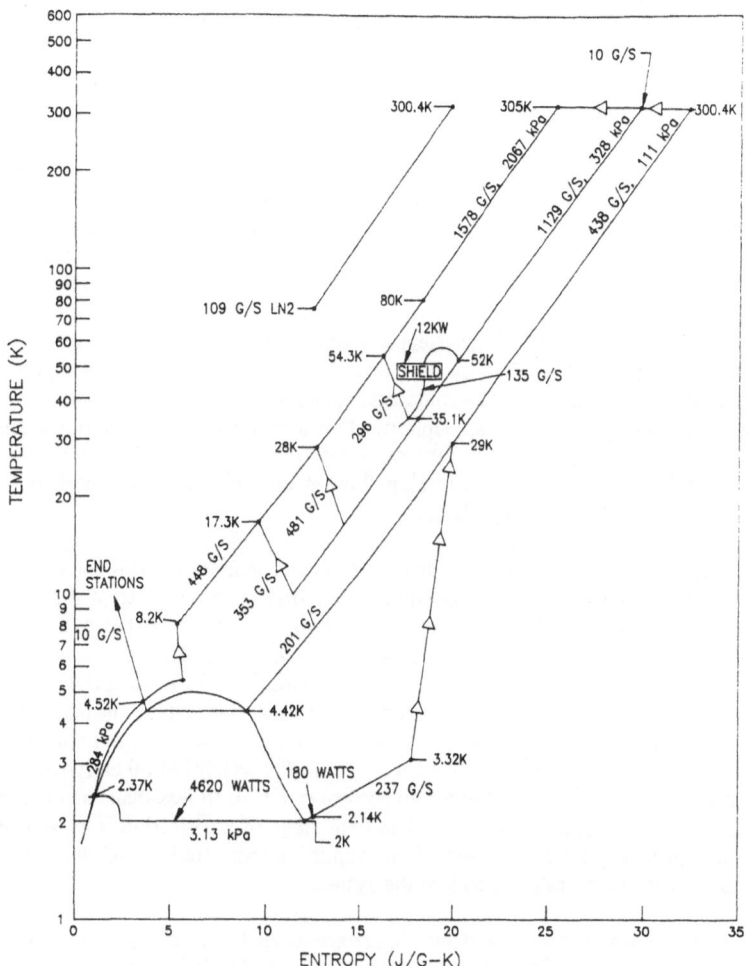

Fig. 2 T-S Diagram for CEBAF Refrigerator

The exhaust from the fourth stage expander is further cooled to approximately 4.5 K by liquid in heat exchanger HX-9 to form a cold, supercritical fluid. Of this stream, 10 g/s go to cool the end stations, approximately 199 g/s go to the return stream of HX-9 (after expansion into the dewar) for cooling of the supercritical stream and 237 g/s go to the 2 K module.

2.0 K Module. In the 2.0 K module, the supercritical stream is cooled to 2.37 K in a counter flow heat exchanger and expanded in an isenthalpic process to 2.0 K and 3.13 kPa (0.031 atm) in multiple parallel JT valves (supplied by CEBAF) associated with the various loads. The refrigeration stream returns from the loads at 2.14 K and is warmed in the counter flow heat exchanger to 3.32 K before entering the first of four CCC arranged in series that boost the pressure to approximately 116 kPa. The exhaust from the fourth stage of cold compression enters the heat exchanger stack at a temperature of approximately 29 K, thus effectively recovering much of the available refrigeration.

Purification. There are purifiers at two temperature levels in the cold box. Dual adsorbers at 80 K provide for the removal of trace amounts of air components while a single adsorber at 20 K primarily removes neon. The 80 K adsorbers for the removal of air components are of conventional design. The 20 K adsorber was designed on the basis of experimental work that had been performed at CVI[2].

Component Descriptions

The refrigerator/liquefier's components were packaged to maximize shop construction and testing and to minimize the field construction requirements. The components were designed into skid and cold box assemblies manufactured at CVI's facilities.

Warm Compression System. The helium compression system consists of two (2) stages of compressors plus final oil removal components. The compressors are an oil injected screw type, manufactured by Howden, individually skid packaged by CVI. Each skid is an independent compressor system with drive motor, gas cooler and oil circulating system for maximum overall system reliability. Only the second stage final oil removal system is common to the individual compressor skids. Each oil circulating system consists of a bulk separator, sump reservoir, pump, cooler, filters and control valves. The oil sump/separator reservoirs are horizontal pressure vessels which are gusseted to base frames. The compressors and drive motors are mounted on top of these vessels which provide a rigid structural support. This design was developed in concert with that of the associated piping to assure that oil would not collect at low points in the piping during shutdown and subsequently be ejected in slug flow at the next start up. Each bulk separator includes two (2) stages of coalescing elements.

Each stage of compression is provided by parallel compressor skids for enhanced system reliability and for efficient capacity reduction. The first stage consists of three (3) identical compressor skids, with 429 kW drive motors. Only two of the three first stage compressor skids are required for the Principal Mode of operation. The third compressor is a complete, operable, standby spare. The second stage consists of three (3) identical 1679 kW compressor skids, any two of which can provide 85% of the Principal Mode refrigeration capacity. The parallel compressor skids are connected between common suction and discharge headers. Each compressor skid is designed to discharge helium with less than ten (10) ppm (by weight) of oil carryover. The second stage discharge flow is subjected to further oil aerosol removal, to less than one (1) ppb, by three (3) serial stages of microglass coalescers. The final stage of purification is an activated carbon adsorption bed, primarily for the removal of any vapors as trace products of oil decomposition or impurities present in the oil as charged into the system.

The six (6) compressor skids and switchgear are located in a sound insulated room. Water piping and conduit are buried in the floor and the process piping is in trenches, providing a clean look and maximum accessibility (see Figure 3). The coalescer skid is located outdoors and the adsorber for the final oil removal system is located in the cold box room.

4.5 K Cold Box. The vacuum insulating vessel supporting all of the cold components which produce 4.5 K refrigeration is a vertical cylinder with a horizontal cylindrical appendage attached to its center section. This configuration was partly dictated by the volume of the components and shipping limitations and permits vertical mounting of the heat exchangers and purifiers while the horizontal section provides convenient access to the turbine expanders and valves. The vertical center section is supported by floor stanchions and the horizontal section is supported by saddles. The top and bottom of the vessel are flanged removable end covers which permit internal access.

Turbine Expanders. The four (4) cold gas expanders required for the refrigeration cycle are supplied by the Advanced Technology Division of L'Air Liquide. These expanders are centripetal turbines shaft coupled to centrifugal compressors, all supported by static gas bearings.

2.0 K Cold Box. The vacuum insulating vessel supporting the cold components which produce 2.0 K refrigeration is a cylindrical vertical vessel with a flanged removable top section for internal access. Valves and piping interfaces are mounted in the bottom stationary section. Internally, this cold box supports the CCC, a heat exchanger, a test heater, valves and piping. These internal components are thermally insulated by dynamic pumping, radiation shielding, a liquid nitrogen shield and heat intercepts. The CCC are mounted, in a clover leaf arrangement, on a structural frame which is thermally insulated.

Fig. 3 Warm Compressor Room

Centrifugal Cryogenic Compressors. The four (4) CCC required for the refrigeration cycle are supplied by L'Air Liquide. The machines are centrifugal compressors supported by magnetic bearings and driven by variable speed electric motors. The motors are cooled by liquid nitrogen. Each cryogenic compressor is supplied with a frequency converter cabinet, enabling motor speed variations, and a control cabinet which contains bearing and motor operative and protective circuits. These cabinets are provided with controls for local operation. The compressors are normally operated through the CEBAF computer which is programmed with polynomial expressions to regulate rotation speeds, in order to avoid surging conditions. The compressor train is initially cooled by 4.5 K helium gas, while rotating the compressors at low speeds. Once cooled, each compressor is accelerated, in succession, according to the suction pressure and speed limitations. As speeds increase the compression ratio of the compressor train increases until the desired pressure is attained. Technical characteristics of the CCC are listed in Table 2.

Table 2. Centrifugal Cryogenic Compressors Characteristics

Cryogenic Compressor	C1	C2	C3	C4
Inlet Press (kPa)	2.84	7.19	18.04	45.6
Inlet Temp (K)	3.32	5.70	9.80	16.9
Outlet Press (kPa)	7.40	18.54	46.6	116.5
Wheel Speed (RPM)	6600	12690	24540	33900
Mass flow (G/S)	237	237	237	237

<u>Controls</u>. The helium refrigeration system is controlled and monitored by CEBAF's central computer system. All process variables and equipment operative status are transmitted for display at the computer console. The screw compressors, turbine expanders and CCC are each provided with local protective interlocks. These components are additionally supplied with local control stations for starting and stopping. The control functions performed by CEBAF's computer include the following:

> Equipment starting/stopping
> Valve positioning
> Temperature controls
> Pressure controls
> Speed controls

HISTORICAL BACKGROUND ON CENTRIFUGAL CRYOGENIC COMPRESSORS

The CCC are clearly the most technically interesting feature of the CEBAF refrigerator. Therefore, the following sections present some historical background on the evolution of these machines, culminating in the design of the CEBAF Centrifugal Cryogenic Compressors.

Development of Centrifugal Cryogenic Compressors

When EURATOM agreed to build the Tore Supra Tokamak, the French Atomic Energy Commission made the decision to operate the toroidal coils at 1.8 K[3]. The specification called for the refrigerator to supply 300 watts at 1.75 K. The conventional approach would have been to provide a multistage Roots type blower system, operating at ambient temperature. Such a system has several drawbacks and the difficulties tend to compound as the scale of the system increases. The multiplicity of blowers presents problems in terms of system reliability and cost. The heat exchangers and return piping become very large due to the extremely high volumetric flow rate and the requirement for very low pressure drop. L'Air Liquide decided to develop a CCC. An advantage of the CCC approach is that it becomes increasingly attractive as the cryogenic power increases.

Cooling liquid helium at 1.75 K requires that the bath pressure be at 1.30 kPa, resulting in a total compression ratio to atmospheric pressure of about 100 to 1. It was decided to provide two stages of cryogenic compression in a series arrangement, the remaining compression to be provided by room temperature liquid ring pumps. The minimum isentropic efficiency selected as a design goal was 40%.

This was a challenging development project in several respects:

1)   the mass flow rate was small
2)   the compression ratio was high[4]
3)   the reliability requirements were quite high

A unique solution was chosen for the bearings featuring an active magnetic system for the support of the rotating parts which consist of the drive rotor, shaft and compressor wheel. The bearings and the motor operate at liquid nitrogen temperature in order to minimize heat leak. At the same time, any contamination problem was avoided because all of the equipment was totally enclosed in the insulation vacuum.

L'Air Liquide also had to adapt centrifugal compressor theory to a machine operating at cryogenic temperatures.

It was decided that state-of-the-art equipment involving a significant development effort should be tested under actual operating conditions rather than, for example, at room or liquid nitrogen temperatures. Consequently, a special test bench was built. The development work subsequently confirmed that this was the correct approach. An early prototype test resulted in an isentropic efficiency of 42%.

The design operating conditions for the Tore Supra machines were established as indicated in Table 3.

Subsequent testing, both at design and off design conditions showed efficiencies as high as 57% and compression ratios as high as 4.1 to 1.

Prior Operating Experience with Centrifugal Cryogenic Compressors

Prior to CEBAF, the only large scale refrigerator operating at or below 2.0 K with CCC was Tore Supra. Since it's start up in 1988, that refrigerator has operated essentially continuously (the torus has been kept cold for 54 consecutive weeks) in an industrial environment[5]. As of February 28th 1991, each CCC had logged 24700 operating hours. There has not been any interruption of the refrigerator operation due to the CCC.

After the two Tore Supra machines, L'Air Liquide built a smaller CCC which processes 6 g/s at the same operating conditions as the first stage of the Tore Supra train. Isentropic efficiencies as high as 63.2% for a compression ratio of 2.22 to 1 and 60.6% for compression ratios as high as 4.27 to 1 were demonstrated on this very small machine. Higher efficiencies could be expected on a larger machine.

CEBAF Centrifugal Cryogenic Compressors

The experience gained with the Tore Supra CCC provided an excellent base of experience and data to tackle the challenges posed by the CEBAF requirements:

1.   four stages in series
2.   discharge pressure above atmospheric pressure
3.   large machines

New theoretical work was launched in order to improve the calculation model using test data collected on the earlier machines.

The four CCC designed and built for the CEBAF Refrigerator are shown in Figure 4. The largest machine is approximately 132 cm in overall length.

Testing of these large machines presented a new difficulty. As with the cryogenic expansion turbines, L'Air Liquide considers it to be very important to check the behavior of a CCC at operating temperature and rotational speed. An agreement was made with the Atomic Energy Commission Center of Grenoble, France to build an experimental loop for superfluid helium

Table 3. Design Conditions for Tore Supra Centrifugal Cryogenic Compressors

	Stage 1	Stage 2
Suction Pressure (kPa)	1.1	3.5
Suction Temperature (K)	4.5	9.3
Discharge Pressure (kPa)	3.5	8.0
Discharge Temperature (K)	9.5	15.5
Mass Flow Rate (g/s)	14.0	14.0
Design Efficiency (%)	50.0	50.0

Fig. 4 Centrifugal Cryogenic Compressors for CEBAF

circulation investigations. This facility made it possible to test each of the CEBAF CCC at or below their design operating temperature and up to the maximum speed of rotation but the flow rates in the test facility were limited to approximately 10% of the design value. An important result of these tests was to demonstrate that these CCC have remarkably good flexibility of operation, being able to provide large pressure ratios at very low flows with surprisingly high efficiency. Some of this test data is included in Table 4.

STARTUP

Startup Experience

The startup of the CEBAF 4.5 K module refrigerator/liquefier was, for the most part, without difficulty.

The integration of the CVI refrigerator/liquefier system with the computer based control system for all the CEBAF cryogenics was very straight forward. After completion of the initial compressor system operation, flow through the 4.5 K cold box was initiated to demonstrate the integrated controls system as well as to filter any particulates from the system prior to installation of the expanders. The turbine expanders were installed by L'Air Liquide and checked and set for proper operation. The 1st, 2nd, and 3rd stage expanders started and performed satisfactorily but the 4th stage expander ran for about 30 minutes before it failed in such a way that it was not rotatable. The cause of the failure is under investigation. However, it should be noted that the 4.5 K module is fully functional and is continuing in service while awaiting the return of the 4th stage expander.

Table 4.   Comparison of Bench Test Data to Design for the Third Stage CCC

Design Variable	Nominal	Test
Suction Pressure (kPa)	20.9	6.04
Suction Temperature (K)	10.41	9.10
Discharge Pressure (kPa)	59.6	16.85
Compression Ratio (NA)	2.85/1	2.79/1
Mass Flow Rate (g/s)	237	30
Isentropic Efficiency (%)	66.5	55.8

Startup Status

The CEBAF refrigerator/liquefier 4.5 K module was operated in mid February 1991, turned over to the CEBAF staff and is available for the support of "front end" testing at 4.5 K. Installation of the 2 K cold box is underway. Acceptance testing of the 2 K cold box is scheduled for mid-April, 1991.

CONCLUSIONS

The heart of the CEBAF refrigerator is the state-of-the-art centrifugal cryogenic compressor system. Extensive development, testing and operation of two CCC for over 24700 hours each at Tore Supra have provided a very high level of confidence in the CCC supplied for CEBAF.

The modular design for the 4.5 K and 2.0 K systems has already proven useful as the 4.5 K system is operating and available to liquefy helium and/or provide refrigeration at 4.5 K as required while awaiting completion of installation of the 2.0 K system.

The mechanical design of the warm compression system has provided a clean appearance and good accessibility in the compressor room.

REFERENCES

1. 2.0 K CEBAF CRYOGENICS, C. H. Rode, D. Arenius, W. C. Chronis, D. Kashy and M. Keesee, Advances in Cryogenic Engineering, Vol. 35, (1990), P. 275.

2. EVALUATION OF ADSORPTION COEFFICIENTS FOR NEON AT LOW TEMPERATURE, H. Parish, Advances in Cryogenic Engineering, Vol. 35, (1990), P. 1541.

3. CONCEPTUAL DESIGN OF A SUPERCONDUCTING TOKAMAK: TORUS II SUPRA, R. Aymar et al. IEEE transactions on magnetics, vol MAG. 15 (1979).

4. APPLICATION RANGE OF CRYOGENIC CENTRIFUGAL COMPRESSORS, G. M. Gistau, J. C. Villard and F. Turcat, Advances in Cryogenic Engineering vol. 35, CEC 1989.

5. TWO YEARS OF OPERATION OF THE TORE SUPRA CRYOGENIC SYSTEM, B. Gravil, G. Bon Mardion, G. Ferlin, B. Jager and L. Tavian, Proceedings of SOFT, London Sept 1990.

A HELIUM VENTING MODEL FOR A SSC HALF CELL

R. H. Carcagno, M. S. McAshan, and W. E. Schiesser

Accelerator Division
Superconducting Super Collider Laboratory*
2550 Beckleymeade Avenue
Dallas, Texas 75237

Abstract: When a Superconducting Super Collider (SSC) dipole magnet quenches, the quench protection system will intentionally quench other magnets in the half cell. The result is that the stored energy of all of these quenched magnets will be absorbed equally among them. These simultaneous quenches produce heat, which diffuses from the magnet coils to the main helium (He) coolant channels and thereby eventually causes an increase in the He pressure. When the quench is detected, vent valves open to minimize the He pressure increase and thus prevent damage to the magnets.

The performance of the He venting system has been modeled and simulated to establish whether the venting will take place as required. The model consists of partial differential equation energy balances written radially for the magnet coils, collar, and yoke; and ordinary differential equations of energy and mass balance written for the He in the magnets and relief header. The basic algorithm is the numerical method of lines, with finite difference approximation of the spatial derivatives, and time integration by LSODES.

Simulation results will be presented for an SSC half cell of the Accelerator Systems String Test (ASST) facility. The results will be also compared with recent string quench measurements performed at the Fermilab String Test Facility.

INTRODUCTION

If a superconductor is warmed sufficiently, current no longer flows through it without resistance. The phenomenon of going from the superconductive state to the normal resistive

---

*Operated by the Universities Research Association, Inc., for the U.S. Department of Energy under Contract No. DE-AC02-89ER40486.

state is called quenching. The electrical aspects of a quench and of quench protection systems are described elsewhere.[1,2] From the cryogenics point of view, an SSC magnet quench can be divided into the following three stages:

Stage One

While current is still flowing in a quenching magnet, Joule heating warms the He within and around the coils. As the temperature of the coils rises, trapped He in the cable expands and is expelled into the annular passage between the coils and the beam pipe (Figure 1). The He in this passage also receives heat from the coils by convective heat transfer. As a result of this mass and energy transfer, the He in the annular passage is expelled very rapidly into passages within the magnet cold mass and into the magnet interconnect region. The time constant of the first stage is less than a second because of the rapid current decay time constant in the coil. Assuming that the quench protection system works properly, the heat deposited in the coil is equal to the initial stored magnetic energy in the magnet. The energy stored in an SSC dipole at 6.6 T is 1.6 MJ. If this amount of heat is deposited evenly in the coil, the temperature increases to 70°K. However, the heat distribution in the coils immediately after the field decay is not uniform and is a complication in modeling the quench process during this stage.

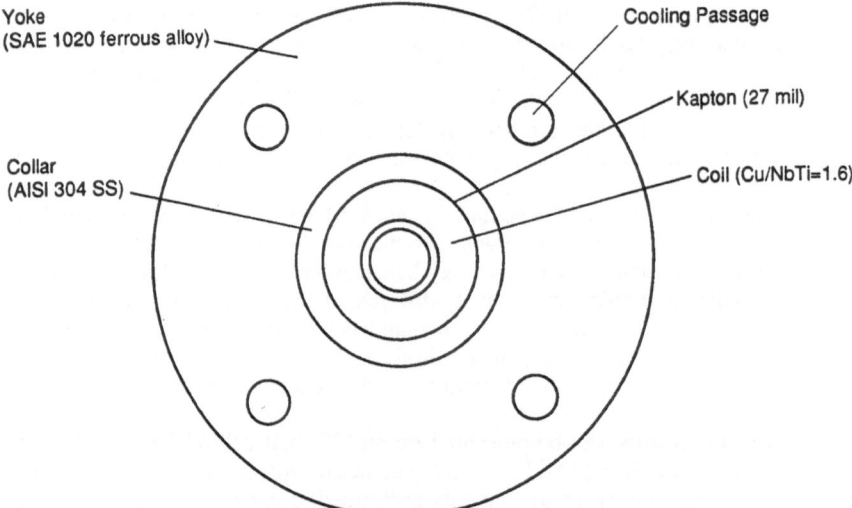

Figure 1. SSC Dipole Cold Mass Model for Heat Conduction.

Stage Two

During the second stage, most of the He will be vented from the half cell as the heat from the quench diffuses outward from the coils. This venting occurs through valves located in the spool pieces at each end of the quenching string (Figure 2). These valves open on signal from the quench detection electronics and automatically on pressure rise to vent into the 20K shield line. The pressure rise and venting rate on quench is determined in large part by the rate of diffusion of heat from the coils into the bulk of the He that is in the coolant channels in the iron. The heat must be conducted through the coil insulation, the stainless steel collars, and the iron itself, and the time constant of this process is on the order of several seconds. It is essential to provide an adequate venting system during this stage. For example, if no venting path is provided, the pressure will rise beyond 10 MPa as a result of the transfer of quench energy into the He at constant density. The time constant of the second stage is of the order of several seconds.

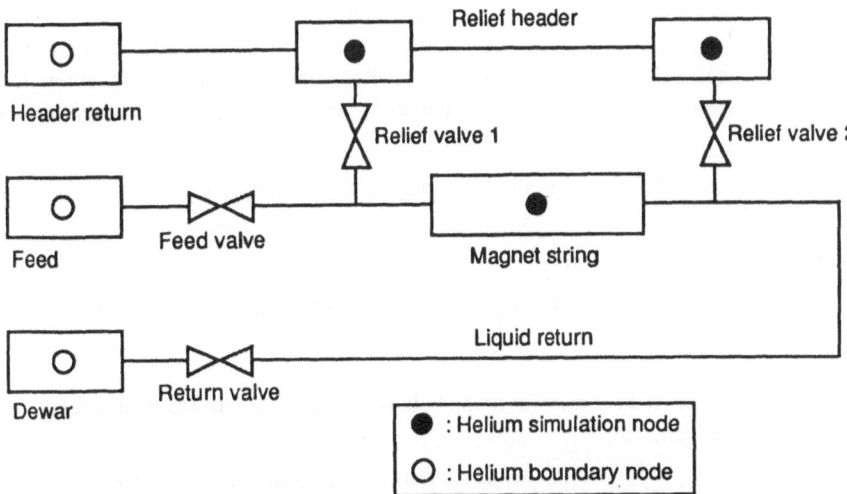

Figure 2. He Network for the SSC String Quench Venting Simulation Model.

Stage Three

In the third stage, which has a time constant of several minutes, all the He inventory has been vented, and the magnet is being cooled to the normal operating temperature.

In this paper we are mostly interested in a quench model for stage two. The pressure rise in stage one is too fast to be relieved by opening relief valves, and the pressure and temperature characteristics of stage three depend strongly on how the quench recovery strategy is implemented.

The performance of the venting system will affect primarily the pressure rise during the second stage. It is essential to understand the venting process and the time constant associated with it in order to design an appropriate venting system for the SSC. In this paper we present a model to simulate the SSC He venting system. This model was used to simulate the quench of a SSC half cell in the ASST.

The investigation of quench behavior is one of the primary goals of the half-cell test at Fermilab. The first quench measurements of a string of two magnets became available in early 1991. These measurements, together with modeling, are providing a quantitative basis for understanding the various aspects of the quench processes. We also present comparisons of simulation and measurements for a string quench in the Fermilab String Test facility.

MODEL DESCRIPTION

The venting model presented here includes an approximate model for the first quench stage and a detailed model for the second quench stage. The third quench stage relates to quench recovery, and its analysis is beyond the scope of this work. However, we are including the venting model in the process simulator SSCDYSIM[3] to study quench recovery in an SSC sector of 4 km of magnets.

Model for the First Stage

A simplified model is used to approximate the overall He pressure rise after the first quench stage. We assume that all the He mass trapped in the cable and in the annular passage is expelled into the bulk of the He in the main coolant channels and interconnect regions. The resulting temperature and pressure is calculated by assuming adiabatic compression of this volume.

For a more detailed treatment of this stage, see References 4 and 5.

Model for the Second Stage

The model for the second stage consists of the following three models:

(1) A model for heat conduction in the cold mass;

(2) A model for heat transfer to the bulk of the He that is in the main coolant channels in the iron; and

(3) A model for He venting through relief valves into a relief header.

Each of these models is discussed below.

Heat Conduction in the Cold Mass

Heat transfer in the cold mass is by conduction. The basic equation is then Fourier's second law for variable thermal properties, which in cylindrical coordinates is

$$\rho C_p \frac{\partial T}{\partial t} = \frac{1}{r} \frac{\partial}{\partial r}\left(rk\frac{\partial T}{\partial r}\right) + \frac{1}{r^2 \partial \theta}\left(k\frac{\partial T}{\partial \theta}\right) + \frac{\partial}{\partial z}\left(k\frac{\partial T}{\partial z}\right).$$

$$(1)$$

If angular and axial variations in temperature are neglected, Equation (1) reduces to

$$\rho C_p \frac{\partial T}{\partial t} = \frac{1}{r} \frac{\partial}{\partial r}\left(rk\frac{\partial T}{\partial r}\right).$$

$$(2)$$

Equation (2) is the basic heat conduction equation for the cold mass; it is applied to the coils, with temperature $T_1(r,t)$, to the collar with temperature $T_2(r,t)$, and the yoke with temperature $T_3(r,t)$. Thus, we have the system of three PDEs:

$$\rho_1 C_{p1} \frac{\partial T_1}{\partial t} = \frac{1}{r} \frac{\partial}{\partial r}\left(rk_1\frac{\partial T_1}{\partial r}\right)$$

$$(3)$$

$$\rho_2 C_{p2} \frac{\partial T_2}{\partial t} = \frac{1}{r} \frac{\partial}{\partial r} \left(r k_2 \frac{\partial T_2}{\partial r} \right) \tag{4}$$

$$\rho_3 C_{p3} \frac{\partial T_3}{\partial t} = \frac{1}{r} \frac{\partial}{\partial r} \left(r k_3 \frac{\partial T_3}{\partial r} \right) + \frac{Q(r,t)}{V} . \tag{5}$$

Note that Equation (5) contains the term $Q(r,t)$, which reflects the heat transfer with the He flowing through the main coolant channels. Variations of temperature with θ and z are neglected because (1) the axial conduction along the magnet is small (the collar and yoke are made of laminations), (2) calculation of the angular heat conduction indicates that it is also small compared to the radial conduction, and (3) the reduction of the model from three dimensions to one significantly reduces the complexity of the model and the associated computation of a solution. Variations of the physical properties have been included in the computer implementation of Equations (3), (4) and (5), i.e., $\rho = \rho(T)$, $C_p = C_p(T)$ and $k = k(T)$.

Equations (3), (4) and (5) each require one initial condition and two boundary conditions. Considering the boundary conditions first, we assume a zero gradient (no heat conduction) condition at the inner surface of the coils ($r = r_0$) and outer surface of the yoke ($r = r_3$)

$$k_1 \frac{\partial T_1(r_0,t)}{\partial r} = 0 \tag{6}$$

$$k_3 \frac{\partial T_3(r_3,t)}{\partial r} = 0 . \tag{7}$$

At the interface between the coils and the collar ($r = r_1$), and between the collar and the yoke ($r = r_2$), we assume that the rate of heat transfer is expressed in terms of contact resistances (C_{12}, C_{23}), i.e., there is a finite jump in temperature at these two interfaces:

$$k_1 \frac{\partial T_1(r_1,t)}{\partial r} = C_{12} (T_2 (r_1,t) - T_1(r_1,t)) \tag{8}$$

$$k_2 \frac{\partial T_2(r_1,t)}{\partial r} = C_{12} (T_2 (r_1,t) - T_1(r_1,t)) \tag{9}$$

$$k_2 \frac{\partial T_2(r_2,t)}{\partial r} = C_{23} (T_3 (r_2,t) - T_2(r_2,t)) \tag{10}$$

$$k_3 \frac{\partial T_2(r_2,t)}{\partial r} = C_{23} (T_3 (r_2,t) - T_2(r_2,t)). \tag{11}$$

Equations (6) to (11) are the required six boundary conditions.

For initial conditions, we use

$$T_1(r,0) = T_{10}, \; T_2(r,0) = T_{20}, \; T_3(r,0) = T_{30} \qquad (12),(13),(14)$$

where T_{10}, T_{20}, and T_{30} are prescribed initial conditions. Equations (3) to (14) are then the complete cold mass model (along with temperature-dependent properties).

Since this model is highly nonlinear (because of the strong dependence of the cold mass properties on temperature), we must use a numerical procedure for the solution of the model equations. We use a finite difference approximation of the right hand sides (RHS) of Equations (3) to (5), defined on a radial grid with index i. The finite difference approximation of the RHS conduction group at the interior grid points is then

$$\frac{1}{r}\frac{\partial}{\partial r}\left(rk\frac{\partial T}{\partial r}\right) \approx \frac{1}{r_i}\left(\frac{\left(\frac{r_{i+1}+r_i}{2}\right)\left(\frac{k(T_{i+1})+k(T_i)}{2}\right)\left(\frac{T_{i+1}-T_i}{\Delta r}\right) - \left(\frac{r_i+r_{i-1}}{2}\right)\left(\frac{k(T_i)+k(T_{i-1})}{2}\right)\left(\frac{T_i-T_{i-1}}{\Delta r}\right)}{\Delta r}\right)$$

$$(15)$$

Boundary conditions (6) and (7) are approximated as

$$T_1(r_0 - \Delta r) = T_1(r_0 + \Delta r) \qquad (16)$$

$$T_3(r_3 - \Delta r) = T_3(r_3 + \Delta r) \qquad (17)$$

which are then substituted in Equation (15) as required, e.g., for the inner surface of the coils, i=1 in Equation (15) and T_{i-1} (= T_1 (r_0 - Δr)) is therefore a fictitious point which can be eliminated by substitution of Equation (16).

Boundary conditions (8) to (11) are substituted directly into Equation (15) with the grid index i set at values corresponding to the coils-collar ($r = r_1$) and collar-yoke ($r = r_2$) interfaces. Consequently, the three time derivatives in Equations (3), (4) and (5) can be computed at all of the radial grid points, including the boundaries and interfaces. This leads to a system of ordinary differential equations (ODEs), one ODE for each radial grid point; Equations (12), (13) and (14) are the required initial conditions. This system of ODEs is then added to the ODEs for the He to make up the complete magnet model.

We used the following materials in our cold mass model: for the yoke, SAE 1020 ferrous alloy; for the collar, AISI 304 Stainless Steel; and for the coils with a copper to superconductor (NbTi) ratio of 1.6. The contact thermal resistance between the coils and the collar (C_{12}) is equivalent to the conductivity of kapton divided by the kapton thickness between the coils and the collar (27 mil). Each region (coils, collar, and yoke) was discretized with seven radial grid points.

Heat Transfer to He

The term Q(r,t) in Equation (5) was calculated as

$$Q(r,t) = -hA \left(T_{wall} - T_{He}\right) . \qquad (18)$$

In Equation (18), h is the heat transfer coefficient for rough walls. The main coolant channel is a pipe of rough walls because it is made out of iron laminations. The heat transfer coefficient h of a rough pipe can be obtained from the following expression[6]:

$$\frac{h}{h_{smooth}} = \sqrt{\frac{f}{f_{smooth}}}.$$

(19)

Equation (19) is valid for Pr ~ 1. The heat transfer coefficient for smooth pipes Re > 10000) can be obtained from Reference 7:

$$h_{smooth} = 0.023 \, C_p \, G \, Re^{-0.2} Pr^{-2/3}.$$

(20)

A general expression for the Fanning friction factor is:[8]

$$\frac{1}{\sqrt{4f}} = -2 \log\left(\frac{\varepsilon/d}{3.7} + \frac{2.51}{Re \sqrt{4f}}\right).$$

(21)

For smooth pipes, $\varepsilon = 0$. We used Equation (21) to calculate the ratio f/fsmooth, and then we used Equations (20) and (19) to calculate h.

He Venting Model

A basic assumption is that the lumped-parameter approach is adequate. A lump or node represents the capacitance or volume of an individual process vessel or group of vessels plus interconnecting piping. Within a lump or node, spatial variations of density, pressure, and temperature are assumed to be negligible. Therefore, homogeneous lumps can be modeled with first-order nonlinear ordinary differential equations. The lumped-parameter approach is commonly used for process simulation studies, and it appears to be adequate for the SSC quench simulation presented here; i.e., we do not expect spatial variation of the state variables to be significant.

The specific lumping assumptions made are illustrated in Figure 2. The magnet string was considered one node, and the relief header was split into two simulation nodes, one at each relief location.

Mass and energy balances were written for all nodes. The mass balance is

$$\frac{d\rho}{dt} = \frac{\left[\sum_{I=1}^{N} W_I - \sum_{J=N+1}^{M} W_J\right]}{V}.$$

(22)

with N input streams amd M output streams.

Assuming that no work is done on the surroundings and that there are negligible changes in kinetic and potential energies, the energy balance is

$$\frac{dU}{dt} = \frac{\left[\sum_{I=1}^{N} W_I(H_I - U) - \sum_{J=N+1}^{M} W_J(H_J - U) + Q\right]}{\rho V}.$$

(23)

Flow between nodes was assumed to be proportional to the square root of the product of the pressure difference and the density of the upstream node:

$$W = C\sqrt{\rho\Delta p} \; . \tag{24}$$

The conductance C of the venting line is the only free parameter in our quench venting model. The value of this conductance was adjusted for agreement with the measured pressure level during venting for the Fermilab string quench.

For He properties, we used the international standard for the thermodynamic properties of He, the model developed by R. D. McCarty of NIST (formerly NBS) [9]

ASST HALF CELL QUENCH SIMULATION

A SSC half cell consists of a spool piece, five dipoles, one quadrupole, and seven interconnects. The total stored energy is about 8 MJ at 6.6 T. The He inventory in the half cell is about 0.6 m^3. We assume that 5% of the cable volume is filled with He, and that the beam pipe o.d. is 37 mm.

We assume that the stored energy is deposited evenly in the coils, raising the coil temperature to 70°K. The relief valves open immediately following a quench, venting He into the 20°K relief header. During venting, the temperature and pressure of the relief header change, affecting the peak pressure in the magnet during the process.

A description of the ASST quench model is given in Table 1.

Table 1. Boundary Conditions, Initial Conditions, and Parameters for the ASST Quench Model.

BOUNDARY CONDITIONS	PARAMETERS
Feed T = 4.5 K	Coil Inner Radius = 0.025 m
Feed P = 0.4 MPa	Coil Outer Radius = 0.050 m
Dewar T = 4.5 K	Kapton Thickness between coil and collar =
Dewar P = 0.12 MPa	6.85 × 10^{-4} m (27 mil)
Header Return T = 20 K	Collar Outer Radius = 0.068 m
Header Return P = 0.25 MPa	Yoke Outer Radius = 0.170 m
	Coolant Channel Diameter = 0.03175 m
	String Volume = 0.6 m^3
INITIAL CONDITIONS	He Mass in coils and in annular passage =
Coil T = 70 K	2.1 Kg/dipole
Collar, Yoke, and He T = 4.5 K	Relief Header Node Volume = 0.24 m^3
String P = 0.4 MPa	Relief Header Conductance = 0.00162 m^2
String Flow = 0.1 Kg/s	Relief Valve Conductance = 0.00068 m^2

Simulation Results

Figure 3 shows the string pressure as a function of time for two scenarios: venting through both relief valves and venting through relief valve 1 only. The first peak of about 1.5 MPa occurs as a consequence of adiabatic compression of the bulk of the He as the He mass in the coils and annular passage is expelled. As expected, this peak is the same for both scenarios. After the relief valves open, there is a rapid decrease in pressure. Eventually, the quench heat reaches the bulk of the He (Figure 4), and the pressure rises again to a second peak that depends on the number of relief valves open and is higher for the case of only one relief valve open. The second peak occurs during the stage two mentioned in the introduction. Most of the inventory venting will occur during the second peak, as shown by Figure 5. The peak power input to the He is about 17 KW at 18 seconds, with a venting flow rate of about 2 Kg/s. Ninety percent of the inventory is vented in about 30 seconds, carrying away 23% of the quench energy.

Figure 6 shows the model mid-temperature of the coils, collar, yoke, and He as a function of time. The cold mass reaches an equilibrium temperature of 29°K at about 200 seconds.

Figure 7 shows the 20°K header pressure at the relief valve locations. Immediately after opening the relief valves the pressure decreases to almost 0.1 MPa because of the mixing of cold He with warm He. Later, the pressure rises to a peak of about 0.4 MPa at 30 seconds because of the He inventory added to the relief header during venting.

Figure 8 shows the 20°K header temperature at the relief valve locations. These temperatures drop rapidly from 20°K to about 7°K.

COMPARISON WITH MEASUREMENTS

In early 1991, the first measurements of quench of a string of two SSC magnets became available. These measurements were performed at the Fermilab String Test Facility,[10] and they provided pressure data at the magnet location and temperature data at the relief valve location. In order to test our model, we simulated the quench venting process for the Fermilab String.

The main differences with the ASST simulation are that the Fermilab string consists of two 40 mm aperture dipole magnets instead of five 50 mm aperture dipole magnets, and that the relief header of the Fermilab string is at room temperature instead of at 20°K.

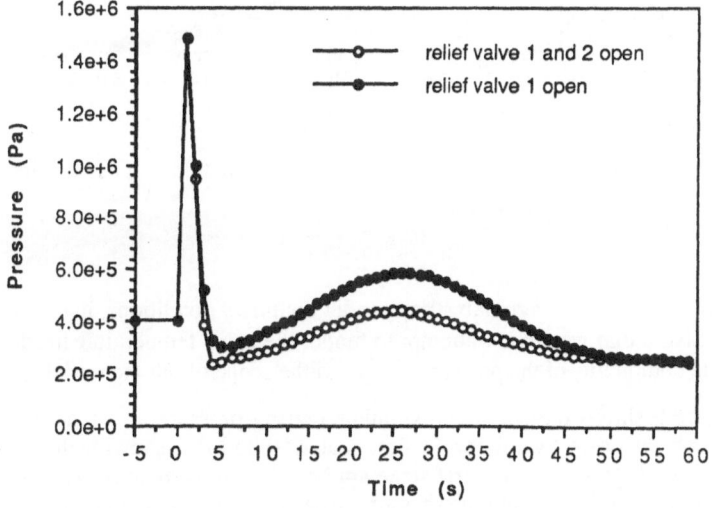

Figure 3. Pressure as a Function of Time for the ASST Quench.

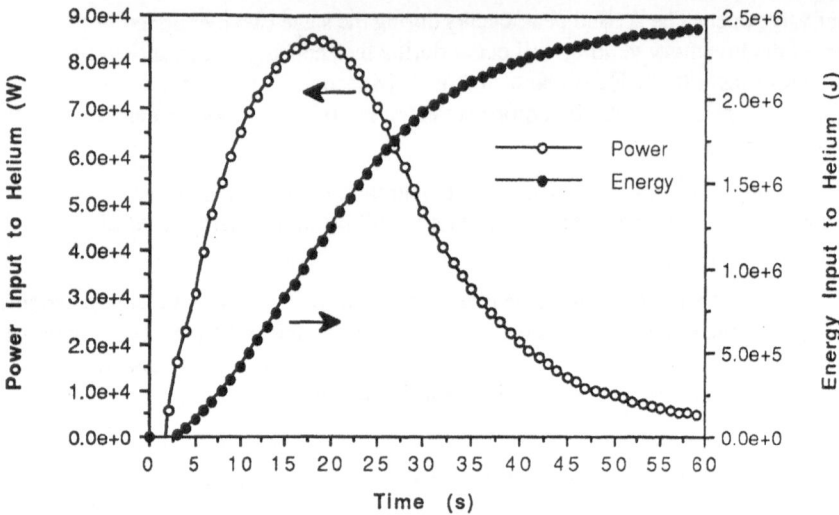

Figure 4. Power and Energy Input to the He as a Function of Time for the
ASST Quench.

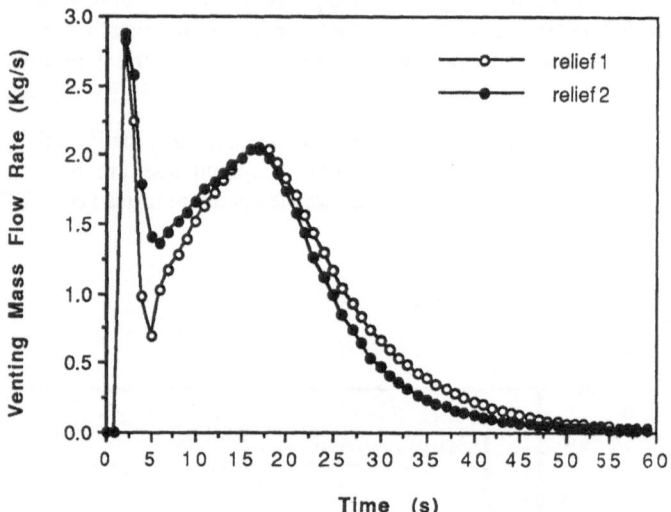

Figure 5. Venting Mass Flow Rate as a Function of Time for the
ASST Quench.

Table 2 shows the Fermilab string model boundary conditions, initial conditions, and parameters. Note that we did not attempt to model the room temperature header because the considerable complexity of the problem is beyond the scope of our He model.

Figure 9 shows measured and simulated string pressure as a function of time. The model predicts the correct value for the first peak of 0.85 MPa, showing that the assumption of adiabatic compression for the first stage applies. The agreement between simulation and measurement is excellent during the pressure decay after the second pressure peak, showing that the model predicts the time constant associated with the venting process very well. However, the model does not agree well with the pressure measured during the first and second pressure peaks. Possible reasons for this disagreement are unknown pressure rise in the warm header, vent line cool down effect, and faster venting of the fraction of He inventory located between laminations.

Figure 10 shows measured and simulated temperature as a function of time. Note that the simulation results are for the He temperature in the magnet, and the measurements were done at the relief valve location. We had no temperature measurements for the magnet. The relief was initially at room temperature, and the temperature drops rapidly as liquid He is being vented. It takes more than 5 seconds to approach the level of the liquid He temperature. The simulated venting He temperature when 90% of the He inventory has been vented is about 9°K, and the temperature measured at the relief valve location shows 8°K.

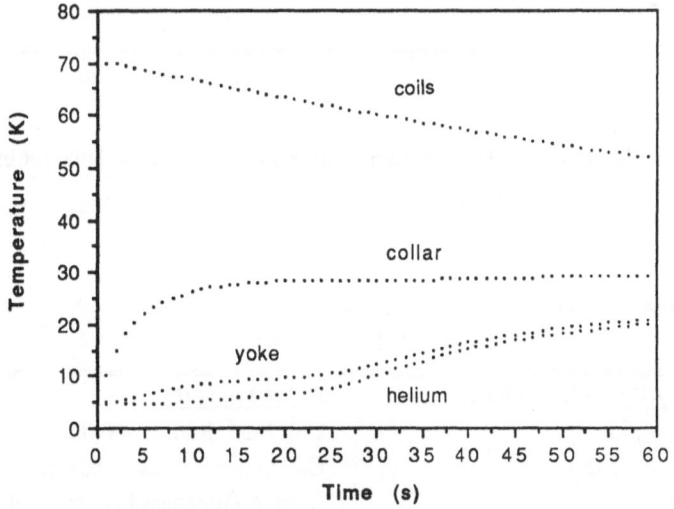

Figure 6. Mid-temperature of the Coils, Collar, Yoke, and He as a Function of Time for the ASST Quench.

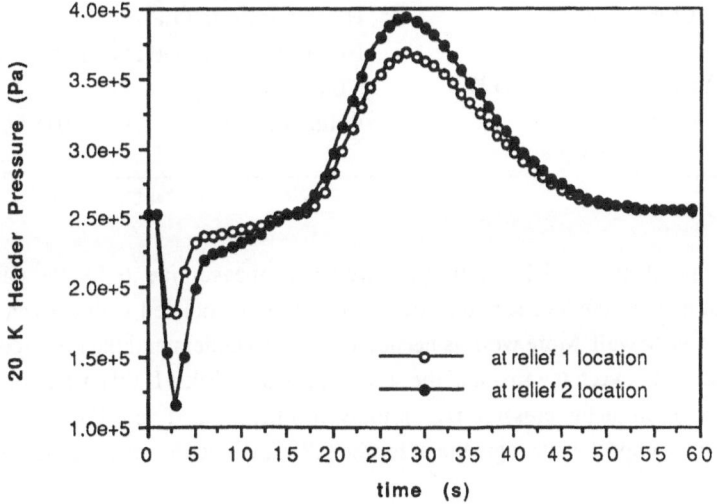

Figure 7. 20K Header Pressure as a Function of Time for the ASST Quench.

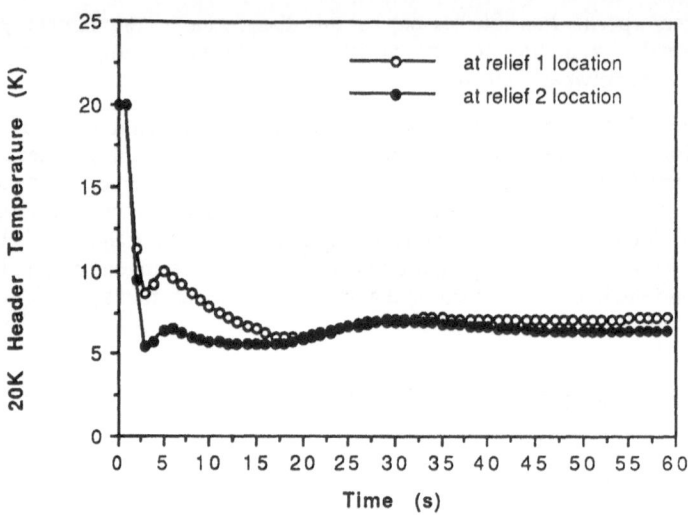

Figure 8. 20K Header Temperature as a Function of Time for the
ASST Quench.

Table 2. Boundary Conditions, Initial Conditions, and Parameters for the
Fermilab String Test Quench Model.

BOUNDARY CONDITIONS	PARAMETERS
Feed T = 4.5 K	Coil Inner Radius = 0.02 m
Feed P = 0.5 MPa	Coil Outer Radius = 0.04 m
Dewar T = 4.5 K	Kapton Thickness between coil and collar =
Dewar P = 0.12 MPa	6.0×10^{-4} m (24 mil)
Relief Header T = 300 K	Collar Outer Radius = 0.0554 m
Relief Header P = 0.27 MPa	Yoke Outer Radius = 0.133 m
	Coolant Channel Diameter = 0.03175 m
INITIAL CONDITIONS	He Volume = 0.3 m^3
Coil T = 70 K	He Mass in coils and in annular passage =
Collar, Yoke, and He T = 4.5 K	0.8 Kg/dipole
String P = 0.5 MPa	Relief Valve Conductance = 0.00068 m^2
String Flow = 0.05 Kg/s	

In general, the model results agree well with measurements for the parameters of
interest; that is, pressure levels, temperature, and timing associated with the venting process
are predicted very well. More work is needed in order to understand the disagreement of the
pressure change between the first and the second pressure peaks. If this behavior is primarily
due to effects such as the pressure rise in the warm header and vent line cooldown during
venting, then we expect better agreement for the ASST, where the relief header is at 20°K.

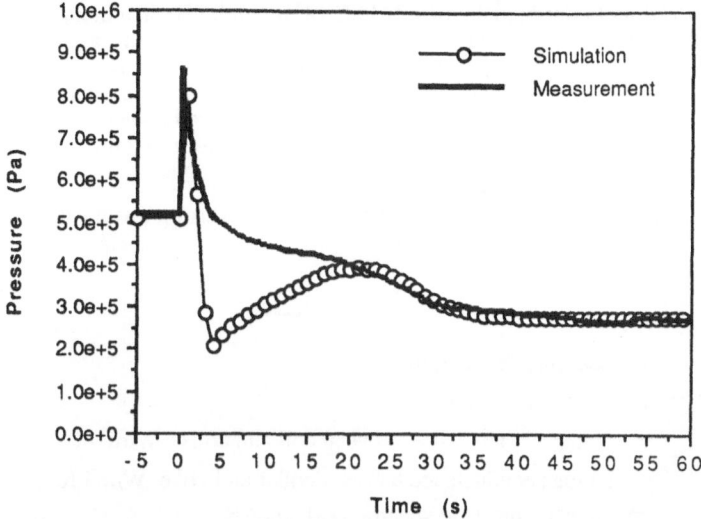

Figure 9. Measured and Simulated String Pressure as a Function of Time for the
Fermilab String Test Quench.

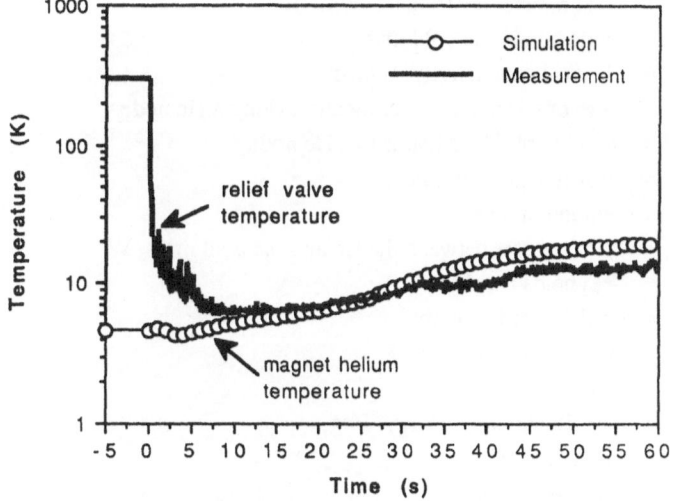

Figure 10. Measured and Simulated Temperature as a Function of Time for the
Fermilab String Quench.

CONCLUSIONS

The He venting model presented in this paper was developed primarily to predict the pressure level, temperature level, venting rate, and venting process time constant after an SSC half-cell quench. This is determined in large part by the rate of diffusion of heat from the coils into the bulk of the He that is in the main coolant channels in the iron yoke.

According to our model typically 90% of the initial He inventory is vented in about 30 seconds. A typical peak venting rate is about 2 Kg/s at 17 seconds, and a typical peak power into the He is about 20 KW per dipole magnet at 17 seconds. At 30 seconds, the He temperature is about 9°K. Apart from the initial pressure spike after He expulsion from the coils and in the annular passage, typical pressure levels are between 0.4 and 0.6 MPa for the relief system considered. The cold mass reaches an equilibrium temperature of about 29°K at approximately 200 seconds.

Comparisons of the model results with Fermilab string test measurements show good agreement for the parameters of interest.

NOMENCLATURE

A	=	heat transfer area, m^2
C	=	conductance, m^2
C1	=	contact conductance between coil and collar, W/m^2-K
C23	=	contact conductance between collar and yoke, W/m^2-K
C_p	=	constant pressure specific heat, J/kg-K
f	=	Fanning friction factor
G	=	mass flow rate per unit area, kg/(m^2-s)
H	=	specific enthalpy, J/kg
h	=	heat transfer coefficient, W/m^2-K
i	=	node number in cold mass
k	=	thermal conductivity, W/m-K
M	=	sum of all streams entering and exiting a He node
N	=	number of inlet streams to a He node
P	=	absolute pressure, Pa
Pr	=	Prandtl number
Q	=	heat transfer between the He and the cold mass, W
Re	=	Reynolds number
r	=	radial coordinate, m
T	=	absolute temperature, K
t	=	time, s
U	=	specific internal energy, J/kg
V	=	volume, m^3
W	=	flow rate, kg/s
z	=	axial coordinate, m
ρ	=	density, kg/m^3
θ	=	angular coordinate, rad
Δr	=	distance between nodes in cold mass, m
Δp	=	pressure difference between upstream and downstream He node, Pa

Subscripts

I	=	inlet stream counter
i	=	inlet stream, cold mass node number
J	=	outlet stream counter

1,2,3: coils, collar, and yoke, respectively

REFERENCES

1. Conceptual Design of the Superconducting Super Collider, SSC-SR-2020, March 1986.

2. Site-Specific Conceptual Design of the Superconducting Super Collider, SSCL-SR-1056, July 1990.

3. D. G. Hartzog et al., "Dynamic Modeling and Simulation of the Superconducting Super Collider Cryogenic Helium System," Supercollider 1, Proceedings of the IISSC, New Orleans, 1989.

4. M. A. Hilal et al, "Helium Pressure Rise of Superconducting Super Collider Dipole Magnets Following a Quench," Adv. Cryo. Engr., Vol. 33, Plenum Press, New York (1988).

5. T.J. Fagan and P.W. Eckels, "Process Control Oriented Quench Analysis of a SSC Magnet," Super Collider 2, Proceedings of the IISSC, Miami Beach, 1990.

6. W. M. Kays, Convective Heat and Mass Transfer, McGraw-Hill, 1966.

7. K. Timmerhaus and T. Flynn, Cryogenic Process Engineering, Plenum Press, New York, 1989.

8. C.F. Colebrook and C.M. White, J. Inst. Civil Eng., Vol. 10, No. 1, pp.99-118, (1937-1938).

9. R.D. McCarty, "Thermodynamic Properties of Helium-4 from 2 to 1500 K at Pressures to 10^8 Pa," J. Phys. Chem. Ref. Data, 2(41), 923-958 (1973).

10. J.C. Theilacker, "Cryogenic Testing at the SSC String Test Facility," Adv. Cryo. Engr., Vol. 33, Plenum Press, New York, 1988.

DESIGN OF THE MULTILAYER INSULATION SYSTEM FOR THE SUPERCONDUCTING SUPER

COLLIDER 50 mm DIPOLE CRYOSTAT

W.N. Boroski, T.H. Nicol, and C.J. Schoo

Fermi National Accelerator Laboratory
Batavia, Illinois

ABSTRACT

The development of the multilayer insulation (MLI) system for the Superconducting Super Collider (SSC) 50 mm collider dipole cryostat is an ongoing extension of work conducted during the 40 mm cryostat program. While the basic design of the MLI system for the 50 mm cryostat resembles that of the 40 mm cryostat, results from measurements of MLI thermal performance below 80K have prompted a re-design of the MLI system for the 20K thermal radiation shield. Presented is the design of the MLI system for the 50 mm collider dipole cryostat, with discussion focusing on system performance, blanket geometry, cost-effective fabrication techniques, and built-in quality control measures that assure consistent thermal performance throughout the SSC accelerator.

INTRODUCTION

The design requirements for the thermal insulation system in the SSC collider dipole cryostat dictate that heat leak from thermal radiation and residual gas conduction be limited to 0.677 W/m^2 to 80K, 0.093 W/m^2 to 20K, and 0.003 W/m^2 to 4K. Moreover, the insulation system must perform reliably throughout the 25-year operating life of the accelerator in an environment which includes severe thermal cycles, radiation exposure, and upset insulating vacuum conditions. Essential to meeting these requirements is an insulation system design that 1) limits radiant heat transfer to design limits; 2) has sufficient volumetric heat capacity to reduce the effects of thermal transients; 3) has sufficient layer density for improved gas conduction shielding; and 4) is comprised of materials suitable for extended use in a high radiation environment. Finally, the system design must be such that cost-effective fabrication and installation techniques establish consistent thermal performance throughout the entire accelerator.

The thermal insulation system for the 50 mm collider dipole cryostat is comprised of cryostat-length assemblies of multilayer insulation fabricated and installed as blankets on the 4.5K cold mass and the 20K and 80K thermal radiation shields.

Table 1. Structural Properties of Candidate Materials

ORGANIC MATERIAL	POLYESTER	POLYIMIDE	POLYAMIDE
TRADE NAME[†]	MYLAR, REEMAY, DACRON	KAPTON	NYLON, CEREX
MOISTURE ABSORPTION	0.4% @ 50% Rh REEMAY SPUNBONDED 0.5% @ 98% Rh	1.3% @ 50% Rh ————	8.0% @ 50% Rh CEREX SPUNBONDED 3-5% @ 95% Rh
OUTGASSED MASS PER UNIT MATERIAL MASS	DACRON NET 1.1×10^{-4} g/g DOUBLE ALUMINIZED MYLAR 2.6×10^{-3} g/g	DOUBLE ALUMINIZED KAPTON 3.1×10^{-3} g/g 3.9×10^{-3} g/g	NYLON NET 4.0×10^{-2} g/g ————
IONIZING RADIATION DAMAGE	5.7×10^{8} RAD (50% MAX. MECHANICAL)	5.0×10^{9} RAD (50% MAX. MECHANICAL)	7.0×10^{7} RAD (50% MAX. MECHANICAL)
IRRADIATION EVOLVED GASES	3-5 ml/g @ 10^{9} RAD H_2(70%), CO_2(20%), CO(10%)	————	20-25 ml/g @ 10^{9} RAD H_2(52%), CO(20%), CO_2(12%), N_2(8%), O_2(3%)

† DUPONT DE NEMOURS & CO; REEMAY INC; JAMES RIVER CORP.

MATERIALS SELECTION

The selection of materials for the MLI system was formulated after consideration of such design parameters as mechanical strength, radiation tolerance, hygroscopicity, vacuum outgassing rates, and thermal performance measurements[1-6]. Candidate materials that were considered for use include polyesters, polyimides, and polyamides; structural properties of these materials are tabulated in Table 1. The material of choice for each of the MLI blanket components is polyethylene terepthalate (PET), a polyester. The material was chosen for its high strength, low moisture absorption, low outgassing rate, and high resistance to ionizing radiation damage. Table 2 outlines the materials selected for use in the 50 mm cryostat insulation system.

Table 2. Materials selected for the 50 mm. cryostat insulation system

BLANKET COMPONENT	MATERIAL	DESCRIPTION	COMPANY
REFLECTOR	ALUMINUM VIA VACUUM DEPOSITION	ALUMINIZED METAL COATINGS; BOTH SIDES; EMISSIVITY <0.03	MULTIPLE VENDORS
REFLECTOR SUBSTRATE	POLYETHYLENE TEREPHTHALATE	FLAT FILM, 0.03 mm THICK, NO PERFORATIONS	DUPONT & CO.
SPACER	POLYETHYLENE TEREPHTHALATE	SPUNBONDED POLYESTER, 0.1mm THICK, 17 g/m^2	REEMAY, INC.
COVER LAYERS	POLYETHYLENE TEREPHTHALATE	SPUNBONDED POLYESTER, 0.23mm THICK, 46 g/m^2	REEMAY, INC.
HOOK FASTENER	POLYESTER	HOOK #80, WHITE #012 WIDTHS: 25mm & 50mm	VELCRO USA, INC.
LOOP FASTENER	POLYESTER	LOOP #2000, WHITE #012 WIDTH: 25mm	VELCRO USA, INC.
THREAD	POLYESTER	V125 WHITE	BELDING CORTICELLI THREAD CO.

BLANKET DESIGN GEOMETRY

<u>80K MLI System Configuration</u> - The insulation system for the 80K thermal shield must have a mean apparent thermal conductivity of 0.76 x 10^{-6} W/cm-K to meet the design heat load budget to 80K. This is achieved by using a multilayer insulation system comprised of reflective layers of aluminized polyester separated by spunbonded polyester spacer layers. The apparent thermal conductivity of an MLI blanket comprised of these materials has been measured to be 0.52 x10^{-6} W/cm-K in the temperature range 300K to 80K[6].

The 80K MLI system consists of two 32-reflective-layer blanket assemblies for a total build of 64 reflective layers. The blankets are designated as the inner 80K and outer 80K blanket, respectively. Each 32-layer blanket has a nominal stack height of 8.86 mm which equates to a mean layer density of 3.61 layers/mm. The design geometry of the 80K outer MLI blanket is shown in Figure 1. The design geometry for both 80K MLI blankets is the same with the exception of the emissivity flap. The emissivity flap consists of three layers of DAM which are folded over the seam area of the blanket during cryostat assembly. The flap serves to maintain the low emissivity of the blanket over the entire outer surface area. The inner 80K MLI blanket does not require an emissivity flap as the outer 80K MLI blanket covers the stepped joint / stepped seam area of the inner blanket.

The reflective layers of the 80K MLI system are comprised of 0.03 mm thick PET film, aluminized with a nominal coating thickness of 350 angstroms, or 0.9 ohms/square, per side. The reflective layers are held apart by single spacer layers of 0.10 mm spunbonded PET material. Single layers of 0.23 mm spunbonded PET cover the blanket top and bottom, and serve to position the polyester hook and loop fasteners at the blanket edges. The fasteners are affixed to the cover layers by sewing. A third 0.23 mm PET layer is located midway through the blanket assembly to separate the upper and lower 16 reflective layers. The multiple blanket layers are sewn together as an assembly along both edges of the blanket. Non-lubricated polyester thread is used in all sewing operations.

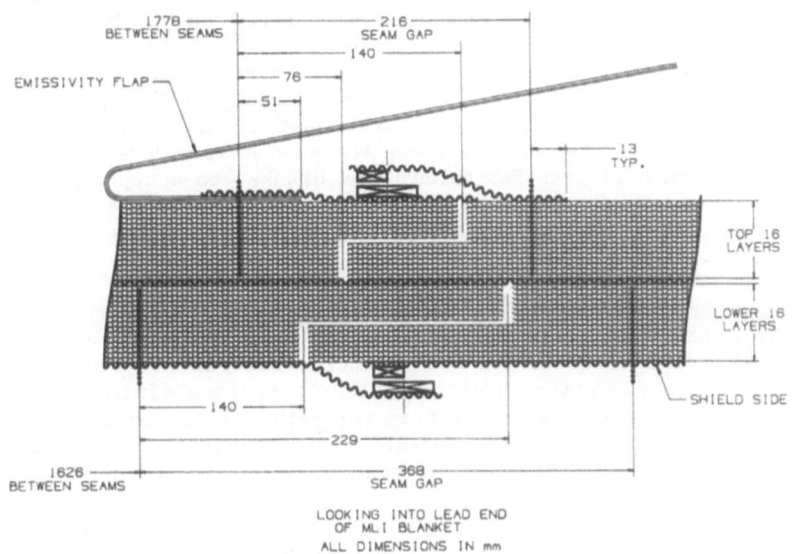

Figure 1. Outer 80K MLI blanket geometry

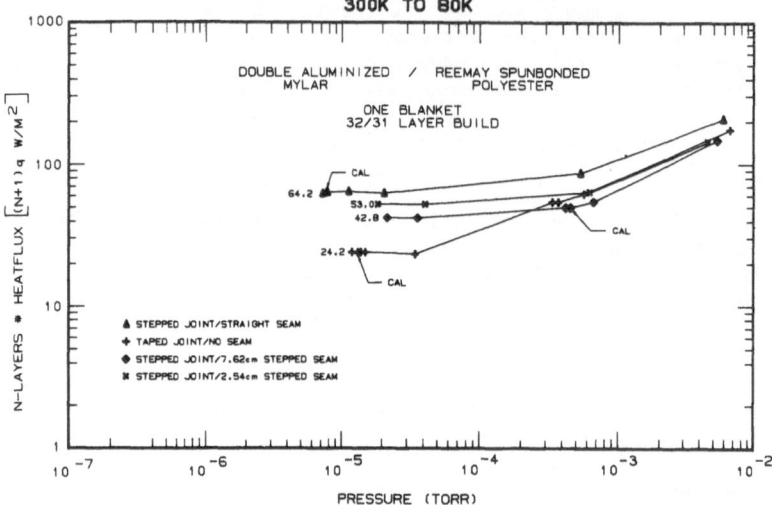

Figure 2. Thermal performance of the stepped seam geometry

At each blanket edge, the upper MLI layers are sewn together from the upper cover layer through to the midlayer, with the thread terminated in the heavier midlayer material. The seam location is then incremented 7.62 cm laterally along the midlayer, and the lower MLI layers are sewn together from the heavy midlayer through to the lower cover layer. The resulting stepped-seam geometry is shown in Figure 1. Sewn seams are advantageous in that they hold the many layers together in a package that can be treated as a single entity during shipping, handling, and cryostat installation. The trade-off comes through increased conduction through the blanket.

Thermal performance measurements have been conducted on various sewn seam geometries to study the effect of sewn seams on MLI blanket thermal performance[6]. Results from these measurements that illustrate the effect of various sewn seam geometries on MLI blanket performance are presented in Figure 2. As evidenced by the data, solid conduction through a straight-through seam significantly increases the heat transfer rate through the blanket. However, a step of 7.62 cm in the sewn seam reduces the amount of solid conduction heat transfer through the blanket to acceptable levels by interrupting and lengthening the conductive heat path of the thread. With the stepped seam geometry, the MLI system design meets the infrared heat load budget to 80K.

20K MLI System Configuration - The MLI system for the 20K thermal shield must have a heat flux less than 93 mW/m^2 to meet the heat load budget to 20K. Calculation of radiation heat transfer between two aluminum dipole cryostat thermal radiation shields of "ideal" surface finish, with no MLI installed, predicts that the heat transfer rate will be 29 mW/m^2, which is well below the design heat load budget. This prediction is substantiated by experimental data which shows low heat transfer rates between aluminum-taped surfaces for temperatures below 80K[7,8]. Furthermore, evidence has been presented that suggests that MLI should not be installed on surfaces below 80K[8,9]. However, several conditions and assumptions exist which must be addressed when considering these statements.

Experimental results which show low heat transfer rates between two aluminum surfaces are made under ideal laboratory conditions where 1) surface contamination is controlled to assure low surface emissivity; and 2) insulating vacuum levels are in the region where residual gas conduction is negligible. In practice, the thermal shields of the cryostat will be used as is, with no special actions taken to achieve as low a value of surface emissivity as theoretically possible. The result will be surface emissivities significantly higher than those experienced under ideal conditions. In fact, assuming "working" shields with surface emissivities approaching 0.1, the heat flux to 20K increases to 129 mW/m^2, substantially higher than the design budget. With respect to insulating vacuum, it is a reality of accelerator operation that the system be operable during periods of insulating vacuum considerably above 10^{-6} torr. Heat leak into the 80K shield is rather insensitive to fluctuations in insulating vacuum as it is driven primarily by radiation heat transfer. However, heat leak into the 20K shield is highly sensitive to pressure, and increases rapidly above 10^{-5} torr. The addition of an MLI blanket on the 20K shield serves not only as an impedance to infrared heat transfer but to gas conduction as well, as the radiation shields also function as gas conduction shields[10,11].

With respect to measurement results which suggest that MLI should not be installed on surfaces below 80K, one must consider the effects of aluminum coating thickness on MLI performance at very low temperatures. If the addition of an MLI blanket on a low temperature surface (in the region 4K - 20K) causes an increase in the heat transfer rate between boundary surfaces, then one must consider what has changed. Given the same degree of insulating vacuum, the only change is in emissivity of the cold boundary.

Since aluminum has electrical resistivity, an electromagnetic wave will penetrate some distance into the aluminum before being absorbed or reflected. This phenomenon, known as radiation tunneling, explains why the thickness of a metallized coating has a considerable effect on the emissivity of metallized films. The thicker the metallized coating, the higher the reflectivity and lower the emissivity. And as electromagnetic wavelengths increase with decreasing temperature, then the radiation tunneling depth also increases, causing a subsequent increase in emissivity[10,12]. This effect has been seen experimentally by Obert, et al.[13], during surface emissivity measurements. When aluminized Mylar was placed over a stainless steel surface, the reflectivity for the 300K to 77K case improved by a factor of 5. However, for the 77K to 4K case the emissivity was found to "deteriorate drastically", becoming worse than the stainless steel surface itself.

Of further concern is the increase in transmissivity of thin aluminum coatings at low temperatures. Transmissivity can be defined as the percent of radiant energy striking an object that passes through without restriction. Figure 3 is a theoretical calculation[14] of the transmissivity through various aluminized coating thicknesses as a function of temperature. Note that a coating thickness of 300 angstroms serves well to impede the passage of radiant energy above 100K. However, a rapid increase in transmittance is seen with decreasing temperature until at 10K approximately 55% of all radiant energy striking the surface will pass through unhindered. It is interesting to note that several researchers have compared the performance of aluminum tape to NRC-2 insulation at 4K; the nominal coating thickness for NRC-2 is 250 angstroms per side. Some increase is seen in performance as the metallized film thickness increases, although it becomes rather small above 500 angstroms. The nominal aluminized coating thickness specified for the 20K MLI system is 600 angstroms per side.

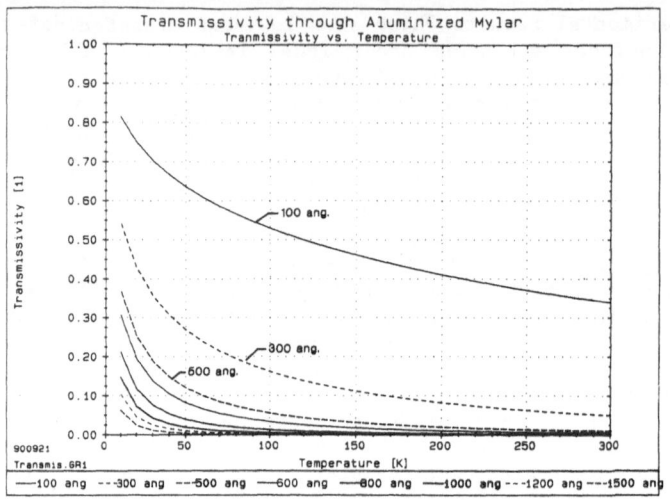

Figure 3. Transmissivity as a function of coating thickness

The final issue which must be addressed in MLI performance below 80K is that of solid conduction through the blanket layers. Below 80K, solid conduction becomes the dominant mode of heat transfer through an MLI blanket. Since infrared heat transfer drops off rapidly below 80K, the 20K MLI system does not require a large number of reflective layers. Hence, the total number of reflective layers in the 20K blanket has been decreased from 32 on the 40 mm design to 10 layers for the 50 mm design. To further limit solid conduction through the blanket, three spacer layers have been added between each set of reflective layers to decrease layer density.

A series of thermal performance measurements between 80K and 20K are underway to experimentally evaluate the design of the MLI system for the 20K shield. The first measurement completed was on a 32-layer MLI blanket modeled after the SSC 40 mm dipole cryostat design. With the warm surface operating near 80K, the measured heat flux through the blanket was 231 mW/m^2, a factor of 2.5 higher than the design budget. The aluminized Mylar had a nominal coating thickness of 350 angstroms per side. Thermometers installed on the individual blanket layers showed evidence of thermal shorting between adjacent layers. A graph of the data plotting individual layer temperatures through the body of the blanket at steady-state is presented in Figure 4. The high degree of thermal shorting between layers, evident in the data, offers one explanation for the excessive heat load.

Performance measurements on a MLI blanket modeled after the present design of 10 reflective layers each separated by three spacer layers are currently underway. The test blanket will have sewn seams comparable to those specified in the cryostat design. Finally, the blanket will contain thermometers to evaluate the interlayer temperature distribution under various operating conditions.

At present, the blanket design specified for the 20K MLI system incorporates 10 reflective layers of PET film aluminized with a nominal coating thickness of 600 angstroms, or 0.45 ohms/square, per side. The reflective layers are held apart by sets of three spacer layers of 0.10 mm spunbonded PET material. Single layers of 0.23 mm spunbonded PET cover the blanket top and bottom to position the polyester hook and loop fasteners. The multiple layers are sewn together along both edges of the blanket.

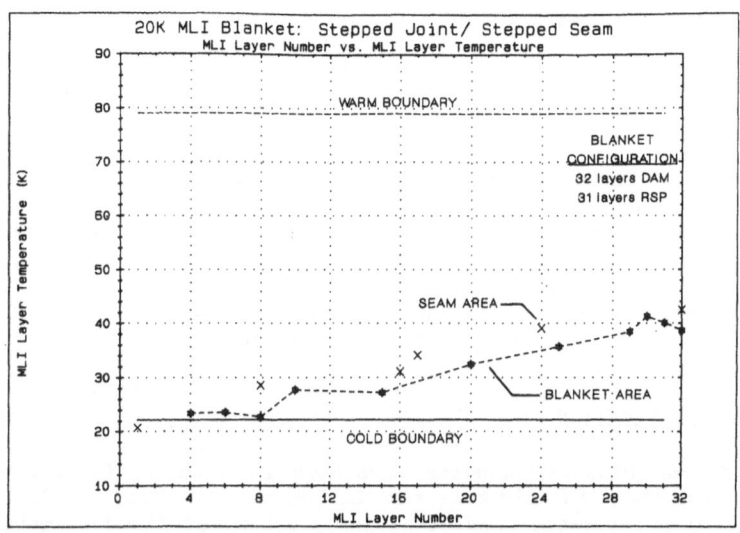

Figure 4. Temperature profile through a 32-layer MLI blanket

At each blanket edge, the upper MLI layers are sewn together from the upper cover layer through to the middle three layers of 0.10 mm PET material, with the thread terminated in the three layers. The seam location is then incremented 7.62 cm laterally along the midlayer, and the lower MLI layers are sewn together from the middle three layers through to the lower cover layer. The resulting stepped-seam geometry is shown in Figure 5.

Cold Mass MLI Configuration - While radiation heat transfer between 20K and 4K is almost negligible, the region is very sensitive to changes in insulating vacuum. Results from computer modeling have shown that the presence of a 10-layer MLI blanket on the cold mass significantly reduces the heat leak into the cold mass at pressures above 10^{-6} torr[15]. Thus, the primary function of the cold mass blanket is to impede residual gas conduction between 20K and 4K.

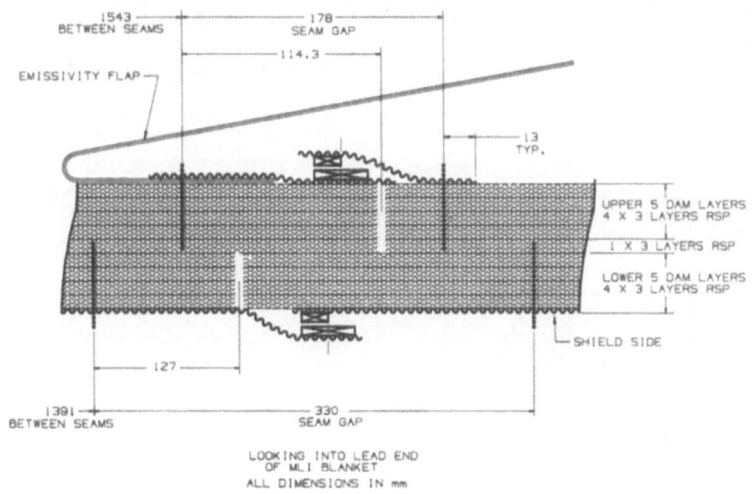

Figure 5. 20K MLI blanket geometry

The MLI system for the 4K cold mass consists of a single 5-reflective-layer blanket assembly, spirally wrapped twice around the cold mass surface for a total of 10 reflective layers. The MLI is installed on the cold mass between support locations; there is no MLI around the cold mass cradle assemblies. The blanket design incorporates 5 reflective layers of double-aluminized PET film, separated by single spacer layers of 0.10 mm spunbonded PET material. The cold mass blankets do not employ heavy PET cover layers or hook and loop fasteners, but are held in place with 7.62 cm wide reflective adhesive-backed tape. The MLI layers are sewn together near both edges of the blanket during blanket fabrication. During cryostat installation, one sewn seam is cut off as the blanket is spirally-wrapped onto the cold mass. The remaining sewn seam serves to maintain layer registration and hold the blanket assembly together.

METHOD OF FABRICATION

A large diameter winding apparatus is used to fabricate the MLI blankets for the 4.5K cold mass and the 20K and 80K shields. The apparatus consists of a rotatable mandrel having a 5.5 meter fixed diameter with an outer surface that is crowned with a convex cross-section. A cross-section view of the mandrel is illustrated in Figure 6. Adjacent supply spools hold the blanket reflective and spacer materials. The function of the apparatus is to wrap the appropriate number of MLI blanket layers around the fixed mandrel. Since each wrap of MLI material increases the circumference of the mandrel, each successive layer of material is slightly greater in length than the preceding layer. Additionally, the convex surface of the mandrel causes the amount of material between sewn seams in the width direction to increase as successive layers are wrapped. Thus, the blanket fabrication method provides integral material in its length and width dimensions to accommodate thermal contraction to cryogenic temperatures. The last layer wrapped onto the mandrel has the greatest length; therefore, it becomes the first layer against the cryogenic surface.

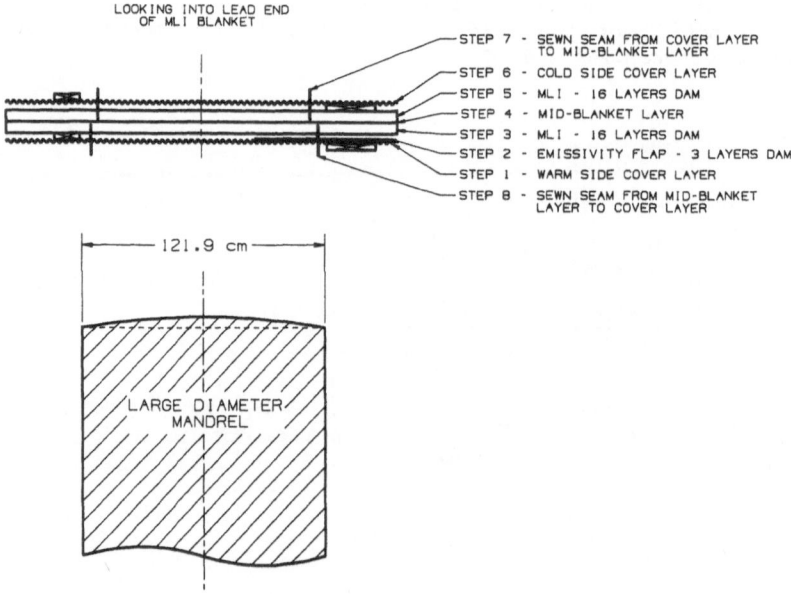

Figure 6. MLI blanket fabrication procedure

Each finished blanket is bound together at its edges by sewing through the blanket with a lock-stitch set at 1.5 stitches per cm. The sewn seams are made by rotating the blanket through a stationary sewing machine while the blanket remains on the mandrel. This method serves to "lock in" the extra material for thermal contraction, as well as fixing the mean layer density of the blanket. Once the sewing operation is completed, a single cut is made across the width of the blanket parallel to the mandrel axis and the assembly removed from the mandrel.

The resulting MLI blanket has sufficient length and width for an SSC shield or cold mass assembly. In addition, the sewn seams serve to control layer density while insuring interlayer cleanliness, three-dimensional stability, and extra material to accommodate thermal contraction. Finally, the sewn seams fix interlayer registration to maintain support post penetrations and blanket edge alignment during shipping, handling, and installation onto the cryostat.

The final stage in blanket fabrication is to locate and cut support post penetrations and alignment fiducials along the blanket length. As an additional allowance for thermal contraction of the blanket assemblies during thermal cycling, the support post penetrations are located 3.81 cm farther apart than the actual support post locations. This builds into the blanket sufficient material to compensate for local thermal contraction. Alignment fiducials marked on the inner layer of the blanket perpendicular to the blanket length function as quality control measures during cryostat installation. Alignment of the fiducials during cryostat assembly assure that the blanket has been installed parallel to the cryostat axis.

CRYOSTAT INSTALLATION

Locations of the MLI blankets in the 50 mm dipole cryostat are shown in Figure 7. During cryostat assembly, the MLI blanket is wrapped around the shield such that the edges of the blanket overlap to form a stepped, butt-joint connection as illustrated in Figure 1. The effect of the joint connection on overall thermal performance is significantly decreased by staggering the blanket penetrations; there are 24 uninterrupted layers at any point along the 80K shield, and 5 uninterrupted layers along the 20K shield length. Thermal measurements made on MLI joint configurations prove that connections made in this manner approach the performance of a MLI blanket with no seams[16]. To further reduce the effects of the blanket joint on thermal performance, the inner and outer 80K blanket seams are staggered on the cryostat.

Blanket installation begins by securing opposite ends of the inner blanket lower cover layer to each other by full engagement of the hook and loop fasteners. As the lower cover layers are overlapped and secured, the perpendicular alignment marks are superimposed, thereby confirming a cylindrical blanket assembly along the cryostat length. The MLI layers between the cover layers are then joined along the cryostat length using the stepped-butted joint. As the blanket edges are drawn together, tension on the blanket is taken by the sewn seams and cover layers. The MLI material located in the greater blanket area between sewn seams is isolated from.the tension by the seams. The joint configuration is completed by full engagement of the upper cover layer hook and loop fasteners over the joint area. The resulting blanket installation is secured from opening by the closure of the two hook and loop pairs. Thermal testing of the hook and loop fasteners disclosed that the connection became more solid during cooldown due to the thermal contraction of the hook/loop connection.

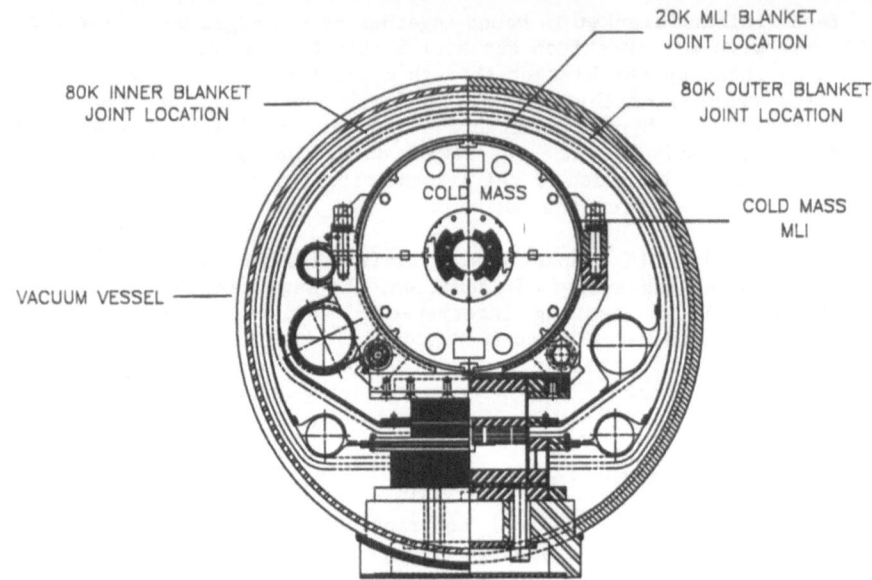

Figure 7. MLI locations in the 50 mm cryostat

The outer blanket is installed in a like manner. However, the outer cover layer of the outer blanket must be removed to uncover the outermost, or warm, layer of reflective DAM. Additionally, the emissivity flap, which consists of several layers of DAM, must be folded over the heavy spacer material covering the seam area to provide a low emissivity surface to thermal radiation.

QUALITY CONTROL

Quality control features inherent in the blanket design permit quick and easy assessment of the quality of each MLI blanket installation during cryostat production.

Uniform distance between sewn seams after blanket installation is a concise indication of proper blanket installation. As a fixed amount of material is contained between sewn seams, and as the amount of material remains constant from assembly to assembly as a function of the fabrication method, then maintaining a fixed distance between sewn seams maintains a fixed layer density from cryostat to cryostat. The result of controlled layer density is consistent thermal performance throughout the accelerator. During the 40 mm cryostat program, the preferred distance between seams across the joint connection was empirically determined for each blanket connection by trial fitting sections of MLI onto actual shields. The separation of the hook and loop fasteners was adjusted until the desired blanket fit was obtained. When the fasteners are superimposed during cryostat installation, the blanket installation is correct. A like procedure will be employed during the initial phase of 50 mm cryostat production.

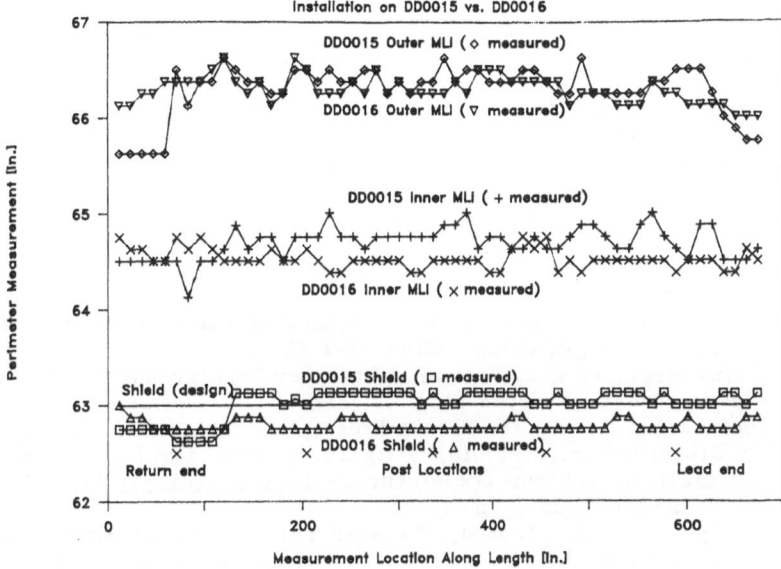

Figure 8. Quality control of MLI installations

During blanket fabrication, the seams are sewn parallel to within 32.0 mm along the entire blanket length. Hence, the distance between sewn seams across the body of the blanket is a fixed and known parameter. As a quality control procedure, measurements of the distance between sewn seams are made along the length of each blanket immediately after cryostat installation. These measurements provide a clear indicator of the quality of each blanket installation. Furthermore, the results, when compared with data from other cryostat installations, allow assessment of installation consistency from cryostat to cryostat. Figure 8 illustrates the consistency of blanket installations between cryostats; shown is data taken from the MLI installations on two SSC 40 mm dipole cryostats.

Finally, a visual confirmation that alignment marks, placed on the MLI blanket perpendicular to the cryostat axis, are superimposed during cryostat installation assures that the blanket is installed parallel to the cryostat axis. Moreover, it is a positive indication that sufficient material is contained between support locations to accommodate thermal contraction.

SUMMARY

The MLI system for the SSC 50 mm collider dipole cryostat closely resembles the MLI system for the 40 mm cryostat. The performance of the 80K MLI system has been experimentally evaluated, with results indicating that the design geometry meets the heat load budget for the 80K thermal shield. Thermal performance measurements are underway to quantitatively define the thermal performance of the MLI system design to 20K. All materials selected for use in the MLI system have properties appropriate for use in the SSC accelerator environment. Finally, the MLI blanket design allows for cost effective fabrication techniques which assure consistent thermal performance throughout the SSC accelerator.

ACKNOWLEDGEMENTS

The development of the MLI system for the SSC dipole cryostat has involved the talents and efforts of many individuals. In particular, the design and development of the system to its present state has been accomplished through the efforts of J. D. Gonczy, J.G. Otavka and M.K. Ruschman. Further appreciation is given to J. Kerby for analytical contributions and to D. Arnold, K. Sarna and K. Swanson for their assistance in preparing the illustrations for this work.

REFERENCES

1. M.H. Van de Voorde and C. Restat, 'Selection guide to organic materials for nuclear engineering,' CERN 72-7 (1972).
2. H. Burmeister, et al., Test of multilayer insulations for use in the superconducting proton-ring of HERA, in: "Advances in Cryogenic Engineering," Vol. 33, Plenum Press, New York (1988), p. 313.
3. A.P.M Glassford and C.K. Liu, 'Outgassing rate of multilayer insulation materials at ambient temperature,' J. Vac. Sci. Technol., Vol. 17, No. 3, May/June 1980, p. 696.
4. W. Burgess and Ph. Lebrun, Compared performance of Kapton and Mylar based superinsulation, in : "Proceedings of the Tenth International Cryogenic Engineering Conference," Helsinki, Finland (1984), p. 664.
5. T. Ohmori, et al., Thermal performance of candidate SSC magnet thermal insulation systems, in: "Advances in Cryogenic Engineering," Vol. 33, Plenum Press, New York (1988), p. 323.
6. J.D. Gonczy, W.N. Boroski, and R.C Niemann, Thermal performance measurements of a 100 percent polyester MLI system for the Superconducting Super Collider: Part II, Laboratory Results (300K-80K), in: "Advances in Cryogenic Engineering," Vol. 35, Plenum Press, New York (1989), p. 497.
7. E.M.W. Leung, et al., Techniques for reducing radiation heat transfer between 77K and 4.2K, in: "Advances in Cryogenic Engineering,", Vol. 25, Plenum Press, New York (1980), p. 489.
8. T.R. Gathright and P.A. Reeve, Effects of multilayer insulation on radiation heat transfer from 77K to 4.2K, Proc. MT-9 Zurich, 1985, p. 696.
9. K. Kutzner, F. Schmidt, and I. Wietzke, Radiative and conductive heat transmission through superinsulations - experimental results for aluminum coated plastic foils, Cryogenics, July 1973, p. 396.
10. C.L. Tien and G.R. Cunnington, Cryogenic insulation heat transfer, in: "Advances in Heat Transfer," Vol. 9, Academic Press, New York (1973), p. 349.
11. P.E. Glaser, et al., Thermal insulation systems, a survey, NASA SP-5027.
12. R.G. Scurlock and B Sauli, Development of multilayer insulations with thermal conductivities below 0.1 μW cm^{-1} K^{-1}, Cryogenics, May 1976, p. 303.
13. W. Obert, et al., Emissivity measurements of metallic surfaces used in cryogenic applications, in: "Advances in Cryogenic Engineering,", Vol. 27, Plenum Press, New York (1982), p. 293.
14. R.P. Schutt, Some thoughts on superinsulation, Isabelle Division, Technical Note No. 21, Brookhaven National Laboratory (1976).
15. "Integrated thermal math model of the production dipole magnets for the Superconducting Super Collider," Report No. SSC-CDG-364-SDP.
16. Q.S. Shu, R.W. Fast, and H.L Hart, Theory and technique for reducing the effect of cracks in multilayer insulation from room temperature to 77K, in: "Advances in Cryogenic Engineering," Vol. 33, Plenum Press, New York (1988), p. 291.

INTEGRATED CRYOGENIC SENSORS

D. B. Juanarena

M. G. Rao

Pressure Systems, Inc.
34 Research Drive
Hampton, VA 23666

Aniga Enterprises
110 Harlan Drive
Grafton, VA 23692

ABSTRACT

Integrated cryogenic pressure-temperature, level-temperature, and flow-temperature sensors have several advantages over the conventional single parameter sensors. Such integrated sensors were not available until recently. Pressure Systems, Inc. (PSI) of Hampton, Virginia, has introduced precalibrated precision cryogenic pressure sensors at the Los Angeles Cryogenic Engineering Conference in 1989. Recently, PSI has successfully completed the development of integrated pressure-temperature and level-temperature sensors for use in the temperature range 1.5 - 375 K. In this paper, performance characteristics of these integrated sensors will be presented. Further, the effects of irradiation and magnetic fields on these integrated sensors will also be reviewed.

INTRODUCTION

Integrated cryogenic pressure-temperature, level-temperature, and flow-temperature sensors have several advantages. Integrated pressure-temperature and level-temperature sensors for cryogenic use in the temperature range 1.5 - 375 K have been recently developed at PSI. Integrated pressure-temperature sensors will not only eliminate the disadvantages of limited frequency response, thermo-acoustic oscillations, added heat leaks and failure due to leakage or blockage associated with the gauge lines of the conventional pressure sensors, but also provide the precise thermodynamic state of the cryogenic system. Integrated level-temperature and flow-temperature sensors will provide an accurate cryogen quantity, quality, and mass flow measurements.

The selection criteria and the thermometric characteristics of the temperature sensors and the performance characteristics of the pressure and level sensors of the integrated cryogenic sensors will be presented in this paper.

Moreover, the effects of irradiation and magnetic fields on these integrated sensors will also be reviewed.

INTEGRATED CRYOGENIC PRESSURE-TEMPERATURE SENSORS

Integrated pressure-temperature sensors are the basic building blocks of the level-temperature and flow-temperature sensors. The selection criteria and the thermometric characteristics of the integrated temperature sensors are presented in the following section.

Integrated Temperature Sensor

The major criteria in selecting a cryogenic temperature sensor for the integrated sensors are interchangeability, wide operating temperature range, high reproducibility, two wire configuration, and cost. Table I summarizes these characteristics for the likely candidates: carbon, carbon-glass, cryogenic linear temperature sensor (CLTS), germanium, platinum, rhodium-iron resistance and Si diode thermometers.

Table 1. Comparison of Cryogenic Thermometers

Sensor	Interchange-ability	Temperature Range	Signal Level	Reproducibility	Number of leads	Cost (uncalibrated)
Carbon	no	0.1-100K	10mV	Poor (~20/charge)	4	<$1
Carbonglass	no	1-300K	1-10mV	1mK	4	$155
CLTS	no	4-300K	20mV	0.2K	4	$310[1]
Germanium	no	0.02-100K	1-10mV	<1mK	4	$150
Platinum	no[2]	15-700K	1-100mV	10mK	4	$100
Rhodium-iron	no	1-800K	1-50mV	10mK	4	$395
Si diode	yes	1-475K	0.1-1.8V	20mK	2	$180[3]

[1] Ruggedized
[2] With two wire configuration
[3] Calibration not required

Though carbon and germanium resistance thermometers have high sensitivity in the low temperature range (1.5 -100K), they have limited operating range. By using them in conjunction with a platinum resistance thermometer, the operating temperature range can be extended but at the expense of additional cable, connectors, feedthroughs, and installation. Except for Si diodes, all the other sensors provide low-level signals, and transmitting such low-level signals over cable lengths of 200 feet or more present difficulties with respect to thermal emf's and electrical noise.(1) Further, with the exception of Si diodes, all the other sensors need to be calibrated for achieving temperature measurement accuracies of better than 0.5 K and this is prohibitively expensive.(2) The Si diodes have small thermal mass, are very rugged, and relatively low cost. For these reasons, we chose to incorporate the Si diode sensor as our basic temperature sensing element.

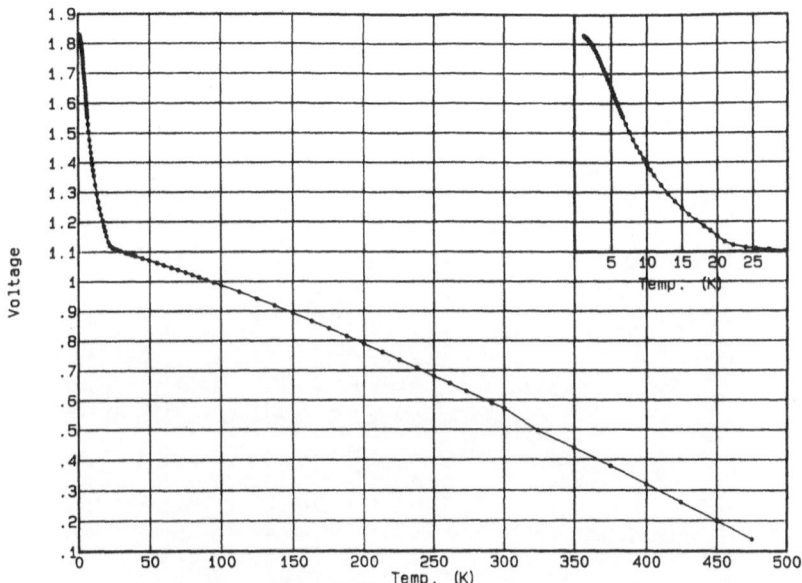

Fig. 1. Forward Diode Voltage vs. Temperature

Figure 1 gives the thermometric characteristic (forward voltage with 10 μA current) of Si diodes as a function of temperature. These sensors have linear thermometric characteristics in the temperature range 25-475 K with ~2.2 mV/K sensitivity. The sensitivity increases below 25 K and is ~55 mV/K at 4.2 K. These Si diode temperature sensors have virtually identical thermometric characteristics, as the maximum deviation between any two sensors is less than 0.05 K over the operating temperature range (1.5 - 475 K). The reproducibility of these sensors is better than 20 mK.

Integral Pressure Sensor

Precision cryogenic pressure sensor technology is incorporated into the integrated pressure-temperature sensor. During the past three years, we have developed cryogenic pressure sensors and the ability to accurately calibrate these sensors between 1.5 and 475 K(3). The pressure sensors as shown in Figure 2 consist of a highly doped silicon piezoresistive pressure sensor mounted in an all-welded stainless steel package. Differential and absolute devices have been produced between the ranges of 1.4 to 3400 kPA (0.2 to 500 psi; 0.02 to 33 bar). The pressure sensors require 1 mA constant current excitation and have a bridge resistance of 6000 ohms nominally. The low power dissipation (~6mW) makes this sensor ideal for applications where power dissipation is critical. At 4.2 K and 1 mA excitation, the pressure transducers provide between 50 and 500 mV output depending on the pressure sensors' full scale range.

Calibration data at 4.2 K and 300 K for 50 kPA (0.5 bar) differential, and 200 and 1000 kPA (2 and 10 bar) absolute pressure sensors are presented in Figures 3, 4, and 5 respectively. Figure 6 gives the full scale output of 200 and 1000 kPA (2 and 10 bar) absolute sensors as a function of temperature, and Figure 7 presents their normalized sensitivities, again as a function of temperature.

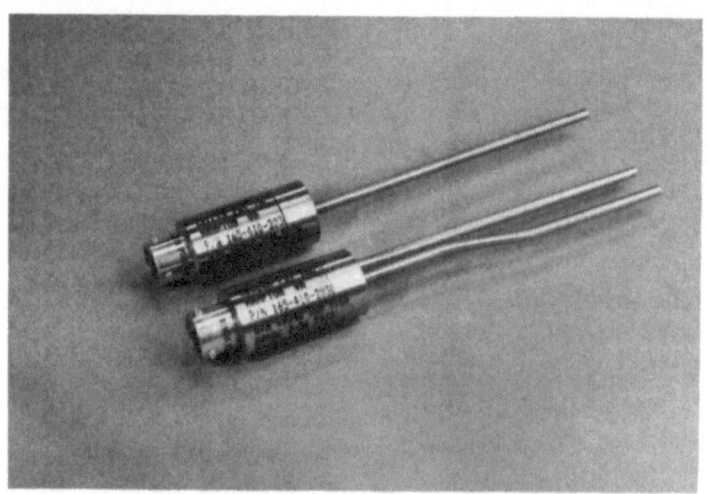

Fig. 2. Cryogenic Pressure Transducers

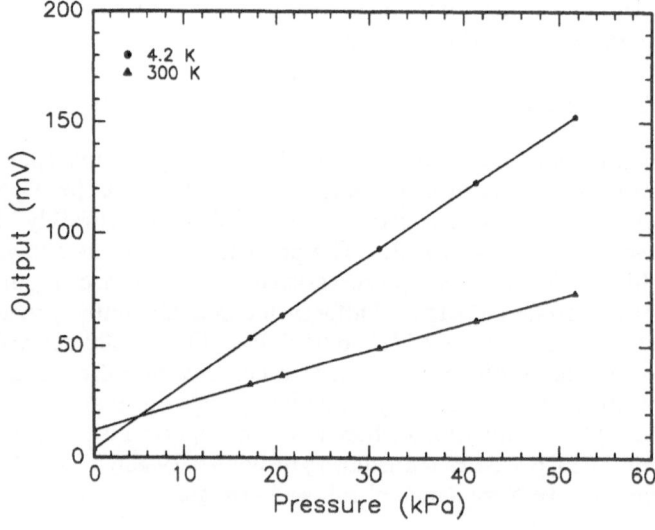

Fig. 3. Voltage Output vs. Pressure for 50 kPA Pressure Sensors

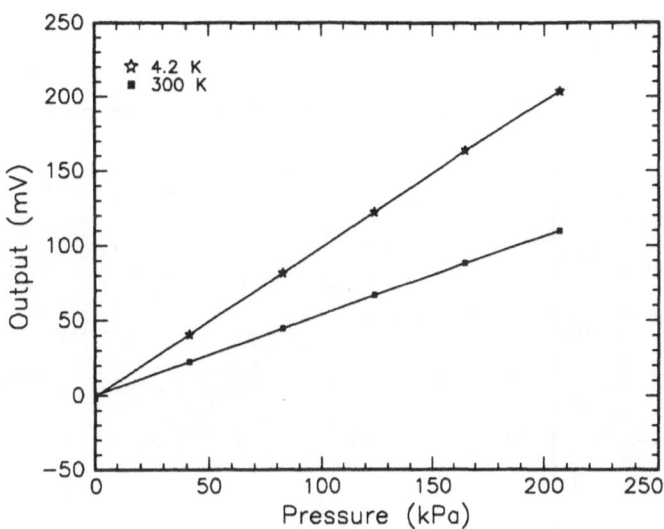

Fig. 4. Voltage Output vs. Pressure for 200 kPA Pressure Sensors

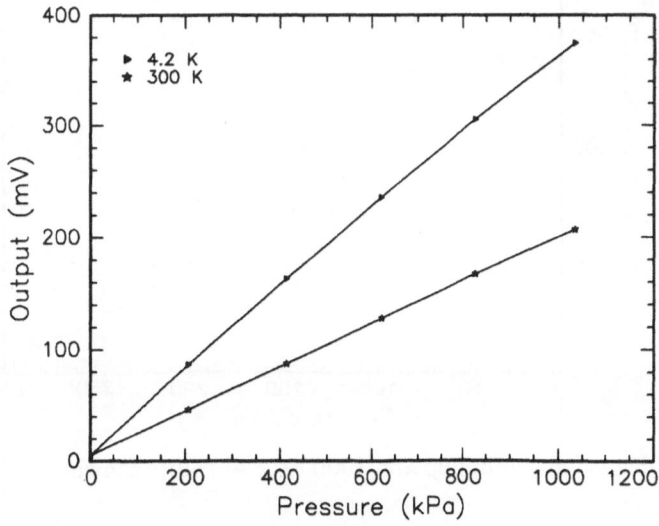

Fig. 5. Voltage Output vs. Pressure for 1000 kPA Pressure Sensors

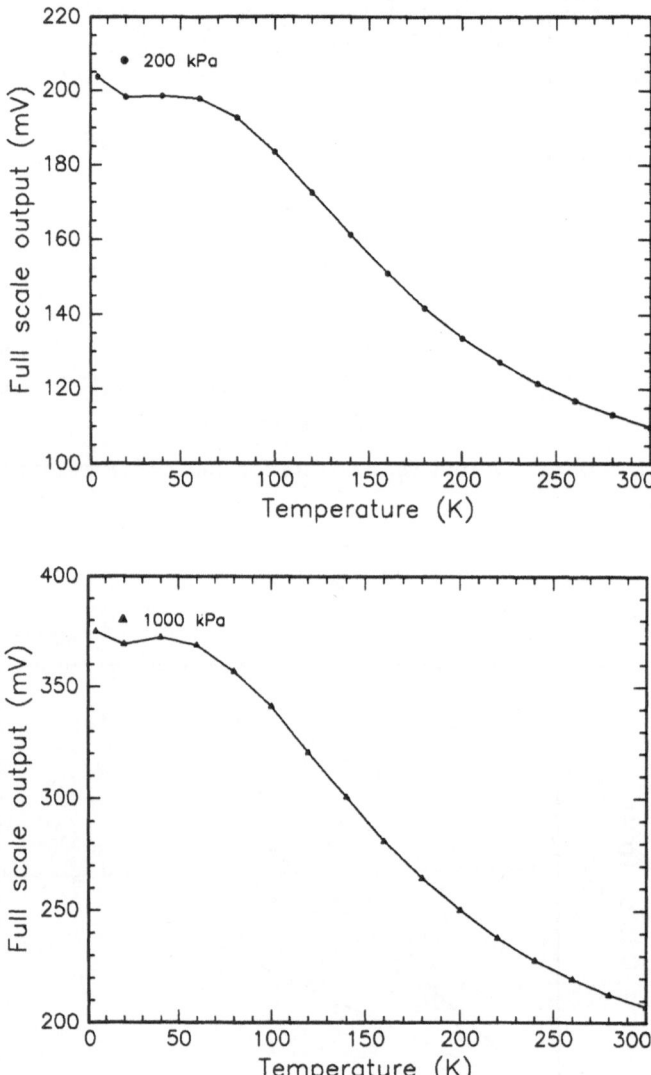

Fig. 6. Full Scale Output of 200 and 1000 kPA Absolute Sensors vs. Temperature

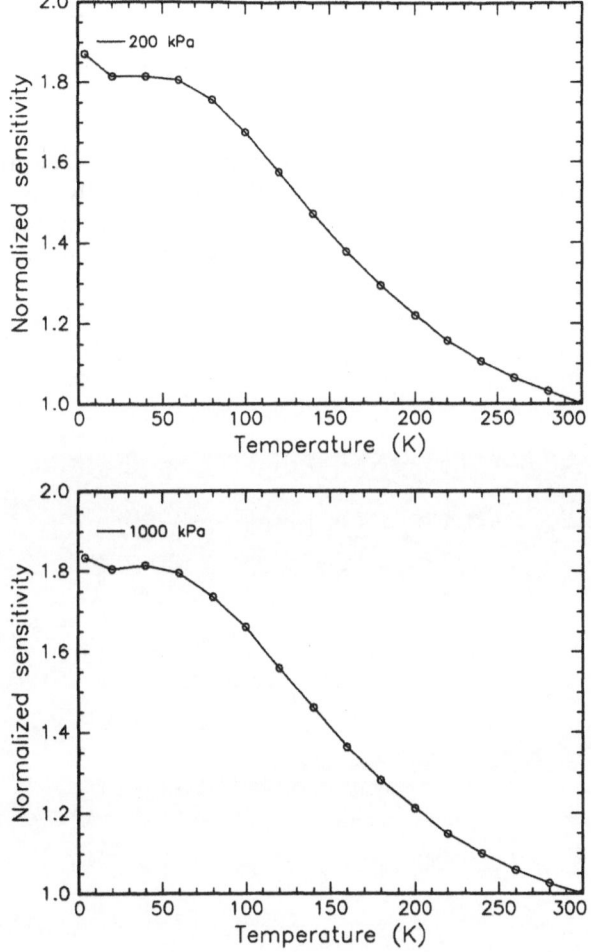

Fig. 7. Normalized Sensitivities for 200 and 1000 kPA
Absolute Sensors vs. Temperature

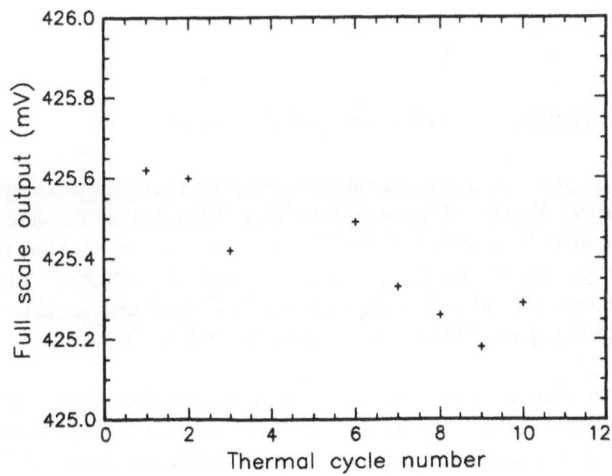

Fig. 8. Full Scale Output at 78 K as a Function of Thermal Cyclings
Between 78 and 300 K for 1000 kPA Sensors

Figure 8 illustrates the repeatability of a 1000 kPA (10 bar) pressure sensor at full scale pressure over a number of thermal cyclings between 300 and 78 K. This data was taken over a period of several weeks and is representative. Typical thermal hysteresis is less than 0.1% F.S.

For applications not requiring very long term vacuum or pressure integrity, the pressure sensors are produced with commercially available electrical connectors. For applications requiring extremely long term hermeticity or high line pressure, special ceramic/metal feedthroughs may be provided.

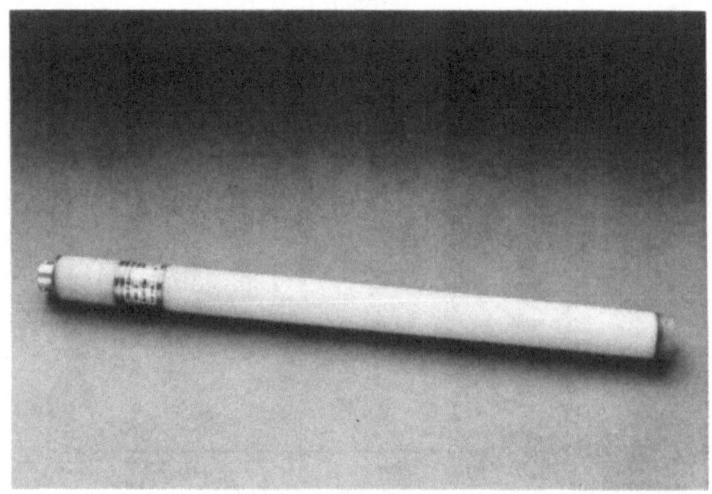

Fig. 9. Cryogenic Liquid Level Sensor

INTEGRATED LEVEL-TEMPERATURE SENSOR

A new cryogenic liquid level sensor has been recently developed at CEBAF in Newport News, Va.(4). Using a sensitive differential pressure sensor, the cryogen is measured by sensing the differential pressure exerted by the column height of the cryogen across the pressure sensor diaphragm. This sensor has been designed to operate at LHe temperature (1.5 - 4 K) and incorporates features which eliminate cryogen condensate problems within the sensor.

PSI has further developed the packaging and calibration techniques for the level sensor. Additionally, a Si diode temperature sensor has been combined to provide level and temperature measurement integrated into one housing. The new level-temperature sensor is shown in Figure 9.

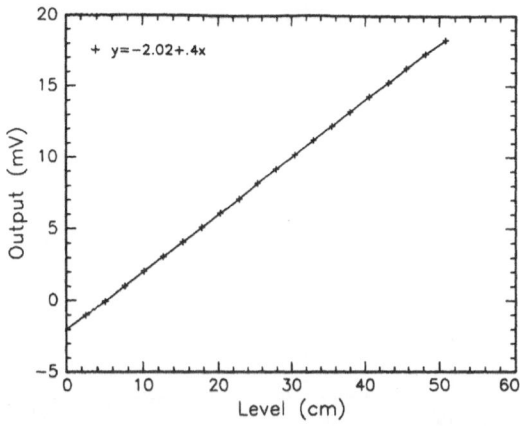

Fig. 10. Voltage Output vs. Level for Cryogenic
Level Sensor Operating at 4.2 K

Figure 10 gives the output signal of a typical 20" LHe level sensor as a function of LHe level. These sensors can be used to measure the level of any cryogenic liquid and over the entire temperature range of any cryogen. These sensors have a resolution of better than ±1 mm of cryogen and can be calibrated to provide uncertainties of less than ±0.5% F.S. range of the level sensor.

IRRADIATION AND MAGNETIC FIELD EFFECTS

The understanding of the effects of irradiation and magnetic fields on sensors is important if the sensors are to be used in particle accelerators. In the following, the irradiation and magnetic field effects on the integrated sensors are reviewed in light of the available data on Si diodes and cold pressure sensors.

Irradiation Effects

Si diode thermometers from Southampton University (SMDT), Lake Shore Cryotronics and CEBAF have been irradiated with gamma rays at room temperature in steps of decades of rad for determining irradiation effects in them(2). Figure 11 gives the difference between the indicated temperature at 4.2 K before the sensors were irradiated and the indicated temperature after each decade irradiation dose. This temperature difference includes the variation of both temperature due to ambient pressure changes during the 1-1/2 year test period. The temperature indication was lower with each decade of irradiation and a maximum temperature difference of -0.2 K is indicated by Lake Shore diode. Although irradiation had a small effect on the temperature sensors at 4.2 K, the effect can be expected to be considerable above 30 K, in the intrinsic range of the sensors, due to the increase in surface leakage currents as a result of irradiation above 1 Mrad.

Figure 12 gives the effect of irradiation on the fullscale output of a cold sensor at 4.2 K. The pressure sensor output variation is less than ±5 mbar at 4.2 K due to irradiation and this is within the calibration accuracy of the sensor.

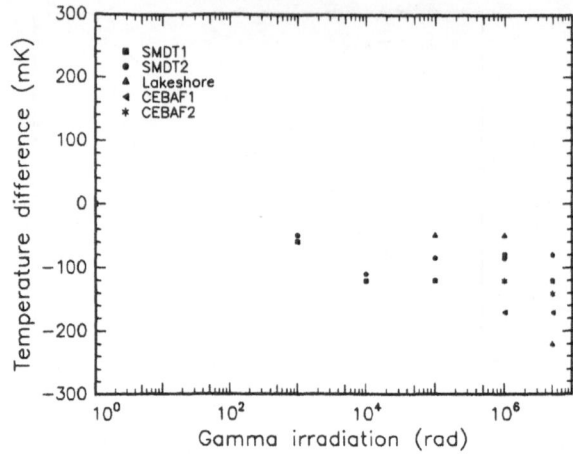

Fig. 11. The Effect of Gamma Irradiation on Si Diode Thermometers

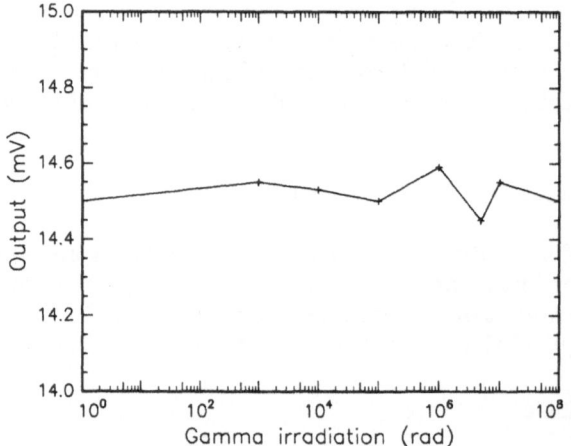

Fig. 12. The Effect of Gamma Irradiation on Cold Pressure Sensors

Magnetic Field Effects

The effect of high magnetic fields on semiconductor junctions have been reviewed elsewhere (1). Si diodes are known to be highly affected by the strong magnetic fields and are not recommended for use in such environments. However, if the magnetic fields are under 1.2 Tesla and perpendicular to the sensor, the effect of the magnetic field is found to be less than 0.2%(5).

The cold pressure sensor consists of four piezoresistors forming a wheatstone bridge configuration. Since the effect of magnetic field on the four closely matched resistors of the sensor can be expected to be the same, it can be assumed that the output of the bridge will not be affected by the magnetic field. This has been confirmed during the preliminary studies and further experiments are planned in the future.

APPLICATIONS

Applications for integrated cryogenic sensors include aerospace, particle physics research, industrial cooling, etc. To date, these sensors have been successfully employed in advanced propulsion studies, in LHe level measurements for particle accelerator development, and level measurement for cryogenic tank gaging.

Small size, low power dissipation, and rugged design suggest that there are a variety of research and commercial applications.

CONCLUSIONS

Integrated cryogenic pressure-temperature, level-temperature, and flow-temperature sensors have several advantages. PSI has recently developed such integrated sensors and is in the process of introducing them into the market. These integrated sensors have low measurement uncertainties:

o temperature - ±20 mK
o pressure - ±0.25% F.S.
o level - ±0.5% F.S.

The accurate measurement of mass flow requires the simultaneous measurement of several fluid parameters (fluid flow and density and/or temperature). PSI is currently developing a cryogenic mass flow meter based on the integrated cryogenic pressure-temperature sensor. These integrated sensors can be used in radiation and magnetic field environments. However, the integral temperature sensor will have unacceptable uncertainties in temperature measurement in magnetic fields over 2 Tesla.

REFERENCES

1. Rao, M. G., "Semiconductor Junctions as Cryogenic Temperature Sensors," Temperature It's Measurement and Control in Science and Industry, Vol. 5, American Institute of Physics, 1982, pp. 1205-1211.
2. Rao, M. G., "CEBAF LINAC Cryogenic Instrumentation Requirements and a Review of the Available Sensors," CEBAF Report 89-001. Linear Accelerator Conference Proceedings, Newport News, VA, June 1989.
3. Boyd, C., Juanarena, D., Rao, M. G., "Cryogenic Pressure Sensor Calibration Facility," Advances in Cryogenic Engineering, Vol. 35B, Plenum Press, New York City, NY, 1990, pp. 1573-1581.
4. Rao, M. G., "New Liquid Helium Level Sensor," CEBAF TN-0184, CEBAF, Newport News, VA, 1989.
5. Bahaj, A. S., et al, "The Effect of Magnetic Field and Orientation on The Southampton Diode Thermometer in the Temperature Range 4.2 - 300 K," Proc. of ICEC 12, Southampton U.K., Butterworths, U.K., 1988, p. 384.

PROCESS DESIGN FEATURES OF THE SSC MTL CRYOGENIC SYSTEM

Udo Wagner   Willem Keyer

Sulzer Cryogenics   Koch Process Systems, Inc.
CH-8401 Winterthur   20 Walkup Drive
Switzerland   Westborough, MA 01581

ABSTRACT

SSC magnets will be performance tested at the Magnet Test Laboratory (MTL) under conditions simulating the environment of the SSC main ring. The cryogenic system consists of five test stands and the associated equipment including cryogen storage, purification, thermal conditioning and helium refrigeration necessary to support the test program. The helium refrigerator must be capable of cooling up to ten magnets in various stages of clean-up, cooldown, warm-up and operation.

This paper discusses some of the process design considerations employed to achieve an easily controlled and thermodynamically efficient system suitable for the widely varying operation conditions.

GENERAL REQUIREMENTS FOR A SYSTEM SUITABLE FOR MAGNET TESTING

Testing magnets means, among other things, to operate them deliberately under conditions which are to be avoided under normal continuous operation, in order to experience the limits of their mechanical and electrical design.

For the process design, this means special emphasis must be placed on the following points:

Capability to Cover a Wide Range of Operating Modes

Operating modes will vary from nearly all liquefaction to all refrigeration, under maximum power as well as under reduced part-load conditions. Even "overload" conditions, characterized by an excessive flow of cold gas after a quench occurred, has to be taken into consideration.

Easily Controllable

Even with a multitude of pre-designed operating modes one has to bear in mind that the actual operation will only be met by chance. Additionally, it can never be expected that any operating mode will be static over an extended period of time.

The controls should not try to cover pre-calculated operating modes but should be suitable to adapt the plant, quickly and with greatest possible simplicity, to meet the changing requirements at the cold end.

Availability and Reliability

The SSC technical specification has defined minimum requirements for redundancy and technical standards.

All points considered, a suitable refrigeration system for the SSC-MTL must not be a system that tries to be most efficient by using a complex process design to achieve a specified capacity.

- It has to be a flexible plant capable of covering times of over and undercapacity.

- It must not be disturbed by varying loads at the cold end. On the contrary, it should be able to adapt easily to varying cooling demands.

- It should have a simple and straight forward control philosophy instead of a complicated control mechanism with interconnections in operating modes, multiple set points for each mode, etc.

- It must be capable to handle a certain amount of gas contamination.

In summary, it should be a system operating mostly unnoticed in the background, enabling everyone to concentrate on the task at hand, and that is testing magnets. Achieving a good plant efficiency is of minor priority but need not be completely neglected, providing that it will not impact plant simplicity and reliability.

PROCESS DESIGN

Boundary Conditions Outlined by the Technical Specification

The SSC specification for the MTL shows that its designers have considered all possible hazards for the plant which might result from the above mentioned special application. This resulted in requirements which established a frame for the process design.

- A multiple pressure cycle with staged compressors and a seperate expander loop, to minimize the potential endangerment of the expanders caused by fluctuations in the low pressure line due to quenches.

- Liquid nitrogen precooling.
- Oil-free turboexpanders.

General Plant Layout

For descriptive purposes the plant can be divided into three major systems:

System I

System I, directly associated with the magnets, is comprised of the following components: the end boxes providing all connecting lines and a subcooler for each individual test stand; the pump box providing the high circulating flow of supercritical subcooled helium; all the valving incorporated in these units for distributing the coolant flow and handling warm-up, cooldown and quenches; the interconnecting transfer lines to and from the other main systems.

System II

System II consists of: the warm helium storage buffer; the compressor set for the refrigerator/liquefier; the refrigerator/liquefier; and the liquid helium Dewar.

The system delivers the required cooling capacity for the cold end and is capable of re-introducing the returning flow, at any appropriate temperature level, into the low pressure return. The system can run independently, isolated from the other systems. It is designed for continuous smooth operation, adapting to the changing load requirements without need for much attention.

System III

This system contains all the necessary auxiliary subsystems: the CCWP system for Cleanup, Cooldown (to 80K), Warmup, and Purge; the refrigeration recovery system, which recovers the cooling capacity of the cold gas from the subcoolers in each test stand; the liquid nitrogen system.

REFRIGERATOR/LIQUEFIER PROCESS DESIGN DESCRIPTION

General Description

Re: P&F Diagram Fig. 1
    T-s Diagram Fig. 2

The helium is compressed from 1.05 bara to 17.9 bara in a two-stage compressor system. Each stage consists of two parallel screw compressors with identical capacity. The intermediate pressure is approximately 3.0 bara. The helium enters the coldbox at 17.1 bara and 308K. It is then cooled in the coldbox by heat exchange with the low pressure helium flowing back to the compressor set, by precooling with LN_2 , and by means of dynamic gas bearing expansion turbines.

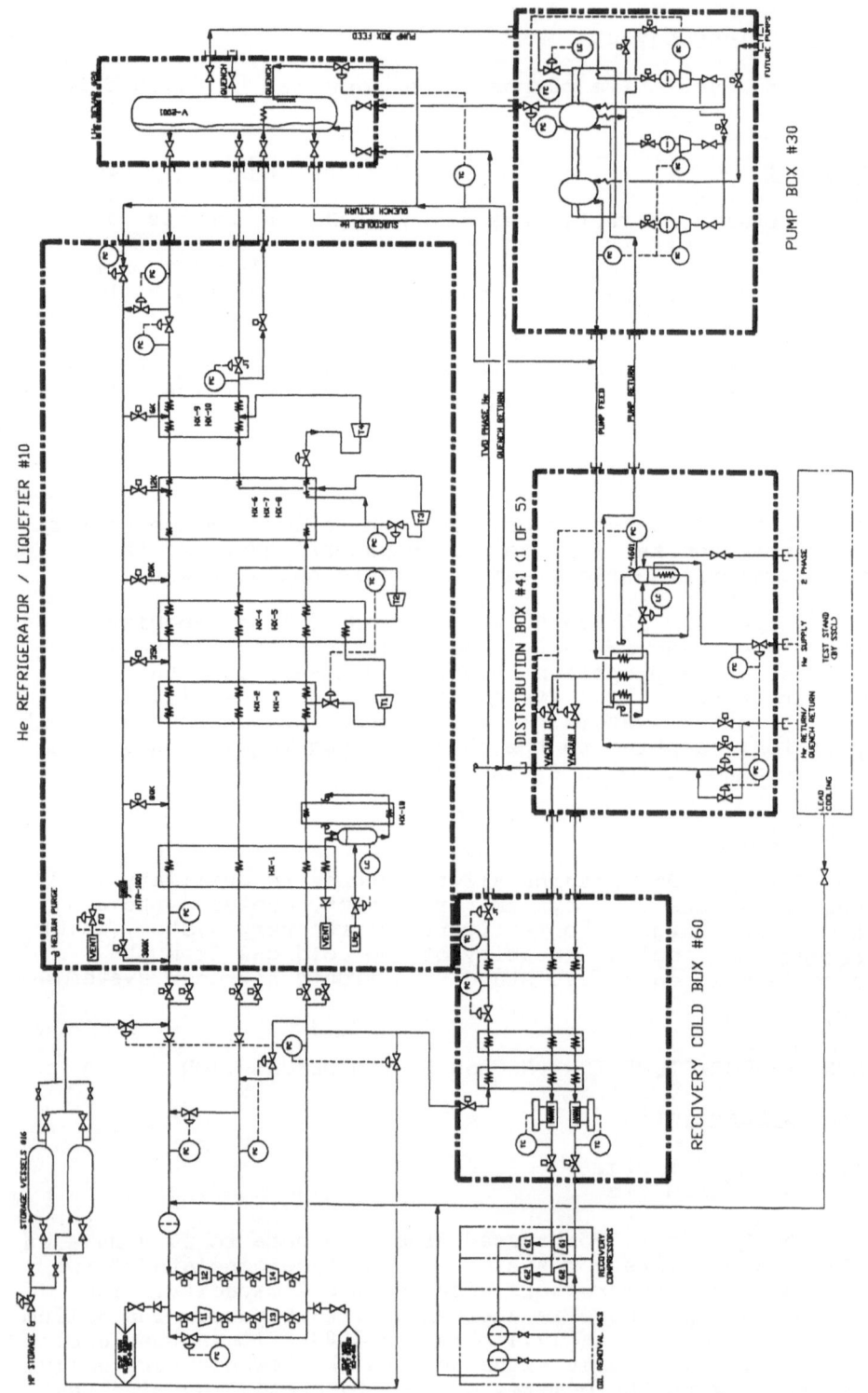

FIGURE 1. SSC MTL System Layout

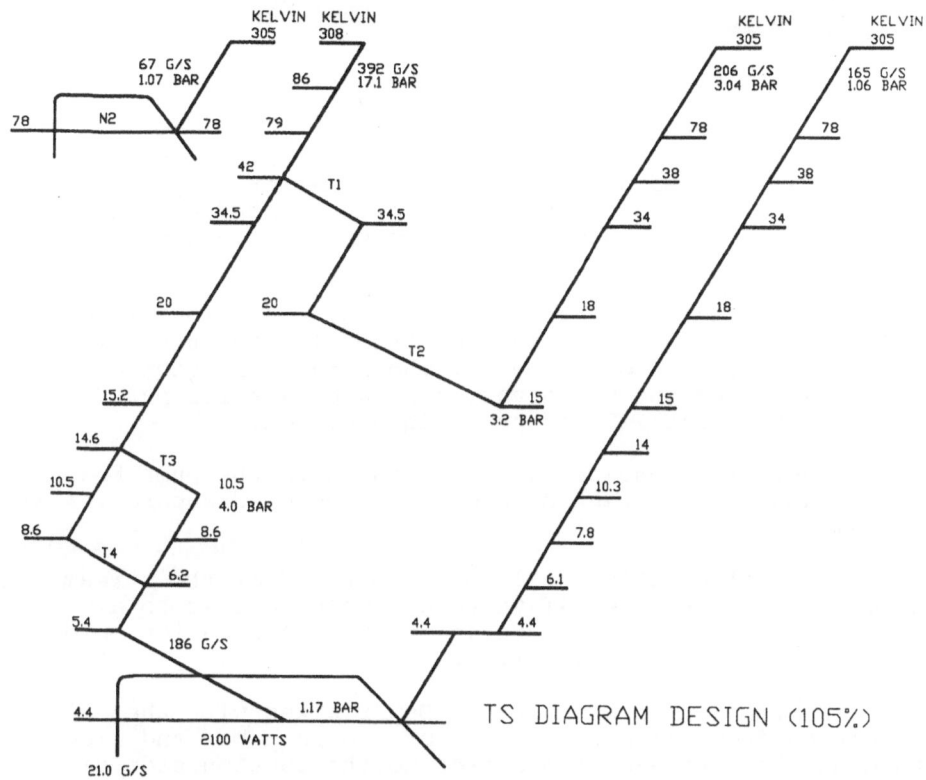

FIGURE 2. T-S Diagram for Refrigerator/Liquefier Design Mode.

The cooling cycle can be divided into three stages of cooling: LN2 precooling, the turbine cycle, and the Joule-Thomson (J-T) cycle.

The LN2 Precooling

Precooling with liquid nitrogen is achieved by means of a natural circulation loop, which drives the LN2 from the bottom of the boiler through the second plate fin heat exchanger core (HX-1B). The gas leaving the heat exchanger re-enters the boiler, displacing saturated vapor through HX-1. The high pressure helium gas leaving the second heat exchanger will have a temperature of approximately 80K.

The Turbine Cycle

The gas flow through the turbine cycle is separated from the main flow at a temperature level of about 40K. It passes through two in series expansion turbines (T1 and T2) with one heat exchanger (HX-4) between them. The flow exits the second turbine and enters the cold end of HX-5 at a pressure of 3.2 bara and 15K and continues to provide cooling until it exits the cold box to the intermediate pressure stage of the compressor set.

The J-T Cycle

The J-T stream incorporates two J-T expansion turbines in a staged parallel arrangement. 37% of the stream enters the first of the two J-T turbines where the helium is expanded to 4.0 bar and approximately 10.5K. Both J-T streams are cooled down in HX-8, where after the remaining high pressure stream is expanded in the second J-T turbine. The combined stream has then a pressure of 4.0 bar and a temperature of 6.2K.

In the last heat exchanger, HX-10, the stream is cooled to 5.4K. The J-T stream is then throttled through a J-T expansion valve into the Dewar. The liquid fraction of the helium entering the Dewar varies between 91% for pure refrigeration and 22% for pure liquefaction.

Liquid from the Dewar is passed on to the pump box which supplies the test stands with subcooled supercritical helium.

As an alternative, all (or a portion) of the stream normally entering the J-T valve can be by-passed to a subcooler in the Dewar, then on to the distribution boxes as subcooled supercritical helium.

The vapor generated in the Dewar mixes with the helium returning from the surge vessel of the pump box and flows through the heat exchangers back to the suction side of the compressor set.

REFRIGERATOR/LIQUEFIER EQUIPMENT DESCRIPTION

Compressors

Oil lubricated screw compressors from Sullair are used. The first stage consists of two parallel machines type C25LB704-22, the second stage of two parallel machines type C25SA704-26.

This compressor set is able to compress:

First Stage: 225 g/s helium from 1.05 bara and 308K to 3.4 bara.

Second Stage: 450 g/s helium from 3.2 bara and 308K to a maximum pressure of 19.5 bara.

The isothermal efficiencies of the compressors related to the shaft power are: 56% for the first stage and 53% for the second stage, at 100% capacity.

Turbines

As expansion machines centripetal turbines with dynamic gas bearings from Sulzer are used. In the turbine cycle, two machines type TGL32-28 are arranged in series; in the J-T cycle two machines type TGL22-12 are arranged in parallel.

The upper turbines are designed to provide full capacity in the liquefaction case where they achieve an isentropic efficiency of 75% (T1) and 78% (T2). The J-T turbines provide the cooling capacity near the temperature level which helps to minimize the necessary J-T flow. The parallel arrangement provides a good overall plant efficiency and has an advantage during part load operation, where the upper turbine (T3) is throttled to keep the efficiency of the lower turbine (T4) high.

The isentropic efficiencies of these turbines are 61% (T3) and 63% (T4), for the design case.

All indicated turbine efficiencies, which were used for the cycle design, are conservative compared to the measured performance of similar refrigerators.

Heat Exchangers

The heat exchangers are of plate fin construction. The cores for the low pressure helium stream are designed to provide a maximum overall pressure drop of only 75 mbar, as specified by SSC, so that a low Dewar pressure/temperature can be maintained with compressor suction pressure above 1 atma.

The total NTU's is approximately 130 for the whole coldbox set.

CONTROLS

The process is controlled by eight control loops which remain unchanged for all modes. Some of these controllers are designed to function only when a limiting set point is reached and do not control during normal operation.

Normal Operation

The slide valves of the compressors are only operated for startup and will be at a fixed capacity position during operation.

Table 1. Compressor Controls

Controlled Value:	Controlling Device:	Set Point	Remark
Suction pressure LP	By-pass valve LP-stage	1.05 bara	Constantly controlling
Suction pressure HP	By-pass valve HP-stage	2.70 bara	Limiting function
Delivery pressure to coldbox	Make-up valves to and from buffer (split range)	18.5 bara	Constantly controlling or limiting function

Table 2. Coldbox Controls

Controlled Value:	Controlling Device:	Set Point	Remark
Level in LN_2 phase separator	LN_2 feed valve	75% level	Constantly controlling
Outlet temperature out of turbine T2 (turbine cycle capacity)	Inlet valve into turbine T1	15.0 K	Limiting function
Pressure before J-T expansion valve	J-T valve	4.0 bara	Constantly controlling
Pressure of the LP stream at the last heat exchanger	Outlet valve of the Dewar	1.2 bara	Limiting function
Capacity of turbine T3 under part load conditions	Inlet valve of turbine T3 trying to control its up-stream pressure	12.0 bara	Limiting function

These control loops are sufficient to adapt the process according to the demands of the cold end. With the delivery pressure at a constant value, the coldbox capacity is fixed and the plant output will vary between liquefaction and refrigeration. As long as the capacity needs at the cold end are below the plant capacity, any excess capacity will result in liquefaction of helium into the Dewar. When the cold end capacity demand rises above the plant capacity, the result will be depletion of helium in the Dewar and a pressure rise in the warm buffer.

The capacity balance between refrigerator and cold end is visible by the position of the make-up valves, to and from the warm buffer.

The turndown of the plant capacity is achieved by lowering the set point of the inlet pressure of the high pressure helium supply to the coldbox. Thus the compression ratio, as well as the mass flow through the coldbox, are diminished in a very efficient and simple way.

Instead of maintaining different set points at the controller, the adaption of the high pressure can simply be made self-regulating by fixing the load valve from the warm helium buffer to the suction side at its closed position. Thus, the delivery pressure floats according to the capacity demands. Overcapacity means liquefaction into the storage Dewar and dropping of the delivery pressure. Lack of capacity means a higher return flow from Dewar and surge vessel and an increasing delivery pressure. The delivery pressure control limits the pressure to a maximum value.

Once the plant is cooled down and a certain amount of helium is stored in the Dewar, the make-up valve can be closed.

This kind of capacity control is not only simple, reliable, and quick adapting, but also a very efficient way of controlling the plant capacity. During turndown operation of 30% or less of nominal capacity, it is more efficient to operate only one of the two J-T turbines. Turbine T3 is simply switched off by controlling the turbine inlet valve on high pressure. If the pressure drops below 12.0 bara, the value is throttled to hold this pressure, and totally closed when a delivery pressure of about 11.5 bara is reached.

Quench Recovery

In this context, quench recovery means the reaction of the refrigerator/liquefier during and after a quench. According to the SSC specification, the majority of quench gas is directed to the liquid helium storage Dewar. When the quench vapor has reached a temperature of about 12K, it is switched to the coldbox quench return line at a comparable temperature level. With rising temperature, the gas is returned at higher temperature levels. For the refrigerator/liquefier this means a rise in the cold gas return stream, and an imbalance between the high pressure stream and the low pressure stream. The plant reaction will be:

-    The compressor bypass valves close to enable full capacity.

-    If the mass flowrate exceeds the compressor capacity, the outlet valve of the liquid helium storage Dewar will start to close, trying to maintain an acceptable value at the low pressure return.

-    The delivery pressure will rise, and depending on the plant capacity prior to the quench, will reach a maximum value.

-    With maximum delivery pressure and a high cold return flow, the plant reaches a maximum liquefaction rate.

The increase in LP helium flow results in a greater cooling capacity, consequently LN_2 precooling will be reduced. The mass flow through turbines T1 and T2 will diminish according to the decreasing outlet temperature of T2.

As explained earlier, with floating delivery pressure the plant will adapt itself to the quench condition, without deviating from the normal control philosophy.

Flexibility

Flexibility of the plant means the ability to shift between refrigeration and liquefaction while maintaining maximum possible capacity for both modes.

Refrigeration is mainly limited by the first stage compressor capacity and the heat exchanger effectiveness, especially at the cold end. For this mode, the arrangement

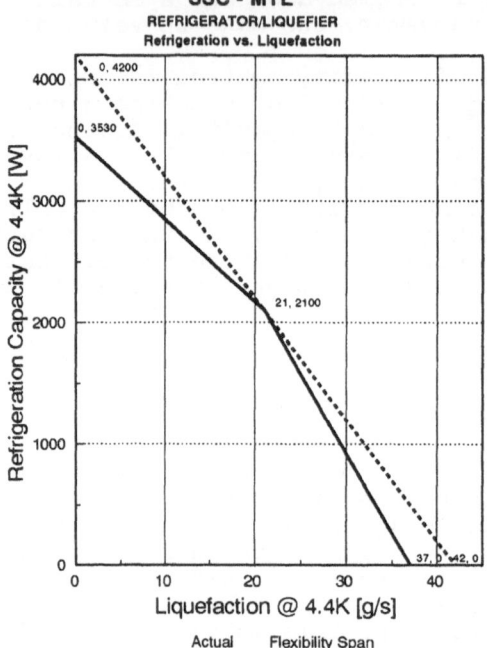

FIGURE 3. Refrigeration vs. Liquefaction

of the J-T turbines provides maximized capacity characteristics.

Liquefaction is mainly limited by the precooling capacity. This mode presents the design bases which determines the liquid nitrogen precooling and the sizing of turbines T1 and T2, which are able to operate with high flow rates at relatively high temperatures.

For the mixed mode, the plant has the following design data: 2100 W   refrigeration at 4.4K, and
            21.0 g/s liquefaction at 4.4K

As pure refrigerator or pure liquefier, the plant will reach: 3530 W   maximum refrigeration at 4.4K, or
            37.0 g/s maximum liquefaction at 4.4K

(See Figure 3, Refrigeration Capacity Versus Liquefaction Capacity)

Efficiency

To calculate the efficiency of the process and compare such different capacities as liquefaction, refrigeration, gas compression, nitrogen consumption, and electric motor input power, it is necessary to agree on a basis of comparison. For this application, it is common practice to calculate the exergy of the single loads and add them to each other.

Furthermore, it is necessary to make an assumption concerning the equivalent power value of the liquid nitrogen consumption. An overall efficiency of 33% relative to Carnot for the liquid nitrogen production seems realistic and defines a basis for comparison.

For the design mode, the exergies (related to ambient temperature of 308K) are:

Output:

Item	Exergy
2000 W refrigeration at 4.4K	134.9 kW
20 g/s liquefaction at 4.4K	135.2 kW
30 g/s He gas from 1.05 to 17.1 bar	53.1 kW
Total Carnot Power Output	323.2 kW

Input:

Item	Exergy
Expected motor lead power	1198.0 kW
LN_2 consumption of 70 g/s at 33% of Carnot	162.4 kW
Total Input	1360.4 kW

These figures result in a plant efficiency of: **23.8%**

To make the process efficiency more transparent, it can be split into separate parts to compare how the major losses are distributed.

Process Part	Efficiency(Expected Values)
J-T Loop up to 15K	68.3%
Turbine Cycle T1 and T2	75.1%
LN_2 Cooling in the First HX	91.2%
Total Coldbox Process	46.8%

The total coldbox efficiency of 46.8% is characteristic for modern multi-purpose plants, where a number of compromises have to be made to ensure flexibility. Single purpose plants without LN_2 precooling can have higher efficiency.

CONCLUSION

The SSC MTL's refrigerator/liquefier plant was designed to be simple in operation while still meeting a multitude of operating modes in a very efficient manner. The self-adapting control philosophy of the plant allows the operator to focus on what is required, and that is testing magnets.

CONTROL SYSTEM DESIGN CONSIDERATIONS FOR THE

SSC MAGNET TEST LABORATORY CRYOGENIC SYSTEM

Herb Peters and Max Kozyrczak

Koch Process Systems
20 Walkup Drive
Westborough, MA 01581

ABSTRACT

The Magnet Test Laboratory (MTL) Cryogenic System will
be used to support prototype and production testing of
dipole and quadropole magnets for the SSC. The MTL
Cryogenic Control System must be capable of providing
remote and unattended operation of the cryogenic system.
This paper discusses some of the design considerations
involved in the selection and design of a control system
that will have changing requirements over time; and must
eventually interface with the overall MTL control system
for which the requirements have yet to be determined.

INTRODUCTION

The MTL Cryogenic Systems will be operating prior to
the overall implementation of the control system at SSCL.
Even though the MTL Cryogenic System controls must be able
to stand alone and carry out all of the modes of operation
required to safely operate the Cryogenic Systems for
testing of SSC magnets, it is still only one of several
computer controlled systems that will be needed for magnet
testing. These other computer controls are for power
supply, quench protection, and magnet measuring. All of
these systems must be integrated under a supervisory
control system managing the overall process of magnet
testing. Therefore, the MTL Cryogenic Control System must
satisfy the additional requirement of interfacing with an
as yet undetermined supervisory control system.

DESIGN APPROACH

The purpose of the MTL Cryogenic System and its
controls is to qualify a large number of magnets in the
most expeditious manner. It must be highly reliable so as
not to be the limiting factor in the magnet production
schedule.

The Cryogenic Control system must be simple to operate but provide the necessary flexibility to handle the MTL transient or off-design conditions. The design for operational flexibility and controllability utilizes the following factors in determining the control system philosophy:

o Provide for remote operation and data acquisition from a central control console for ease of operator interface and system controllability.

o Provide sufficient process instrumentation to understand and optimize the cryogenic system performance.

o To avoid interactions between systems each cryogenic subsystem must be decoupled as much as possible through the use of appropriate controls.

o Control loops must be easily reconfigurable. It is unrealistic to expect to fully understand the operational and control requirements of the cryogenic system up front or to expect them to be invariant with time. Therefore, control loops must be part of the control system software, not hardware, to allow for this flexibility.

o Distribute the front-end control and input/output interface to the local areas where the actual process equipment is installed.

o Safety shutdown circuits will remain hardwired at process equipment.

o Provide isolation between systems to allow independent maintenance of each subsystem.

CONTROL SYSTEM CRITERIA

Utilizing the above mentioned design approach, control system criteria were developed for both the hardware and software as follows:

o The control system must be able to carry out all modes of operation required to safely operate the cryogenic system for testing SSC magnets.

o The control system must be able to expand easily to handle future test stands and possible alternate modes of operation for the MTL as determined by the SSCL.

o Ease of programming and initial definition of the configuration for the cryogenic process control requirements is essential. Since the MTL control system must be operational in a relatively short timeframe and must also maintain a high degree of flexibility, developing any original software for process control techniques cannot be advocated because of the unpredictable amount of time required for completion.

o   The control system must have proven performance and
dependability for both the hardware and software in either
cryogenic, industrial or process control applications.

o   Software ability to communicate with future MTL
supervisory computer control systems in a manner which will
enhance the involvement of the SSCL personnel in the
development of the controls software for the cryogenic
systems is also required.

COMPUTER CONTROL SYSTEM SELECTION

      At this point in the procedure of the design
consideration, specific technical information from detailed
Process and Instrumentation Diagrams (P&ID's) concerning
the number, location and duty of the various analog and
digital input and output signals were gathered and
tabulated. This information included the number of process
control loops, the input signal conditioning requirements
for process instrumentation, and the number and type of
equipment status indicators.

      The above mentioned technical information and the
general design approach coupled with the SSCL MTL design
specification formed the basis for a design and procurement
specification for the Cryogenic Control System. This
specification was furnished to ten (10) computer control
system suppliers whose systems were characterized into
three groups:

o   Programmable Logic Controllers (PLC), generally configured
as stand-alone systems.

o   Distributed Control Systems (DCS), closed-loop analog
control based.

o   Mini or Microcomputer based systems with open-ended
configuration or central processor/PLC.

EVALUATION CRITERIA

      Koch Process Systems (KPS) recognized that control
system selection is based on many objective as well as
subjective issues. Some of the items are technical in
nature, while others depend upon user familiarity and
preference. All control system designs were evaluated with
only one thing in mind, the potential to meet the cryogenic
computer system specification. The evaluation was
performed without any preconceived vendor preferences; and
the various methods of accomplishing the control
requirements of the MTL Cryogenic System were considered
carefully.

      The computer control system proposals were then reviewed
and consideration given to the following criteria:

o   Process control performance: Speed, input/output
capacity, memory capacity, and interface with process
instrumentation.

o Operating systems experience: The total number of installed systems plus the variety of applications. The design personnel and other individuals responsible for operating the systems were contacted to get their impressions regarding overall system performance.

o Ease of programming and configuration of process control parameters.

o Software performance and adaptability to future MTL control requirements.

o Corporate strength of the vendor: Size and financial stability.

o Years of operation (longevity).

o Flexibility and changeability: Such that next year's improved processor would readily adapt to last years hardware.

o Ease of expansion in consideration of the inevitable growth of any intelligent system. It must have the ability to communicate with a supervisory computer control/data acquisition system.

o Technical approach and support critical to before, during and after installation.

o Cost.

Based on careful consideration of the above criteria, the Texas Instrument Model 70 TISTAR Computer Control System has been selected as the control system for the MTL Cryogenic System.

TEXAS INSTRUMENTS

The TISTAR control system is a process controller-based distributed control system which provides centralized control and reporting capabilities along with the ability for future communication to and from a supervisory computer via an optional Ethernet.

The TISTAR system performs the following:

o Automatically suppresses and enables auxiliary alarms.

o Logs all operator actions.

o Integrates sequential and regulatory tasks.

o Provides connectivity between components, resulting in economical and reliable control.

o Enhances overall system performance by distributing a greater percentage of control to the process controller level.

The TISTAR system consists of three major subsystems for quick, simple and seamless integration of discrete and continuous control functions. Description and features of these major subsystems are as follows:

1. The Command Center is the supervisory control center of the system. This is where the operator interface and engineering functions are performed.

Here the communications and reporting software packages generate, process and distribute information and commands bi-directionally up-from and down-to the Process Controllers, and via an optional link up-to and down-from, a supervisory computer or personal computers.

The nucleus of the Tistar Command Center is the Relational Database Management System which receives and stores data from the Process Controllers in addition to receiving and storing data entered manually from the workstations.

2. Process Controllers use a parallel processor architecture to execute multiple control programs. The main CPU performs relay ladder type programs and input/output scans, while a special function CPU performs PID loops and high level statement driven programs. Interlocks and status are made available between both CPU's by using the highspeed, multi-processor architecture of a VME-type bus.

The Process Controllers for the MTL Cryogenic Control System are the Model 545 and 565 Controllers. These controllers interface with a wide variety of functions over a high speed input/output bus.

Development of programs for the Process Controllers is accomplished with standard menu-driven software called TISOFT or with an Applications Productivity Tool (APT). TISOFT is a controller level programming tool which allows for both online and offline Relay Ladder Logic (RLL) programming. The Applications Productivity Tool (APT) is a package which provides structured designs of continuous, discrete and sequential control schemes using graphical techniques.

For the MTL Cryogenic Control System, all software configuration and provision of the documentation package will be performed by Koch Process Systems, Inc.

3. The Communications Network is TIWAY I which accesses, evaluates and modifies control data stored in the separate memories of the interconnected Process Controllers. TIWAY I provides a central collection point for information and automatically corrects missing and duplicate messages which is a feature essential to safety control. For the MTL Cryogenic Control System, a redundant TIWAY I data highway is provided.

CONCLUSION

The TISTAR distributed control system has many more detailed operational features which have not been addressed in this paper due to the narrow focus of the topic. The key features of the TISTAR distributed control system, which directly relate to the MTL Cryogenic Control System design, have been highlighted.

The TISTAR control system is a commercially available unit with standard menu driven software, which is easily configured at the Process Controller level in relay ladder logic format. The TIWAY I communications network links the Process Controllers with the TISTAR Command Center data exchange unit. It is within this unit that third-party independently developed software compatible with the TISTAR system protocol can be utilized by SSCL in the future.

The reliability and flexibility of these system features provide the SSCL with the ability to operate the MTL Cryogenic Control system and acquire a database of operational experience which may be integrated into the development of control software systems for subsequent cryogenic plant controls.

REFERENCES

1. RFP-SSC-90A-01107 Superconducting Supercollider, Request for Proposal MTL Cryogenic System, Section 2, Technical Specification.

2. Supercollider 1, Proceedings of the International Industrialization of Superconducting Superconducting Super Collider, New Orleans, 1989, "SSC Refrigeration System Design Studies", M. S. McAshan.

TRANSIENT SHIELDED CRYOGENIC CONTAINERS

A. P. Varghese and B. X. Zhang

Air Products and Chemicals, Inc.
Gardner Cryogenic Department
2136 City Line Road
Lehigh Valley, Pennsylvania 18001, USA

ABSTRACT

Transporting and storing industrial gases in the liquid phase is safer and more economical than in the high pressure gas phase. However, due to the extremely low liquid phase temperatures of the industrial gases and the present state of insulation techniques, the storage of these gases as liquids has been restricted to relatively short durations.

A totally new cryogenic container, utilizing a transient shield developed by engineers at Gardner Cryogenics, has improved the storage duration considerably and has been granted a patent. For cryogenic containers used for the storage of cryogenic liquids for occasional use as gas at ambient temperature, the heat leak can be substantially reduced by the use of a transient thermal shield. This transient thermal shield, through variations in its temperature with time, will extract, retain and release refrigeration from occasional or cyclical flow of fluid being warmed up. In such a process, the shield will intercept the heat leak into the storage and thus extend the storage time without the use of any energy to produce refrigeration for this purpose. Testing of a prototype container for liquid hydrogen storage has indicated an improvement of five hundred percent in storage time. Cryogenic containers with transient thermal shields are already finding applications in replacing high pressure gas storage, very low volume usage, storage of high purity cryogenic fluids, and in localities where venting of stored product is not allowed. This paper will describe the construction and the test results of a series of transient shielded cryogenic containers.

INTRODUCTION

Fluids can be transported and stored as a compressed gas, or liquid. For example, compressed gas helium generally requires about fifty-three times its own weight in transportation or storage equipment, while liquefied helium needs only about five times of its own weight in transportation or storage equipment. Present highway regulations on weight and size limit the maximum quantity of helium in one package to about 15,000 *lbs.* (6,804 *kgs*) which obviously has to be in the liquid phase. The cost advantages in transportation coupled with purity considerations have made liquefied gases the prime choice for industrial and scientific applications where gases are unavailable from production facilities or from pipe lines. However, storage of cryogenic fluids for low consumption applications has been a problem due to the unavailability of high performance storage equipment. Engineers at Gardner Cryogenics, Department of Air Products and Chemicals, have developed and patented a Transient Shielded Storage System for cryogenic fluids to extract the refrigeration from stored cryogenic fluid being warmed up for eventual use as gas at ambient temperature.

BACKGROUND

In commercial and scientific applications where cryogenic fluids are eventually used in their gaseous phase, the refrigeration in the cryogenic fluids is wasted to the ambient in the vaporizers of present user storage facilities. In addition, an amount of refrigeration equivalent to the total latent heat at atmospheric pressure of one to three percent of the capacity of an average conventional storage and transportation container is generally lost to the atmosphere as normal heat leak into cryogenic fluids.

In applications where the gas consumption is lower than the gas generated by the normal heat leak into the container, the extra amount of gas has to be vented out to maintain the storage pressure below the limits specified by the design of the storage system. Loss of gas in this manner can be minimized by using the transient thermal shield system developed by Gardner Cryogenics.

Normally, a conventional cryogenic container consists of two vessels (see Figure 1). The inner vessel, which contains the cryogen, is made of austenitic stainless steels, high nickel steels, or aluminum alloys due to the loss of ductibility of most other comparably or lower priced materials at cryogenic temperatures. It is usually covered with multilayer or powder types of insulation. The inner vessel and the insulation will be enclosed in a vacuum jacket made of carbon steel or stainless steel; steel being used for its ability to protect the container and the insulation in case of a fire. The insulation reduces the radiative heat leak into the inner vessel, while the high vacuum system accomplishes the reduction of gas conduction and convection heat leaks. Conventional containers of capacities up to 33,000 gallons have a normal evaporation rate between 0.5 and 3 percent of the total capacity per day, the 3 percent being the rate for containers of capacities around 1,000 gallons for inexpensive cryogenic fluids.

GAS SHIELDING

For a conventional cryogenic container, certain amount of heat, depending on the construction of its insulation system, will penetrate the insulation system reaching the inner vessel. The internal support system as well as the piping system also introduce heat leak into the inner vessel. The contained liquid cryogen will be vaporized at a rate inversely proportional to its latent heat, and the gas generated has to be consumed immediately or vented to the atmosphere, in order to prevent a pressure build-up within the inner vessel.

In principle, the gas shielding utilizes the vaporized gas to refrigerate one or more thermal shields surrounding the inner vessel before the gas is consumed or vented. Being properly designed, the shield(s) will intercept part of the total heat leak, and hence reduce the vaporization rate. The gas shielding operates on the basis of a nearly uniform, continuous flow of boiled-off gas through the shield.

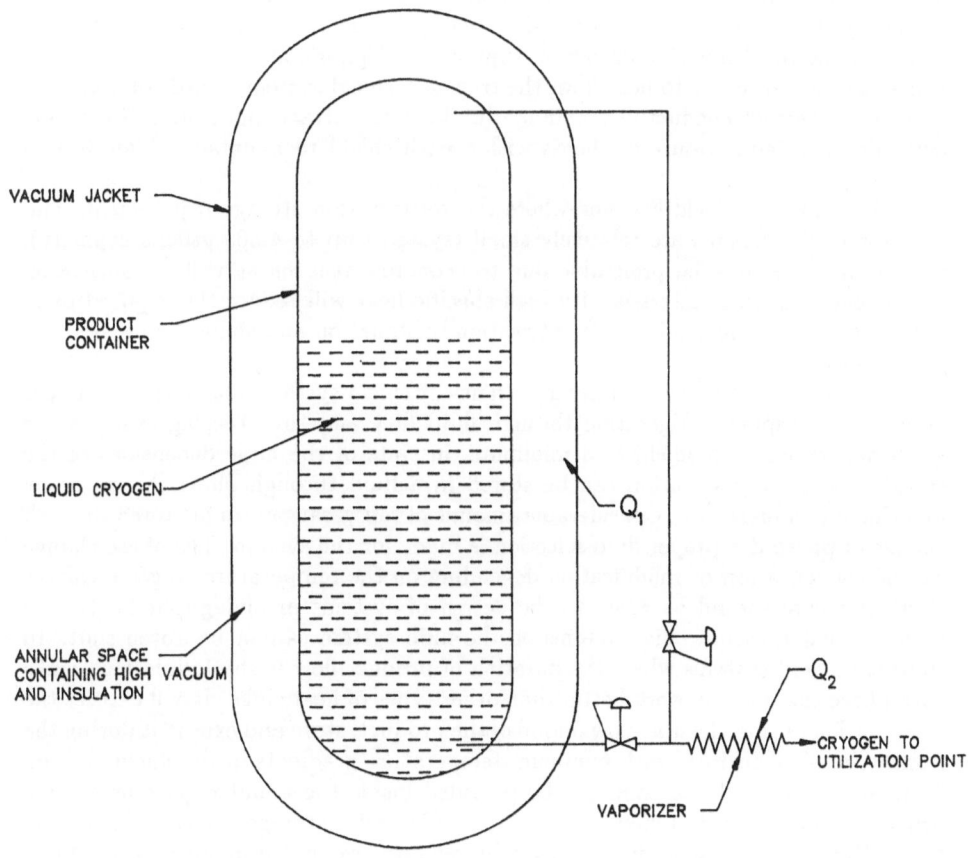

VACUUM JACKET

PRODUCT
CONTAINER

LIQUID CRYOGEN

ANNULAR SPACE
CONTAINING HIGH VACUUM
AND INSULATION

Q_1

Q_2

CRYOGEN TO
UTILIZATION POINT

VAPORIZER

TOTAL REFRIGERATION LOST = $Q_1 + Q_2$

Figure 1. Conventional cryogenic storage system.

Rarely does the condition of a nearly uniform continuous flow ever exist with a customer station, wherein, the product consumption can be erratic and often dominated by a work week requirement. Normally, fluid product is drawn and used only 40 hours or less out of a 168 hour week or at best 24 percent of the time. The transient shields reduce the effect of these variations in the flow of fluid product on the interception of heat leak into the cryogenic container. This is accomplished through the extraction and storage of refrigeration from the flow of product from the container. The transient shields are placed in the annular space between the inner vessel and the vacuum jacket at optimum locations dictated by the service pattern and the type of insulation used. While the insulation, supports and piping are made from materials which are very resistive to heat flow, the transient shield system is made of materials with high thermal conductivity and specific heat to extract and store refrigeration and minimize temperature gradients within each shield to maintain a high system efficiency.

In transient shield systems where the refrigeration storage requirements and the system dimensions are relatively small (systems up to 4,500 gallons capacity), solid heat sinks may be preferable due to economic reasons as well as fabrication convenience. Again, materials with high specific heat will reduce the required mass of the heat sinks and render their fabrication, installation and supporting easier and less expensive.

Fluid heat sinks are used in large transient shield systems due to their inherent ability to transport refrigeration through fluid flow and thus keeping temperature gradients within each shield to a minimum, in spite of the large dimensions of the shield system. Refrigeration can be stored in a fluid through phase change or by lowering its temperature. Considerable amount of refrigeration can be stored through change of phase if a proper fluid selection is made for the system. The phase change can be condensation or solidification depending on the temperature ranges involved. Particular care should be given to the design of systems involving solidification of fluids so that pressure relief systems on the fluid system cannot be frozen shut. In transient shield systems where the duration of product flow is short, fluid heat sinks with phase change may work better than other types of heat sinks. It will extract the refrigeration during the brief flow period of the product, store and expend it during the lengthy, no-flow periods, with minimum temperature gradients in the shield system. Storage for the fluid heat sink can be provided inside the annular space or outside depending on its mode of operation and size. The inside storage is more efficient due to a better thermal insulation system as the vacuum level in the annular space will be lowered by the extremely low temperature of the stored product. For liquefied helium storage, helium is the fluid heat sink of choice due to its high specific heat and its ability to stay in the gas phase at liquid helium temperatures, if the mode of operation preferred for the fluid heat sink is change in sensible heat. Nitrogen will be a good choice for change of phase operation of the fluid heat sink due to its low condensing and solidifying temperatures to facilitate efficient operation of the transient shield system. Latent heat of vaporization, latent heat of fusion and changes in sensible heat of nitrogen can be used to store the refrigeration in a transient shield system for liquid helium. Extreme care shall be exercised in the design and operation of fluid heat sink systems to maintain high purity of the fluid heat sink so that moisture or other high freezing point substances will not enter the system and freeze in the fluid passages disabling the operation of the transient shield system.

SYSTEM OPERATION

To describe the operation of the transient shield system, consider an operation cycle of a week on the liquid helium system shown in Figure 2. As illustrated in this figure, an amount of heat Q_3 flows in through the vacuum jacket, and part of the insulation system, reaching the transient shield. A part of this heat inflow, Q_4, is retained in the transient shield to increase the enthalpy or the internal energy of the transient shield system depending on whether it works on an isobaric or an isochoric basis. The remaining heat inflow

$$Q_5 = Q_3 - Q_4 \tag{1}$$

goes into the liquid helium container to increase the enthalpy, internal energy or both of the contained helium depending on whether the container is maintained at constant pressure, sealed up or at varying pressures with occasional flow which is the most common case. In a balanced system, a helium mass flow W_g from the vapor side will meet all customer requirements and bring the container pressure down within a certain specified pressure cycle range. In cases when the vapor generated by the heat leak exceeds the consumption requirements, pressure in the inner vessel will keep building up, and within certain time, it will reach the threshold of the pressure relief device. The excessive vapor will be vented to the atmosphere to maintain the pressure setting. This indicates an under usage of helium with respect to the design conditions. Conversely, a final pressure below the specified range will be an indication of a higher helium usage than the design consumption rate, resulting in eventual vaporization of liquid to meet customer helium requirement. In such a scenario, the back pressure regulator on the vapor side will close and the pressure regulator on the liquid side will open and a flow of liquid helium, W_l, will occur. Both gas flow W_g, and liquid flow W_l will pass through the transient shield system transferring its refrigeration and carrying out the amount of heat Q_l, which is accumulated over the week and possibly subcooling the system if the liquid flow W_l is substantial. While final pressures outside the specified range can affect the efficiency of storage of the system, it still will be much better than the conventional non-shielded systems available in the market.

Tests were conducted on a transient shielded 3,000 gallon liquid helium customer station designed for a cyclic withdrawal of 28,000 standard cubic feet per month at an operating pressure of 135 $psia$. The cyclic test consisting of 40 hour on out of 168 hours with a flow equivalent of 28,000 standard cubic feet per month yielded an average heat leak of 6 Btu/hr compared to 30 Btu/hr for a conventional gas-shielded 3,000 gallon customer station.

1,500 gallon transient shielded customer stations with an average heat leak as low as 8.4Btu/hr with a cyclic flow of 7,854 standard cubic feet per month of hydrogen is now available. With the required hydrogen consumption as low as 7,854 standard cubic feet per month or 7 cylinders per week, hydrogen has become accessible even to the low volume users.

SAFETY CONSIDERATIONS

As an illustration of a typical transient shielded cryogenic system, the flow diagram is shown in Figure 3. If for any reason, the container pressure rises above the specified limits, a back pressure regulator will relieve it with the flow going through the transient shield system, leaving the refrigeration to reduce the heat leak and thus slow down or eliminate the pressure rise. A redundant pressure relief system which

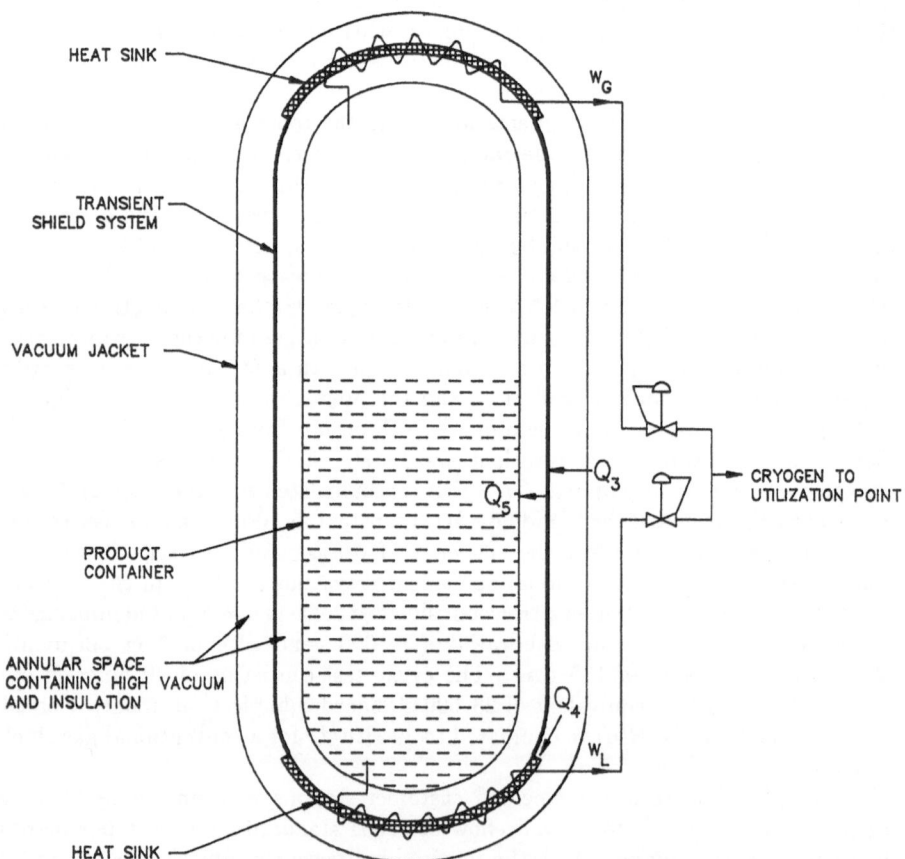

Figure 2. Transient shielded cryogenic storage system.

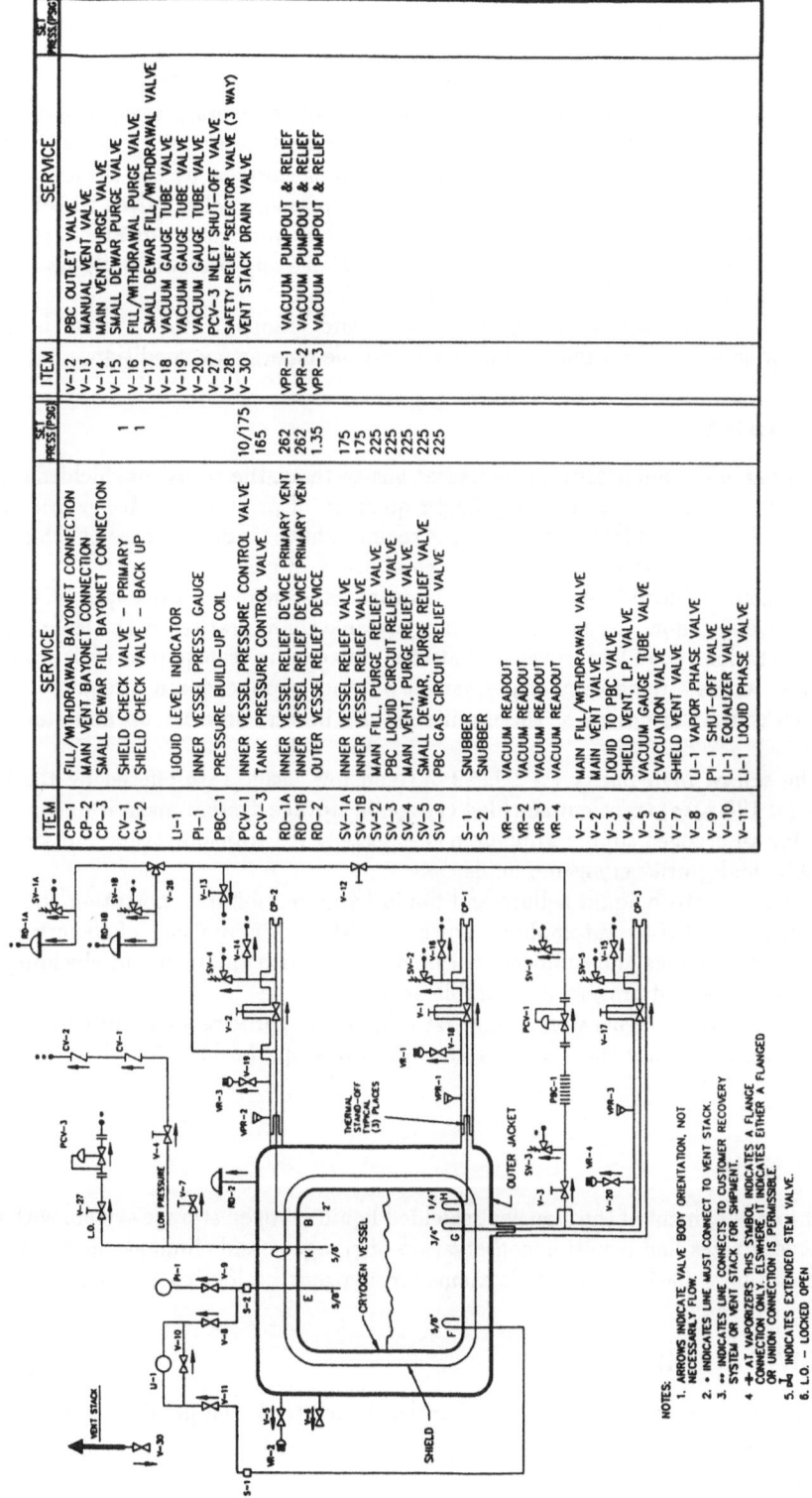

Figure 3. Flow diagram of liquid cryogen transient shielded storage system.

will always provide at least one pressure relief valve and a parallel rupture disc open to the liquid helium storage system is installed on each transient shielded helium customer station. These devices are designed to keep the maximum pressure under the test pressure of the system even when the storage system is engulfed in fire with a loss of vacuum. An optional rupture disc set at a pressure between 150 and 200 percent of the maximum allowable working pressure of the storage system could also be provided. It could be connected to a separate vent stack as a back-up. Any portions of the piping which can trap liquid helium or cold gas by the closing of two or more valves or other devices are protected with pressure relief valves to keep the pressure in this portion of the piping below their test pressure. An annular space relief device is provided on the vacuum jacket to relieve any pressure build-up in the vacuum space due to any highly unlikely internal leaks.

The transient shielded cryogenic storage and transportation system is thus designed for safe operation under almost all possible emergency conditions.

APPLICATIONS

Compared to high pressure compressed gas systems, the transient shielded cryogenic system can transport and store larger quantity of product at a lower price per cubic feet. The low heat leak and usage rate required have made the transient shielded cryogenic systems economical to even the small users of gaseous products.

Transient shielded liquid helium or hydrogen systems with capacities of $1,500$, $3,000$ and $4,500$ gallons are available in transportable configurations with Department of Transportation (DOT) approval so that the units can be transported with product to locations where a temporary or permanent product supply source is required. The totally self-contained design of these units results in very low set up and start up times.

The safety, high purity (> 99.999 %) and low usage rate offered by the low pressure (< 175 $psig$) transient shielded cryogenic storage system make it universally suitable for all cryogen supply applications especially for transportation and storage of valuable, high purity cryogenic fluids.

Systems to store liquid helium and liquid hydrogen for transportation applications such as MAGLEV, automobiles, trucks and ships, where usage of the cryogens may not be continuous and uniform, can be designed with the transient shielding to conserve cryogens and prevent undesirable venting.

Slush hydrogen generators using the evaporation process and slush hydrogen storage systems can use the transient shielding to keep the heat leak low for high slush yield.

CONCLUSION

The development of the transient shielded liquid cryogen storage system with its low heat leak rates and resulting conservation of cryogens and elimination of venting will open up new markets and applications for cryogens besides conserving energy.

ACKNOWLEDGEMENT

The effort devoted by P. F. Kraihanzel in preparing the graphics is cordially acknowledged.

REFERENCES

1. R. F. Barron, "Cryogenic Systems", 2nd Ed., Oxford University Press, New York Clarendon Press, Oxford (1985).

2. "Handbook on Materials for Superconducting Machinery, Mechanical, Thermal, Electrical, and Magnetic Properties of Structural Materials, Metals and Ceramics Information Center", MCIC-HB-04, Battelle's Colombus Laboratories (1974).

3. Cryodata, *GASPAK* user's guide, version 2.2, Niwot (1989).

4. R. H. Herring and A. P. Varghese, "Method and Apparatus For Storing Cryogenic Fluids", United States Patent No. 4,877,153, (October, 1989).

14. Detectors II

CHARGE CORRELATION MEASUREMENTS OF DOUBLE-SIDED DIRECT-COUPLED SILICON MICROSTRIPE DETECTORS

M. L. Wood, J. F. Kuehler, G. R. Kalbfleisch,

D. H. Kaplan, and P. Skubic

Department of Physics and Astronomy
University of Oklahoma, Norman, Oklahoma 73019

A. D. Lucas and C. D. Wilburn
Micron Semi-Conductor, Ltd.
Lancing, Sussex BN158UN, England

ABSTRACT

Charge correlation measurements of several Micron 38 mm by 58 mm by 300 micron thick double-sided DC-coupled microstripe detectors have been made. They have been bench tested with a Sr-90 source, with the detectors operated at -22 degrees Celsius. The correlation of the charges collected from both the diode ("holes") and the ohmic ("electrons") stripes are equal within a signal to noise resolution of 20:1 (i.e., 1200 electrons noise) using common-mode subtracted double-correlated sampling with the Berkeley SVXD readout chip.

INTRODUCTION

It is becoming increasingly clear that double-sided silicon microstripe detector systems can play a crucial role in the future of high energy physics. With this belief in mind, the University of Oklahoma has been investigating these devices for some time.[1,2]

During recent months we have been conducting a series of in-house tests of the performance and charge correlation characteristics of certain double-sided DC coupled Silicon devices in tandem with an associated Berkeley designed VLSI readout system. This work is part of an ongoing research and development program that has involved both beam tests with collaborators[3,4,5] and source tests at Oklahoma, the goal of which is the development of double sided Silicon microstripe detector systems suitable for deployment at the SSC.

Advantages of Double Sided Silicon as SSC Tracking Detector Elements

The need for both high quality inner and outer tracking at the SSC has been established many times.[5,6] Very briefly, the general advantages of the use of Silicon microstripes for SSC tracking include very high resolution (5 microns has been achieved), very high efficiency with good signal to noise characteristics, transparency, the absence of complicated associated gas systems, reliabilty and compactness. The use of VLSI associated readout with a high resolution large scale system allows the efficient handling of the resulting large amount of data and will be *de rigeur* at the SSC.

Currently, most silicon devices are about 300 microns in thickness and produce about 24,000 electron-hole pairs per normally incident Minimum Ionizing Particle (mip). Typical signal-to-noise (s/n) ratios for single sided devices operated at room temperature under near optimum conditions have been in the range of 10 - 30; in general s/n ratios are expected to improve with decreasing temperature, although until recently this feature has largely remained unexploited.

Interestingly, the use of double sided devices offers several significant advantages over the use of "equivalent" multiple layers of single sided devices. The most obvious such advantage lies in the reduction of the necessary thickness of silicon needed for good signals for both coordinates, thus reducing conversions and multiple scattering. Associated with this is also a shortening of the necessary extrapolation length for vertexing due to the shortened dimensions of the device; this feature will be crucial for efficient "close-in" reconstruction at the high multiplicity environment of the SSC.

The inclusion of pulse height information, easily available on modern VLSI readout chips, would also allow for the local resolution of the pattern recognition "ghost problem" that occurs in instances of multiple apparent hits in the same detector without the use of a third view and the resulting decreased transparency. This is accomplished through the comparison and identification of x and y pulse heights with each other. The results will substantially reduce the critical combinatoric problems that otherwise threaten to plague inner SSC detectors and should allow for the crucially important relatively clean reconstruction of tracks through several layers. Further, one could also improve the signal to noise ratio for a given "point" on a track through the combining of the signals from the two sides. For reasons such as these it has become important to investigate the charge correlation characteristics of double sided silicon devices.

EXPERIMENTAL METHOD

Detectors

The four detectors reported on in this paper were double-sided DC coupled devices manufactured by Micron Semiconductor, Ltd. (Sussex, England). Each device was made from a 3" silicon wafer and had overall dimensions 38mm x 58 mm x 300μm thickness, although, for these tests only a fraction of the available area (the same for each device) was instrumented for readout. Both sides of each device were divided into three regions according to pitch (see Figure 1) - a central region of 25 micron pitch was surrounded on each side by a region of 50 micron pitch. This division was effected so that the variation of performance characteristics with pitch (charge sharing, resolution, etc.) could be studied conveniently on a single device.

Stripes on each of the diode and ohmic sides of each detector were read out with the Berkeley SVXD IC (see below). These chips were mounted on fan-in cards with an approximately 50 micron pitch to receive wirebonded detector channels. A total of two SVX chips were used to read out each side (diode and ohmic).

On the diode sides, one chip ("SVX1") was placed on one end of the detector straddling the boundary of the central 25μm region and one of the 50μm regions. This chip received signals from 64 consecutive stripes in the 50μm region and from 64 stripes spaced apart by one stripe (i.e. every other stripe) in the 25μm region. Another chip ("SVX2") was mounted on the opposite end of the diode side wholly

Ohmic Side

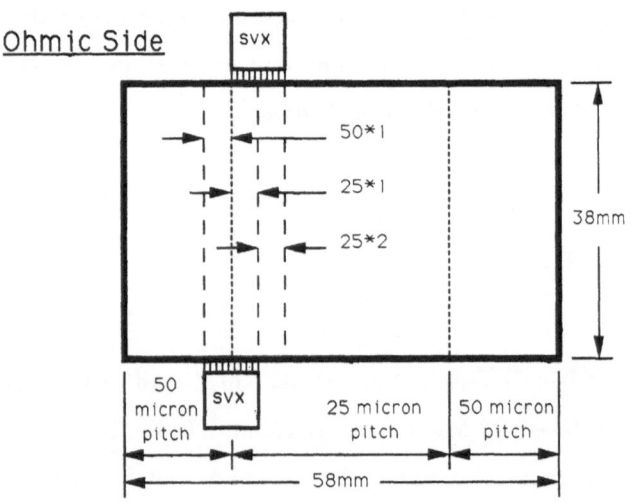

Diode side

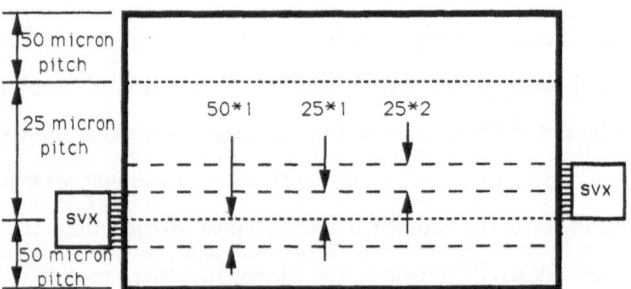

Fig. 1 Ohmic and Diode Side Detector Layouts.

in the 25μm region in such a manner as to receive signals from the 64 interleaved stripes that were between the stripes readout by SVX1. This chip also received signals from another 64 stripes (again "every other") stripes in the area next to that covered by SVX1 (see Figure 1).

Readout

The readout of the microstripe planes was accomplished with the Berkeley SVX version D (SVXD) IC and Berkeley microprogrammable SVX readout Sequencer ("SRS") and fast ADC ("SDA") modules which were built and tested at Oklahoma. The SVXD and the Berkeley modules have been described in detail elsewhere;[7,8] the discussion here will be brief. For the layout we refer the reader to Figure 2.

SVX Readout IC

The SVXD readout chip is a CMOS IC with multiplexed serial readout and sparsification capability. The chip incorporates 128 channels of analog amplification and storage and associated integrated digital address generating circuitry for handling input from wire bonded microstripe channels. Each analog channel consists of a high gain charge sensitive amplifier (CSA) followed by a set of sample and hold and threshold storage capacitors, a comparator to compare signal to threshold and a latch. Outputs of the SVX comprise up to 128 channels of sparsified analog pulse height information brought out serially on a single line ("analog out" (AO)) and 8 bits of associated address information (chip and channel ID) output on 8 address lines. A number of SVXD chips may be daisy chained together.

The operation of the SVX chip is normally controlled by a series of signal patterns input simultaneously on bidirectional lines. In the "write" mode, input clocking signals open and close various switches which pass charge along through the analog section chain or load (or clear) sample and hold capacitors. During only a brief interval corresponding to part of this "clocking cycle" the chip is sensitive to charge from the detector ("Integration Window" ("IW")). In the "read" mode, the same lines return digital addresses; coincident with this, the channel by channel analog pulse heights are placed on an output bus ("AO", for "Analog Out").

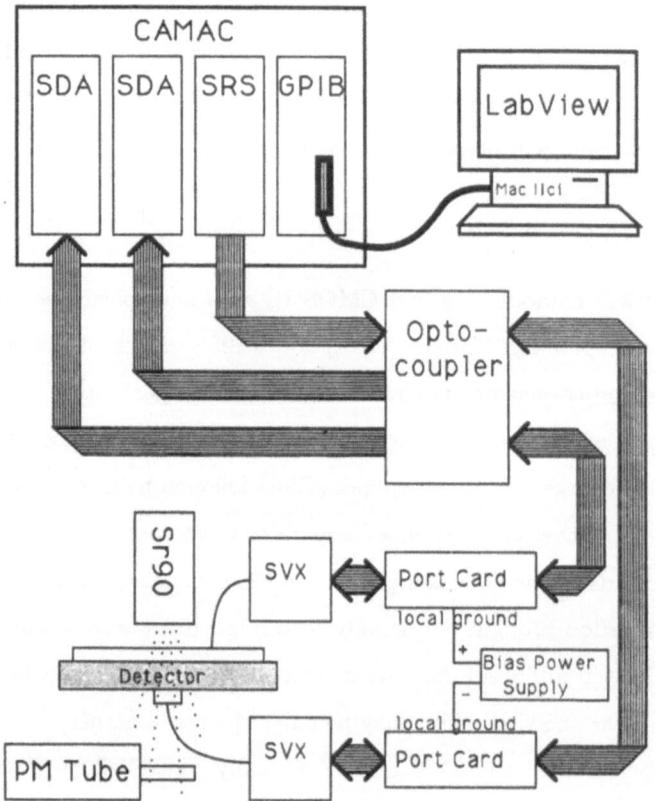

Fig. 2 Experimental Setup.

a. SRS

The SRS as built at Oklahoma is a CAMAC based module used to generate the signal patterns for clocking and reading out of the SVX. The pattern generator is based on a programmable AMD 2910 micro-sequencer IC and three physical memories into which the associated assembled microcode is downloaded. The micro-sequencer can be programmed, e.g., to execute loops, cause time delays and to branch.

b. SDA

The SDA receives and stores data from the SVX. It contains an 8-bit flash ADC for digitizing analog signals and two digital memories for storing the addresses and digitized analog data. The timing signals necessary for strobing the ADC and the digital memories, as well as setting the LAM, are created by the SRS microcode program and transmitted from the SRS to the SDA. The analog signals arriving from the SVX have a span of about 2 volts; the SDA has an adjustable offset to accomodate this. The SDA analog input stage also has an adjustable gain which can be set to 1,2,4 or 8. "Gain 1" corresponds to about $(1/256)(2V) \approx 8mV$ per ADC count.

Microcode and Triggering

The microcode consisted essentially of three independent loops - a clocking loop, a readout loop and a "quiesence loop". As mentioned above, during the clocking loop switches are opened and closed so that the SVX may accept charge and pass it on. Charge from ionizing particles incident during the brief integration window only is passed on to the SDA. In typical running the integration window was an interval about 600 ns in length (corresponding to two SVX clock pulses).

Prior to the arrival of an experimental trigger the clocking loop was executed repeatedly, the SVX thus being repeatedly loaded with and subsequently cleared of "noise" signals. The arrival of a photomultiplier signal from a 1 cm^2 x 2mm thick slab of scintillator provided a trigger signal, which, if coincident with the "open" IW, caused the executing microcode to initiate a readout cycle. Upon the completion of this, the code would branch to the quiescent loop while the SDA memories were interrogated. Subsequent to this the clocking loop would be reactivated. For simplicity we used a clocking cycle with no explicit noise thresholds.

Data Acquisition

Data from the SDA modules were collected by an Apple MacIntosh IIci computer with National Instruments "Labview" data acquisition software via GPIB bus connection. Data analysis was also done on MacIntosh II computers.

Biasing of DC Coupled Detectors and Optical Couplers

To ensure collection of charge, silicon detectors must have a DC (bias) voltage applied from diode to ohmic side that fully depletes the active silicon volume. With DC coupled devices, the absence of a coupling capacitor precludes the direct application of this bias across the detector since that would load the amplifier inputs heavily relative to their local grounds. To circumvent this difficulty, we applied the bias voltage across the amplifier "grounds" from ohmic to diode side (see Figure). Since the ohmic and diode sides were read out with distinct SVX chips, this meant that the local "grounds" of ohmic and diode SVX chips differed by the bias voltage.

Signals returned from the ohmic side SVX chip and from the diode side SVX chip thus differed from each other nominally by the full bias voltage. Although the ohmic and diode signals were received by distinct SDA modules, these modules were mounted in the same crate and thus used a common ("true") ground (see Figure 2). Accordingly, digital and analog signals passing from one of the SVX chips had to be optically (or capacitively) coupled, respectively, to the SRS and the appropriate SDA module.

DC Leakage Currents and Cooling of Detectors

One consequence of DC coupling of detectors to amplifiers is the continual presentation of reverse-bias leakage current at the amplifier input. In this mode of operation it is therefore important that the accumulation of leakage charge over the IW time interval not be so large as to saturate a considerable portion of the usable dynamic range of the readout IC. In addition, large channel-to-channel differences in leakage current can also have bad effects on the useable dynamic range.*

We observed experimentally that typical leakage currents were about 10 μA integrated over the whole of one side of the detector ($\sim$ 1000 channels); as implied in the footnote, this is not optimal. Interestingly, we also observed that channel-to-channel variations in the leakage current were often almost as large as the typical current itself. This would have further complicated the extraction of the signal. It thus became important either to cancel or to directly eliminate the leakage current.

One method that is commonly employed to this end involves the use of so-called "quadruple-sampling". In this scheme the noise charge is sampled twice so as to establish a slope (essentially the leakage current). The resulting expected charge integration is subtracted from the two signal samples before final storage on the sample and hold capacitors; thus what is ultimately sent on is the signal minus the expected leakage current contribution. As this involves four separate samplings, the noise is expected to be roughly $\sqrt{2}$ worse than for ordinary "double-correlated sampling". Another, potentially superior method that is showing signs of promise involves cutting off the leakage current by effecting an AC capacitive coupling. We have been studying the characteristics of a variety of AC-coupled detectors; some preliminary results have already been reported.[3,4]

Yet another method for improving the noise characteristic is to eliminate the leakage current directly at the source by cooling the device. This method was used for the results reported on in this paper.

RESULTS

Temperature Dependence of Leakage Currents

We found experimentally that cooling from room temperature to $\approx$ -22C re-

---

*To understand this roughly, note that a 10 nA leakage current per channel and a 0.6 μs integration window implies a charge of 6 fC accumulated. A mip is roughly 4fC. At "gain 8" (desirable for best signal to noise) the linear dynamic range of the SVX-SDA combination is about 0.25 volt of output at the typical chip gain of about 14mV/fC. This then implies a usable dynamic range of the order of 20fC, and we conclude that 10 nA of unsubtracted leakage current per channel is too high.

duced both the absolute leakage currents and the channel to channel variations by a factor of about twenty, as is shown for one of the detectors in Figure 3. All data reported on in this paper was taken at about -22C, unless otherwise indicated.

Noise Characteristics

In Figure 4 we show the noise pulse height spectrum of a typical 50 micron region ohmic channel. In this Figure and those of the same sort following, the abscissae correspond to ADC counts. We initially refer the reader to the lower, broader distribution. This represents the raw noise pedestal as directly observed. However, such a plot is not directly usable in the data analysis due to its relatively large width. One of the major contributions to the apparent broadening of the noise peak results from the so-called "common-mode" noise which one often observes in the use of silicon microstripe detectors. This refers to event-to-event fluctuations in which all of the usable stripes change pulse height by the same number of counts. Such common-mode fluctuations can easily be evaluated in the data analysis and must be subtracted from the raw data on an event-by-event basis. The taller, narrower distribution of Figure 4 shows an equivalent number of "noise events" for the same detector channel, but with the common-mode correction included. It is seen that the standard deviation σ has decreased to about 2.7 ADC counts (gain 8). Results for the 25 micron region strips are similar and will be discussed below. All Figures subsequent to Fig. 4 (except for Fig. 9) showing data have had the common-mode correction included.

Figure 4 refers to data taken with the integration window (IW) set to a width of two clock pulses (i.e., 600 ns.). We also studied the variation of the width of the noise distribution with IW width; the results of this are shown in Figure 5 and are compared to an (expected) square-root fit, the fit being shown as the solid curve. The conclusion is that this fit is not unreasonable.

Uncorrelated Diode and Ohmic Signals

In Figures 6 and 7 we show signal and pedestal distributions in logarithmic histograms. Our convention, reflected in the Figures, is that diode signals are shown as a negative displacement in counts, while ohmic signals, since they correspond to the opposite sign of collected charge, are shown as a positive displacement in counts.

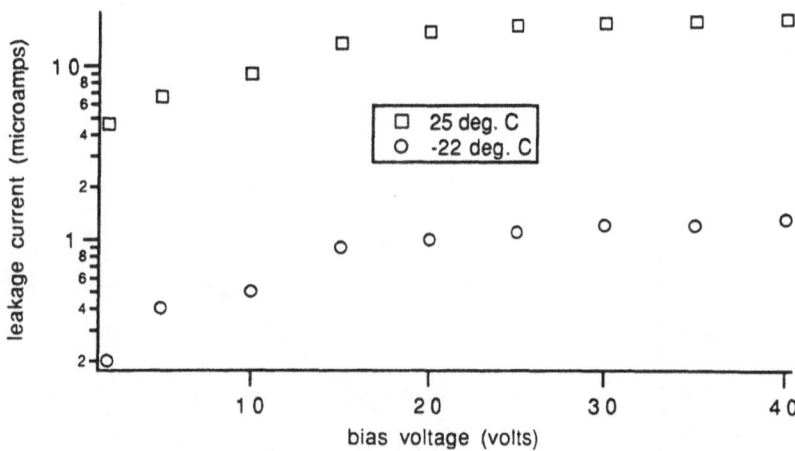

Fig. 3 Leakage Current versus Bias Voltage.

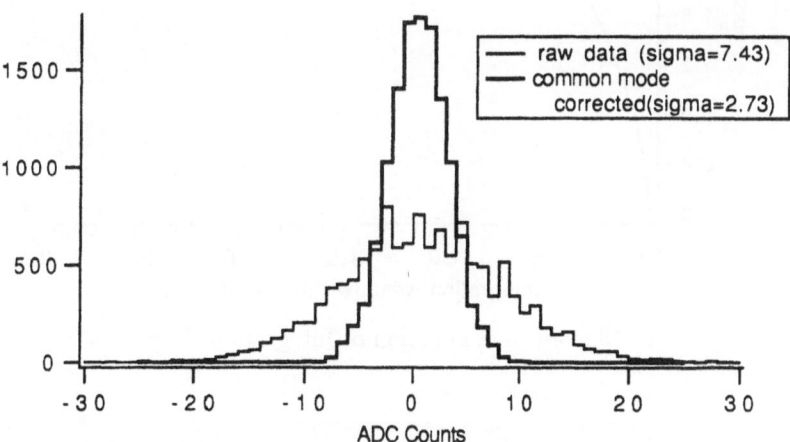

Fig. 4 The Effect of Common Mode Correction on "Noise Events" (raw sigma = 7.43, corrected sigma = 2.73).

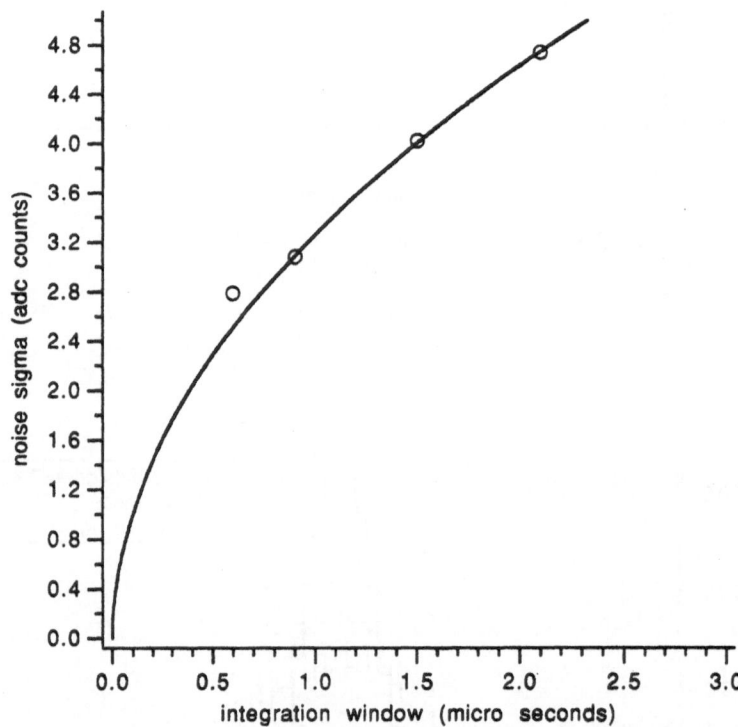

Fig. 5 RMS Noise as a Function of Integration Window Width.

Let us begin by referring, for example, to Fig. 6a. The dominant peak to the right represents the noise pedestal which results primarily from triggers where an incident particle did not traverse the stripe under observation. The logarithmic nature of the plot also shows more clearly the signals due to hit stripes. The corresponding histogram for the ohmic side is shown in Figure 7a. These Figures show data summed over all useful channels in the 50 micron region of a detector.

Interestingly, however, histograms in this form are still not optimally useful, since the detectors show "charge sharing" between stripes - for a useful analysis one must combine signals from neighboring channels. (This apparent charge sharing is believed to be mostly due to the incident track angles and multiple scattering of low energy β-rays in the detector).

Figure 6b is a histogram of the pulse height of the "biggest double" per event in the 50 μm pitch regions of the diode side of one of the detectors. In finding the "biggest double" per event we scanned through all of the useful channels in the relevant region of the detector under investigation and histogrammed the summed pulse height of the pair of neighboring channels that had the largest absolute value of summed pulse height for the event. The quantity histogrammed, then, was the signed value (+ or -) from the pedestal mean of the double hit cluster with the biggest absolute pulse height in the event. We note from this plot that the mean pulse height for the "biggest double" is about 70 counts. The corresponding data for the 50 micron pitch region of the ohmic side of one of the detectors is shown in Figure 7b; here the mean biggest double pulse height is essentially similar to that for the diode side.

[The reader may wonder why the noise pedestals seem split around zero in Figures 6 and 7. This is because noise pulse heights can be either positive or negative relative to their mean. (Recall that what is plotted is the signed value of the cluster with the biggest absolute value.)]

In Figures 6c and 7c we show the "biggest triples" per event for the 50 micron regions of the diode and ohmic sides, respectively. These are defined by an algorithm similar to that used for doubles (plot signed summed pulse height for case of biggest absolute value of summed pulse height) for triples of adjacent strips. We note here that the means are about 79 for each of the two sides, respectively, which is not much bigger than the corresponding mean values for the biggest doubles. This indicates

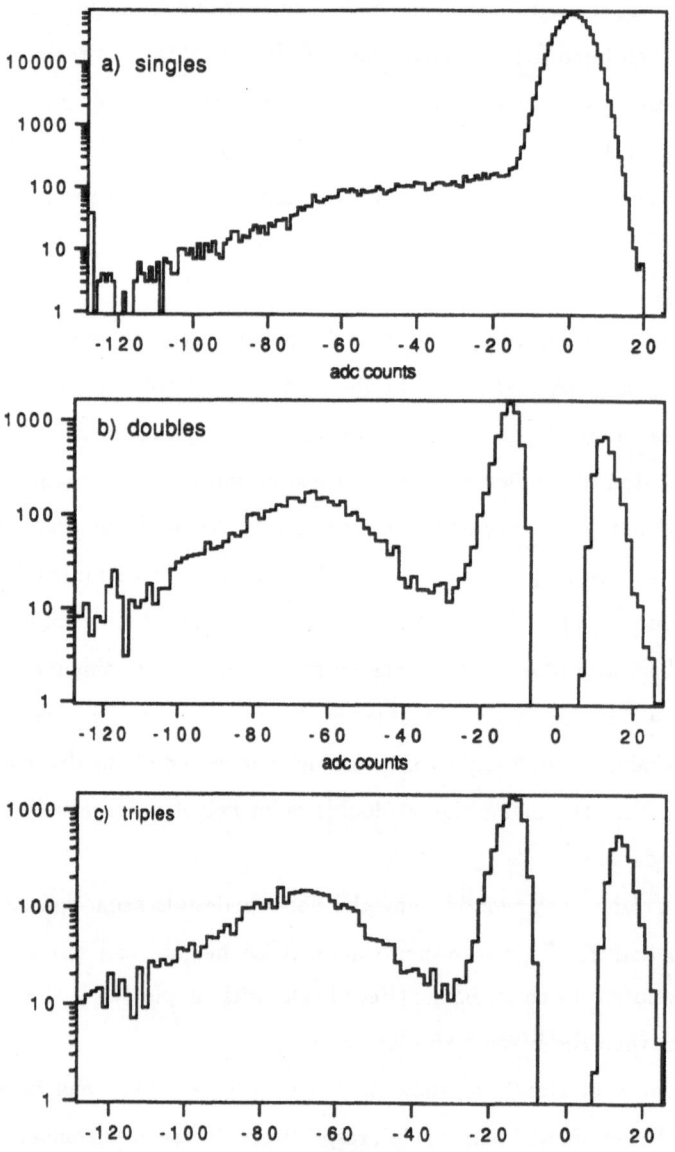

Fig. 6 Diode signals for the 50*1 region.

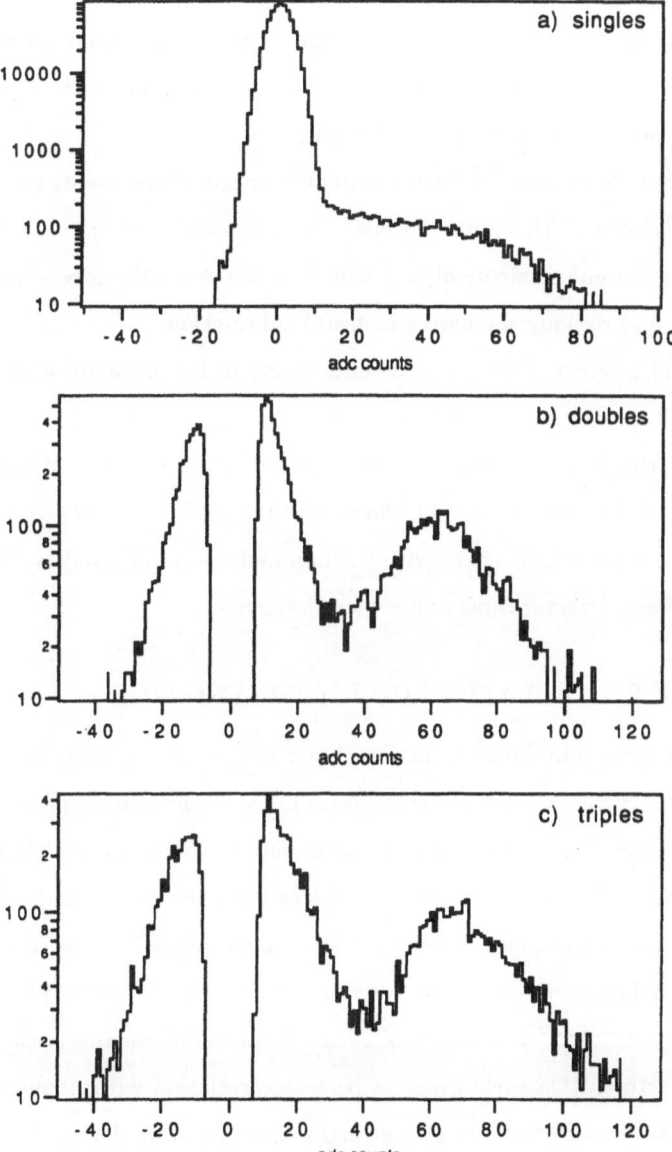

Fig. 7 Ohmic side signals.

that the charge sharing is mostly limited to just two adjacent strips, which as we remarked above, was expected from the angular distribution of tracks from the source. In Figure 8 we show a histogram of cluster sizes. This also shows a peak at about 2. Assuming that the typical events involve cluster pulse heights of about 79 ADC counts, the single stripe signal to noise figures can be approximated by 79/4.0 $\approx$20 and 79/3.0 $\approx$27 (recall diode and ohmic noises were about 4.0 and 3.0 resp.), for the diode and ohmic sides respectively - *i.e.* an average of $\approx 23 \pm 3$. Thus the signal-to-noise for an "n-cluster" is $23/\sqrt{n}$.

In Figure 9 we show an "event display" for one of the events that had a single "triple-hit cluster." This is an "on-line" display made by a channel-by-channel subtraction of the pulse heights of an event from those for the preceeding event so as to quickly and roughly simulate a pedestal subtraction.

Also of interest to us was a determination of the behavior of the signal with biasing voltage. This is shown for the cases of singles, "biggest double" and "biggest triple" for the 50 micron pitch region of the diode side of one of the detectors in Figure 10. From this we see that effective full depletion appears to occur at a bias of about 15 volts, which agrees with the nominal value indicated by Micron Ltd. as obtained from detector capacitance measurements.

CHARGE CORRELATION OF DIODE AND OHMIC SIGNALS

As mentioned in the introduction above, of special interest to us was the possibility of observing a systematic correlation between the ohmic and diode side pulse heights for the same events with the same detector. (Such a correlation would be expected since the ionizing particle produces both electrons, which migrate to the ohmic side, and an equal number of holes, which migrate to the diode side of the device.) In Figure 11b we show a diode versus ohmic "scatter plot" of "biggest triple" pulse heights for a sample of events from the 50 micron pitch region of one of the detectors. (On such a plot, a perfect correlation with no noise spread and equal "gains" would show up as a straight line with a 45 degree slope.) It is seen that there is a strong correlation between ohmic and diode pulse heights.

Figures 11a and 11d show the projected ohmic and diode side pulse height histograms as projected from the scatter plot of Fig. 11a with their means of 79.8

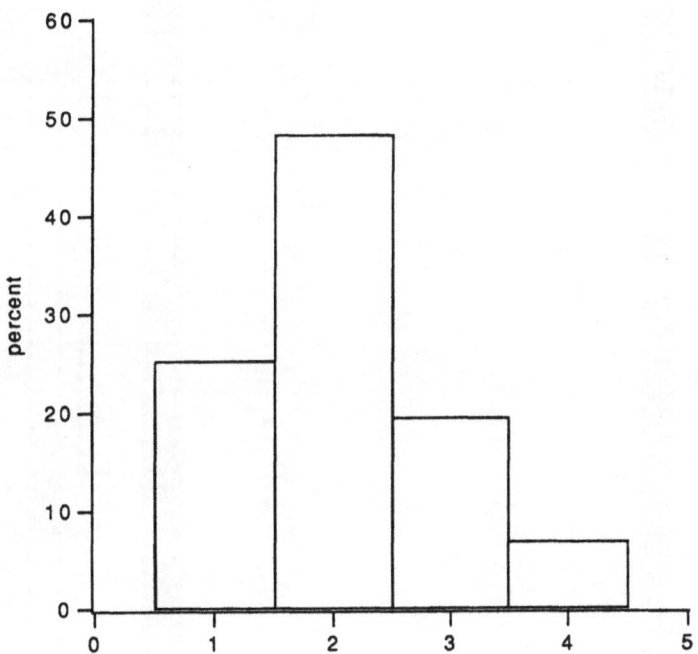

Fig.   8 Distribution of singles, doubles, triples, quadruples for diode side.

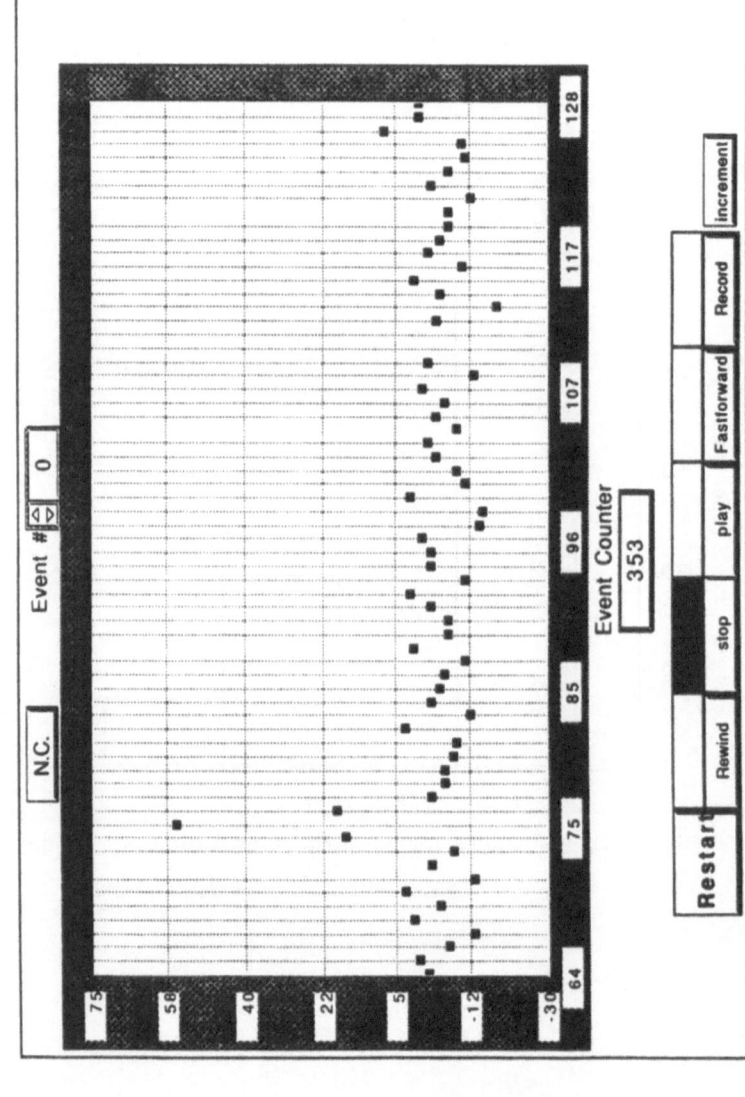

Fig. 9 "On-Line" Event Display from "Labview" panel showing a 3-hit cluster. Previous event has been subtracted for rough noise suppression.

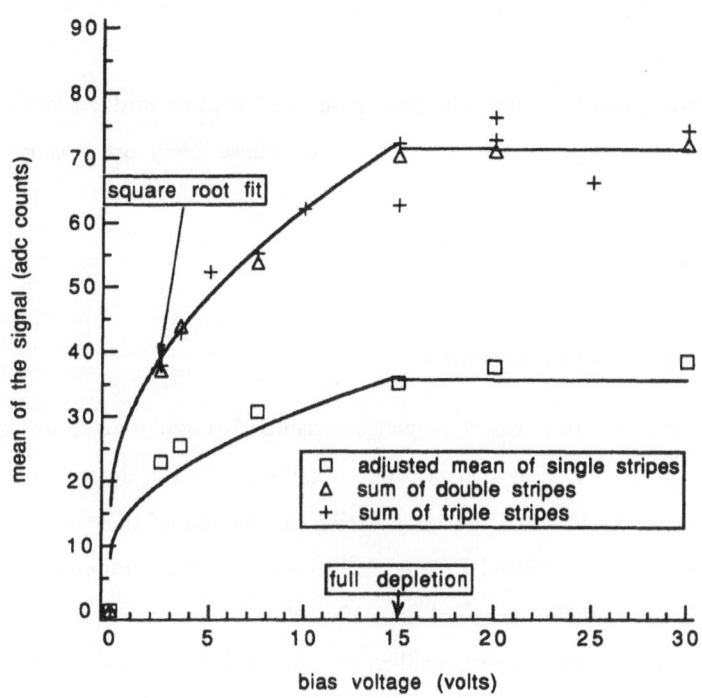

Fig. 10 Signal Size versus Bias Voltage.

and 79.2 ADC counts respectively. The rms noise width can be determined by projecting out a pulse height histogram at 45 degrees to the ohmic and diode axes of the scatter plot. This corresponds to the "diagonal width" of the central band of the plot and is shown in Figure 11c.

We note that the rms σ indicated in this histogram is ≈ 6.0 ADC counts. That this is consistent with previously mentioned results can be seen roughly as follows: We indicated earlier that typical σ's for the ohmic and diode sides were about 3.0 and 4.0 counts respectively. Since the "diagonal noise" has contributions from both sides, we take an "average" figure of 3.5 counts. Since the triple-hit clusters involve "three noises", we would estimate a composite "noise σ" of $3.5(\sqrt{3}) \approx 6.0$, which agrees with the data.

Figures 12 and 13 show the analogous scatter-plots and projected histograms for the 25 micron pitch regions of the cases where every other stripe is read out ("25*2") and where every consecutive stripe is read out ("25*1"). The conclusions are similar.

ABSOLUTE CALIBRATION

Energy losses of charged particles passing through matter are theoretically described by the "Vavilov distribution". In Figure 14 the dotted curve shows the values assumed by this distribution function for the case of the "typical" 1.5 MeV/c momentum electron emitted by our Sr^{90} β source. For comparison with this the solid curve shows "50*1 diagonal projection data" (see above) converted to femto-Coulombs by comparison with calibration signals fed into the detector through a (assumed nominal) 66 femtoF capacitor.[9] The agreement is reasonable, considering the errors. From this Figure we see that the mean signal is $\approx 4.2fC$, so that a single stripe signal-to-noise figure of 21 corresponds to a single stripe $\sigma \approx 0.2fC$, or ≈ 1200 electrons.

CONCLUSIONS AND PLANS

We have measured the noise and signal characteristics of ohmic and diode signals and have shown pulse height correlations of these signals for double sided direct-coupled detectors with the Berkeley SVXD IC. It is found that our results agree well with expectations.

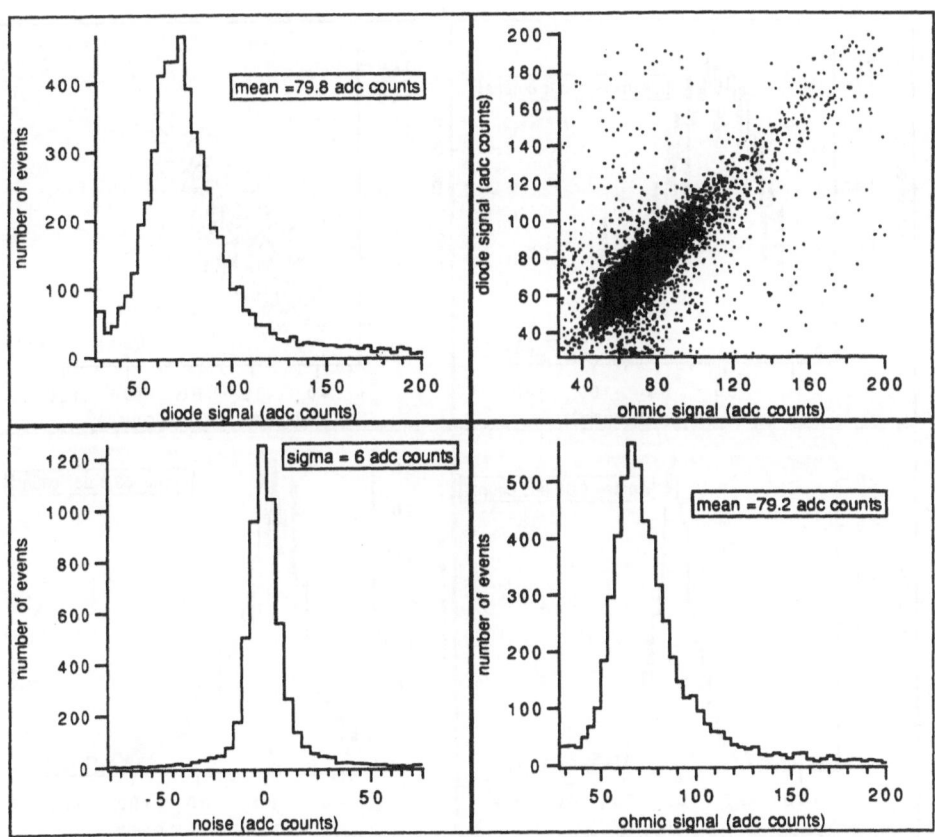

Fig. 11 Scattergrams and histograms of the diode and ohmic signals and noise in the correlation band for the 50*1 region.

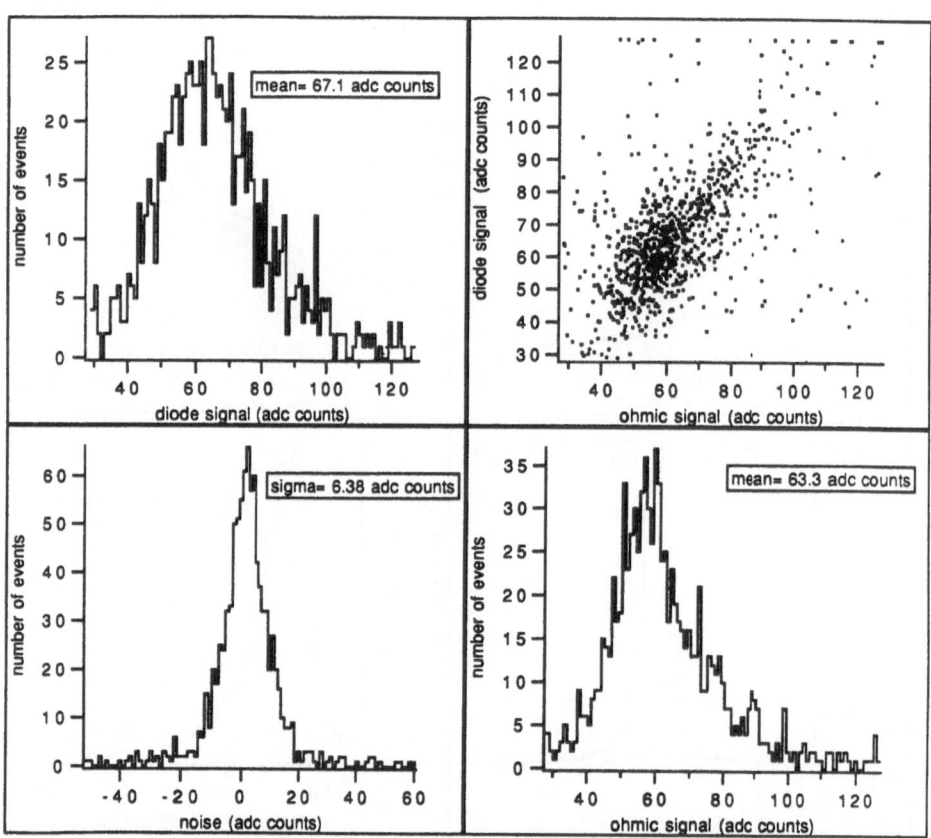

Fig. 12 Scattergrams and histograms of the diode and ohmic signals and the noise in the correlation band for the 25*2 region.

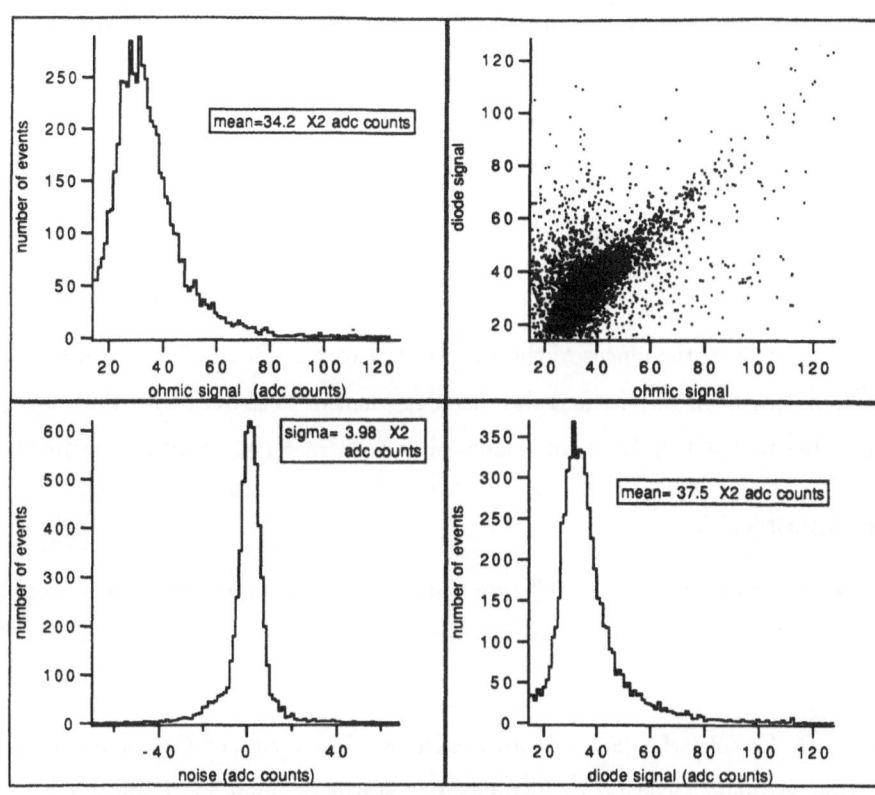

Fig. 13 Scattergrams and histograms of the diode and ohmic signals and the noise in the correlation band for the 25*1 region.

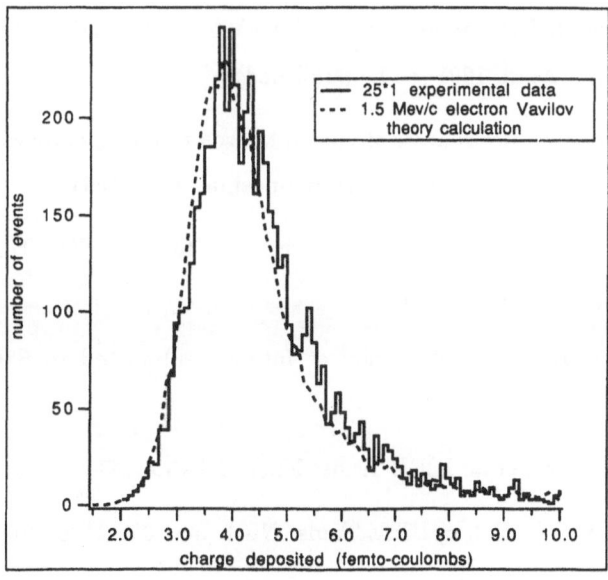

Fig. 14 Comparison of the experimental data from the 25*1 correlation band and a theoretical Vavilov calculation for charge deposited in 300 microns of silicon.

In the immediate future we plan to extend this work to study the effects of integration window timing and the use of quadruple-correlated sampling in cooled and uncooled detectors and to conduct high energy pion beam test studies of both DC and AC coupled detectors at Fermilab.

ACKNOWLEDGEMENTS

The authors thank the University of Oklahoma for strong financial support for this project from the offices of the President, Provost, Dean of Arts and Sciences, the Department of Physics and the Vice President of Research. This work was also supported by the U.S. Dept. of Energy under contract DE-AS-05-80-ER-10629.

REFERENCES

1. G. R. Kalbfleisch, et al., "Charge Correlation Measurements From a Double-Sided Solid State MINI-stripe Detector," IEEE Trans. Nucl. Sci., **36**, 272 (1989).

2. G. R. Kalbfleisch, "Solid State Detectors" University of Oklahoma Internal Report, Feb. 1988; G. R. Kalbfleisch, "Status of Some of the Silicon Detector Development in the United States," Proceedings of a Workshop held at Fermilab (T. Ferbel, ed., Oct. 5-16, 1981), pg. 45.

3. M. Lambrecht, et al, "Evaluation of AC-Coupled Silicon Microstripe Detectors and Berkeley SVXD readout with 227 GeV/c Pions," IEEE Trans. on Nucl. Sci., **38** No.2, April, 1991 (to be published).

4. H. Attias, et al., "Beam Tests of Silicon Microstripe Detectors with VLSI readout," Proc. Fort Worth SSC R&D Symposium, Oct. 1991.

5. P. Skubic, et al, "Vertex Detector Technology for the SSC", Fort Worth SSC R&D Symposium, Oct. 1991.

6. "The BCD Detector", Expresion of Interest submitted to SSC Laboratory (1990).

7. S. A. Kleinfelder, et al, IEEE Trans. Nucl. Sci. **35**, 171 (1989).

8. F. Kirsten and C. Haber, IEEE Trans. Nucl. Sci. **37**, 2188 (1990).

9. S. Tkaczyk, Fermilab, private communication.

APPLICATIONS FOR A NEW TIME DIGITIZER IN SSC EXPERIMENTS

George J. Blanar and Richard Sumner

LeCroy Corporation
Chestnut Ridge, New York 10977-6499

ABSTRACT

A new Time-to-Digital Converter, Application Specific Integrated Circuit (TDC ASIC), the LeCroy Model MTD132 Octal Time Digitizer, has been developed for use in elementary particle and heavy ion physics experiments. The TDC was designed for high multiplicity forward spectrometers' drift chamber readout and will be packaged in high density CAMAC (IEEE-583) and FASTBUS (IEEE-960) instrumentation. However, the possibility of chamber mounted applications for hadron collider experiments has always been included in the design.

Tracking chambers for the SSC will require low power, on-chamber TDCs with sub-nanosecond timing resolution and multi-hit capability. The implementation of the MTD132 has included these attributes in an eight channel circuit requiring little peripheral support, in a low power technology.

In addition to being used as a simple timing device, this high resolution multi-hit TDC can digitize both leading and trailing edges of pulses. Therefore, the MTD132 can be applied to the problem of total charge measurement of an analog signal as a timer in an ADC or via the "time-over-threshold" technique. Monte Carlo studies of the readout of several types of detectors with an MTD132-like device will be presented. Finally, the use of this type of instrument in several data acquisition architectures for SSC experiments will be discussed.

INTRODUCTION

LeCroy Corporation has recently announced a new Time-to-Digital Converter (TDC) integrated circuit designed specifically for Elementary Particle and Heavy Ion Physics experiments. The MTD132 combines high density (8 channels per monolithic), low power (60 mw per channel), and high resolution (less than 1 nanosecond) in a single integrated circuit. The LeCroy MTD132 has been designed with drift chamber and time projection chamber applications in mind. It is the natural next step in the 15 year long evolution of LeCroy time digitizers for particle physics. The high resolution, multi-hit capability, automatic zero suppression (self-sparsification) and excellent double hit resolution of the MTD132 TDC are particularly well suited for drift chamber readout under conditions of high track multiplicity and translate directly into good vertex resolution.

The TDC will be initially available as a 32-channel CAMAC module. A 96-channel FASTBUS module is being designed for large scale applications. On-chamber mounting is being considered for special situations.

This high resolution multi-hit TDC, available for the first time in a low-cost high density system, allows many new applications beyond the traditional drift chambers used in most high energy physics experiments. This paper will discuss several of these new applications, and a novel data acquisition architecture made possible by the characteristics of this new TDC.

THE MTD132 MONOLITHIC INTEGRATED CIRCUIT

The MTD132 uses a high speed clock, a continuously counting scaler, and an interpolator. A conceptual schematic is shown in Figure 1. When a signal arrives the contents of the scaler and a 4 state interpolator are latched and stored in a 16 word first in first out (FIFO) buffer memory. The quantizing step size is 0.75 nanoseconds (nsec). Later signals on the same channel simply store another word in the FIFO. A special input, the COMMON HIT (either START or STOP) input signal is also latched using similar circuits, but without the FIFO memory. External circuitry suspends the latching process depending on whether COMMON START or COMMON STOP mode is in use. During readout, each of the stored signal times is automatically subtracted from the COMMON HIT time. The final output is the time difference between the input signal and the COMMON HIT signal. Hits older than the 49 microsecond full scale are automatically discarded and not read out. If more than 16 hits occurred within the full scale time interval, only the most recent 16 are saved. In COMMON START mode, a time out interval must be specified to define the full scale time (when the latching process is suspended). All hits, until the end of the time out interval, are stored in the FIFO. Again, if there are more than 16 hits, only the most recent 16 are saved.

A short list of the MTD132 specifications includes:

> Wide Dynamic Range, 49 Microseconds Full Scale (16 bits)
> Least Count of 0.75 nsec, Resolution of 0.3 nsec rms
> Differential Linearity of better than 20%
> Pipelined Multi-hit Operation, Up to 16 Hits Per Channel
> Multi-hit Resolution of less than 20 nsec
> COMMON STOP or COMMON START, Selectable
> Leading or Trailing Edge (or both) Recorded
> Output Data Corrected For COMMON HIT
> Fast Readout, No Conversion Time or Searching for Hits
> Digital CMOS, Low Power Technology
> High Density, 8 Channels Per Monolithic

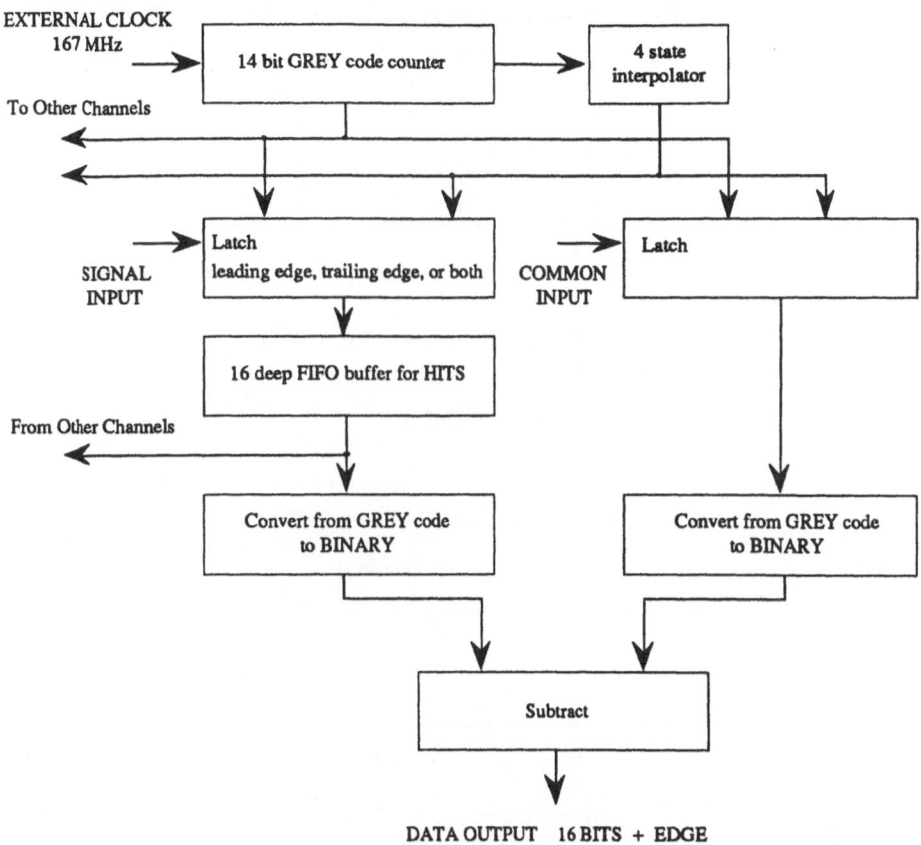

EXTERNAL CLOCK
167 MHz

14 bit GREY code counter

4 state
interpolator

To Other Channels

SIGNAL
INPUT

Latch
leading edge, trailing edge, or both

COMMON
INPUT

Latch

16 deep FIFO buffer for HITS

From Other Channels

Convert from GREY code
to BINARY

Convert from GREY code
to BINARY

Subtract

DATA OUTPUT   16 BITS + EDGE

Figure 1 MTD132 Conceptual Schematic

MEASURING TIME

A typical drift chamber geometry is shown in Figures 2 and 3. The standard drift chamber parameters are drift velocity of 5 centimeters/microsecond and spatial resolution for a single wire (limited by electron statistics and diffusion) of about 200 microns rms. This represents a time resolution of 4 nsec rms. Since the MTD132 quantizing resolution (0.3 nsec rms) is sufficiently smaller than the intrinsic drift chamber resolution, its contribution to the final system error is negligible. Such high resolution is not needed for ordinary drift chambers. However, this fine resolution can be put to good use. Small drift chambers, and those using fast gases can benefit from the improved resolution. More sophisticated drift chamber systems, those that average the signals from many wires for example, demand the highest possible time resolution.

A low precision measurement parallel to the sense wire (Z position) is often needed to resolve the stereo ambiguities when multiple tracks occur in the same set of chambers. The Z coordinate is usually measured by using resistive wires and charge division with flash ADCs on each end of the wire. Precision track information can then be obtained from the stereo chambers themselves.

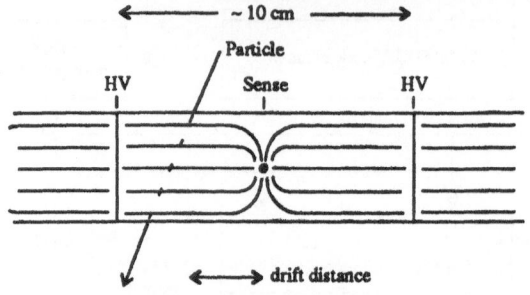

drift time = drift distance/drift velocity

t_0 = drift time + $t_{particle}$

Figure 2

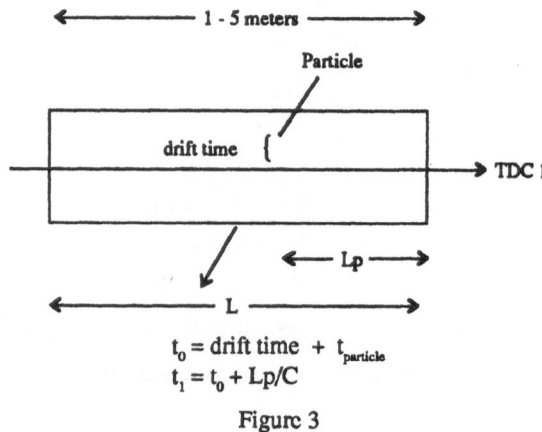

t_0 = drift time + $t_{particle}$

$t_1 = t_0 + Lp/C$

Figure 3

If a TDC is used at each end of a drift wire, the difference in arrival times of the two signals can be used to determine the approximate position along the wire as shown in Figure 4. Using two TDC channels, the time sum can be used to measure the time of arrival of the pulse at the wire (automatically correcting for the propagation along the wire). The arrival time is measured with a resolution of 0.42 nsec, independent of the Z position. This assumes independent clocks for the two TDCs. If we use a common clock, the stop error is correlated and the total error on the sum is 0.52 nsec. The time difference can be used to measure the position along the wire (the Z coordinate). For the time difference, the correlated common stop error drops out, giving an error of only 0.3 nsec. This translates into a position error of 4.5 cm rms (since the time difference changes twice as fast as either time measurement). The worst case error is only 11.5 cm (one half of one least significant bit of the TDC). Note that this error is constant, independent of the wire length, and not a constant fraction of the wire length, unlike the charge division methods. These error calculations assume that the time slewing error at the discriminator is negligible. Naturally, in order to take advantage of the improved resolution of this new TDC, all system components must be equal to the task!

The method of using two TDCs per wire will give the highest hit rate capability without ambiguities. If rates per wire are modest, the second TDC can be created by delaying one of the two signals by twice the wire propagation delay, plus the minimum double hit resolution

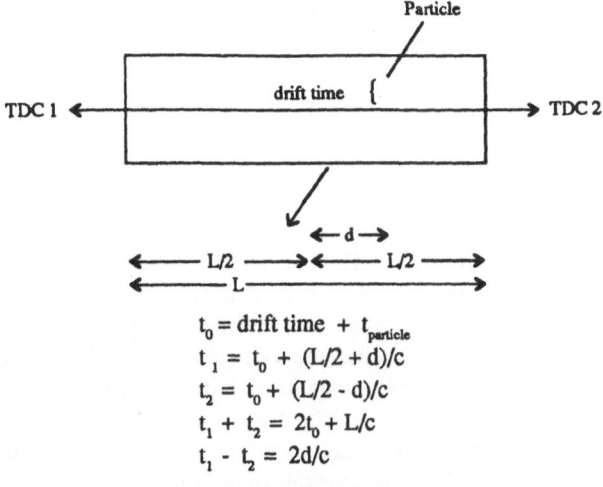

$$t_0 = \text{drift time} + t_{particle}$$
$$t_1 = t_0 + (L/2 + d)/c$$
$$t_2 = t_0 + (L/2 - d)/c$$
$$t_1 + t_2 = 2t_0 + L/c$$
$$t_1 - t_2 = 2d/c$$

Figure 4

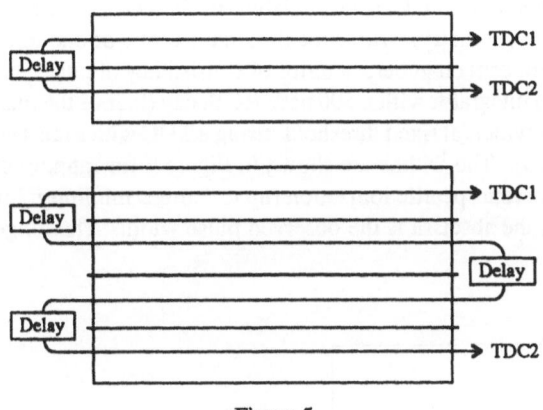

Figure 5

of the TDC, and ORing the two signals into one MTD132. This exploits the multi-hit capability and excellent double hit resolution of the MTD132. For a 5 meter wire, the delay required is less than 50 nsec. One still needs an amplifier and discriminator at each end of the wire, however.

If the rates per wire are low and the probability of hits on adjacent wires is also low, the same result can be accomplished with one TDC per wire and connecting the wires together at the far end, as shown in Figure 5. The same pulse arrives on both TDCs, the time difference tells the Z position and which wire was struck. As above, the time sum automatically corrects for the propagation along the wire.

This concept can be extended further by combining more than two wires. By daisy chaining four sense wires into one long wire, the number of TDCs required is halved. There is no ambiguity as to where the hit was, either in Z or wire number, as long as the hit rate is low enough that the pulse pairs from different hits are separated. Note that the long wire should be correctly terminated at each end by the amplifiers, and the links between wires should maintain the wire impedance. There is an ambiguity in the wire identification when the hit occurs close to the end of a wire. This can be completely eliminated by using a delay element of a few nanoseconds (longer than the TDC time resolution) to interconnect the wires. As above, the two TDCs required can be combined into one multi-hit TDC, if high quality delay lines of sufficient length are available.

MEASURING CHARGE

The existence of a true pipelined TDC with high resolution makes a true pipelined ADC possible. In fact, any signal which can be converted into a time difference can now be measured in pipeline fashion. If the signal rates are low, rather simple pulse shaping and a comparator (time-over-threshold) will produce a pulse, the width of which is a function of the original pulse area. In fact, if the shaped pulse has an exponential decay, the response is logarithmic.

As an example, consider a drift chamber pulse. If we can measure the charge collected, in addition to the time, we can determine the rate of ionization. This identifies the particle type in some kinematic regions, and works particularly well for heavily ionizing particles such as low energy protons.

We have made a Monte Carlo calculation of the charge collection for a circular drift chamber (straw tube geometry), with a maximum drift time of one microsecond. This is one of the worst cases for drift chambers in terms of consistency of pulse shape. After shaping the pulse with a simple integrator with a 500 nsec RC decay (half of the maximum drift time), we measured the pulse width (at fixed threshold) using a TDC with a least significant bit (LSB) of 1 nsec (see Figure 6). The widths are shown in Figure 7 for random drift distances (impact parameters) and random specific ionization (up to 5 times minimum ionizing). The ordinate is the total charge, the abscissa is the observed pulse width. The longest pulses are about 6

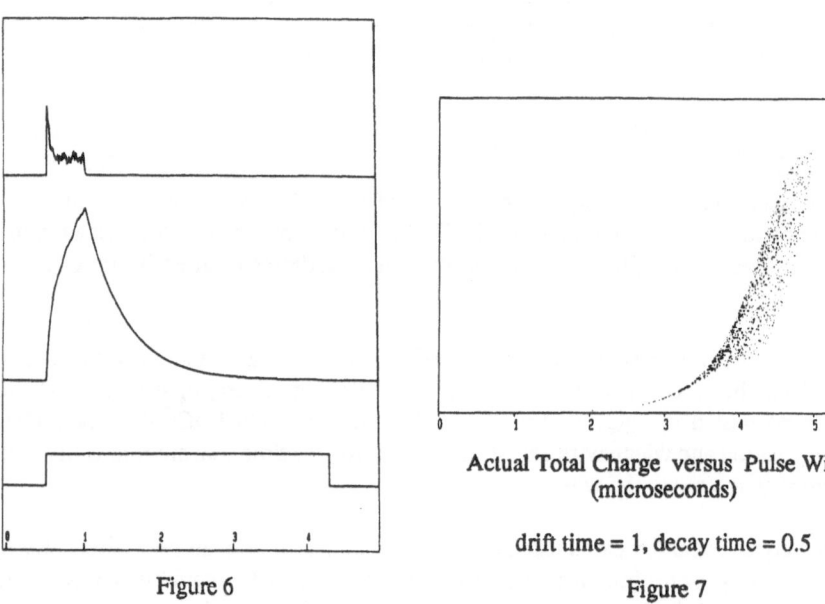

Actual Total Charge versus Pulse Width
(microseconds)

drift time = 1, decay time = 0.5

Figure 6   Figure 7

microseconds. This is a particularly messy plot because of the short shaping time since the length and shape of the charge pulse depends on the impact parameter of the particle. For small impact parameters (the distance of closest approach to the sense wire), the input pulse is 1 microsecond long and constant in amplitude. For the largest impact parameter, the charge is collected in a very short spike after the maximum drift time of one microsecond. However, by fitting a function to the pulse width and correcting for the time of arrival of the leading edge (the time when the first electron arrived at the wire), the total charge can be recovered with an rms error of 3.7% as shown in Figure 8.

If the shaping time is long compared to the drift time, the resolution can be improved (Figure 9) to a final resolution of less than 1% (Figure 10). In this case, however, the shaped pulses are 15-20 times the maximum drift time! Obviously, this is only suitable for very low hit rate situations.

The time-over-threshold method is easily extrapolated to other signal inputs. For a fast scintillator and photomultiplier, the shaping times can be short since the input pulse shape is more nearly constant. In this case, the resolution will be limited only by the TDC resolution. A shaping time of 75 nsec will give 1% resolution with the MTD132.

The method of simple shaping produces a logarithmic response and is suitable for relatively low rates. The data is pipelined with no deadtime or analog storage required. It does have the traditional problems of logarithmic converters however, problems with small signals and calibration difficulties. In addition, the simple shaping is not independent of the input pulse shape. By using active shaping, these problems can be avoided and a linear response obtained.

By using the input signal to open and close a GATE, the signal can be integrated. Then if the integrator is discharged with a constant current, in the manner of an traditional Wilkinson converter, the output width (the run down time) is independent of the input pulse shape.

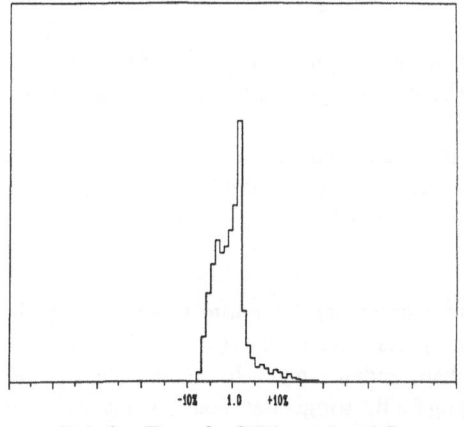

Relative Error for 2-Dimensional fit

mean = 1.000, rms = 0.037

Figure 8

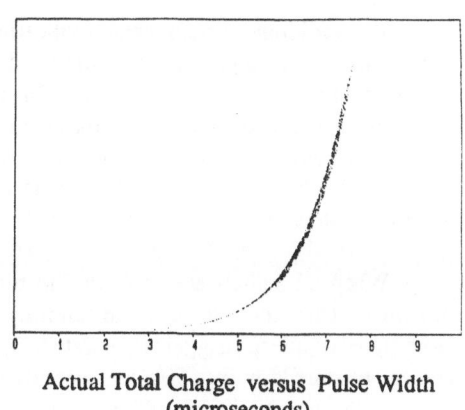

Actual Total Charge versus Pulse Width
(microseconds)

drift time = 0.25, decay time = 1

Figure 9

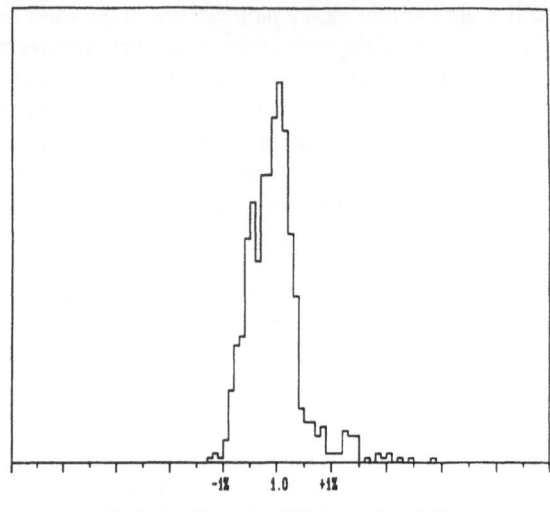

Relative Error for 2-Dimensional fit

mean = 1.000, rms = 0.005

Figure 10

Although this is pipelined and no analog storage is needed, there is a deadtime associated with the run down (indeed, the long shaping time is also a deadtime, but it does not shut off the input as the GATE will). The resolution can be quite good; 12 bits for example, with only a 3 microsecond maximum run down time.

MULTI-HIT PIPELINES

Modern high energy physics experiments use many thousands of channels of time and amplitude measurement. If the ADC requirements can be met by the pipelined techniques described above, some very interesting data acquisition and trigger configurations are possible.

The traditional system separates the trigger and data acquisition systems. After the initial trigger, all the detector elements are digitized and collected in a buffer memory to form an event record for that particular event. While this is taking place, the trigger system may refine its decision and either keep or discard the event while it is still in the buffer memory. The goal of the system architecture is to throw away as many event records as possible as soon as possible. This leaves only the "good" physics events to be read out by the host computer and transferred to archival storage.

When all signals are converted to time measurements, the entire experiment can be pipelined. That means there are no carefully timed gates or delay lines (or capacitor arrays) to store signals while the trigger is generated. Drift chamber data is naturally pipelined by the TDC and can be stored for up to 50 microseconds waiting for the trigger decision. Analog data must be converted to pulse widths and then is treated exactly as the drift chamber data. The MTD132 must be stopped to read it out and its pipeline is only 16 hits deep. Any hits after 16 cause the loss of the oldest data. Therefore, this scheme is suitable for experiments in which the trigger delay is short, or the detector occupancy is low enough that the data of interest is not overwritten by later events.

As a concrete example of this new architecture, we propose the following data acquisition system for an SSC experiment which uses the existing MTD132 TDC. This system, called FRAte TRAinSSC (Fast Readout Architecture for Time Resolved Acquisition in the SSC), can be applied to almost any high speed triggering and data acquisition problem. However, it is particularly well suited to the problem of the future hadron colliders, the SSC RHIC, LHC and other high rate experiments such as the Tau-Charm Factory. The FRAte TRAinSSC concept is similar to an early bubble chamber experiment, i.e., simply collect all the data (take the picture) and trigger off line (on the scanning table). In this case, the scanning table is replaced by a network of high speed computers (the processor farm or ranch). In this system the hardware trigger is completely eliminated. All trigger decisions are made in software by the processor ranch.

The FRAte TRAinSSC architecture takes advantage of the inherent structure of the beam. Each beam consists of a particle bunch separated from the next by 16 nsec. There are about 2000 bunches forming a continuous train in each beam pipe followed by a gap of about 1.7 microseconds (used in the machine/accelerator program). All together there are 8 trains in each of the 54 mile long beam pipes. The beams counter rotate and are made to collide in the center of the experimental detector system. The beams act like two 2000-car "freight trains" that collide for about 32 microseconds followed by a 1.7 microsecond pause before the next freight train - freight train interaction. Since there about 1.5 interactions per beam crossing, there will be about 3000 interactions in the 32 microseconds train - train crossing period.

In the FRAte TRAinSSC scheme, all detector readout is reduced to reading out time intervals. For drift chambers this very easy since the data is already "time like". Analog data is converted to a time interval using the concepts discussed above. Pixel detectors can indicate the presence of a hit by turning on a bit for a time fixed period. All readout channels have a time digitizer like the MTD132 connected to them and all the TDCs are started at the beginning of the train - train interaction and are read out at the end of the train - train interaction 32 micro-seconds later. With an occupancy in the detector of 0.1% per interaction per channel, the 3000 interactions will yield an average of 3 hits per TDC. With its 16 hit capacity, the present design would permit 99.7% efficiency for leading and trailing edge measurements (time-over-threshold) and nearly 100% for single edge data (drift chamber timings).

During the 1.7 microseconds "between trains", the data is converted by the MTD132 and the digitized, sparsified data is transferred to a local readout buffer. Since all timings are made with respect to the start of the train - train interaction, the information about which beam crossing produced the particular data is never lost. It is an integral part of the data itself. Overlapping events are handled in a very simple way, since everything is recorded.

During the next 32 microseconds of train - train interaction, the data is pulled out of the local readout buffers and transferred via 100 1GB/sec fiber optic links to a "cross bar" switch. This switch then delivers the sparsified, compacted and time ordered data to the computer ranch where it is processed with architectures like the IBM™ CSP (Common Signal Processor) or a neural network system.

The FRAte TRAinSSC system eliminates the expensive and restrictive pipelines, intermediate storage and multi-level processing engines that are required by the traditional SSC

™ IBM is a registered trademark of International Business Machines.

readout system. It substitutes a pipelined, multi-hit time digitizer instrument, the MTD132, and takes advantage of the rapid decline in the price of processing power and the cost of high speed networks. There are few things on Earth that are getting cheaper faster than the cost of MIPs or MFLOPs.

The SSC is still many years off, but there are some low-rate, fixed target experiments being planned which could easily be designed for a triggerless data acquisition system. Neutrino experiments using a pulsed horn beam are possible candidates. The rates per detector element are low, the live time is short (a few milliseconds pulse length on the horn) with a low duty cycle. The average data rates are quite manageable, and the computer requirements to perform the software trigger are within the range of feasibility.

SUMMARY

Precise, multi-hit, pipelined time measurement can be the key to effective detector instrumentation in the high speed, high flux nuclear, heavy ion and elementary particle physics experiments of the future. The LeCroy MTD132 is the first of this new generation of data acquisition instruments.

Such devices can be applied to traditional applications (such as drift chambers) to provide a new level of performance. The MTD132-based instruments will be able to read out time projection chambers and drift chambers up to 400 times faster with 3 times better resolution and with much less data handling overhead compared to the instrumentation in use today.

The availability of this level of performance and low cost makes the pipelined measurement of analog signals practical for the first time. Applications requiring only modest precision such as dE/dx measurements with drift chambers and time projection chambers are easily accommodated. Sampling calorimeters are also possible areas of application. Electromagnetic calorimeters, with their requirements of precision and dynamic range, are also being considered.

Time digitizers like the MTD132 easily fit into traditional data acquisition schemes. Since this new TDC self-sparsifies, data transfer is simplified. However, the new instrument permits radically new architectures like FRAte TRAinSSC to be considered for hadron colliders and certain fixed target experiments. Studies have begun with a group at FNAL to use this concept in a neutrino experiment now being planned for the mid 90's. Informal discussions have also started with IBM and others to understand how to best get the very large data stream from a FRAte TRAinSSC data acquisition architecture into a processing ranch.

The first generation of such a time digitizer exists, works and is available for experiments today. This is not a paper product or a design study. The MTD132 is presently packaged in a 32-channel CAMAC module (LeCroy Model 2277) for about $3,000. This unit contains four 8-channel MTD132s and is being used in detector and architecture studies in the U.S. and in Europe. A high density, low cost 96-channel version of this instrument is under design in FASTBUS (LeCroy Model 1877). It is presently scheduled for formal introduction at the 1991 IEEE/NSS Conference in Santa Fe, and will be suitable for use in many large experiments ranging from CEBAF to LEP upgrades.

RAD-HARD ELECTRONICS STUDY FOR SSC DETECTORS

T. Ekenberg, J. Dawson, A. Stevens, and W. Haberichter

Argonne National Laboratory
High Energy Physics Division
Argonne, IL 60439

ABSTRACT

The radiation environment in a SSC detector operating at a luminosity of 10^{33} cm^{-2}s^{-1} will put stringent requirements on radiation hardness of the electronics. Over the expected 10 year life-time of a large detector, ionizing radiation doses of up to 20 MRad and neutron fluences of 10^{16} neutrons/cm^2 are projected. At a luminosity of 10^{34} cm^{-2}s^{-1} even higher total doses are expected. The effect of this environment have been simulated by exposing CMOS/bulk and CMOS/SOS devices from monolithic processes to neutrons and ionizing radiation. Leakage currents, noise variations, and DC characteristics have been measured before and after exposure in order to evaluate the effects of the irradiations. As expected the device characteristics remained virtually unchanged by neutron irradiation, while ionizing radiation caused moderate degradation of performance.

INTRODUCTION

At a luminosity of 10^{33} cm^{-2}s^{-1} the total dose inside a SSC detector during its projected ten year life-time can be as high as 20 MRad of ionizing radiation and 10^{16} neutrons/cm^2. A change in luminosity to 10^{34} cm^{-2}s^{-1} will further increase the dose. Radiation damage is strongly dependent on detector geometry and distance from interaction point [1]. The dependence of the dose on distance from interaction region and pseudo-rapidity in a SSC detector [4] is shown in Fig. 1.

The read-out electronics is one of the most radiation-sensitive systems of a detector. Response time and number of channels dictate that the electronics for many detector elements must be placed closed to the interaction point (such as the silicon tracking system and liquid argon calorimeter). The challenge to the electronics designer posed at the SSC by high event rate and large channel count is then further compounded by the severe radiation damage experienced by some detector elements.

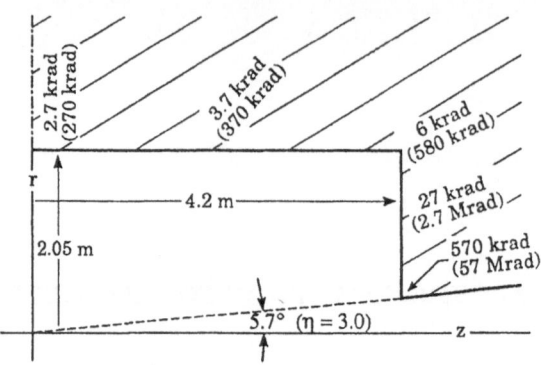

Figure 1. Total dose at several locations in a SSC detector. This figure is for an electromagnetic calorimeter of the SDC collaboration. Maximum dose occurs at shower maximum and dose indicated is for a standard SSC year at 10^{33} cm^{-2}s^{-1}, and (in parenthesis) for 10 years at 10^{34} cm^{-2}s^{-1}. Figure is taken from [4].

The effects of radiation damage on electronics have been studied by scientists associated with the military and space science community for the last 30 years, and a large body of data has been accumulated. Commercial rad-hard processes have also been developed, so the technological know-how to fabricate rad-hard electronics already exists. The purpose of this study have been to evaluate radiation resistant processes from several vendors and previous results have been published on JFETs [2] and BiCMOS processes [3].

RADIATION DAMAGE MECHANISM IN FETS

Radiation damage at the SSC is expected to be primarily caused by neutrons and ionizing radiation. A calculation of the expected neutron spectrum from a uranium/scintillator calorimeter is shown in Fig. 2. Neutrons cause damage in semiconductor materials mainly by kinematic displacements of silicon atoms from their sites in the crystal lattice. This creates recombination centers that decrease the life-time of minority carriers. Since MOSFET devices are majority carriers, they are effectively uneffected by neutrons. Some secondary processes, like ionization from recoil ions, will cause small damage in MOSFETs.

The main mechanism for radiation damage in MOSFET devices is ionizing radiation [5]. Ionizing radiation degrades the performance of FETs in two ways:

- Shift in the threshold voltage V_{th}, and

- Decrease in carrier mobility.

When an ionizing particle pass through the gate oxide it generates electron-hole pairs. As long as the device is biased during irradiation, the electrons are swept out of the oxide by the applied electric field, due to the higher mobility of the electrons, while the holes are trapped. This "hole trapping" in the gate oxide cause the the

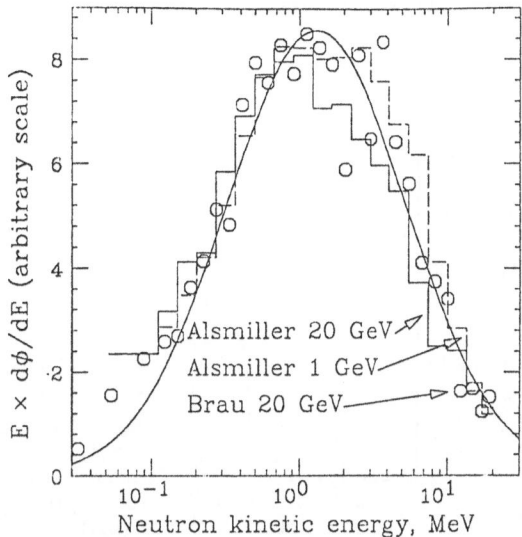

Figure 2. The calculated neutron spectrum for a uranium/scintillating calorimeter at the SSC. Figure is taken from [1].

threshold voltage to shift towards more negative values, making it easier to invert a NMOS and more difficult to invert a PMOS device. This effect can be minimized by using devices with a thin gate oxide since

$$\Delta V_{th} = -\frac{Q_{ss}}{C_{ox}} = -\frac{Q_{ss}t_{ox}}{\epsilon_{ox}}. \tag{1}$$

The threshold shift will be smaller for smaller t_{ox} since Q_{ss} is volume effect, decreasing with decreasing t_{ox}. The threshold shift caused by charge trapping in the oxide will saturarate at a maximum value, after which this mechanism will cause no further shift. Ionizing radiation also causes a build-up of interface states at the silicon/oxide interface. These states traps carriers and further shifts the threshold voltage. In NMOS devices the interface states cause a shift towards more positive V_{th} and in PMOS-devices the shift is negative.

The total effect on V_{th} in NMOS devices is a shift towards negative voltages followed by a turn-around when the "hole trapping" saturates and then a shift towards positive voltages due to the interface state build-up. In PMOS-devices the two effects are cumulative and we see a continous shift towards more negative voltages.

The interface states created by ionizing radiation also decreases the carrier mobility by providing additional scattering centers for carriers. This will translate into higher thermal noise of the channel since the effective resistance, R_{eff} is given by

$$R_{eff} = \frac{1}{g_m} \propto \frac{1}{\mu}, \tag{2}$$

where μ is the mobility.

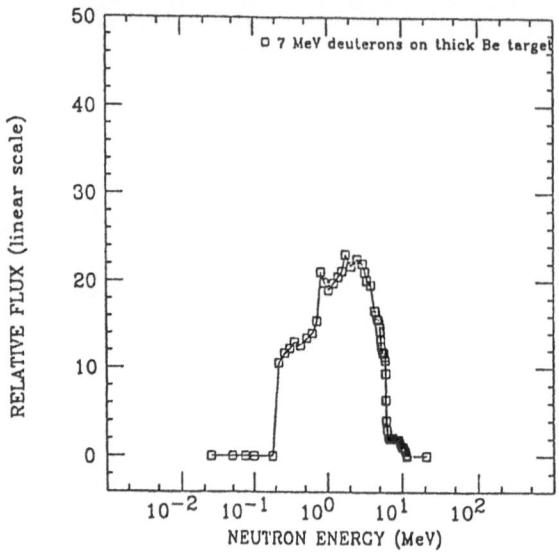

Figure 3. Neutron spectrum at fast neutron generator at Argonne, used in neutron irradiations.

MEASUREMENT TECHNIQUE

We performed irradiations on individual transitors from monolithic rad-hard CMOS/bulk and CMOS/SOS processes with both neutrons and ionizing radiation. The CMOS/SOS process is "Hughes 1 Mega Rad Hard SOS Process" and the CMOS/bulk process is UTMC 1.5 μm rad-hard UTDR process. The neutrons doses were between 3.5×10^{13} and 2×10^{14} neutrons/cm^2. Neutron irradiations were done at the fast neutron generator at the Engineering Physics Division at Argonne which generates neutrons by accelerating 7 MeV deuterons on a beryllium target. The spectrum (see Fig. 3) peaks at 2.5 MeV and produces a flux of 10^{10} neut/cm^2/sec. The devices were irradiated with ionizing radiation doses of 1 Mrad(Si), 3 Mrad(Si), and 10 MRad(Si). Irradiations were done at the high level gamma room at the Biology and Medical Research Division at Argonne. The source is a Cobalt-60 source rated up to 2 MRad(Si)/hr. A detailed description of the radiation sources at Argonne is given in [2]. The high dose rate at both the neutron and ionizing radiation source enabled us to simulate a ten year SSC dose in 4 – 8 hours, four orders of magnitude difference in dose rate. We have not addressed the issue of dose rate effects in this paper.

Parameters measured before and after irradiations include leakage currents through gate oxide, I-V curves ($I_D = f(V_{DS}, V_{GS})$), transconductance (g_m), and series noise at frequencies up to 100 kHz. Noise measurements were done using a Quantech Noise Analyzer and a HP 4145B Parametric Analyzer was used to aquire the other device characteristics. All devices were biased at $V_{DS} = 5$ V and $I_D = 100$ μA, corresponding to the digital low state of an inverter, both during measurements and irradiations. All devices were kept at room temperature during and after irradiations. During all neutron irradiations the chip-packages were activated and all post-radiation

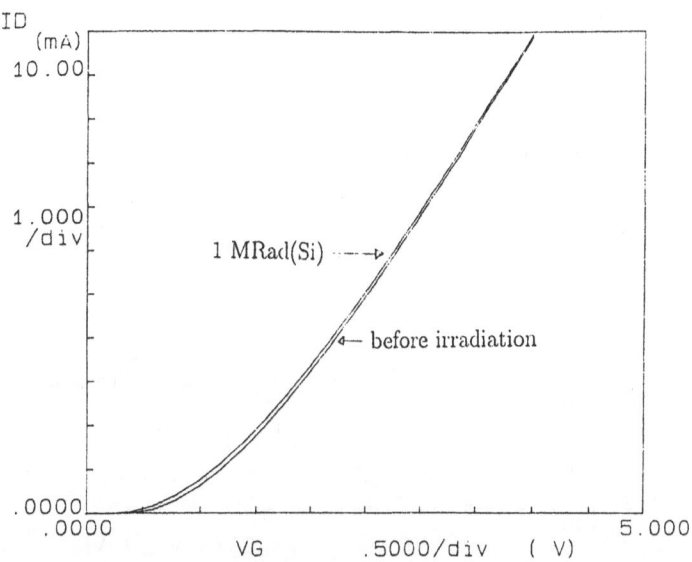

Figure 4: Threshold shift in NMOS-device from "Hughes 1 Mega Rad Hard SOS Process" after 1 MRad(Si) of ionizing radiation.

measurements were performed between three and four weeks after exposure. The devices irradiated by ionizing radiation were all measured approximately three to four days after irradiation, but several were re-measured four to six months later and showed no additional degradation or annealing effects during that time.

RESULTS

The DC-characteristics of CMOS/SOS devices showed practically no change after irradiation with up to 10^{14} neutrons/cm^2 as expected. However the low-frequency series noise increased by a factor of two at 2×10^{14} neutrons/cm^2. The transconductance remained unchanged indicating that the series noise increase is limited to the low-frequency flicker noise, not effecting the thermal noise. 1 MRad(Si) of ionizing radiation introduced a small negative shift in V_{th}, see Fig. 4, and also caused an increase in series noise by a factor of two, but these changes are within wafer to wafer variations. Transconductance was reduced by ~ 10 percent indicating a slight increase in thermal noise of the devices. Although this process is not rated up to 10 MRad(Si) of ionizing radiation, we still investigated the effects of this dose. The threshold voltage shifted with several volts and the low-frequency noise increased with a factor of two. This noise increase was accompanied by a reduction in transconductance of approximately 30 percent. No increase in gate leakage currents were noted for any of the devices.

CMOS/bulk devices were irradiated with neutron fluences between 3.5×10^{13} and 2×10^{14} neutrons/cm^2, and ionizing radiation between 1 MRad(Si) and 10 MRad(Si). At 3.5×10^{13} neutrons/cm^2 both V_{th} and noise characteristics were unchanged, while at the higher dose a small shift in the V_{th} of the PMOS device was noticed. The series noise of the p-channel device increased with a factor of two.

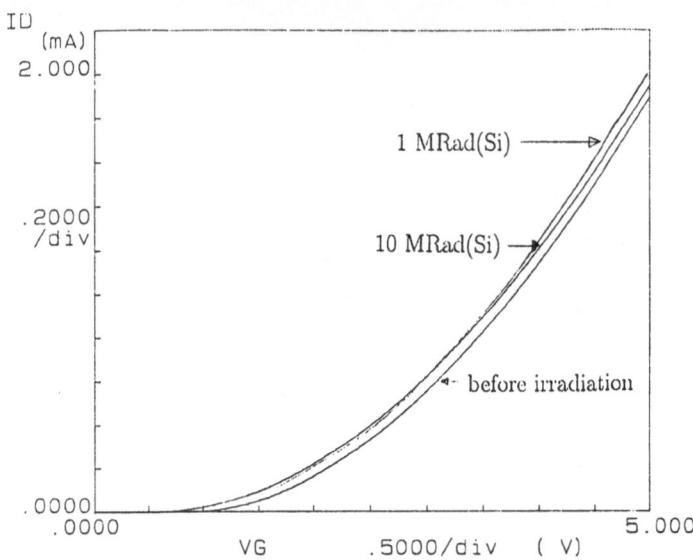

Figure 5. Threshold voltages in UTMC 1.5 μm rad-hard UTDR process for 1 MRad(Si) and 10 MRad(Si) of ionizing radiation compared to pre-rad value.

1 MRad(Si) of ionizing radiation caused a negative threshold shift of ~150 mV in both the n-channel and the p-channel device. A small increase in series noise were noted for both devices as well. With increasing radiation doses the n-channel devices went through the saturation point of "hole trapping" in the oxide and the threshold started to shift positive, partly cancelling the initial negative shift. At a dose of 10 MRad(Si) the turn-around can be seen but the positive going shift due to the inteface states is not enough to bring V_{th} more positive than it was in the pre-rad device, see Fig. 5. In the PMOS the shift is negative for all doses and the effect of additional irradiation is just to shift V_{th} even further away from the pre-rad value, making the total shift for 10 MRad(Si) approximately 200 mV. The series noise of the n-channel devices shows the effect of the continous build-up of interface states with increasing noise for each additional dose, Fig. 6. There was no change in transconductance by ionizing radiation, so the series noise increase was confined to the low-frequency flicker noise.

CONCLUSIONS

We have irradiated transistor level devices from two monolithic rad-hard MOSFET processes: UTMC 1.5 μm rad-hard UTDR CMOS/bulk process, and "Hughes 1 Mega Rad Hard SOS Process." Irradiations were performed at sources at Argonne and the doses were between 3.5×10^{13} and 2×10^{14} neutrons/cm^2 and between 1 MRad(Si) and 10 MRad(Si) of ionizing radiation.

MOSFETs are majority carrier and are not effected by neutron irradiation and we found that to be the case with our test devices. Threshold shifts caused by neutrons did not exceed 50 mV and series noise increased by a factor of two. These changes are within the variations seen in devices between different wafers and from

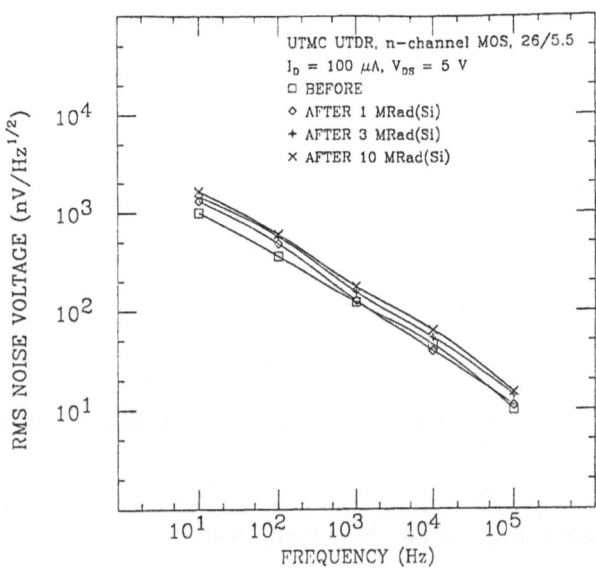

Figure 6. Series noise in UTMC 1.5 μm rad-hard UTDR process in 26/5.5 $(\mu m)^2$ n-channel device pre-rad and after 1 Mrad(Si), 3 MRad(Si), and 10 MRad(Si).

lot to lot in production. There was no change in transconductance indicating that the noise increase was limited to the low-frequency flicker noise and did not cause an increase in thermal noise, the main noise source at the fast shaping times of SSC signal processing.

Ionizing radiation is the main damage mechanism in MOSFETs thru the charge trapping in the gate oxide and the increase in surface state in the silicon/oxide interface. "Hughes 1 Mega Rad Hard SOS Process" is rated to 1 MRad(Si) of ionizing radiation and performed very well up to that dose: threshold shift was less than 100 mV and the noise increase was by a factor of two. At 10 MRad(Si) the device was still functional but threshold shift were large and noise had increased further.

UTMC 1.5 μm rad-hard UTDR CMOS/bulk process performed well up to the maximum dose of 10 MRad(Si) with threshold shifts confined to 200 mV in the worst case: p-channel threshold shift at 10 MRad(Si). The series noise increased by a factor of two at 10 MRad(Si) but no change in transconductance was noted. This indicates that the degradation in noise performance was limited to the low-frequency flicker noise.

ACKNOWLEDGMENTS

The authors would like to thank A. Smith, D. Smith, and G. Holmblad for assistance with irradiations and dosimetry. We would also like to thank E. Streets for help with neutron dosimetry. Discussions with H. Kraner, V. Radeka, and P. O'Connor of Brookhaven National Laboratory were helpful in the analysis of the results.

Work supported by the U.S. Department of Energy, Division of High Energy Physics, under contract W-31-109-ENG-38.

REFERENCES

[1] D. E. Groom, ed., "Radiation Levels in the SSC Interaction Regions," *SSC-SR-1033*, June 10, 1988.

[2] A. Stevens, J. Dawson, et al. "Rad-Hard Electronics Development Program for SSC Liquid-Argon Calorimeters," in: *The 1990 International Industrial Symposium on The Supercollider 2*, Plenum Press, New York.

[3] T. Ekenberg, J. Dawson, et. al. "Radiation Damage Testing of Transistors for SSC Front-End Electronics," *Conference Record of the 1990 IEEE Nuclear Science Symposium*, pp.843 – 845, (1990).

[4] G. H. Trilling, et al. "Letter of Intent by the Solenoidal Detector Collaboration," SDC note *SDC-90-00151*, (1990).

[5] G. C. Messenger, M. S. Ash, *The Effects of Radiation on Electronics Systems*, Van Nostrand Reinhold Co., New York, 1986.

DEVELOPMENT OF A CUSTOM MONOLITHIC DEVICE FOR DATA ACQUISITION FROM A SCINTILLATING CALORIMETER AT THE SUPERCONDUCTING SUPER COLLIDER

T. Ekenberg, J. W. Dawson, R. L. Talaga,
A. E. Stevens, and W. N. Haberichter

Argonne National Laboratory
High Energy Physics Division
Argonne, IL 60439

ABSTRACT

A clock-driven continuous sequential write/random read data acquisition architecture for a scintillating calorimeter at the SSC is presented. Simplicity of design and operation as well as potentially dead time-less operation are the motivations of this effort. The architecture minimizes the number of fast control signals, thereby reducing pickup from digital control lines by sensitive analog circuits in the front-end device. This architecture also reduces the logic necessary on the front-end device improving reliability and easing design and operation. Operation and design of the front-end device are discussed.

INTRODUCTION

A major challenge in the SSC detector development is the design of efficient data acquisition and trigger electronics under the severe constraints imposed by the high event-rate, stringent power consumption budget, and the radiation damage. The SSC will be operating with a luminosity between 10^{33} cm^{-2}s^{-1} and 10^{34} cm^{-2}s^{-1} and with a beam crossing frequency of 60 MHz, generating 10^8 events per second. Total number of read-out channels will be on the order of 5×10^5. Considerations of response time and hermeticity of the detector forces much of the calorimeter electronics to be located deep inside the detector where they are suspect to high radiation doses as well as forced to be minimized with respect to physical space and power.

In order to design the electronics to be placed in the immediate vicinity of the detector elements one has to optimize with respect to several constraints: power consumption, speed, noise performance, packing density, radiation damage, and a high degree of multiplexing to reduce cable volume. Many of these constraints are inversely related to each other, and there are always trade-offs to be made. Another important design consideration is the effect of a large number of high speed digital control signals switching in the vicinity of sensitive analog devices.

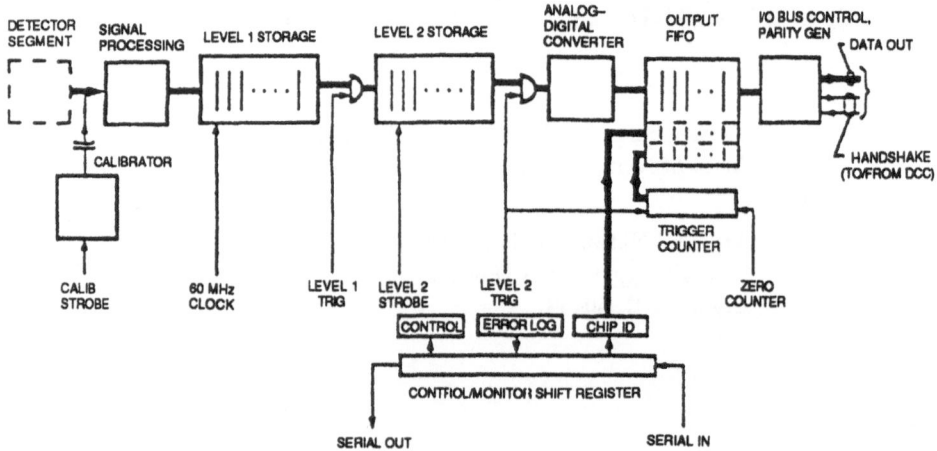

Figure 1. Generic DAQ design for SSC detector. Figure from [1].

A generic DAQ architecture for SSC detectors has been developed in the High Energy Physics community and is outlined in Fig. 1. In this architecture the detector signal is routed into a signal processing unit and through the level 1 pipeline/storage device and is transferred to level 2 storage on a level 1 trigger. The signal is held until a level 2 trigger accept arrives. At this time it is digitized and stored in the output FIFO, after which the signal is shipped to the level 3 processing. The high event-rate of the detectors (10^8 events/second) dictates that the event reduction done by the trigger must be on the order of $10^3 - 10^4$ for the level 1 trigger, $10^1 - 10^2$ for the level 2 trigger, and another factor of $10^1 - 10^2$ in the level 3 processing. This results in approximately 100 events/second written to permanent storage.

In this paper we propose an architecture for a scintillating calorimeter DAQ. It consists of a front-end device, which may be implemented as a VLSI circuit, that stores the data and digitizes valid level 2 events with an on-board ADC, and a central control unit that interfaces the front-end device to rest of the DAQ and to the trigger system. This architecture limits the number of fast control signals to the front-end device to two, and minimizes the logic on the front-end device.

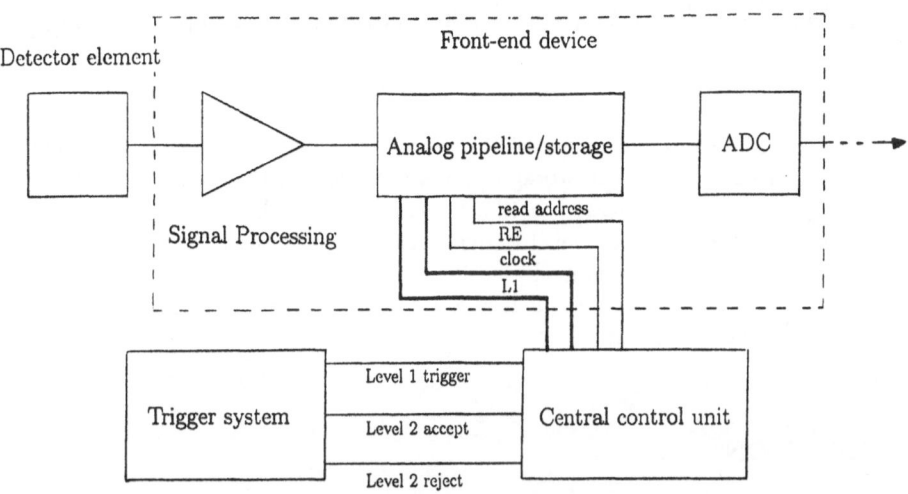

Figure 2. Schematic of proposed architecture. Border of front-end device outlined as well as the flow of control signals from the central control unit (CCU) and trigger system. Fast control signals are marked with heavy lines.

ARCHITECTURE

The motivation for the proposed architecture is simplicity. In order to ease design, trouble-shooting, and operation of the entire calorimeter system, an architecture has been developed that minimizes the number of fast and slow control signals, and reduces the amount of logic on the front-end device. The only fast signals necessary are the global clock and the level 1 trigger. Logic at the front-end device is limited to an array write-selector, decoding logic for the loading of the read addresses, and some logic necessary for diagnostics.

The reduction in total number of control signals is important in order to minimize the cable count, but the implementation of a front-end device with only 2 signals that must be driven at 60 MHz also improves noise performance of the system. This is the case since mixed analog/digital signal systems are highly sensitive to cross-talk between the high-frequency digital signals and sensitive small-signal analog circuits. The motivation of reducing the controlling logic on the front-end device is two-fold. Since all front-end devices for the calorimeter would be the same, significant savings in silicon area can be accomplished by collecting all controlling logic in one place. Reliability is also improved by implementing only very simple functions on the front-end device which will be largely inaccessible except during special circumstances. This system will be easier to operate, trouble-shoot, and repair with all control logic in a central location.

The architecture is implemented with a front-end device and a central control unit (CCU), as shown in Fig. 2. The front-end device is a mixed analog/digital signal IC, or set of ICs. The controlling signals from the CCU to the front-end device are the Level 1 (L1) trigger, global clock, Read Enable (RE), and the read address word. Included in the front-end device is the signal processing, the analog pipeline/storage, and the ADC.

The front-end device of the proposed architecture consists of eight arrays of analog storage sites. The storage sites are capacitors that record and store charge or voltage signals. Each array is a 128 elements deep pipeline, *i.e.* a 2 μs deep memory that holds the data until the level 1 trigger makes its decision. Each array is controlled locally for writing and reading by a 128 bit long loadable shift register. This shift register selects elements for sequential write or random read operations as instructed by the central control unit.

OPERATION

All interfaces between the trigger system and the DAQ system, as well as between level 3 processing and the front-end device of the DAQ, are handled by the CCU. When the calorimeter system is enabled for data acquisition, the CCU enables the front-end devices by write enabling the first array. The shift register associated with the write selected capacitor array is clocked by the 62.5 MHz global clock, writing each array element in sequence. When the write pointer reaches the end of the array, it wraps around and starts at the beginning of the array again. This continues until a level 1 trigger arrives at the CCU. The CCU logs the trigger and issues a L1 trigger to the front-end devices. At this time the array selector is incremented, write enabling the second array, and writing commences with the first element of the second array. The analog data is continuously written to the second array until the next L1 trigger arrives. At this time the array-selector is incremented again and writing continues at the beginning of the third array. In this manner each array is written in order, and when the last array contains a L1 trigger, the write pointer wraps around to the first array again. See Fig. 3 for functional lay-out of front-end device.

The L1 trigger is a fast signal and its arrival must be precisely defined and timed since it indicates valid data a fixed-time interval preceding its arrival. The CCU maintains the correct timing relationship of the L1 trigger with respect to the level 1 trigger signals. When the level 2 trigger arrives at the CCU it is processed and the storage site address word corresponding to the triggering event is generated. This address word is presented to the front-end device and is latched at the front-end device when the CCU issues a read enable (RE) to the front-end device. RE and the read address lines are slow signals. When the address is latched the selected site is connected to the ADC and the value digitized and outputted through to the level 3 processing. This system is flexible with respect to the number of storage sites to be digitized since this can be changed in the CCU and the front-end devices remain uneffected. This is important since the actual number of samples needed for baseline subtraction is not yet fixed.

This operation is dead time-less in our application. An event rejection ratio in the level 1 trigger system of 10^3 to 10^4 produces level 1 triggers on the average between every 16 to 160 μs [3]. Level 2 trigger decisions are made approximately every

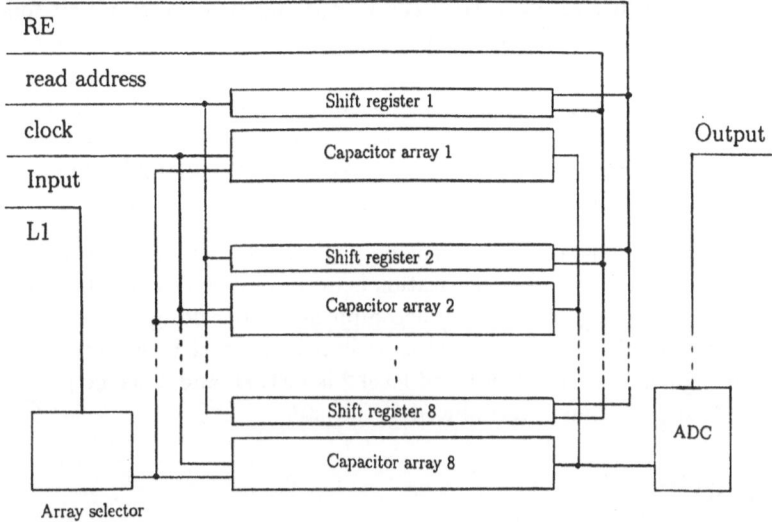

RE

read address

clock

Input

L1

Shift register 1

Capacitor array 1

Output

Shift register 2

Capacitor array 2

Shift register 8

Capacitor array 8

ADC

Array selector

Figure 3. Functional lay-out of front-end device. Each capacitor array contains a write select decoder and a read address decoder. Input is a analog signal from the detector and output is the digitized signal going to level 3 processing.

10 μs with a maximum latency of 50 μs. This will allow continuous operation with eight storage arrays since level 2 decisions will be made before the array comes up in the write queue again. No special level 2 reject is necessary since the storage arrays are just over-written every time they come up in the write queue.

Since the local shift registers are cleared on every L1 trigger the device is automatically re-synchronized at that time. A local counter on the front-end device tracking the global clock outputs its value for each valid event, allowing for a global check on the timing integrity of the data at the level 3 processing. If corrupt timing data is found the event is discarded without any need for re-synchronizing of the front-end devices. This means that corrupt data is always isolated to one event and does not effect subsequent events.

DESIGN

Several different schemes are currently being investigated for implementing the front-end device. One way is to fold the signal processing and storage unit into one by integrating directly on the storage capacitors. This approach has been investigated previously by Anghinolfi, et.al. [2]. Integrating directly on a storage capacitor is a very attractive option in a scintillating calorimeter where the current pulses from the photo multiplier tubes are expected to be confined to less than 16 ns, the time between beam-crossings. This can be implemented with a high speed operational amplifier with the storage capacitors switched in and out of its feedback path every 16 ns. The inverting input of the op amp is connected directly to the base of the photo multiplier tube.

One can also use a time continuous signal processing unit followed by a sample and hold circuit. In this approach a pre-amplifier converts the current signal from the detector to a voltage signal which is processed through a shaping amplifier. The output of the shaping amplifier is then sampled at discrete time-intervals and stored on a analog storage device pending digitization. This is especially well suited for systems with slow signals that will be extended over many beam-crossings at the SSC, like liquid Argon calorimeters.

To keep the data on the same physical storage location during the entire time between initial write and digitization is important since every transfer introduces a degradation of analog data. This is accomplished in our design by using the capacitor arrays as the initial memory element as well as both level 1 and level 2 storage. The transfer of signals between level 1 and level 2 is virtual, and since no physical signal is transferred no dead time is introduced at this step.

SUMMARY

A clock-driven continuous write/random read DAQ architecture that limits the number of fast control signals to two and minimizes the logic at the front-end is presented. This improves reliability and makes design and operation simpler.

The interface between the front-end devices and the trigger system and the level 3 processing units is the central control unit (CCU). The CCU controls all front-end device of the calorimeter system. Each front-end device consists of eight capacitor arrays and a few simple logic functions. During data acquisition the data is being written to the write selected array every 16 ns. The array is 2 μs deep and is continuously being over written until a first level trigger arrives. The next array is then write enabled and writing continuous at the beginning of that array. This is repeated for every level 1 trigger until all eight arrays are full. At this time the writing continues on the first array again.

The operation of this architecture is dead time-less in our application and the front-end devices are automatically re-synchronized on every level 1 trigger by virtue of always start writing at the beginning of each array. This isolates events with corrupted timing information from subsequent events, and no special re-synchronization must be carried out.

ACKNOWLEDGMENTS

Work supported by the U.S. Department of Energy, Division of High Energy Physics, under contract W-31-109-ENG-38.

REFERENCES

[1] H. H. Williams, et al., "Front End Electronics Development for SSC Detectors, " *SSC-PC-031*, 1989.

[2] Anghinolfi, F., et al., "Monolithic CMOS Front-End Electronics with Analog Pipelining," *Conference Record of the 1990 IEEE Nuclear Science Symposium,* pp. 543 – 552, (1990).

[3] G. H. Trilling, et al., "Solenoidal Detector Collaboration Expression of Interest," SDC note *SDC-90-00085,* 1990.

DEVELOPMENT OF PIXEL DETECTORS FOR SSC VERTEX TRACKING*

Gordon Kramer
Hughes Electro-Optical Data Systems Group
El Segundo, CA 90245

Eugene L. Atlas, F. Augustine, Ozdal Barkan, T. Collins,
Wayne L. Marking, Stuart Worley, and Ghassan Y. Yacoub
Hughes Technology Center, Carlsbad, CA 92009

Stephen L. Shapiro
Stanford Linear Accelerator Center
Stanford University, Stanford, CA 94309

John F. Arens and J. Garrett Jernigan
Space Sciences Laboratory
University of California, Berkeley, CA 94720

David Nygren, Helmuth Spieler, and Michael Wright
Lawrence Berkeley Laboratory
Berkeley, CA 94720

P. Skubic
University of Oklahoma
Norman, OK 73019

ABSTRACT

A description of hybrid PIN diode arrays and a readout architecture for their use as a vertex detector in the SSC environment is presented. Test results obtained with arrays having 256×256 pixels, each 30 μm square, are also presented. The development of a custom readout for the SSC will be discussed, which supports a mechanism for time stamping hit pixels, storing their xy coordinates, and storing the analog information within the pixel. The peripheral logic located on the array, permits the selection of those pixels containing interesting data and their coordinates to be selectively read out. This same logic also resolves ambiguous pixel ghost locations and controls the pixel neighbor read out necessary to achieve high spatial resolution. The mechanical and thermal design of the vertex tracker and the proposed signal processing architecture will also be discussed.

INTRODUCTION

Pixel detectors are uniquely suited for vertex tracking close to the beam-line in high energy accelerators such as the Superconducting Super Collider. Their unique characteristics are a result of their small sensitive area, which is generally less than 10^4 square microns. The detectors described herein are silicon PIN diode arrays.

---

* Work supported by Department of Energy contract DE–AC03–76SF00515.

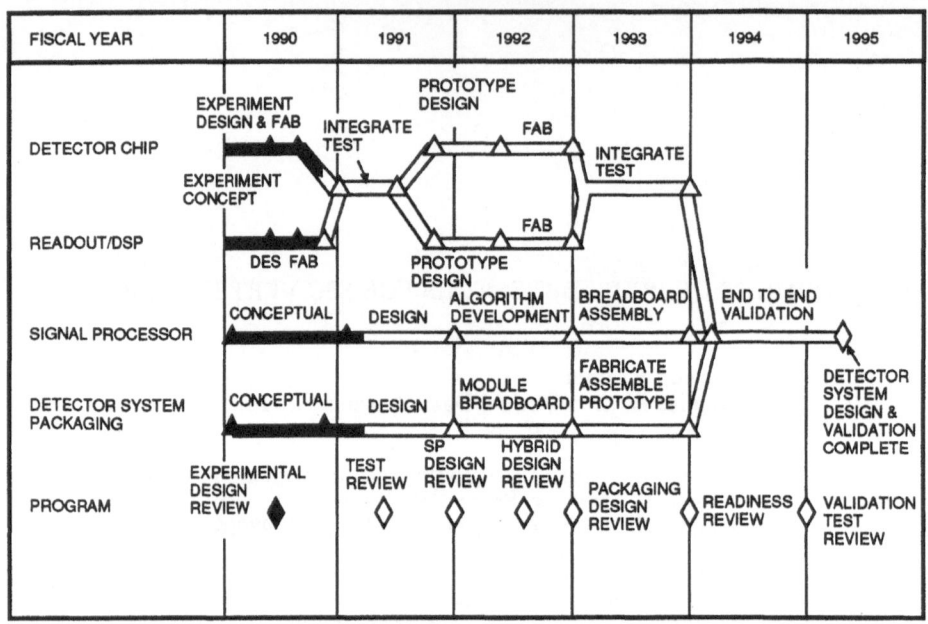

Fig. 1. SSC pixel development time line.

A principal benefit is high resolution position determination to better than 10 μm accuracy. The resulting high resolution tracking provides for accurate momentum measurements and the determination of secondary vertices. Interesting events in the Super Collider will come from the decay of short lived particles which travel a short distance from the primary interaction point and then decay, creating a secondary vertex. Additionally, the pixel's small area allows high track densities and multiple interactions to be resolved, since the probability of multiple hits within the same detector during an event is small. Lastly, the low pixel capacitance results in a signal to noise ratio from minimum ionizing particles of greater than fifty to one. This allows for improvements in system design and a reduction in the effects of radiation. The signal-to-noise ratio degrades due to an increase in noise from radiation induced increases in dark current. The small pixel area results in an exceptionally small initial dark current. Pixel detectors are, thus, well suited to applications close to the beam-line where radiation dosages are highest.

Hughes Aircraft Company is participating in a pixel detector development collaboration with high energy physicists from various universities and national laboratories.[1] This collaboration has been in place for the last two years. Many of the tests and results described in this paper, however, are attributable to the earlier effort funded by the SSC Generic Research Program.[2] The collaboration hopes to capitalize on the technology developed by Hughes Aircraft Company through defense department contracts related to infrared sensor technology over the last 20 to 30 years. The long range plan of the collaboration is the development of a validated end-to-end pixel detector system which can serve as a vertex detector at the SSC. This plan is shown in Fig. 1. The collaboration is presently on schedule, in having demonstrated a working room-temperature hybrid pixel array,[3] and is presently designing and testing the first pixel detectors designed specifically for SSC application.[4] Together with the design of the individual detector chips, the collaboration is pursuing the overall system design. This design includes the detector, the electrical interconnections, the cable packaging, the mechanical and thermal design, and the signal processing. The successful implementation of a vertex detector requires consideration of these system issues in the early stages of development.

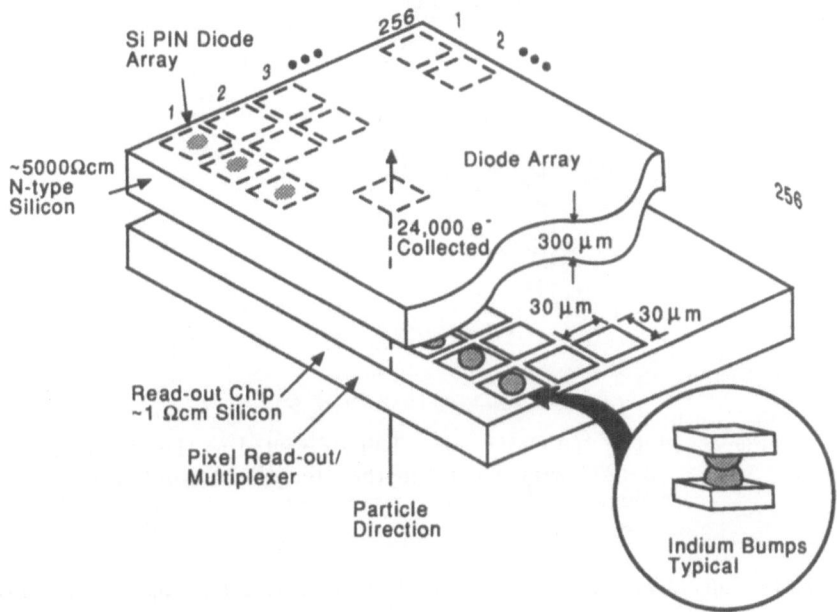

Fig. 2. Schematic Representation of pixel detector hybrid.

PIXEL ARRAYS

The individual detector arrays employ the hybrid approach which has proved extremely successful for infrared sensor arrays. A detector chip, which is an array of silicon PIN diodes, is connected via indium bump interconnects to another silicon array of readout unit cells.

The hybrid approach allows each array to be processed separately and individually optimized. Thus, one can change either the detector design or the readout design more readily. From a manufacturing point of view the yield losses in the detector and readout processing are not compounded because they can each be individually selected prior to hybridization. This is shown schematically in Fig. 2. Figure 3 is a photo of a pixel detector array tested last year. It uses an existing x-y scanned readout. This is a 256 × 256 array with 30 μm square pixels . A detector of this type was exposed to

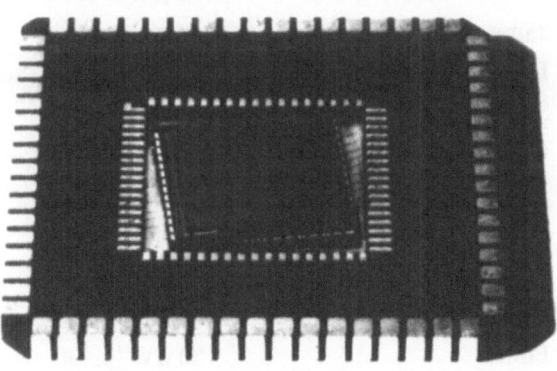

Fig. 3. Photograph of 256 × 256 Array having 30 μm square pixels.

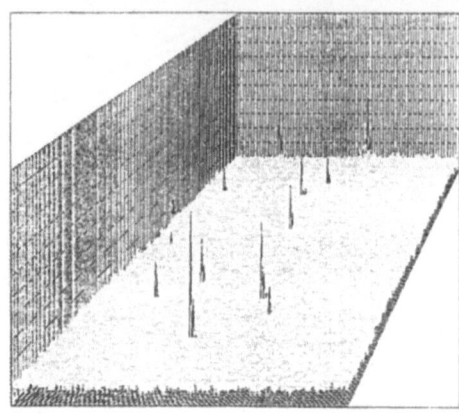

Fig. 4. Plot of 250 GeV/c pions being detected as they traverse a 256 × 256 array illustrating the strengths of two-dimensional readout to aid in pattern recognition.

250 GeV/c pions at Fermilab, and produced signals with a high signal to noise ratio. A plot of the output of one of these pixel detector arrays is shown in Fig. 4. This plot can be viewed as a proof-of-concept that a significant improvement in pattern recognition can be obtained with two dimensional pixel detectors. Accurate x-y positioning of the particle hit within the 30 μm pixel area is shown. High signal to noise ratios, greater than 50:1 were obtained and charge sharing between adjacent pixels is also evident, thus, allowing very accurate determination of the center of the charge cloud created by the incident particle.

The SSC environment imposes unique requirements on the readout electronics due to its high interaction rate. It is not feasible to read out every pixel in a frame by scanning each frame completely since the beams cross every 16 ns. Therefore, a smart readout is required which allows one to record only those pixels which have been hit. A further level of data reduction is the sparse readout feature, which allows the selection of only those hits deemed interesting by an external trigger. Thus the data rate is significantly reduced by reading out only those pixels which have been hit during interesting events.

Table 1.   SSC Prototype Array Parameters

Parameter	Goal
Array format	64 × 32
Pixel dimension	50 μm × 150 μm
Time stamp	50 ns
Digital address storage	2 hits/trigger level 1
Smart/sparse readout	Yes
Ghost elimination	Yes
Analog data storage	Yes
Maximum signal	50,000 $e-$
Minimum signal definition	3,000 $e-$
Noise at room temperature	$\leq$ 300 e^-
Power	20 μW/pixel
Readout time	< 10 μs/array

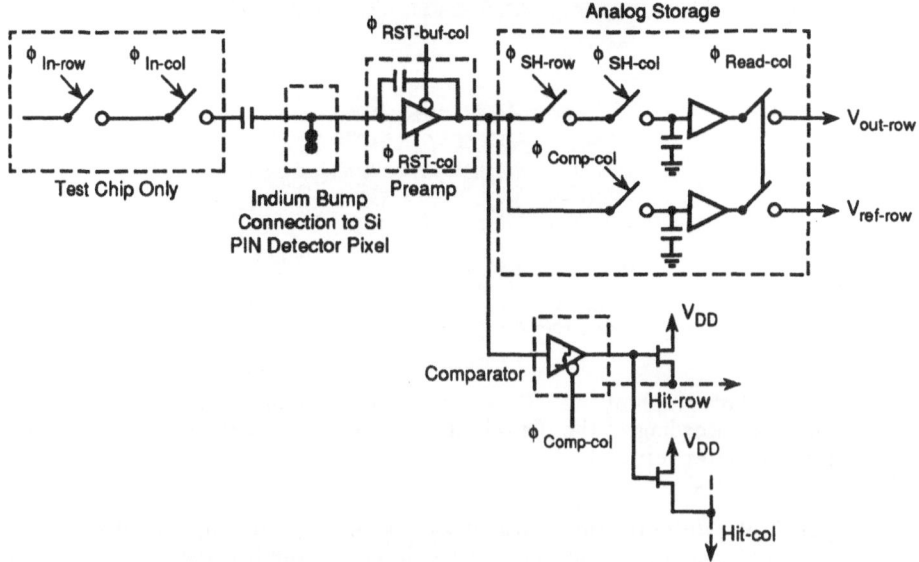

Fig. 5. Schematic of the simplified unit cell architecture.

SSC PROTOTYPE

A readout is under development to meet the SSC requirements. The pixel dimensions are 50 μm × 150 μm in a 64 × 32 array. The goals set for this first prototype are listed in Table 1. The critical parameters are a noise level of less than 300 electrons rms at room temperature, power dissipation of less than 20 μW per pixel, a readout time for the array of less than 10 μs, and the ability to time stamp an event within 50 ns. The unit cell architecture is shown in Fig. 5. The unit cell behind each pixel detector includes an amplifier, analog storage, and an analog comparator. Following a first level trigger, one can return to the pixel and retrieve the stored analog charge within. The test chip, which is a precursor to the prototype array is shown in Fig. 6. This chip allows one to simulate a hit in a pixel by inputting electronically a signal to the unit cell. There is an output signal which indicates that a hit occurred within the array. The hit-row and hit-column signals are read out on serial outputs on the periphery. One can then go back and, using the row and column select lines, read out the analog signal.

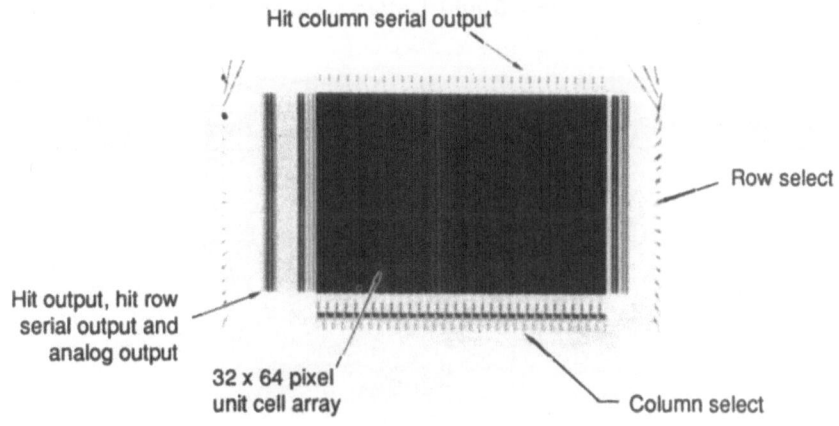

Fig. 6. Photograph of the 32 × 64 SSC array under evaluation.

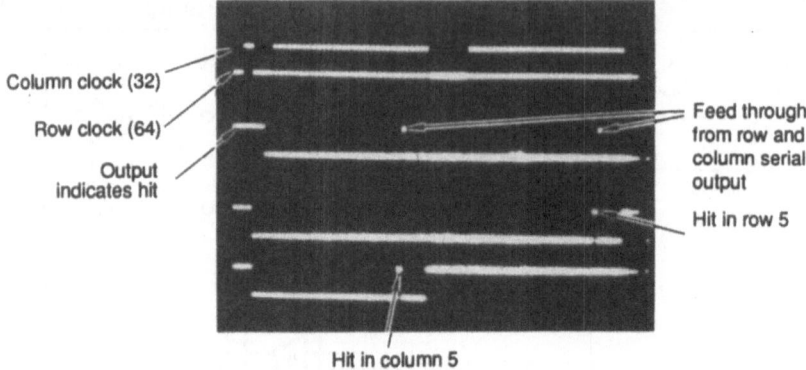

Column clock (32)

Row clock (64)

Output indicates hit

Feed through from row and column serial output

Hit in row 5

Hit in column 5

Fig. 7. Scope photograph of the SSC array readout demonstrating the time stamping and the recording of the digital address. Analog information in stored in the pixel for later retrieval.

Figure 7 illustrates the smart readout operation on a test chip. A signal simulating the deposition of a charge of 5,000 electrons is input via the test row and column inputs to the pixel at the intersection of column 5 and row 5. The scope photograph shows the column and row clocks on the upper trace. The scope trigger is at the extreme left of the trace, and the clock duration is set to two microseconds per clock pulse. The second trace is the output hit, which is indicated by the rising edge of the signal. The hit signal and the strobing of the charge into the pixel are simultaneous to within the resolution of this photo. The spurious pulses on this trace represent cross talk within the prototype array, and are generated by the digital signals on the row and column output lines. Their cause has been determined and is being eliminated on the next prototype array. The row and column outputs can be interpreted as a FIFO and locate the row 5 and column 5 position of the hit. Thus, the array indicates a hit on a specific pixel at a specific time. The hit information is stored in memory, and the analog information is stored in the unit cell for later retrieval. Figure 8 is a photograph showing an analog signal being retrieved from the selected pixel, (in this case, a 25000 electron signal in row 3, column 3). The upper trace is the output of the 32 rows (only 22 shown) of column 3 and the output analog signal is seen in row 3. Thus, the test array has demonstrated the smart and sparse readout capability. The second iteration of this array prototype will be hybridized to a detector array for further testing.

RADIATION HARDNESS

The present array is fabricated in a nonradiation hard, L_{eff} equal 1.2 μm, single poly, double metal (pitch = 3.5 μm) CMOS process. Subsequent processing will

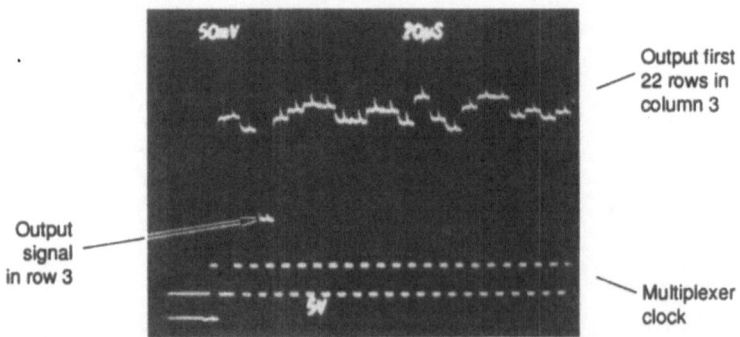

Output first 22 rows in column 3

Output signal in row 3

Multiplexer clock

Fig. 8. Scope photograph of the analog readout from a selected pixel.

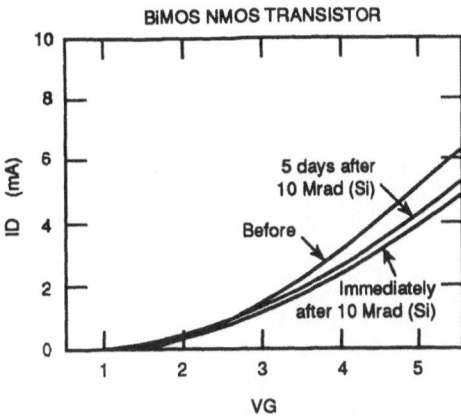

BiMOS NMOS TRANSISTOR

5 days after
10 Mrad (Si)

Before

Immediately
after 10 Mrad (Si)

Fig. 9. Curve of drain current versus gate voltage of an NMOS transistor before and after irradiation with 10 MRad of ionizing radiation.

be done using a radiation hard process. An example of a radiation hard process in hand at Hughes Aircraft Company is illustrated in Fig. 9. The transfer characteristics of an NMOS transistor are measured both before and after a 10 MRad dose of ionizing radiation[5] and a 25% drop in drain current observed. To completely demonstrate radiation hardness, the actual circuit under discussion must be irradiated. To this end, the unit cell readout described in this paper has also been fabricated in a radiation hard SOS/CMOS process and a number of die await testing in the laboratory.

THINNED READOUT

In the design of a complete vertex detector system, one must be concerned with its total mass. Silicon detectors previously used in high energy physics have been 300 μm thick. The proposed detector array of PIN diodes discussed in this paper should also be of the order of 300 μm thick for optimal signal-to-noise. Using the hybrid approach, the readout chip, which is also in the path of the particle, will introduce additional mass to the detector array. One might naively view this technology as a doubling of the mass of the system. This, however, is not the case. Figure 10 illustrates the ability to back-thin the readout from its initial thickness of 300 μm to 50 μm after the hybrid has been created and tested. This technique has been used on a number of Hughes projects in the past producing thinned, working arrays. The hybrid pictured here has a 300 μm thick PIN detector and a 50 μm-thick readout for a total of thickness of 350 μm of silicon.

MECHANICAL AND THERMAL DESIGN

Figure 11 shows a proposed mechanical design for the vertex tracker. A significant feature of this design is the use of a ceramic foam of silicon carbide (SiC) as a structural element. The pixel detector arrays are mounted in a louvered fashion on the barrel cylinders and mounted flat on the end cap disks. The SiC foam which has a density of 3% of bulk SiC has a radiation length of 337 cm, and offers a minimal mass for scattering and a porosity which allows gas cooling of the system. A 1-cm thick foam support structure is equivalent in scattering mass to that of a detector hybrid, about 0.3% of a radiation length each. The ability to machine the SiC foam to provide the necessary reference surfaces is shown in Fig. 12. Boron carbide, not used here, is also available as a foam at 3% of its bulk density. This foam has a radiation length of 693 cm, but has slightly worse thermal characteristics. The use of ceramic foams provides a uniform mass distribution design with minimal frame structure thereby reducing mass concentrations. The gas cooling system adds no contribution to the scattering mass, and requires no assembly or maintenance. Table 2 summarizes the benefits of this unique design.

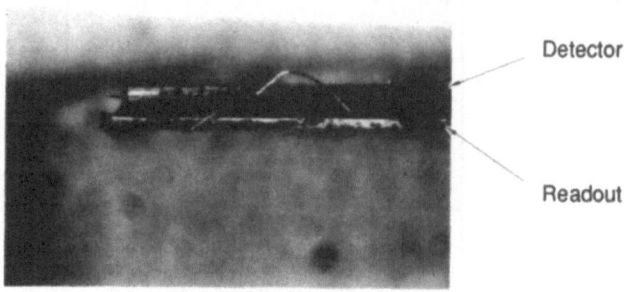

Detector

Readout

Thickness reduced
from 300 μm → 50 μm

Fig. 10. Photograph of a working detector hybrid which has been back-thinned so that the readout electronics is only 50 μm thick.

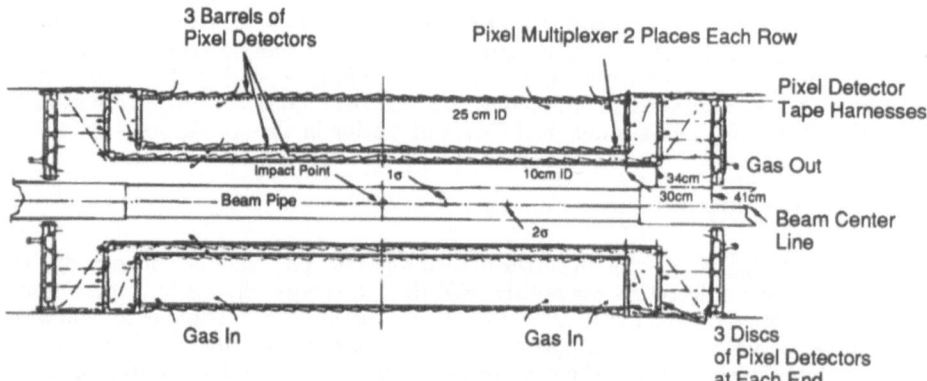

3 Barrels of
Pixel Detectors

Pixel Multiplexer 2 Places Each Row

Pixel Detector
Tape Harnesses

25 cm ID

Impact Point   1σ   10cm ID

Gas Out

Beam Pipe   34cm
30cm 41cm

2σ

Beam Center
Line

Gas In

Gas In

3 Discs
of Pixel Detectors
at Each End

Fig. 11. Schematic representation of the proposed mechanical design of a three layer vertex detector for the SSC, showing the silicon carbide (SiC) foam structural features, including the gas cooling.

Fig. 12. Photograph of a SiC foam substrate demonstrating its machineability.

Table 2. Features of the Present Design Concept Related to the Design Requirements

REQUIREMENT / FEATURES	MINIMUM RADIATION LENGTH	THERMAL DISSIPATION	POSITIONAL ACCURACY	EASE OF ASSEMBLY MAINTENANCE	COVERAGE η =0–2.5 (90° – 10° COVERAGE)	RADIATION HARDNESS	LONG LIFE
STRUCTURE OF CERAMIC FOAM MONOCOQUE	TOTAL MASS <10% RADIATION LENGTH UNIFORM DISTRIBUTION	–COMBINED HEAT TRANSFER & STRUCTURAL FUNCTION	RESOLUTION <10 μm R, θ	–REPEATABLE MACHINING & ASSEMBLY	-OVERLAPPED CHIPS FOR ~100% FILL FACTOR	>10 Mrads	>10 YEARS
GAS COOLING	NEGLIGBLE CONTRIBUTION	-33°C TO 0°C OPERATING RANGE 0.25 W/cm² (average)	-RESOLUTION <25 μm in Z DIRECTION	-NO ASSEMBLY -NO MAINTENANCE	-NO CONCENTRATED MASS	-UNAFFECTED	>>>10 YEARS
MODULAR DESIGN 1 PIXEL CHIP 9 STRIP CHIPS 144 SUBASSYS			-LARGE SUBASSYS REDUCES TOLERANCE BUILDUP	-REMOVABLE SUBASSY -VERTEX DET. REMOVABLE FOR TRACKER CALIBRATION	-FULL COVERAGE ACHIEVED		MODULES CAN BE REPLACED

Laboratory tests have shown that passing nitrogen gas through SiC foam at velocities of about 40 cm/s can dissipate 0.25W/cm² with a temperature differential of about 4°C across one layer.

The mechanical design uses one basic pixel detector array which is 2 cm × 2 cm, and the complete system of barrels and end caps contains 3,004 chips. The louvered construction provides and overlapping of chips for a 100% lapping fill factor. If one were to retain the pixel size at 50 μm ×150 μm, and orient the 150 μm dimension along the beam axis, then one would achieve a spatial resolution of about 10 μm rms in the azimuthal direction, and about 30 μm rms in the z direction. It is anticipated that in the final design, the long dimension can be reduced. Should we elect to do so, the spatial resolution in the z direction will be reduced as well.

SIGNAL PROCESSING

To complete the system design, one needs to consider the system architecture. The concept presently being explored uses an intermediate data acquisition chip to collect the signals from up to 64 pixel arrays for data compression and sequencing. The multiplexed data is then sent to an outside demultiplexer and momentum and vertex processor to determine the track parameters. The decision to send data from the pixel arrays to the data acquisition chip is made by a first level trigger which comes from sources external to the vertex detector, generally from a calorimeter. In some experiments, parameters from the pixel data could be used as part of the first level trigger. The signal processing architecture is shown in Fig. 13. Table 3 is referred to as an N^2 chart. Various hardware elements of the system are shown on the diagonal. The columns contain the inputs to each of the hardware element, and the rows contain the outputs of each hardware element. Thus, for the data acquisition chip, one can see the inputs coming from the pixel detector hit data, and command inputs from the controller. The outputs from that chip would be a bit stream going out to the smart demultiplexer, sequence commands, to the pixel detectors, and status information to the controller.

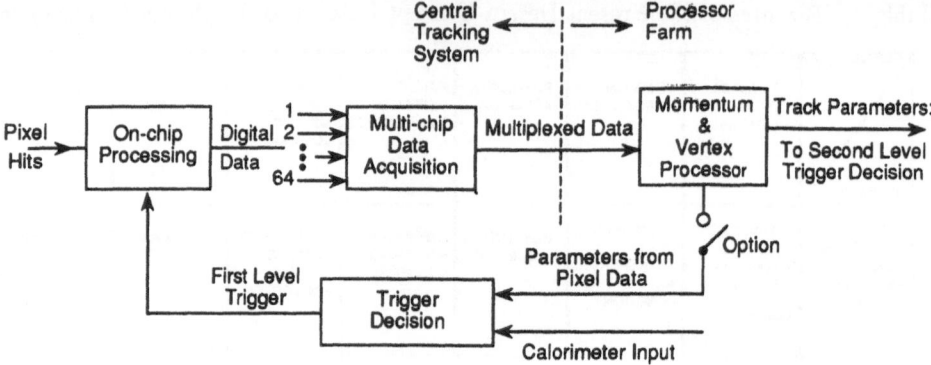

Fig. 13. Schematic representation of a strawman pixel signal and data processor architecture.

SUMMARY

Development is proceeding on schedule for the design of a complete pixel detector subsystem. A proof of principle demonstrating the advantages of room-temperature, hybrid, two-dimensional pixel detectors has been achieved. Similarly, a preprototype array has demonstrated the smart, sparse readout concept under development. Radiation hard processing is available for implementation of the readout electronics. A mechanical and thermal design concept has been established which is a simple low mass, modular, design. The ability to produce thinned hybrids has been demonstrated, and signal processing and architecture issues are being addressed.

Table 3. An N^2 Chart for the Pixel Signal and Data Processor

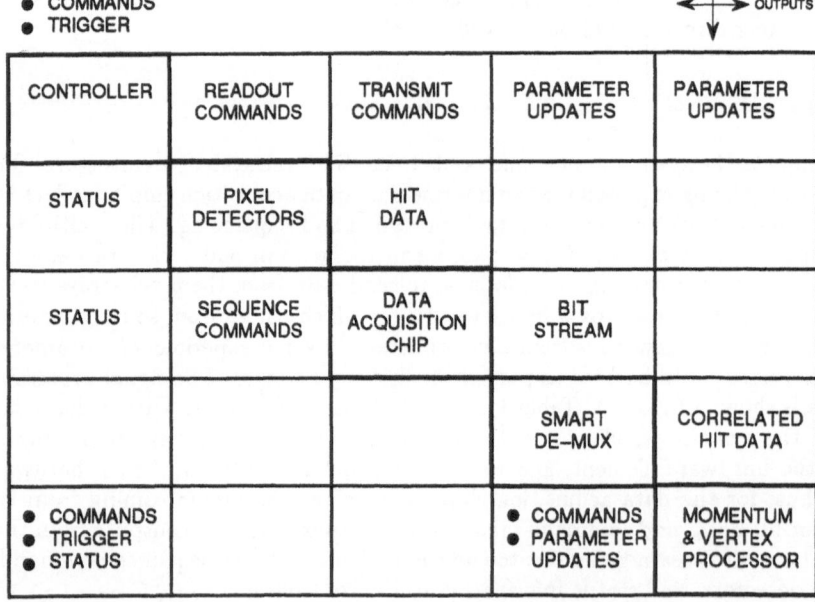

REFERENCES

1. G. Kramer et al., "The Pixel Detector Development Collaboration," Summary Report for FY'90 and Proposed Effort for FY'91, presented to the SSC Laboratory in September 1990.

2. S. Shapiro, W. Dunwoodie, J. Arens, and J.G. Jernigan, "Microdiode Arrays for Charged Particle Detection," Research Proposal Submitted to the SSC Laboratory August 1987 and August 1988.

3. S. Shapiro, J. Arens, J. G. Jernigan, G. Kramer, T. Collins, S. Worley, C. Wilburn, and P. Skubic, "Performance Measurements of Hybrid PIN Diode Arrays," Proc. Symp. on Detector Research and Development for the Superconducting Super Collider, Fort Worth, Texas, 1990; SLAC-PUB-5357.

4. O. Barkan, E. Atlas, W. Marking, S. Worley, G. Yacoub, G. Kramer, J. Arens, J.G Jernigan, S. Shapiro, D. Nygren, H. Spieler, and M. Wright, "Development of a Customized SSC Pixel Detector Readout for Vertex Tracking," ibid.; SLAC preprint SLAC-PUB-5358.

5. T. Ekenberg, J. Dawson, W.Haberichter, R. Talaga, V. Radeka, S. Rescia, and H. Kraner, "Rad-Hard Electronics Study for SSC Detector," these proceedings.

A SUPERCONDUCTING AIR CORE TOROID MUON SPECTROMETER FOR THE

EMPACT/TEXAS DETECTOR

P. Spampinato, T. Brown, J. Connelly, S. Gottesman, J. Klafin,
J. Mueller, and J. Pusateri

Grumman Space and Electronics Division
Grumman Corporation
Bethpage, N.Y.

INTRODUCTION

The primary effort after the submittal of the Expression of
Interest (EOI) [1] in May 1990 was downscoping the EMPACT/TEXAS
detector system to reduce cost. This required parameterizing the
various detector systems to determine performance-cost sensitivities,
and design modifications to determine absolute cost reductions. The
result of this work was the Letter of Intent (LOI) [2] document
submitted to the SSCL in December 1990. In support of the LOI, work
during this phase of conceptual design focussed on expanding the
analysis and design of the superconducting toroid magnets, developing
analysis and design details for the muon detector spectrometers, and
modifying the support structure of the overall muon detector system.

SYSTEM CONFIGURATION

The overall configuration consists of a central toroid made up
of 72 superconducting coils contained in a dewar, and two end toroids
each consisting of 72 superconducting coils. The coils are attached
to overhead support beams through dewar-enclosed tie rods that are
structurally isolated from the dewar vessels. The vacuum vessel
structure for each dewar is also attached to the overhead beams by
means of room temperature tie rods. The toroids are surrounded by
paired assemblies of muon spectrometers that are mounted to common
structures. These are also attached to the beams by means of tie
rods. Figure 1 shows the overall assembly of the detector system in
the underground hall.

The muon spectrometer panels (superlayers) are made up from one
meter wide chamber modules that are oriented in bend and non-bend
planes. Each detector pair consists of two superlayer panels joined
by a rigid truss structure. Each pair is assembled in an alignment
fixture located in the surface assembly facility. Completed
superlayer pairs are then transported to the hall for installation
around the toroids.

SUPERCONDUCTING TOROIDS

The air core toroids were reconfigured and down-sized to
establish an optimum momentum resolution, consistent with the physics

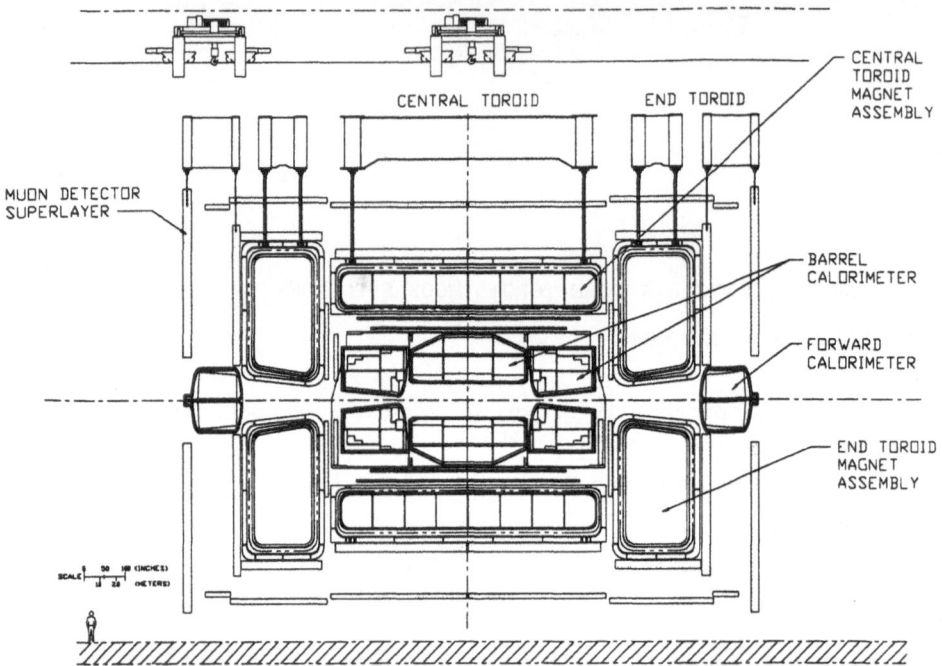

Figure 1. Elevation view of the assembled detector system.

objectives and cost guidelines set by the SSC Laboratory. With
respect to the EOI, the central toroid inner radius was increased by
1.15 m to allow space for the addition of a second muon chamber in
the inner bore and for the enlargement of the calorimeter. The
outside central toroid radii was unchanged and the peak field was
held at 1.5 T. The length of the end toroid was reduced by 0.5 m and
the peak field was reduced from 4.5 T to 3.5 T. The total magnet
system stored energy was reduced from 2.9 GJ to 1.9 GJ. Table 1
summarizes the overall parameters of the superconducting toroid
magnet system presented in the LOI.

Table 1. Toroid System Parameters

Item	Central Toroid	End Toroids
Coil type	discrete	discrete
No. of coils	72	72
Conductor	NbTi/Cu	NbTi/Cu
	(Al stabilized)	(Al stabilized)
Type of stabilization	CICC	CICC
Total number of turns	1276	788
Maximum field (T)	1.5	3.5
Current (KA)	32	32
Current density (A/cm^2)	3048	3048
Stored energy (MJ)	835	514
Coil centering force (M Nt)	0.79	2.6
Conductor length (Km)	44.8	15.8
Conductor weight (m Ton)	125	44
Winding geometry		
inner radius (m)	5.4	1.6 average
outer radius (m)	7.5	8.6
length (m)	15.5	3.9

The design and cost analysis of the toroid magnet system was performed by Grumman, its industrial partners (General Dynamics, and Ansaldo), and Oak Ridge National Laboratory (ORNL) [3,4,5,6]. The work was based on their experience in the design and fabrication of aluminum structures; design, fabrication and winding of large superconducting magnets; and the design and operation of the Large Coil Test Facility (LCTF).

Toroid Magnet Design

The conceptual design of the toroid system followed the original aluminum design set forth in the EOI. This approach uses azimuthal webs that tie the inner and outer winding and also connect the end faces. Connecting the inner and outer windings allows the support of the larger radial inward force acting on the inner winding to be supported, in part, by the outer winding and also improves the stability of the structure. The baseline subelement of the central toroid is a two layer 5 degree segment (compared to the single layer EOI design) shown in perspective in Figure 2.

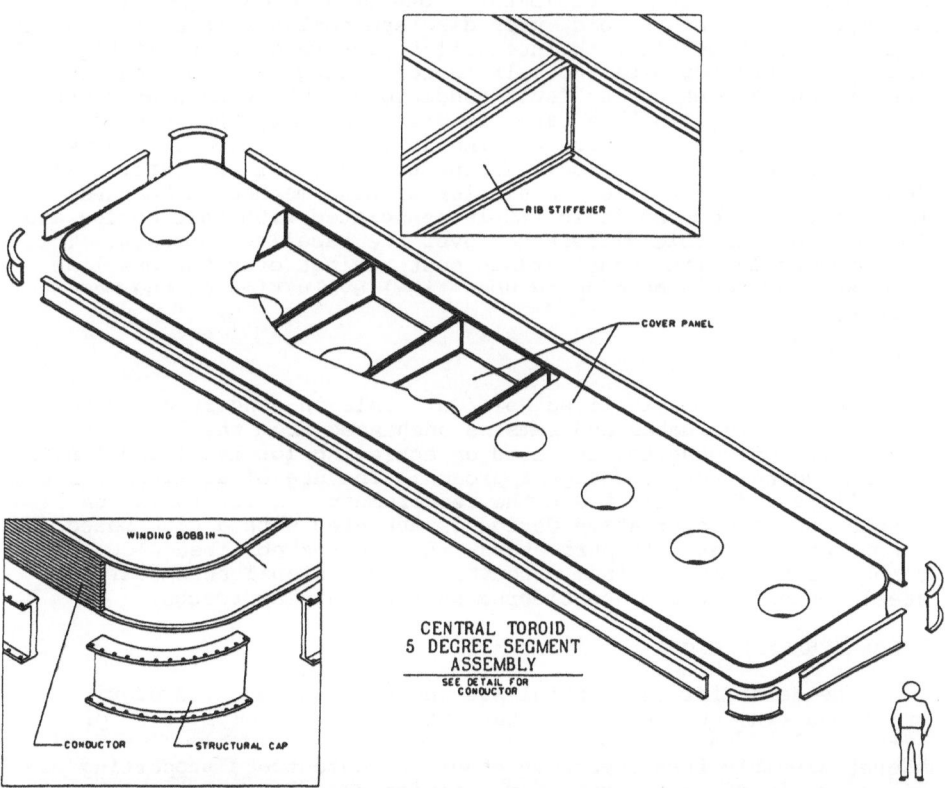

Figure 2. Perspective view of a 5° segment.

One option considered for coil winding has the bobbin stationary while rotating the conductor spool and the cleaning and taping equipment around it. For the central toroid, the conductor transitions from two layers on the inner radius to a single layer on the outer radius. Spacers are used to support the layers in the transition areas. A closure member is then welded to the bobbin to form an integral structure and the complete assembly is vacuum impregnated. A chill bar, located between the winding and structure is used to protect the conductor while welding. The 5° coil subelements are joined together by bolting at a flange interface located along the outer perimeter of the coil. The end toroid follows the same configuration and assembly approach as the central toroid.

Other concepts were also considered in developing the basic 5° winding segment. The approach taken by Ansaldo was to independently wind the conductor on a separate mandrel, remove it from the winding mandrel and then install the conductor into a preassembled 5° structural assembly. Assembly of the 5° magnet segment is a factory operation along with the further build-up of a 45° assembly. The 45° assembly is shipped to the site where the final assembly of the toroid magnet will be completed.

The baseline approach to wind the toroid conductor was to flare the conductor out as the winding transitioned from the smaller toroidal width of the inner radii to the larger winding width at the outer radii. The same approach was taken for the end toroid, transitioning from multiple layers at the inner turn to a single layer at the outer turn. This approach was used for the cost analysis. Other configuration options, however, have been considered and will reduce the coil winding cost and offer greater access for muon detector alignment equipment. One option developed for a 9° segment for the end toroid, uses discrete coil windings of constant width joined together with intercoil structure as shown in Figure 3. This structure supports the self-forces acting between conductor bundles and forces due to fault conditions. The conductor cross section geometry, in this case, is established by the space defined for a rectangular conductor winding at the inner radius. A web structure, as described above, connects the inner and outer coil legs. This option allows a simpler winding method to be used. Ansaldo evaluated the field ripple associated with this option and found no discernable differences over the baseline approach. This approach would offer considerable cost savings over the baseline approach and could be adopted upon final evaluation of the momentum resolution.

Superconductor

The aluminum stabilized NbTi/Cu, cable-in-conduit conductor (CICC) was reevaluated and remains unchanged from the EOI concept. The choice of conductor is based on achieving low radiation length within a conservative design approach. The use of aluminum for the stabilizer and conduit meets the requirement for low radiation length, and a conservative design is achieved with a CICC based on conservative operating parameters. A copper-stabilized CICC conductor with an aluminum conduit, or an optional thin stainless steel conduit could be considered as alternate approachs.

Dewar System Design

The dewar for the central and end toroids consists of rib stiffened aluminum structure that surrounds the complete toroid coil array. All dewar ribs are located in the plane of the 5° toroid magnet assembly face sheets in order to concentrate supporting structure in discrete areas and minimize the structural thickness that a muon would pass though in areas away from the regions of concentrated structural support. The dewar is supported by four tie rod struts, extending down from an overhead structural frame. The tie rods are attached to fittings located near the top end of the dewar, with a pair of circumferential ribs used to stiffen the dewar ends and to provide pick-up points for the tie rods.

The central toroid inner radii was expanded in the LOI design to allow space for two muon detector panels, thereby establishing a line-line measurement with the outer chambers. This expansion of the

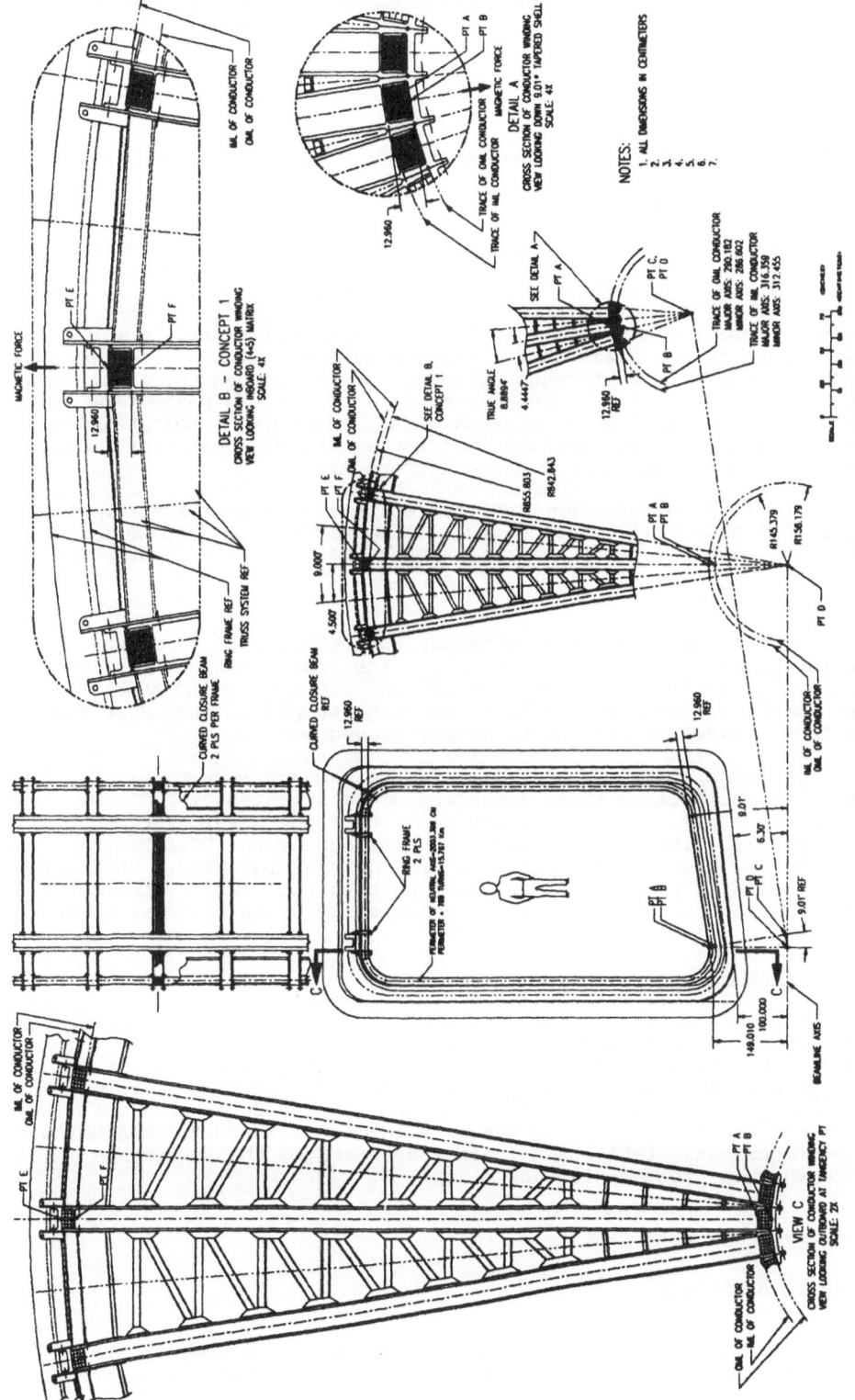

Figure 3. Details of the 9° segment with a constant conductor winding cross-section.

969

inner radius simplified the overall dewar construction and assembly of the central toroid system. The outer dewar cylinder can be completely assembled, a cold wall system installed and tested followed by the insertion of a fully assembled magnet system using the assembly fixture sized to install the calorimeter. The magnet and dewar support can be transferred to the overhead support structure and electrical and coolant lines can be connected prior to installing the dewar end closure structure.

Auxiliary Systems

A cryogenic supply system was designed and costed to supply each of the three superconducting air core toroids with single phase supercritical helium, two phase boiling helium, and liquid nitrogen cooling for the thermal shielding [7]. The supercritical helium is used to cool the conductor joints and electrical transitions and is circulated in parallel channels through the conductor. The two phase helium is circulated through tubes welded to the conductor structure to maintain cryogenic temperatures. Some of the two phase flow boils to cool the current leads of each magnet assembly. Coolant tubes are routed such that all manifold interfaces and tube electrical breaks are accessible through a cryostat access area.

The fundamental electrical operating requirement for the magnets is 32,000 amps at 12 volts DC [8]. This requires cooling water at 30 gpm with a 30 psi pressure drop, 3-phase, 480 volt AC input power from a 500 KVA source, four 8000 amp thyristors for each magnet, and related controls.

Copper and aluminum room temperature electrical leads were investigated for cost-effectiveness. The cross sections required for equivalent resistive losses were 200 and 265 cm^2 respectively, and the conductor lengths were based on power supplies located at the surface. Aluminum was the conductor chosen [4].

Vacuum pumping ports are located on the end structures of each magnet dewar because these areas are not obstructed by the muon detector panels. The initial evacuation of the dewar systems is accomplished with pumping equipment provided by the SCCL Experimental Facilities, and this equipment is used only as required. Thereafter, turbomolecular pumps (TMP) that are part of the dewar system maintain the required level of vacuum. Cryopumps may be an alternative to TMP's.

Magnet Analysis

Finite element analysis (FEA) of the dewar and magnet structures was carried out to augment the design process and refine the component sizing commensurate with material design allowables. Analysis included three dimensional FEA of the 5° magnet segments under magnetic pressure loading, and FEA of the dewar structure under combined gravity and vacuum pressure loads. This work, together with two-dimensional stability calculations provided the confidence necessary to proceed with the design [9].

MUON DETECTOR SYSTEM

General Arrangement

The muon chambers consist of two primary sets of panels (superlayers), horizontal and vertical, that are symmetrically arranged about the detector interaction region. The eleven different superlayers are shown in Figure 4.

A superlayer consists of layers of muon chamber modules oriented in bend-plane and non bend-plane directions for the horizontal type, and at 45° azimuthal displacements for the vertical type. Figure 5 schematically shows the arrangements of each of the superlayers. Note that some chambers have two-wire layers, while most have four-wire layers. Figure 6 illustrates a chamber module, a superlayer assembly, and subsequent installation in the hall.

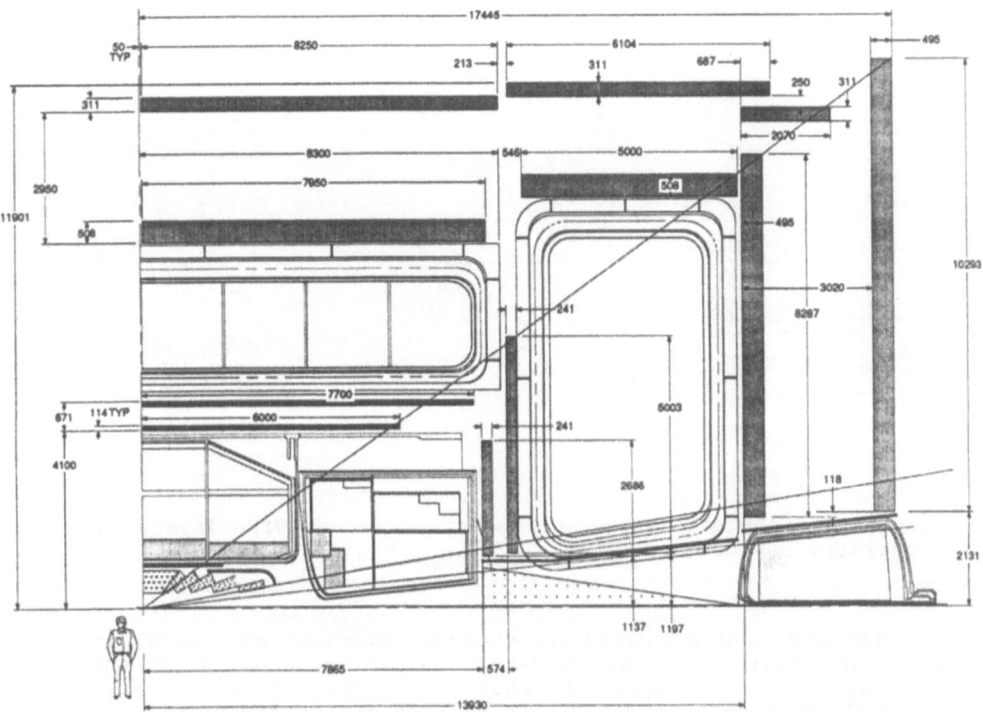

Figure 4.   Geometry of the EMPACT/TEXAS muon spectrometers.

Chamber Design

The basic element of the muon detector system is the chamber module [10]. A chamber consists of a rectangular frame 1 m wide, 12.5 cm high and of varying lengths up to 10 meters. The frame is covered with top and bottom covers that provide stiffness and ionizing gas containment. Most of the chambers contain four wire layers (some contain two) with each wire on a 2.5 cm pitch. Precision bridges hold the wires in place to required tolerances. Sheets of U-shaped cathodes surround each wire. Figure 7 is an isometric view of the muon chamber.

The box-like enclosure of each chamber has external mounting pads with machined surfaces that precisely locate the chambers on the invar support tracks. The tracks also have precisely machined matching surfaces and indexed holes. Since the pads are indexed to the location of the wire bridges, a superlayer assembly containing many chambers can be assembled having accurately predictable wire positions. Wires are positioned in the chamber with respect to external fiducial marks to within plus or minus 25 μm, and chambers are located with respect to each other on the invar tracks with a precision of 25 μm.

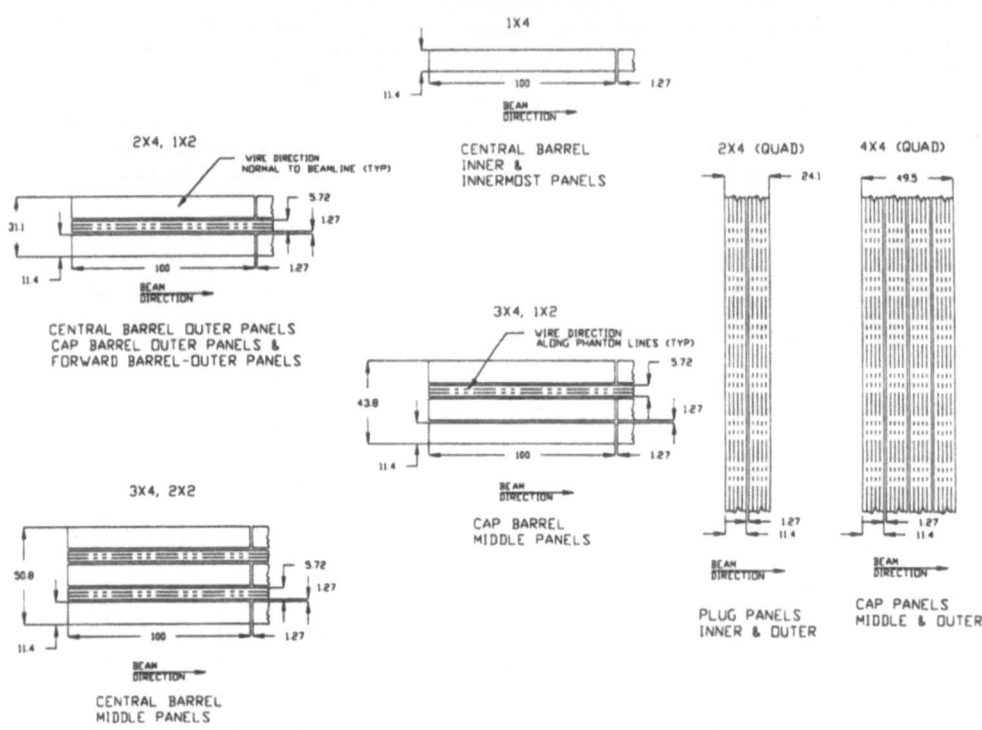

Figure 5. Schematic arangement of chambers and wire layers for each superlayer detector panel.

The side and end walls are extruded aluminum bar stock with the upper and lower inside edge undercut to provide seating for the gas tight covers. Wire support bridges are positioned at 1 meter intervals along the length of the muon chamber. The bridges are made of stable insulating material with bonded invar attachment blocks at each end. The bridge assembly is precision grooved to accept the anode wires. At wire assembly, a second insulating bar is bonded to the grooved wire holder piece. At each bridge location, the side walls are undercut to a precision thickness and locating pads are attached at the bottom and top edge of the side wall. The locating pads are initially oversized, attached to the side wall at the bridge locations, and then surface-machined to define a precise height dimension at each bridge location. A precision pin hole with a tapped end is located in the invar at one end of the bridge and a sliding pin hole located in the opposite invar end. Precision holes are also located in the side walls at each bridge location. The cathodes are made of extruded or roll-formed aluminum to minimize material thickness.

Superlayer Design

Each superlayer consists of two or more layers of chamber modules. For the horizontal type, these are arranged in bend plane and non-bend plane arrays as shown in figure 6. Invar tracks that are arranged as continuous members span the length and width of a superlayer, and are spaced at one meter intervals. The tracks are also stabilized with invar cross-bracing. Because of the low coefficient of thermal expansion, temperature gradients along the straps will have no significant effect on changing the positions of the chambers. The temperature in the hall will also be controlled to plus/minus 5° C in order to ensure that temperature induced changes are negligable.

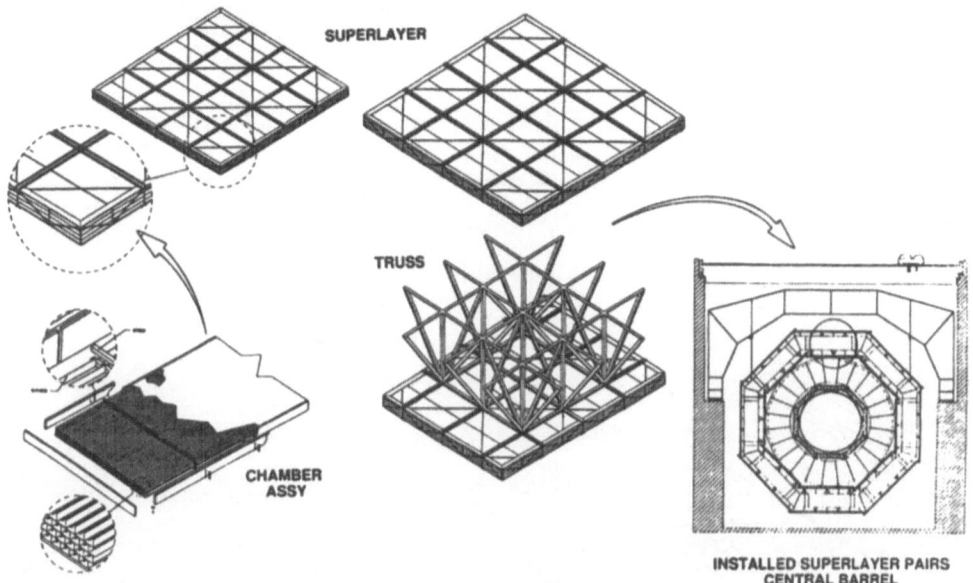

Figure 6. Muon spectrometer fabrication, assembly and installation.

A grid of aluminum channels and edge angles are attached to the invar tracks at the top, bottom, and edges of a superlayer assembly. The upper and lower channels are bolted together through the interstices that are formed by the array of chambers; the bolts are spaced approximately one meter apart. Hence, there is a matrix of attachments to the aluminum strongback structure that provides a stiff support for the superlayer. Since the aluminum structure has a higher coefficient of thermal expansion than the invar, the channel to track attachments are designed with slotted interfaces (except for one central attachment) in the plane of the superlayer to allow for thermal changes.

The vertical type superlayers are assembled in the same manner, using invar and aluminum materials. However, for these the layers of chambers are arranged as quadrants instead of bend and non-bend planes. Figure 8 is a planar view of the cap and plug superlayers. Note that each chamber layer below the top layer is rotated 45° azimuthally to make up the quadrant arrangement.

Assembly and Alignment of Superlayer Pairs

Adjacent superlayers are joined into pairs by attaching the welded truss structure shown in figure 6 to the channel supports. The optimum truss pitch in the x- and y-directions to achieve acceptable deflections will be determined by finite element analysis. For this baseline design, a 1 meter pitch is shown because it coincides with the interstices formed by the assembly of chamber modules. The truss assemblies will vary for each superlayer pair since they are dependent on the orientation of superlayers, i.e. the gravity loads.

Two superlayers and a truss are assembled in a precision alignment fixture located in the surface assembly facility. Each pair is assembled in the fixture in the same orientation that it will have in the hall so that the effect of gravity can be factored into the precise alignment of one superlayer to its mate [11,12].

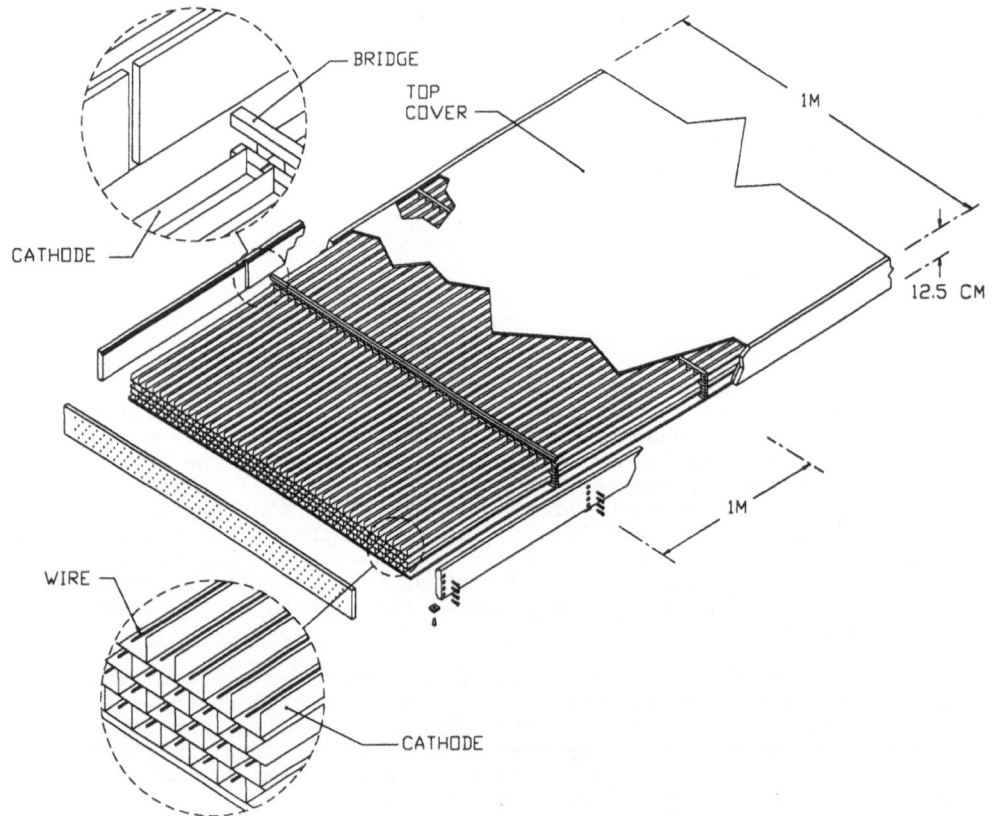

Figure 7. Muon spectrometer chamber module.

In the hall, horizontal and vertical superlayer sets are independently aligned using laser metrology techniques. That is to say, the horizontally oriented panels are aligned relative to each other, and the vertically oriented panels are aligned relative to each other.

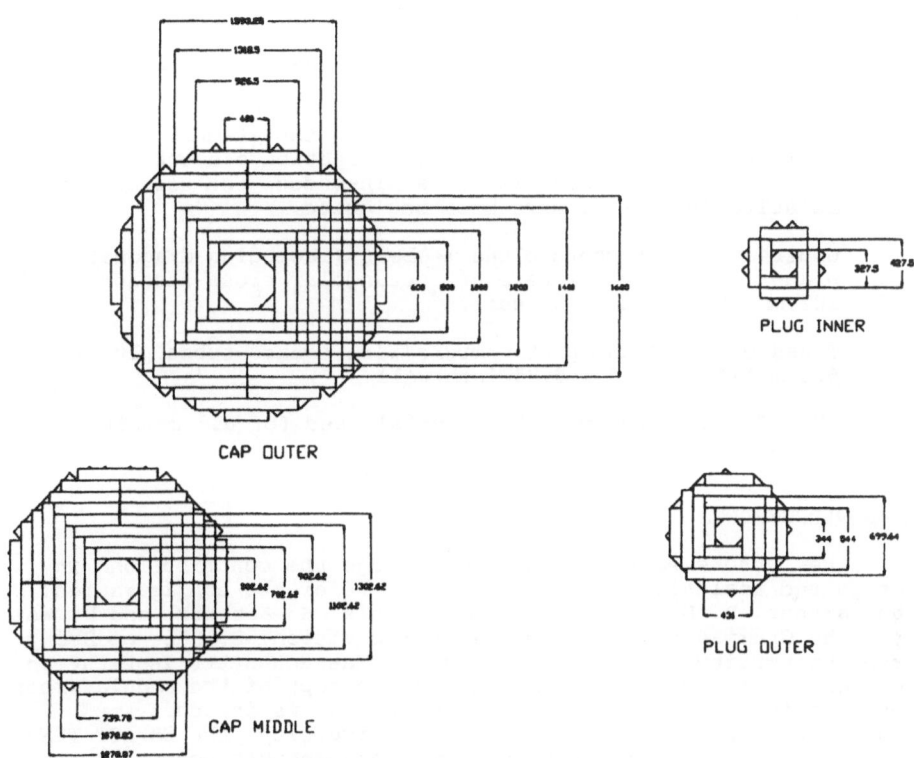

Figure 8.   General arrangement of the four vertical superlayers.

Finite Element Analysis

Finite element analysis of a typical, octagonal sector of the central barrel muon detector outer truss system was performed.   The results provided truss nodal deflections due to gravity acting on a pair of superlayer panels and the supporting truss system.   Figure 6 shows the arrangement of the superlayers and truses that were used for modelling.

Deflection results were obtained for the finite element model supported consecutively in three positions relative to the detector horizontal plane. These positions, as rotated about the beamline axis, were respectively, horizontal (0°), sloped (45°), and vertical (90°).   The maximum nodal deflection normal to the truss system chord plane was 0.346 mm (0.014 in.) at midspan with the truss system in the horizontal (0°) position.   The maximum nodal deflection parallel to the chord plane was 0.295 mm (0.012 in.) at midspan with the truss system in the vertical (90°) position.

It should be noted that the above nodal deflections are the result of a model solution analysis that contained many assumptions regarding element type and sizing, nodal degrees of freedom or restraint, type of reactive force distribution, and assumptions regarding application of superlayer gravity loads.   Subsequent parametric analyses would verify or alter some of the model assumptions.

In this stiffness controlled design, member stresses were considered not to be a design driver until considerations of minimum deflection, layout, etc. were satisfied. Then, member sizing optimization could begin. Hence, member stress results are not presented.

The following criteria are related to the finite elements used in the model:

o   All truss members were modelled using the standard ROD type element. This is a constant cross-section element that carries axial load only. Because it is pin-ended, the effects of relative end fixity have been neglected.

o   Upper and lower chord plane members (including diagonal chord plane cross-bracing) were of rectangular cross-section, 12.7 x 102 mm (0.50 x 4.00 inches).

o   Truss diagonal and post members were square tubes, 102 x 102 x 6.4 mm (4.0 x 4.0 x 0.25 inch) wall.

o   2219-T81 aluminum was the material used for all member elements.

ACKNOWLEDGEMENTS

The authors would like to acknowledge the contributions in design and analysis used to develop the reference configuration: J. Baumgartner, D. De Angelis, R. Karastathis, H. Katcher, and D. Weiss from the Grumman Space and Electronics Division; S. Snee of Stony Brook University; and L. Osborne of the Massachusetts Institute of Technology, for providing the original concept of the modular muon detector chamber. A special acknowledgement is in order for Professor Michael Marx of Stony Brook University, the EMPACT/TEXAS coordinator, for leadership, guidance, and encouragement.

REFERENCES

1.   The EPACT Experiment At The Superconducting Super Collider, Expression of Interest, May 1990.

2.   Letter of Intent for the (EMPACT/TEXAS) Superconducting Super Collider, November 1990.

3.   EMPACT/TEXAS Tech Note 368, "EMPACT/TEXAS Muon Detector System Final Engineering Report", T.G. Brown, P.T Spampinato, February 1991.

4.   EMPACT/TEXAS Tech Note 353, "EMPACT/TEXAS Toroidal Magnet System Cost Report", F. Halfen, December 3, 1990.

5.   EMPACT/TEXAS Tech Note 373, "Final Report For The EMPACT Toroidal Detector, Subcontract 20-98791 To Grumman Aerospace Corporation", H. Gurol, General Dynamics Corporation, January 14, 1991.

6.   EMPACT/TEXAS Tech Note 374, "EMPACT Toroids, Preliminary Design Study", R. Penco, Ansaldo Componenti ABB, January 1991.

7.   EMPACT/TEXAS Tech Note 362, "EMPACT/TEXAS Cryogenic Supply System", Tim Myers.

8.   R.E. Wright and J.N. Luton, Oak Ridge National Laboratory;
        summary table in E/T Tech Note 353, "EMPACT/TEXAS Toroidal
        Magnet System Cost Report", December 3, 1990.

9.   EMPACT/TEXAS Tech Note 313, "Progress Report on Finite
        Element Analyses", S. Snee, October 29, 1990.

10.   Chamber configuration based on the concept by L. Osborne,
        MIT.

11.   EMPACT/TEXAS Tech Note 225, "A System Concept for Aligning
        EMPACT Muon Detector Chambers for the SSC", J. Connelly.

12.   EMPACT/TEXAS Tech Note 363, "November 1990 Addendum No 1:
        A System Concept for Aligning the EMPACT Muon Detector
        Chambers for the Superconducting Super Collider", J.
        Connelly.

15. Environment, Safety and Health

ENVIRONMENTAL IMPACTS OF THE SUPER COLLIDER

Thomas A. Baillieul

Chicago Operations Office
U.S. Department of Energy
9800 S. Cass Avenue
Argonne, IL 60439

William Hasselkus

Office of Superconducting
  Super Collider
U.S. Department of Energy
Washington, D.C. 20585

INTRODUCTION

1. The National Environmental Policy Act of 1969, affectionately referred to as "NEPA", is a simple piece of legislation with far-reaching implications.

    a. It sets a requirement for Federal government decision makers to consider the environmental consequences of their actions <u>before</u> deciding on a course of action.

    b. A decision maker is essentially anyone who causes something to happen; and the action can be just about anything.

    c. NEPA comes into play at the point in time where a proposed action is matched to a physical location.

    d. NEPA implementation is recorded in many ways. The DOE maintains a long list of categorical exclusions for actions which practice has shown to be inconsequential -- such as processing records, or maintaining physical plants. However, in selecting a categorical exclusion for an action, the decision maker/project manager must at least think about the activity to be performed and its possible environmental consequences.

    e. A large project like the SSC, involving an undeveloped site, automatically qualifies for the highest level of environmental analysis under NEPA-- the Environmental Impact Statement (or EIS).

Supercollider 3, Edited by J. Nonte
Plenum Press, New York, 1991

The Environmental Impact Statement

2.   Feasibility studies and conceptual design efforts for the SSC began in the mid-1980's; however, the requirements of NEPA were not triggered until a decision was made to select a site for the facility.

   a.   To select a site to build the SSC, the DOE invited proposals from all interested parties (April 1987). Forty-three responses were received, primarily from state governments, but a few from county and local jurisdictions, and one or two from individuals. The Invitation for Site Proposals requested specific information about the environmental characteristics of the proposed sites -- the first step in environmental consequence analysis.

   b.   Of the 43 original proposals, the DOE deemed 36 to be qualified, and these were sent to the National Academy of Sciences/National Academy of Engineering for an initial screening review. [One state withdrew its proposal part way through the review]. The NAS/NAE selected seven proposals to be placed on its list of Best Qualified candidates (Arizona, Colorado, Illinois, Michigan, North Carolina, Tennessee, and Texas).

   c.   Each of the seven finalists was asked to provide supplementary environmental information about each site as a basis for beginning the required environmental impact analysis.

3.   NEPA requires that in any decision, the Federal manager evaluate the alternatives to the proposed action.

   a.   The DOE determined that the seven finalist sites constituted a reasonable set of alternatives for consideration in the EIS.

   b.   The EIS also reviewed the major design options considered in arriving at the SSC Conceptual Design (magnet type, beam energy).

   c.   NEPA and its implementing regulations require that all EISs prepared include a "no action" alternative.

4.   The AE firm which had prepared the SSC Conceptual Design Report, RTK Associates of Oakland, CA, was tasked to develop the environmental impact analyses for the seven finalist sites and to write the EIS.

   a.   As required by regulation, the EIS contractor had to certify that it had no financial interest in the outcome of the EIS site selection or SSC construction.

   b.   RTK, and its expert team, began to review and confirm the voluminous environmental database provided by each site finalist. Also limited site visits for data collection and clarification were scheduled.

5. The DOE established a dedicated team for the EIS which included: members of the site-selection task force; representatives of the Office of NEPA Oversight, and General Counsel; and project management staff from DOE's Chicago Operations Office (manager of the RTK contract).

   a. Early and regular involvement by the organizations which would have to review and approve the EIS was critical to getting this large a document completed in a short space of time.

6. One of the results of NEPA is to bring the Federal decision-making process out into the public light. NEPA requires specific levels of public involvement in the EIS process.

   a. The first opportunity for the public to get involved in the SSC (outside of grassroots efforts in support of or against a site proposal) was with "scoping" in early 1988. Scoping is a process whereby all interested parties are presented with a description of what a Federal agency proposes to do. The public is then asked to define issues which it considers important to address in the Environmental Impact Statement.

   b. The DOE held scoping meetings in the vicinity of each of the seven finalist sites.

   c. Once the Draft EIS was produced an additional round of public hearings was held during a 45-day open comment period. From this set of hearings, DOE received written comments and oral testimony from approximately 5,700 individuals and groups.

7. The SSC EIS was perhaps the largest environmental effort ever undertaken by the DOE.

   a. RTK and its expert team reviewed and responded to the more than 7,000 comments; conducted additional field work at each site; and finished the Final EIS on a compressed schedule that led to the announcement of a site in January 1989.

The Supplemental Environmental Impact Statement

8. The early planning for the SSC EIS envisioned that the document would be sufficient to address all elements of siting, construction, and operation.

   a. During the DOE internal review of the Draft EIS, it was determined that the generic nature of the conceptual design used in the site proposals and the rather general nature of site environmental information would not be sufficient to support a construction approval.

   b. In its Record of Decision for the 1988 EIS, the DOE announced its intention to proceed with the project and to accept the Texas site proposal. The ROD also committed the agency to prepare a Supplemental EIS which would more fully evaluate the impacts of building and operating the collider at the Texas site, and would look in more detail at alternatives to reduce (mitigate) those impacts.

9. Along with the selection of the site DOE announced the selection of Universities Research Association to establish and manage an SSC Laboratory, and to oversee the construction of the SSC.

    a. A first step was to develop a site-specific design for the collider matched to the geology and conditions of the Texas site.

10. Usually an Environmental Impact Statement is prepared only after a mature design is available to assess. However, in the case of the SSC Supplemental EIS it was decided to initiate the environmental studies in parallel with the design effort.

    a. DOE's Chicago Operations Office was asked to pull together the SEIS team. A partnership was established between the designers and the environmental scientists.

    b. Argonne National Laboratory's Environmental Assessment Division was brought in to help define the scope of the SEIS, to conduct the necessary analyses, and to write the document.

    c. The SSC Laboratory was charged with providing all necessary design information, source terms, and basic field data.

    d. This arrangement caused all the players to focus on the SEIS/design interface in the early phases of the effort and resulted in a design that addressed environmental sensitivities right from the outset.

        1. Scheduling became a problem as the environmental impact analysis had to be performed interactively with the evolving design concept.

        2. Changes in the design of the injector caused an approximate 6 month hiatus in the SEIS while the footprint was reoriented.

        3. Pending approval of the footprint by the Department and the beginning of land acquisition by the State of Texas, SEIS scientists could not study parcels of land that had not been formally announced as part of the revised SSC layout.

        4. Also, with any construction held up until a Record of Decision on the Supplement, there was great project emphasis on the earliest possible ROD.

11. As with the original EIS, a working group with representatives of all interested players was established to coordinate the effort.

   a. The SSC-SEIS Working Group was chaired by the Chicago SEIS Project Manager and involved participants from: Argonne; the SSC Laboratory; the DOE-Washington Office of SSC; the on-site SSC DOE Project Office; DOE's Office of NEPA Oversight; the DOE Office of General Counsel; and the Texas National Research Laboratory Commission.

   b. Each group member brought a unique perspective to the effort. The job of the working group was to assure a common understanding of the SSC project, and the SEIS effort in particular, at all times. Further, this group remained essentially intact throughout the process.

   c. Early involvement of these players--at the stage when analytical approaches and study scope were being defined--saved a lot of time during the review process. There were no surprises to anyone when the document was put together.

12. As with the 1988 EIS, the Supplemental went through an extensive public review and comment process.

   a. During a 45-day public comment period in September and October of 1990, hearings were held in Waxahachie and Ennis, Texas.

   b. Unlike the earlier document, the SEIS received approximately 370 written or oral comments (many of them duplicates).

   c. Certainly a most significant comment came from the Environmental Protection Agency, Region VI, who gave the document its highest rating: lack of objection. This is one of the few environmental documents in DOE to achieve this rating.

13. Upon incorporation of the public comments, the document was finalized. Through close coordination among the Working Group and a lot of hard work by a lot of people, the process was held to a tight schedule.

   a. Notice of availability of the final EIS was filed with EPA on December 21, 1990. A notice appeared in the Federal Register a week later.

   b. After a mandatory 30 day waiting period, a Record of Decision, documenting the decision and action to proceed with construction and operation of the SSC was signed by the Secretary of Energy on February 1, 1991.

14. There is a popular misconception that completion of an environmental document is an end unto itself. One simply puts the published document on the shelf, forgets about it, and proceeds with the project. Many a project has paid a dear price for that attitude. New guidelines from the Council on Environmental Quality clarify what we already knew:

    a. Mitigative action to avoid or control environmental insult, when promised in the SEIS, must be carried out.

    b. Mitigations must be monitored to assure that they, in fact, have brought about the desired environmental impact relief.

    c. If the mitigative actions have not produced the desired effect, additional measures must be taken.

15. For the SSC, as required by Secretary of Energy Notice (SEN) 15-90, the mechanism for assuring adequate mitigation is the Mitigation Action Plan (MAP). The MAP:

    a. Identifies all mitigations promised in the EIS/SEIS.

    b. Pinpoints responsibility for execution, oversight, and verification of each mitigative action.

    c. Describes, to the extent possible, when the mitigations will occur, relative to the project schedule.

    d. The SSC MAP will be a first-of-its-kind document in DOE.

16. In short, we will meet the commitments embodied in the EIS and SEIS. This fits into an overall policy of compliance with all environment and safety and health (ES&H) laws and regulations. The reasons are straightforward:

    a. It is the right thing to do.

    b. The Secretary wants it that way!

    c. We must obey the law.

    d. There are possible criminal/civil penalties for failure to do so.

    e. Individuals and groups in the SSC area are involved.

17. We have a golden opportunity to operate the SSC ES&H program in a manner that reflects the "new DOE." We must:

    a. Leave old attitudes at the door when we enter the SSC program.

b.   Elevate ES&H in everyone's thinking.

c.   Obtain a commitment by all involved to <u>compliance</u>.

d.   Understand that these kinds of attitudes and commitments
     will contribute mightily to SSC project success.

SAFETY IN HANDLING HELIUM AND NITROGEN

G. Schmauch, L. Lansing, T. Santay, and D. Nahmias

Air Products and Chemicals, Inc.
Allentown, Pennsylvania 18195

ABSTRACT

Based upon our industrial experience and practices, we have
provided an overview of safety in storage, handling, and transfer of
both laboratory and bulk quantities of gaseous and liquid forms of
nitrogen and helium. We have addressed the properties and
characteristics of both the gaseous and liquid fluids, typical storage
and transport containers, transfer techniques, and the associated
hazards which include low temperatures, high pressures, and
asphyxiation. Methods and procedures to control and eliminate these
hazards are described, as well as risk remediation through safety
awareness training, personal protective equipment, area ventilation, and
atmosphere monitoring. We have included as an example a recent process
hazards analysis performed by Air Products on the asphyxiation hazard
associated with the use of liquid helium in MRI magnet systems.

INTRODUCTION

Helium and nitrogen are seldom viewed as hazardous materials.
However, when they are used as high pressure gases or cryogenic liquids,
individuals who routinely handle them are subject to a variety of
hazards. This paper is a condensation of a one-day safety orientation
conducted by Air Products for the Magnet Systems Division of the
Superconducting Super Collider Laboratory to highlight these hazards.
The paper focuses on two parts of the agenda from that safety
orientation. First, it discusses the physical properties of gaseous and
liquid helium and nitrogen, and highlights the potential hazards they
create. Argon is included because of its possible use in detectors.
Second, the paper discusses the applicability of process hazards review
methodology, as practiced in the chemical process industry, to designing
safety into SSC cryogenic systems – both above ground and within the
booster and main ring tunnels. As an illustration, one which is
relevant to the SSC program, the paper briefly describes a hazard review
that Air Products performed to evaluate the asphyxiation hazard
associated with the use of helium-cooled Magnetic Resonance Imaging
(MRI) units in confined spaces. The one-day orientation also dealt with
the details of personnel safety in handling gas cylinders and cryogen
dewars.

Supercollider 3, Edited by J. Nonte
Plenum Press, New York, 1991

This paper strives to demonstrate that industry experience and practices in safe handling of helium, nitrogen, and argon are extensive and well proven, and are readily applicable to the SSC. It also shows that the process hazards review techniques can be and should be utilized in designing SSC cryogenic systems.

PROPERTIES AND CHARACTERISTICS

Helium, nitrogen, and argon are colorless, odorless gases at ambient temperature. Helium and argon are noble, inert gases while nitrogen is chemically reactive only under extreme conditions. Nitrogen and argon are present in the air we breathe at 78% and 0.93% by volume, respectively. Helium is present in air in only trace quantities and is typically recovered from natural gas processing. These gases are noncorrosive, nontoxic, and nonflammable. When mixed with air, they can reduce the oxygen concentration to levels that will not support life nor combustion. They are classified as simple asphyxiants.

As shown in Table 1, gas density at ambient temperature and atmospheric pressure ranges from 13% of air for helium to 97% of air for nitrogen, and 138% of air for argon. These low density gases are normally stored and shipped at ambient temperatures under high pressure in steel cylinders to increase their shipping density. Pressures in the range of 2200 to 2600 psig are typical.

Shipping density is substantially increased through liquefaction of the gases. Cryogenic (extremely low) temperatures are needed to accomplish liquefaction; ranging from a Normal Boiling Point (NBP) of -302°F for argon to -320°F for nitrogen and -452°F for helium. Liquid density at NBP compared to water ranges from 12% for helium, 81% for

Table 1 Properties and Characteristics

Property	Helium	Nitrogen	Argon
Molecular Weight	4.00	28.01	39.95
Gas Density, 70°F and 14.7 PSIA in lb./cu. ft.	0.0103	0.0724	0.1034
Gas Density vs. Air (0.0749)	0.138	0.967	1.381
Normal Boiling Point, °F	-452.13	-320.5	-302.6
Liquid Density at NBP in lb./cu.ft.	7.798	50.46	86.98
Liquid Density vs. Water (62.43)	0.125	0.808	1.393
Expansion Ratio, Gas at 70°F and 14.7 PSIA vs. Liquid at NBP	757	697	841
Latent Heat at NBP in BTU/lb.-mole	36	2405	2804

nitrogen, and 139% for argon. Liquefaction provides both a density and pressure incentive for gas storage and shipment. This is illustrated by the expansion ratio of the fluids at the N.B.P. versus S.T.P. These are 757, 697, and 841 for helium, nitrogen, and argon respectively.

POTENTIAL HAZARDS

These properties and characteristics of helium, nitrogen, and argon provide advantages in their storage, handling, and use. However, certain potential hazards are created with these advantages, and these potential hazards must be recognized and addressed. The hazards include the asphyxiant properties of the gases, the very low temperatures, and gases under pressure.

Asphyxiation - The normal concentration of oxygen in air is approximately 21% by volume. Depletion of the quantity of oxygen in a given volume of air by displacement with inert gases is a potential hazard to personnel. When the oxygen content of air is reduced to about 15 to 16% at sea level, the flame of ordinary combustible materials will be extinguished. An individual breathing air at or somewhat below this concentration of oxygen will feel symptoms of fatigue, sleepiness, confusion, loss of coordination. He will be mentally incapable of diagnosing his situation, make errors in judgment, and have a false sense of security and well being. Exposure to atmospheres containing 12% or less oxygen will cause unconsciousness without warning and so quickly that the individual cannot help himself. Death can follow within minutes. Prolonged reduction of oxygen can cause brain damage even if the individual survives.

Areas in which low oxygen concentrations are possible must be well ventilated. Inert gas vents must be piped outside of buildings and enclosures to safe locations. Where low oxygen atmospheres are possible, oxygen analyzers with alarms should be utilized. Continuous area monitors, sniffers, and pocket-sized personnel monitors must be employed to check the atmosphere and maintain surveillance when personnel enter confined spaces and enclosures. When there is any doubt about maintaining safe breathing atmospheres, self-contained breathing apparatus or approved air lines and masks should be employed.

Personnel working in or around potential oxygen-deficient atmospheres rely on the "buddy system" for protection. However, the buddy must be provided with a portable air supply or he is equally liable to asphyxiation if he enters an area to rescue an unconscious partner without it.

Low Temperatures - Cryogenic injuries or "burns" can result from skin contact with cryogenic liquids, cold vapors, or surfaces at cryogenic temperatures. Effects are similar to those of a heat burn but less painful. Severity varies with temperature and time of exposure. Exposed body parts coming in contact with cold surfaces can stick fast due to freezing of available moisture. Skin and flesh can be torn upon removal. The eyes are especially vulnerable to cryogenic liquid droplets and cold vapors. Dry, clean clothing with full length sleeves, gloves, and face shields are used to protect against these hazards.

Certain materials such as plastics, rubber, carbon steel, and ceramics become extremely brittle and lose their strength at cryogenic temperatures. Such materials as brass, stainless steel, aluminum, copper, and special alloys are compatible with cryogenic temperatures.

Cold gases from liquid nitrogen and argon leaks or spills are heavier than air and will tend to settle and flow to low levels by virtue of their density. Depending upon topography and ventilation, hazardous concentrations can be found at low points at considerable distances from the release point. With cold gas from liquid helium, neutral buoyancy is achieved at -387°F, only a few degrees above the normal boiling point.

Gases Under Pressure - When gases are compressed to high pressures, large amounts of potential energy is stored in the container. A sudden release of pressure can cause serious damage to personnel and equipment by propelling a cylinder or whipping a line. Large vessels with a low pressure can be more dangerous than a small vessel with a high pressure since the potential energy is related to the pressure-volume product. A cryogenic liquid or cold vapor trapped in a line between shutoff valves can be particularly dangerous as the liquid vaporizes and warms due to heat leak producing pressures high enough to rupture piping. Pressure relief devices must be provided to protect all pressurized systems and cryogenic systems against excessive pressures that might be generated under various operating conditions. Pressure relief devices must be checked regularly to avoid obstruction by debris and ice, and must be tested periodically to demonstrate functionality.

STORAGE, HANDLING, AND TRANSPORT SYSTEMS

Compressed gas and cryogenic liquid storage and handling systems are engineered and constructed in compliance with a variety of industry codes and standards to insure safety and integrity. Compressed gases are shipped and stored in hollow steel cylinders or tubes which are manufactured in accordance with Department of Transportation specifications. Compressed gases may be stored in ASME coded and stamped high pressure gas storage tubes which are National Board registered if they are part of a stationary storage system. Shutoff valves and relief devices are specified by the Compressed Gas Association and the American National Standards Institute. Small quantities of compressed gases are stored and shipped in individual cylinders common to all laboratories and clinics. Capacities range from 25 to 250 standard cubic feet (SCF). Larger quantities are stored and shipped in tube banks and tube trailers with capacities ranging from a few thousand to 60,000 SCF. A typical tube trailer, equipped for hydrogen transport, is shown in Photo 1.

Liquid nitrogen and argon are stored, shipped, and handed in several ways. Quantities up to 200 liters are handled in liquid cylinders. A typical liquid cylinder and schematic are shown in Figure 1. The cylinder is an insulated, vacuum-jacketed container. Since these cylinders operate at pressures up to 250 psig, their design must comply with DOT specifications. Products can be withdrawn as a gas by passing the liquid through a vaporizing coil, or as a liquid under its own vapor pressure. Safety relief valves and rupture discs protect the cylinders from excess pressure build-up.

Larger quantities are stored and shipped in spherical or cylindrical tanks. They can be mounted on truck chassis for easy transportation or at fixed locations as stationary vessels. Sizes range from 300 gallons to 400,000 gallons for stationary tanks but are weight limited to about 6,000 gallons for over the road transport. A typical customer tank and transport tanker are illustrated in Photo 2.

Photo 1 Typical Compressed Gas Tube Trailer for Over-the-Road Transport

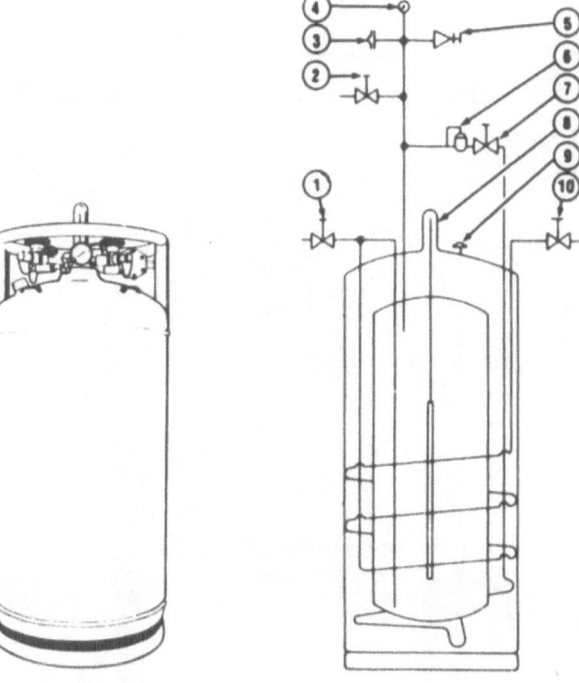

NOMENCLATURE

1. Liquid Valve
2. Vent Valve
3. Burst Disc, 550 PSIG
4. Pressure Gauge
5. Relief Valve, 350 PSIG

6. Pressure Building Regulator
7. Pressure Building Valve
8. Liquid Level Gauge
9. Vacuum Casing Burst Disc
10. Gas Use Valve

Figure 1 Typical Liquid Cylinder and Schematic

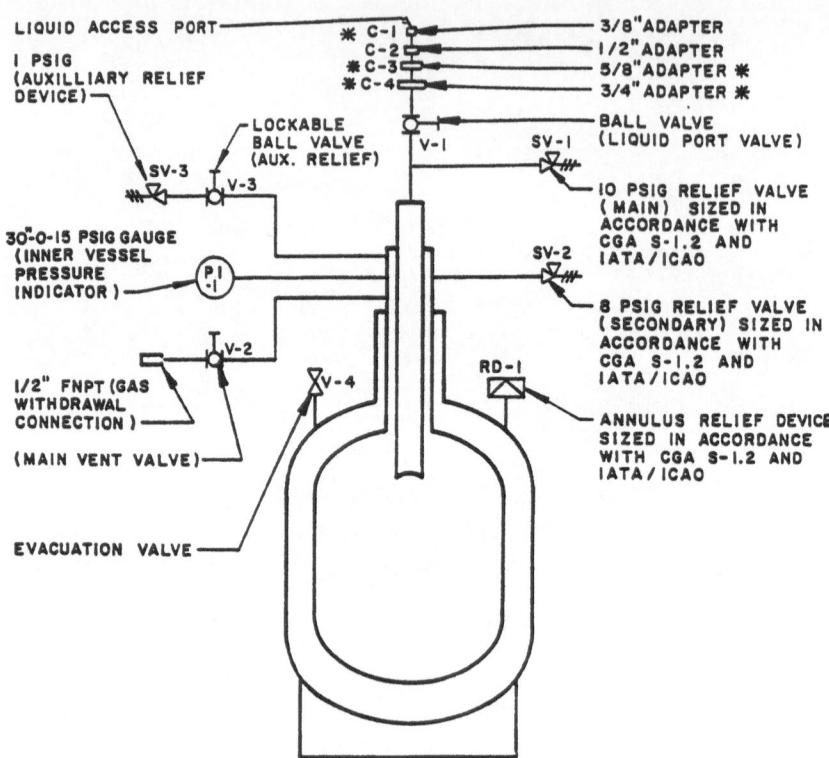

LIQUID ACCESS PORT

3/8"ADAPTER
1/2"ADAPTER
5/8"ADAPTER ✳
3/4"ADAPTER ✳

C-1
C-2
C-3
C-4

I PSIG (AUXILLIARY RELIEF DEVICE)

LOCKABLE BALL VALVE (AUX. RELIEF)

SV-3
V-3
V-1

BALL VALVE (LIQUID PORT VALVE)

SV-1

10 PSIG RELIEF VALVE (MAIN) SIZED IN ACCORDANCE WITH CGA S-1.2 AND IATA/ICAO

30"-0-15 PSIG GAUGE (INNER VESSEL PRESSURE INDICATOR)

P.I -1

SV-2

8 PSIG RELIEF VALVE (SECONDARY) SIZED IN ACCORDANCE WITH CGA S-1.2 AND IATA/ICAO

V-2

1/2" FNPT (GAS WITHDRAWAL CONNECTION)

V-4

RD-1

ANNULUS RELIEF DEVICE SIZED IN ACCORDANCE WITH CGA S-1.2 AND IATA/ICAO

(MAIN VENT VALVE)

EVACUATION VALVE

Figure 2   Schematic of a Typical Liquid Helium Container

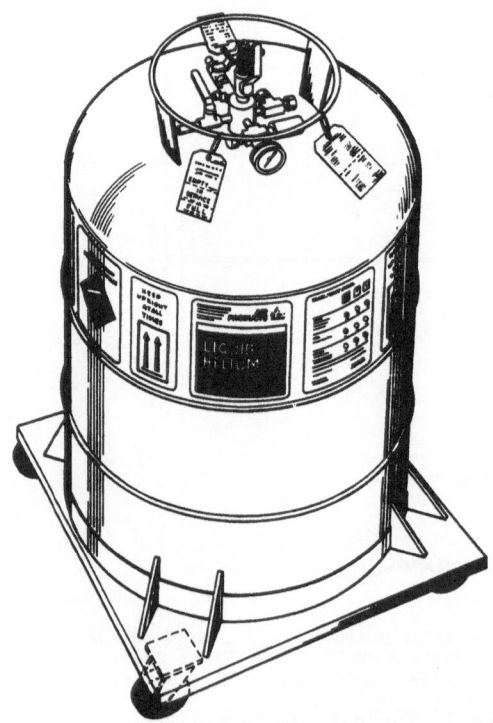

Figure 3   A 500 Liter Liquid Helium Dewar Mounted on Wheels

Liquid helium is stored, shipped, and handled primarily in liquid dewars with capacities ranging from 25 liters to 2000 liters. These containers consist of two cylindrical vessels, one within the other as schematically shown in Figure 2. The annular space is evacuated, and contains multilayer insulation. Multiple pressure relief devices are installed to protect against overpressure. Figure 3 shows one of the larger variety that is mounted on wheels for easy transport.

Larger quantities of liquid helium are transported in tankers with capacities up to 11,000 gallons. These utilize an annular space insulated with vacuum, nitrogen shielding, and multilayer insulation. A typical liquid helium transport container is shown in Photo 3. These tanks can be removed from the trailer chassis for overseas shipments.

Where massive quantities of nitrogen, argon, and helium are used for cryogenic refrigeration purposes, the vaporized gases are recovered and reliquefied.

HAZARDS RECOGNITION, ELIMINATION, AND CONTROL

Fluids like nitrogen, argon, and helium have some inherent potential hazards as were described above. However, these hazards can be amplified and new hazards created as of function of how these materials are used. For each application, the prudent user must review the interaction of the fluids and their ancillary systems, such as storage and handling equipment, with the people and equipment that are using them. Through a Process Hazards Review methodology, potential hazards are recognized and through creative engineering are frequently eliminated from the system. Where elimination is not possible, hazards control can be accomplished through engineered protective systems, operational procedures, administrative controls, and Personal Protective Equipment (PPE).

Photo 2   Typical Customer Tank and Transport Tanker

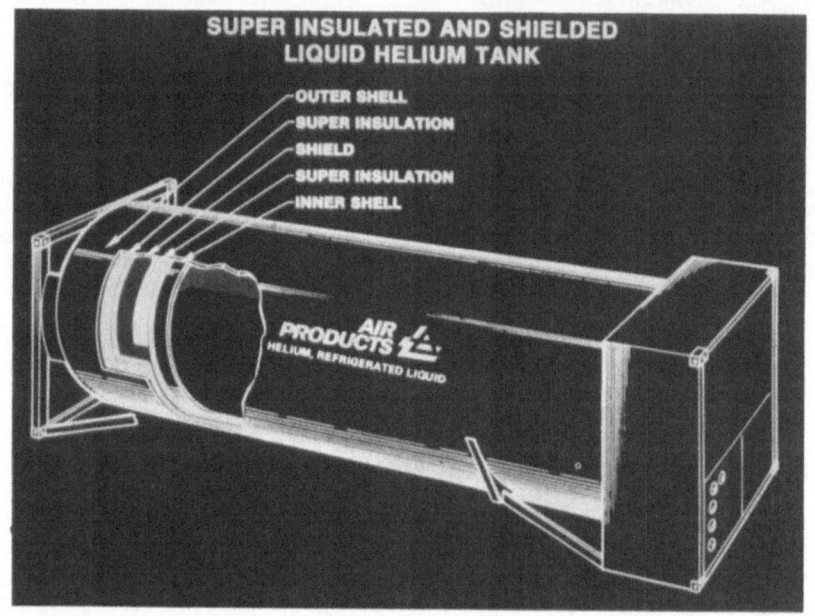

Photo 3 Typical Liquid Helium Transport Container

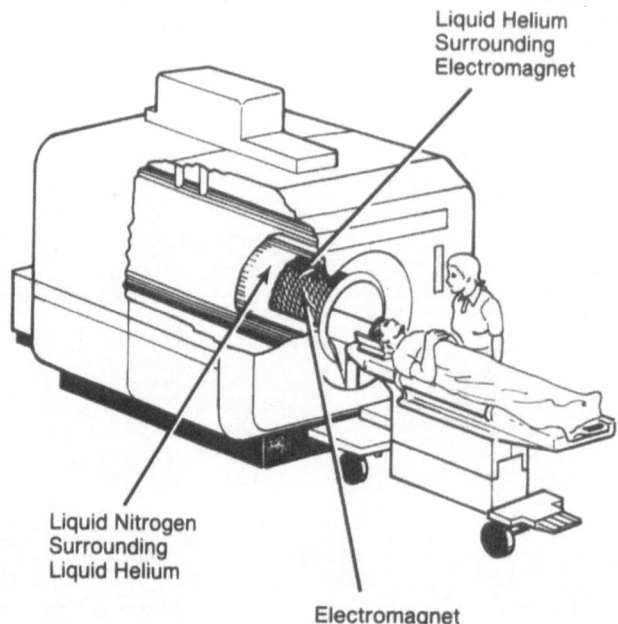

Figure 4 Magnetic Resonance Imaging (MRI) Unit

Magnetic Resonance Imaging – MRI is a fast growing technique that provides the medical profession with a powerful diagnostic tool. A patient is placed inside a strong magnetic field and, via a computer, the molecular makeup is analyzed to produce a detailed, precise digital image. The core of the MR unit is a powerful superconducting electromagnet as shown in Figure 4. Liquid nitrogen and liquid helium are used to provide the low temperature needed for effective magnet performance.

As a supplier of these fluids, Air Products performed hazard reviews of the delivery process, containers, container handling, fluid transfer, and fluid storage at the user's site. Photo 4 illustrates the final design features of the delivery vehicle to insure container cleanliness and handler safety. The closed body vehicle is fitted with louvers to provide ventilation to avoid a potential asphyxiation atmosphere inside the enclosure. A lift is incorporated to raise and lower the containers that weigh several hundred pounds. The containers are fabricated from non-magnetic materials to avoid interaction with the magnet system. Figure 5 shows a variety of magnet hazards that must be avoided. Dewar handlers must wear personal protective equipment which include sturdy leather work gloves, safety glasses with side shields, metatarsal foot protection, and an emergency breathing air pack for use while riding with dewars in an elevator. Safety training on the proper use of PPE and handling of heavy cryogenic containers is necessary to avert strains and bumps.

CONFINED AREA HAZARDS REVIEW AND RISK ASSESSMENT

The rapid growth of Magnetic Resonance Imaging (MRI) in the medical community has placed a great demand on the supply of liquid helium and nitrogen. This growth has resulted in a greater presence of cryogenic storage dewars in MRI facilities. Some MRI units require an annual average replenishment of 8000 liters of helium and 14,000 liters nitrogen, with cryogen resupply of about 250 liters of helium every two weeks and 230 liters of nitrogen per week.

Each dewar is delivered to the facility by the cryogen supplier and placed in a storage area in close proximity to the MRI unit. Full dewars may be stored here for up to three days before use. When required, the MRI technician wheels the dewar up to the MRI cryostat and transfers the cryogen (helium or nitrogen) using a vacuum-jacketed transfer hose. When the transfer is completed, the dewars are wheeled back to the storage room for later pick-up by the cryogen supplier. In most instances, a small heel of product (<5 liters) remains in the dewar.

The most common helium dewar in MRI use is shown in Figure 2 and Photo 4. Each dewar features an inner and outer vessel, with an evacuated annular space. The relief system consists of a 1 psig relief valve, with 8 and 10 psig safety valves. A 3/4" ball valve is provided for liquid withdrawals. Maximum boil-off rates are 1.0% helium per day at 0 psig.

Most nitrogen dewars in MRI service are built as shown in Figure 1. These dewars also feature double-walled construction. The primary relief valve is set at 22 psig. Self-contained pressure-build and economizer circuits are also included. Design boil-off rates are 2.2% nitrogen per day at 0 psig.

Photo 4   Liquid Helium Delivery Vehicle for MRI Users

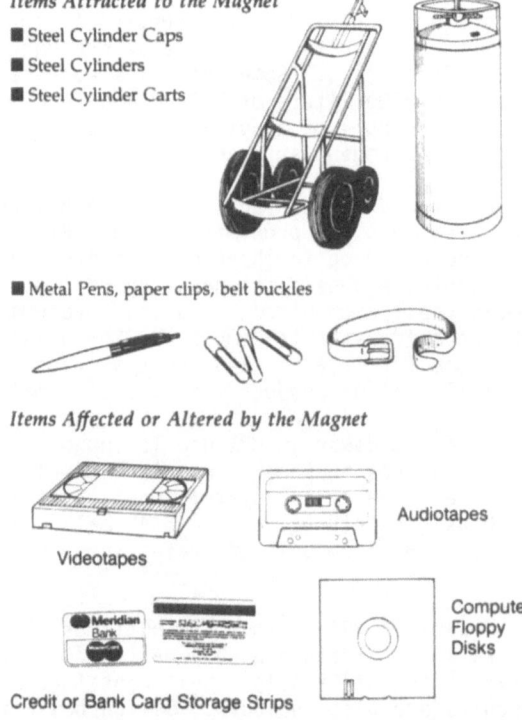

Items Attracted to the Magnet
■ Steel Cylinder Caps
■ Steel Cylinders
■ Steel Cylinder Carts

■ Metal Pens, paper clips, belt buckles

Items Affected or Altered by the Magnet

Videotapes

Audiotapes

Credit or Bank Card Storage Strips

Computer
Floppy
Disks

These objects can also affect the quality of the image if a scan is in progress.

Figure 5   Magnetic Resonance Magnet Hazards

Methodology - Fault tree analysis incorporating some event tree concepts was used in conducting a hazard review of the storage and handling of dewars for the MRI business, but could be expanded to include any operation which requires either liquid helium or nitrogen in dewar quantities. The major objective of the review was to quantitatively assess the risk to the MRI technician due to asphyxiation because of an oxygen-deficient atmosphere.

Because of the many different types of MRI sites, certain simplifying, and in some cases, "worst-case", assumptions were made. These are summarized in the following:

- The MR technician services two sites as primary and assists another technician at two other sites.

- Technician is normally exposed about one hour per dewar handled.

- MR usage rates: 1 LIN dewar/week (52/yr)
                  1 LHe dewar/14 days (26/yr)

- Various storage room sizes were chosen (l x w x h)

    -   5' x 5' x 10' =   250 ft³
    - 10' x 10' x 10' = 1000 ft³
    - 20' x 20' x 10' = 4000 ft³
    - 20' x 40' x 10' = 8000 ft³
       20' x 30' x 15' - 9000 ft³

  The 250 ft³ and 1000 ft³ would be likened to an elevator or broom closet while the higher room volumes might represent storage within the magnet room itself. A uniform distribution of storage rooms among MR users was also assumed.

- The minimum number of air changes initially assumed is 0.5 per hour. This is a conservative estimate and is based on naturally ventilated structures with doors and windows closed. Typically, mechanically ventilated premises that might house MR units would have air changes from four to six per hour, some as high as ten per hour.

- The storage room is well-mixed and the inert (helium and/or nitrogen) concentration is constant throughout.

- The volumetric flow rate of exhaust gas from a dewar release is equal to or greater than the volumetric flow rate of venting inert (i.e., the room does not pressurize).

- An oxygen concentration of 14% was considered to be the threshold for fatality due to an oxygen-deficient atmosphere.

Dewar release rates and durations were calculated for a number of failure scenarios. Most of the failures involve a fixed amount of cryogenic vapor over a relatively short period of time. The ability of a room of a fixed size to adequately vent a given release of vapor can be determined using a simple mathematical model.

Let C be the mole fraction of added inert gas (excluding the nitrogen from the air) in any room. The maximum amount of inert that can be added before a room becomes oxygen-deficient is defined by the equation:

$$(1 - C)(0.21) = \text{target mole fraction } O_2$$

If we assume that 14% oxygen is our target (mole fraction $O_2 = 0.14$), then

$$C = 0.333$$

To solve for the added inert concentration change as a function of time yields the expression:

$$dC/dt = (\text{rate inert in} - \text{rate inert out})/\text{room volume}$$

Or,

$$dC/dt = \frac{[\text{rate inert in} - (\text{mol fr inert})(\text{exhaust rate})]}{\text{room volume}}$$

$$dC/dt = [M1 - (M2)(C)] / VR$$

where:

M1	=	volumetric flow rate of inert discharged into room, SCFM
M2	=	volumetric flow rate of room exhaust, SCFM
VR	=	room volume, ft3
t	=	time, minutes

By our previous assumption, since the room cannot pressurize, if M1 > M2, then M1 = M2.

Integrating this equation, one can then calculate the time it takes for the oxygen concentration to drop to 14% for a given release rate.

$$t = (-Vr/M2) [\ln (C - M1/M2) / -(M1/M2)] \Big|_0^{0.33}$$

Or, this expression can be used to solve for the final room oxidant concentration if the total vent time is known:

$$C = M1/M2 [1 - \exp (-M2 (t)/VR)]$$

Equipment and procedural failures associated with both the LHe and LIN dewars were examined individually. The average probability that a LHe or LIN dewar would be in a "failed state" (or release mode) from any of the potential component failures was calculated for the following events:

- Abnormal boil-off from dewar insulation failure
- Catastrophic failure of pressure or level gauge
- Relief valve opens and fails to reseat
- Improper venting prior to transfer
- Loss of dewar vacuum
- Hose failure during cryogen transfer

Each event was then analyzed, using the specific release rate, to determine the probability of an oxygen-deficient atmosphere as a function of room size and ventilation rate. Given the technician's exposure of one hour per dewar every two weeks (for helium) and one hour per dewar every week (for nitrogen), the total probability of exposure to an O_2-deficient atmosphere per dewar per event was calculated and shown in Table 2. These event probabilities are then summed for each type dewar and multiplied by 100 and 200 dewars respectively for LHe and

Table 2 Dewar Storage Hazard Review - Base Case

EVENT	BASE FAILURE RATE (EVENTS/YR)	GAS FLOW RATE (SCFM)	GAS VOLUME (SCF)	PROB. OF INAD. VENT	HAZARD RATE (EVENTS/YR)	PROB. OF EXPOSURE	PROBABILITY PER DEWAR
LHe DEWAR							
BOIL-OFF WITH SHIELD FAILURE	2.59E-05	0.71	6657.5	0.2	5.17E-06	0.0125	6.47E-08
PI FAILURE AT 8 PSIG	2.35E-07	78.5	634	0.4	9.40E-08	0.0125	1.18E-09
SV-2 RELEASE	2.63E-05	8.3	792	0.4	1.05E-05	0.0125	1.31E-07
V-2 VENTING	4.50E-08	422.67	634	0.4	1.80E-08	1.0000	1.80E-08
DEWAR VACUUM LOSS	4.14E-05	1036	6657.5	1.0	4.14E-05	0.0125	5.18E-07
TRANSFILL HOSE FAILURE	9.36E-08	332.8	6657.5	0.0159	1.49E-09	1.0000	1.49E-09
			TOTAL PROBABILITY (100 DEWARS) =				**7.34E-05**
LIN DEWAR							
PI FAILURE AT 22 PSIG	4.70E-05	0.71	523	0.4	1.88E-05	0.0125	2.35E-07
LEVEL GAUGE FAILURE AT 22 PSIG	1.76E-05	466.7	523	0.4	7.04E-06	0.0125	8.80E-08
OPERATOR VENTS (22 TO 10 PSIG)	7.20E-06	100.7	285	0.2	1.44E-06	1.0000	1.44E-06
DEWAR VACUUM LOSS	2.31E-04	3.07	5658	0.2	4.62E-05	0.0125	5.78E-07
GAS WITHDRAWL VALVE LEAKS	2.60E-07	5.8	5658	0.4	1.04E-07	0.0125	1.30E-09
BURSTING DISC VENTS AT 100 PSIG	1.30E-05	85.7	5658	1.0	1.30E-05	0.0125	1.63E-07
TRANSFILL HOSE FAILURE	9.36E-08	282.9	5658	0.0159	1.49E-09	1.0000	1.49E-09
			TOTAL PROBABILITY (300 DEWARS) =				**5.01E-04**
			TOTAL PROBABILITY =				**5.75E-04**

LIN to provide the total probability of exposure on a yearly basis. The data show that the probable threat associated with the LIN dewars exceeds that of the LHe dewars by nearly an order of magnitude (5.01 x 10⁻⁴ vs. 7.34 x 10⁻⁵). The controlling event for the LHe dewar is dewar loss of vacuum and for the LIN dewar is operator venting. The total probability of exposure to both type dewars on a yearly basis is calculated at 5.75 x 10⁻⁴.

For our purposes, it was felt that improvements could be made in the storage and handling of cryogenic dewars for the MRI business which would have a significant impact on reducing the overall probability. These include:

- install wire seal on ball valve for LHe dewar 1 psig relief line
- establishing a minimum storage room size of 2000 ft³
- ensure reliability of oxygen monitoring of dewar storage areas
- provide technician with five-minute breathing air escape pak when transporting dewars by elevator

Table 3 illustrates the effect of limiting the storage room size to a minimum of 2000 ft³. The probability of inadequate ventilation is reduced to zero for the first four LHe dewar venting cases and for the first five LIN dewar venting cases. Dewar vacuum loss is still the predominant event for the LHe dewar while burst disc venting becomes the dominant event for the LIN dewar. The data shows that with a minimum room size restriction, the probable threat associated with the LHe dewars and the LIN dewars is nearly equal (5.19 x 10⁻⁵ vs.

Table 3   Dewar Storage Hazard Review - Minimum 2000 FT3 Room Size

EVENT	BASE FAILURE RATE (EVENTS/YR)	GAS FLOW RATE (SCFM)	GAS VOLUME (SCF)	PROB. OF INAD. VENT	HAZARD RATE (EVENTS/YR)	PROB. OF EXPOSURE	PROBABILITY PER DEWAR
LHe DEWAR							
BOIL-OFF WITH SHIELD FAILURE	2.59E-05	0.71	6657.5	0	0.00E+00	0.0125	0.00E+00
PI FAILURE AT 8 PSIG	2.35E-07	76.5	634	0	0.00E+00	0.0125	0.00E+00
SV-2 RELEASE	2.63E-06	8.3	792	0	0.00E+00	0.0125	0.00E+00
V-2 VENTING	4.50E-08	422.67	634	0	0.00E+00	1.0000	0.00E+00
DEWAR VACUUM LOSS	4.14E-05	1036	6657.5	1.0	4.14E-05	0.0125	5.18E-07
TRANSFILL HOSE FAILURE	9.36E-08	332.8	6657.5	0.0159	1.49E-09	1.0000	1.49E-09
				TOTAL PROBABILITY (100 DEWARS) =			**5.19E-05**
LIN DEWAR							
PI FAILURE AT 22 PSIG	4.70E-06	0.71	523	0	0.00E+00	0.0125	0.00E+00
LEVEL GAUGE FAILURE AT 22 PSIG	1.76E-06	466.7	523	0	0.00E+00	0.0125	0.00E+00
OPERATOR VENTS (22 TO 10 PSIG)	7.20E-06	100.7	285	0	0.00E+00	1.0000	0.00E+00
DEWAR VACUUM LOSS	2.31E-04	3.07	5658	0	0.00E+00	0.0125	0.00E+00
GAS WITHDRAWL VALVE LEAKS	2.60E-07	5.8	5658	0	0.00E+00	0.0125	0.00E+00
BURSTING DISC VENTS AT 100 PSIG	1.30E-06	85.7	5658	1.0	1.30E-05	0.0125	1.63E-07
TRANSFILL HOSE FAILURE	9.36E-08	282.9	5658	0.0159	1.49E-09	1.0000	1.49E-09
				TOTAL PROBABILITY (200 DEWARS) =			**3.28E-05**
				TOTAL PROBABILITY		=	**8.47E-05**

3.28 x 10^{-5}) and the sum of the two (8.47 x 10^{-5}) is a factor of six (6) lower than the risk predicted in the base case where no room size restriction was used.

Table 4 illustrates the effect of limiting room size to a minimum of 2000 ft^3 and incorporating an oxygen monitor in the room. The dewar vacuum loss is still the controlling event for the LHe dewar with an exposure probability per dewar of 2.77 x 10^{-8} while the burst disc venting remains controlling for the LIN dewar at 8.69 x 10^{-9} per dewar. However, these probabilities are reduced since the analyzer will warn of impending danger and decrease the probability of hazardous exposure. An analyzer failure rate of 0.0535 per year was used. The data shows that with a minimum room size and oxygen analyzer in the room, the probable threat associated with the LHe dewars and the LIN dewars are 2.78 x 10^{-6} and 1.75 x 10^{-6} respectively. The sum of the two (4.53 x 10^{-6}) is a factor of nineteen (19) less than the previous case and a factor of 127 less than the base case.

CONCLUSIONS

Helium and nitrogen are seldom viewed as hazardous materials. However, in certain applications, they create a variety of hazards associated with low temperatures, high pressures, and potential for asphyxiation. Industry experience and practices in safe handling of these fluids are extensive and well proven. The Chemical Process Industries have demonstrated that process hazards review techniques are an effective tool in the identification of hazards and in their

Table 4 Dewar Storage Hazard Review - Minimum 2000 FT3 Room Size and O$_2$ Analyzer

EVENT	BASE FAILURE RATE (EVENTS/YR)	GAS FLOW RATE (SCFM)	GAS VOLUME (SCF)	PROB. OF INAD. VENT	HAZARD RATE (EVENTS/YR)	PROB. OF EXPOSURE	PROBABILITY PER DEWAR
LHe DEWAR							
BOIL-OFF WITH SHIELD FAILURE	2.59E–05	0.71	6657.5	0	0.00E+00	0.0125	0.00E+00
PI FAILURE AT 8 PSIG	2.35E–07	78.5	634	0	0.00E+00	0.0125	0.00E+00
SV–2 RELEASE	2.63E–06	8.3	792	0	0.00E+00	0.0125	0.00E+00
V–2 VENTING	4.50E–08	422.67	634	0	0.00E+00	1.0000	0.00E+00
DEWAR VACUUM LOSS	4.14E–05	1036	6657.5	1.0	2.21E–06	0.0125	2.77E–08
TRANSFILL HOSE FAILURE	9.36E–08	332.8	6657.5	0.0159	7.96E–11	1.0000	7.96E–11
TOTAL PROBABILITY (100 DEWARS) =							**2.78E–06**
LIN DEWAR							
PI FAILURE AT 22 PSIG	4.70E–05	0.71	523	0	0.00E+00	0.0125	0.00E+00
LEVEL GAUGE FAILURE AT 22 PSIG	1.76E–05	466.7	523	0	0.00E+00	0.0125	0.00E+00
OPERATOR VENTS (22 TO 10 PSIG)	7.20E–06	100.7	285	0	0.00E+00	1.0000	0.00E+00
DEWAR VACUUM LOSS	2.31E–04	3.07	5658	0	0.00E+00	0.0125	0.00E+00
GAS WITHDRAWL VALVE LEAKS	2.60E–07	5.8	5658	0	0.00E+00	0.0125	0.00E+00
BURSTING DISC VENTS AT 100 PSIG	1.30E–05	85.7	5658	1.0	6.95E–07	0.0125	8.69E–09
TRANSFILL HOSE FAILURE	9.36E–08	282.9	5658	0.0159	7.96E–11	1.0000	7.96E–11
TOTAL PROBABILITY (200 DEWARS) =							**1.75E–06**
TOTAL PROBABILITY =							**4.53E–06**

elimination and control. These techniques have become a vital part of their safety methodology over the past six to eight years. Air Products and Chemicals, Inc. has used the process hazards review concept for nearly fifteen years and has evolved the process to include hazards quantification, consequence analysis, and risk assessment as illustrated for MRI unit asphyxiation hazards. These techniques permit ranking hazards with respect to frequency and consequence, and provide management with quantitative decision-making data concerning the adequacy of preventive measures.

The practical operating experience and safety analysis work at the Fermilab Tevatron facility at Batavia, Illinois provides a vital background upon which to base helium, nitrogen, argon, and cryogenic safety for the SSC. However, process hazard review techniques can be an effective additional tool in hazards identification, elimination, and control at all levels of the SSC program. Small scale component and magnet testing can be made safer by conducting hazard reviews of proposed systems. The design of the accelerator systems string test (ASST) facility for surface-level testing of magnets and related systems at the SSC Laboratory can reap additional safety benefits from the use of the hazard review process. The design of the prototype installation facility (PIF) for underground testing can benefit greatly from the hazard review process.

APPLICATION OF SYSTEM SAFETY ENGINEERING TECHNIQUES FOR HAZARD PREVENTION AT THE SUPERCONDUCTING SUPER COLLIDER

Barry L. Hendrix

Project Management Division
Superconducting Super Collider Laboratory*
2550 Beckleymeade Avenue
Dallas, Texas 75237

Abstract: A primary goal of the Superconducting Super Collider Laboratory (SSCL) is to establish an exemplary safety program. Achieving this goal requires leadership, planning, coordination, and technical know-how. To ensure that safety is an inherent part of the design, the Environment, Safety and Health Office employs a systems engineering discipline and process known as System Safety. The goal of System Safety—hazard prevention—is accomplished by analyzing systems to identify hazards and to evaluate design and procedural options and countermeasures to prevent, eliminate, mitigate, or control hazards and risks. Establishment of safety and human factors design criteria at the outset of the project prevents unsafe designs and safety violations, reduces risks, and helps in avoiding costly design changes later. This process requires a considerable amount of coordination with a variety of technical disciplines and safety professionals to integrate methods of hazard prevention, mitigation, and risk reduction throughout the system life-cycle.

INTRODUCTION

System Safety is an engineering management science that is being prudently implemented and practiced at the Superconducting Super Collider Laboratory (SSCL). System Safety is the arm of Project Management that works closely with the Environmental, Safety and Health Office on all safety planning and technical safety matters. System Safety is responsible for coordinating safety engineering efforts within each division at the SSCL. Other responsibilities within System Safety include Human Factor Engineering and Safety Risk Management. This paper provides an overview of the philosophy of System Safety and the proven safety engineering and management approaches that are used to influence safe design and operation of the SSC. Proven hazard analysis and risk assessment processes are presented.

---

*Operated by the Universities Research Association, Inc., for the U.S. Department of Energy under Contract No. DE-AC02-89ER40486.

SYSTEM SAFETY

Definition and Purpose

The principal objective of a System Safety program is to ensure that safety, consistent with SSCL mission requirements, is designed into systems, subsystems, equipment, facilities, and their interfaces and operation. A *system* may comprise not only facilities and technical systems hardware, but also the software, interface controls, man-machine interface, operators, training, safety standards, guidelines and operating procedures. In accordance with System Safety philosophy, the formal textbook definition of *system* is a composite, at any level of complexity, of personnel, procedures, materials, tools, equipment, facilities and software. (The elements of this composite entity are used together in the intended operational or support environment to perform a given task or to achieve a specific production, support, or mission requirement). From a System Safety perspective, the word *safety* means freedom from those conditions that can cause death, injury, occupational illness, or damage to or loss destruction of equipment or property, or damage to the environment. *System Safety* can be defined as.the application of engineering and management principles, criteria, and techniques to optimize all aspects of safety within the constraints of operational effectiveness, time, and cost throughout all phases of the system life-cycle. Therefore, a *System Safety program* is the combined management and engineering tasks of planning, implementing, and establishing system-specific requirements to identify system hazards and impose design and management controls to prevent mishaps by eliminating hazards or reducing the associated risk to acceptable levels. For this to be effective, a systematic process is required to accomplish specific safety tasks.

The Process of System Safety

System Safety is not synonymous with Safety Systems, which is a safety design configuration. Instead, one should think of System Safety as the systematic safety process by which those safety systems are prudently and carefully planned and designed to meet their intended purpose. From a systems engineering perspective, System Safety is an effective process that streamlines the safety effort by focusing on essential tasks required to ensure adequate hazard mitigation and risk management. It is a flexible approach used throughout domestic and international industry and is in accord with the current philosophy of the SSC: "to do whatever it takes to ensure systems are safe."

System Safety Engineering is the scientific and engineering approach to solving safety problems of a technical nature. The primary mission is hazard prevention. In order to prevent, eliminate, control, or mitigate specific hazards associated with design and operation, the hazard must first be clearly identified and evaluated for effects, severity, frequency of occurrence, and associated risk. Department of Energy (DOE) and Department of Defense (DOD) experience has shown that risk reduction—through safety analyses and hazard prevention, elimination and control processes—is most effective when implemented early in the system life-cycle.

Early identification and elimination or control of an unacceptable hazard or risk decreases the potential negative impact on the program and the likelihood of a costly retrofit. "After the fact" safety design changes due to negligence, oversight, and the like can be extremely expensive. Designing safety into a system can reduce the overall life-cycle cost and increase the system's reliability and availability. Therefore, it is the responsibility of management to ensure that a formal safety engineering or a System Safety Program is developed so that safety is designed "before the fact" into systems in a logical and efficient manner.

Responsibilities of Safety Professionals

Safety professionals are responsible for preventing injury to personnel and damage to equipment. To ensure safety, these professionals must have the authority to make safety decisions and to influence safe designs and operations. This requires planning and persistence.

One responsibility of all safety professionals is to ensure that engineering and safety standards are not compromised. According to current law, criminal sanctions can now be applied to individuals when willful violations of federal Occupational Health and Safety Administration (OSHA) standards result in a death of an employee. This means that safety professionals are obligated to do not only what is ethical and moral, but what is legally mandatory as well. The phrases "Best Management Practices" or "Most Stringent Guidelines" are not without meaning, but they are not always the right answer. There are practical and budgetary limits to what can be incorporated into a design. Therefore design tradeoff studies may be appropriate, and less stringent approaches may be applicable as long as safety is not compromised. Many design codes and standards may apply to a given project, but taken as a whole, they may yet leave some safety gaps. Codes and standards must be thoroughly evaluated by safety personnel before they are recommended to solve safety problems. In some cases the solution may lie outside the code. It is the responsibility of safety professionals to emphasize the design of systems to operate safely. But they should not blindly adhere to a safety design that yields no safety advantage. While safety professionals cannot violate applicable laws or overlook required minimum standards for safe designs, they should never stand behind a code or regulation without first thoroughly investigating and assessing its applicability.

The degree of safety achieved in a system depends directly on management's emphasis on safety during conceptual planning and design, construction, and operation. Management must see that mishap risk is understood and that risk reduction is always considered in the management review process. The consequences of personnel injury, equipment loss, system downtime, and environmental damage must be fully realized from a legal or liability perspective and from an ethical perspective. DOE and the SSCL are committed to the goal of building the safest system possible within the constraints of time, technology, and budget.

DOE Authorization

DOE Order 5481.1B Change 1, 5-19-87, Safety Analysis and Review System, clearly defines the purpose, policy, general requirements, and guidance for safety analysis, review, and assessment of DOE operations. Section 5, paragraph k of DOE 5481.1B references the use of MIL-STD-882B System Safety Program Requirements "which provides detailed information for organizing, developing, and implementing system safety programs tailored to individual program needs." As authorized by DOE 5481.1B, MIL-STD-882B can be tailored to a broad variety of applications and adapted as appropriate to various types of technologies and industries. It is currently used as a general guideline for implementing an effective System Safety Engineering program at the SSCL.

Organization

Although System Safety is a process, it is also a well-recognized engineering and technical safety management field. At the SSCL the System Safety Manager is assigned to the Project Office and is asked to provide technical interface assistance to the Assistant Director of Environmental, Safety and Health. System Safety, commonly referred to as Project Safety, interfaces closely with the line organization Operational Safety Officers, Safety Coordinators, and safety engineering staff of each division on all technical safety issues. System Safety reports administratively to the Manager of Systems Engineering.

Interfaces

System Safety provides integration assistance on all safety-related matters associated with SSCL design and operation. Provision of this service requires that safety tasks and interfaces be clearly identified. An effective safety program can be accomplished only with the full support and cooperation of upper management and by application of a broad variety of engineering, design, and support disciplines throughout the organization. In addition to direct support to Project Management, technical interface and frequent coordination with the Environmental, Safety and Health Office is required to ensure compliance with DOE orders and federal, state and local safety laws or standards. A close working relationship among System Safety and Systems Engineering Analysts and the support disciplines of Reliability, Maintainability, Availability, and Quality Assurance is required to ensure that the safety process is accomplished and implemented. Continuous interface with the safety officer, safety engineers, and other technical disciplines in all divisions is necessary to coordinate safety efforts and to ensure that the System Safety process of hazard prevention, elimination and control is working. This technical liaison will enhance the overall safety process and will help open channels of communication on safety issues. System Safety provides a technical and managerial interface with all divisions to ensure timely coordination of safety efforts associated with the highly complex and integrated task of developing the SSCL Safety System.

SYSTEM SAFETY TASKS

The following tasks will be performed to ensure that System Safety can effectively influence the safety design of the SSCL.

- Implement a formal System Safety program
- Develop a System Safety and ES&H program plan
- Establish safety liaison with SSCL associates, contractors, vendors
- Review applicable safety policies, procedures, standards and guidelines
- Interface with ES&H and support safety mission
- Develop computerized hazard tracking and risk resolution database system
- Develop test and evaluation safety plan
- Develop System Safety progress summaries (submit to project management)
- Evaluate/recommend requirements for System Safety staff
- Develop preliminary hazard list of top-level safety concerns
- Conduct preliminary safety analysis/train engineers on safety analysis process
- Develop preliminary hazard analysis (PHA) reports (submit to DOE)
- Perform system/subsystem hazard analysis of technical systems design
- Conduct human engineering/human factors studies (as required)
- Perform operating & support hazard analysis of Safety System
- Assist ES&H in occupational health hazard assessments
- Develop safety verification guidelines

- Develop operational safety procedures

- Assist ES&H and divisions in developing and conducting safety training

- Perform assessments of Safety Systems and operations

- Perform safety design criteria compliance assessments

- Review engineering changes, deviation/waiver for safety

- Communicate, coordinate, elevate internal System Safety issues

- Assist safety consultants–university outreach program (as required).

Safety Design Criteria

DOE 6430.1A General Design Criteria is the primary document used to identify acceptable safety criteria for facilities. Technical systems and safety systems in many cases require specific engineering standards. During the development and operation of this project System Safety will assist engineers and engineering support personnel in defining acceptable design criteria for incorporation into requirements documents, specifications, and operating procedures. In some cases, this is being coordinated through the use of appointed safety committees that assess the applicability of various codes, regulations, and engineering standards. In many cases, specific safety design requirements are analyzed and prudent judgment is used to ensure that safety is inherent in the design. System Safety assists systems engineers and designers during the development of specifications. Close liaison with the SSCL Engineering Standards Department is also required.

Many design standards (e.g., ANSI, IEEE, ASME, and NFPA) that are used for a broad variety of facilities, equipment, and technical systems design have safety-related criteria that can be evaluated to determine applicability to the SSCL design. From a safety engineering perspective, many widely used and proven industrial codes and standards are acceptable design criteria, but some sections pertaining to DOE special facilities, such as nuclear or weapons plants, are not applicable to the specific technology of the SSCL. These should be evaluated on a case-by-case basis to determine whether to consider them as requirements or as flexible guidelines. Use of applicable DOE orders and specific commercial engineering standards and guidelines help reduce risks and help ensure that optimal engineering and safety design practices are followed.

Risk

The best safety design is one that has the least risk. The System Safety definition of *risk* is the quantitative or qualitative expression of possible loss in terms of hazard severity and hazard probability that a hazard will cause harm and the consequences of that event. Safety design criteria are intended as acceptable standards to reduce risks to acceptable levels. Although there are other methods of mitigating hazards, such as training, protective equipment, warning devices, and operational safety procedures, they are seldom adequate substitutes for safe design. Whenever possible, within the constraints of technology, time and cost, severe hazards should be prevented, eliminated, or controlled through design. Inherently safe designs don't just happen—they are planned. If safety is not a primary consideration early in the design phase, it is usually a costly afterthought. Experience has proven that unsafe designs are unacceptable. Safety must be built into systems, not added as an afterthought. Residual Risk is that risk associated with significant hazards for which there are no control or mitigation measures.

Hazard Analyses

The hazard analysis process is the backbone of System Safety and is used to identify hazards. A hazard cannot be eliminated or controlled unless it is first clearly identified. Truly effective safety systems and design safeguards cannot be conceptualized, effectively designed, or implemented early in a program planning stage unless system safety processes are employed. Methods for minimizing, mitigating, or controlling hazards either through design or through methods such as placarded warnings, cautions, notes, and safety training are possible only when the specific hazard and all of its effects are thoroughly analyzed and evaluated. For this reason the hazard analysis process must be employed as early in the system design stage as possible. Quantitative hazard analyses are somewhat similar to Reliability Failure Modes Effect and Analysis (FMECA). However, a hazard analysis takes the process a step further and assesses the safety implications, safety effects, hazard severity, and frequency of occurrence classifications and determines which criteria or actions are required to minimize identified hazards and risks. There are many types of hazard analyses including Preliminary Hazard Analysis, System/Subsystem Hazard Analysis, Operating and Support Hazard Analysis, Fault Hazard Analysis, Fault Tree Analysis, Software Hazard Analysis, and Consequences Analysis. Since the Preliminary Hazard Analysis is an initial step in influencing a safety program, it is the most common analysis.

Preliminary Hazard Analysis

The SSCL Project Management Plan (PMP) establishes the requirement for a Preliminary Hazard Analysis (PHA) Report, which is currently being developed. The PMP describes the PHA as.a "document (that) lists and discusses the procedures dealing with the hazards associated with the construction and testing of the SSC." MIL-STD-882B, which is endorsed by DOE 5481.1B, says the purpose of the PHA is to "identify safety critical areas, evaluate hazards, and identify the safety design criteria used."

The PHA is widely recognized as a proven safety approach used by the electronics industry, petrochemical industry, aircraft and aerospace companies, weapon system engineering firms and defense contractors, and various government agencies associated with complex systems. The PHA is an initial risk assessment of a concept or system. To be effective the PHA should begin during the concept exploration phases or earliest life-cycle phase of the program so that safety considerations are included in tradeoff studies and design alternatives. Based upon the best available data, including mishap data from similar systems, hazards associated with the proposed design or functions shall be evaluated for hazard mishap severity, hazard mishap probability, and operational constraint. The PHA also addresses safety provisions and alternatives needed to eliminate hazards or reduce their associated risk to a level acceptable to the appropriate safety-managing agency. At a minimum, the PHA must consider the following for the identification and evaluation of hazards:

- Hazardous components (e.g., flammables, explosives, chemicals, toxic substances, hazardous construction materials, electrical, pressure systems, cryogenics, and all energy sources).

- Safety related interfaces (e.g., material compatibilities, electromagnetic interference, inadvertent activation or lack of control).

- Environmental constraints (e.g., drop, shock, vibration, temperature, noise, exposure to toxic substances, health hazards, fire, electrostatic discharge, lightning, electromagnetic environmental effects, ionizing and non-ionizing radiation including electrical radiation).

- Operating, test, maintenance and emergency procedures (e.g., human factors engineering, human error analysis of operator functions, tasks analyses, task requirements, effects of factors such as equipment layout, lighting requirements, potential exposure to toxic materials and hazardous substances, effects of noise or radiation on human performance, emergency or environmental waste disposal, access and egress, operational safety procedures and safety training).

- Facilities and equipment (e.g., provisions for all phases of operations including daily operations, special operations, maintenance, testing and training).

- Safety related equipment, safeguards, and possible alternative approaches (e.g., interlocks, system redundancy/ back-up devices, hardware or software fail-safe design considerations, subsystem protection, fire detection and suppression systems, personal protective equipment, life support systems and equipment, alert devices, warning devices, detection equipment, ventilation, noise or radiation barriers, placarded warnings).

Hazard Classifications

During the hazards analysis process, risks are determined based upon hazard severity and probability. The following hazard classifications are adapted from MIL-STD-882B and are currently being used at the SSC to determine levels of risk:

Severity of Consequences

    I. (Catastrophic)—death, equipment loss > $500K, > 4 months downtime.

    II. (Critical)—severe injury, Loss $100–500K, > 2 weeks downtime.

    III. (Marginal)—minor injury, Loss $1–100K, 1 day to 2 weeks downtime.

    IV. (Negligible)—no injury, Loss < $1K, < 1 day downtime.

Probability of Mishap

    A (Frequent)—likely to occur repeatedly during the life-cycle of system.

    B (Occasional)—likely to occur several times in the life-cycle of system.

    C (Occasional)—likely to occur sometime in life-cycle of system.

    D (Remote)—not likely to occur in life-cycle of system, but possible.

    E (Improbable)—probability of occurrence cannot be distinguished from zero.

    F (Impossible)—physically impossible to occur.

Risk Zones

    1 (High)—imperative to suppress to a lower level: IA, IB, IC, IIA, IIB are unacceptable (High) risk.

    2 (Medium)—operation requires written, time-limited waiver by management: ID, IIC, IIIA, IIIB require mitigation action.

    3 (Low)—operation permissible: IE, IF, IID, IIE, IIF, IIIC, IIID, IIIE, IIIF and IVA-F are acceptable.

Preliminary Hazard Analysis at the SSC

A Preliminary Hazard Analysis is currently being conducted at the SSC using the System Safety approach. Systems specialists, design engineers, and engineering support personnel assigned to each division are learning how to conduct hazard analyses of the designs for which they are responsible. With oversight from the SSC ES&H Office, the initial PHA was developed by the Project System Safety Office assisted by safety coordinators from the Accelerator, Magnet, Conventional Construction, and Physics Research divisions. In a coordinated effort, 29 engineers and technical personnel conducted first-order hazard analyses associated with over a dozen segments of the SSC design, including Magnet Test Lab (MTL), Magnet Development Lab (MDL), Accelerator Surface String Test (ASST), Injectors, Collider RF Systems, Electrical Systems, Cryogenic Systems, Quench Protection Systems, Conventional Construction, Facilities, and Laboratory Spaces. The PHA is a continuous process, and the document will be updated as new designs and technical requirements evolve. At the end of February 1991, an initial total of 164 hazards were identified and documented. Recommended actions and mitigation methods for each hazard are being planned. Mitigation actions to reduce risk to acceptable levels fall in the following categories:

- Incorporate new or retain current design features

- Develop operational safety procedures

- Conduct specialized safety training

- Develop safety policies

- Conduct inspections

- Comply with OSHA requirements

- Incorporate NFPA codes

- Impose restrictions

- Use protective equipment

- Perform tests

- Plan logistics

- Comply with applicable codes

- Incorporate safeguards

- Provide monitoring equipment

- Enforce policies

- Comply with laws, engineering standards, regulations, and safety rules.

Hazard Tracking

One of the most important areas of an effective System Safety program is documentation of hazards and of methods for resolving risk. A hazard tracking and risk resolution process helps ensure that unacceptable or highly undesirable hazards are documented, evaluated, tracked, and formally proposed to management for action. This systematic process is similar to safety action item identification and assignment of responsibility. An effective hazard tracking system is kept on a centralized database that is accurately updated each time hazard status is changed. A project the size of the SSCL will have literally hundred of identified hazards, many of which must be individually tracked until the associated risks are minimized or reduced to an acceptable level. A paper file is not the most efficient or reliable method for hazard tracking. A computerized database log and track hazards and risks is currently being developed at the SSC.

Risk Assessments

Perhaps one of the most misunderstood areas of safety is risk assessment. Some professionals outside the safety field believe they are qualified to assess risks without all of the facts. A *Risk Assessment* is a comprehensive evaluation of the risk and its associated impact. Risk Assessments, sometimes called safety assessments, are performed as part of the hazard analysis process after hazards are systematically identified and hazardous effects are determined. Risk Assessments are thorough safety evaluations of hazard severity and probability, with a determination of acceptability or unacceptability based upon mitigation actions to reduce risks. Acceptability or unacceptability of a concept, scheme, design or operation is based upon facts such as the environment, conditions, scenarios, human factors, operation, user interface or inherent quality/reliability or safety procedure. Equipment or product integrity, the man-machine interface, potential for human error, types of equipment failures, safety design features, and other factors must be studied to determine risk. To simply determine that a hazard is potentially catastrophic or critical to humans or equipment is not enough. If there is a potential hazard, there must also be a determination of the probability and frequency of mishap occurrence. The methods for determining quantitative risks can be very complex, based as they are upon detailed knowledge of all of the facts and statistics linked to conditions and phases of operation. Often a conservative estimate or prediction, based upon a variety of factors, is the only reasonable way to assess risks. Sometimes research and development studies, computer models, simulations, statistics, reliability data, fault tree analyses, consequence analyses, or historical data associated with similar designs are the best methods for assessing risks. Some risks are reasonably simple to assess; others are very complex and difficult to assess. In the latter cases, extra time and resources may be required. The system/subsystem and all interfaces must be systematically analyzed based upon best available data. Risks are evaluated on the basis of hard facts rather than on an assumption or educated guess. Safety decisions must be justified

Risk Resolution

Once hazards and risks are determined, they must be closely tracked and resolved. An effective process to track hazards and to develop risk resolutions is needed. Recommended actions are based upon the risk assessment.. Safety professionals are responsible for determining the best method of preventing, eliminating, or controlling hazards within the constraints of cost and schedule. Tradeoffs may be required; in some cases "good enough." and "safe" may be appropriate. Specific recommendations for reducing risks must be clearly documented and approved by appropriate management. Options may include:

- Design changes to prevent hazards

- Safety features, interlocks, barriers

- Fail-safe devices

- Monitors, alarms, and control systems

- Redundant systems

- Emergency procedures and precautions

- Operational safety procedures

- Safety training

- Preventive maintenance, inspection, and test

- Warning devices, placards

- Operating limitations, restrictions

- Other prudent mitigation methods.

As previously noted, once a risk has been determined to be borderline or unacceptable in accordance with established criteria, it must be resolved. Hazards and risks do not simply disappear, and sometimes they are hidden. A safety or reliability analyst may fail to identify a hazard or combination of hazardous conditions. A residual risk—a risk that has not been resolved—is usually unacceptable. During the hazard analysis process, hazards are identified and assessed, and actions are recommended to reduce unacceptable hazards to an acceptable level. Risks can sometimes be reduced by adding design features, safeguards, redundancies, back-up systems or safety devices. Such a course may not be economically feasible beyond the preliminary design stage. Therefore the timing of safety analyses and assessments is critical. Analyses and assessments must start early in the design process and continue until design is complete. Once the operational phase begins, safety does not stop. At that time emphasis shifts from the machine design to man-machine interfaces and human errors. If the human factor was not engineered into the safety design, such an omission will show up at this time. Because it is often not practical or possible to redesign the system, safety risks associated with human factor and human errors must be assessed and mitigation measures must be employed to reduce operational hazards and risks. These mitigation methods are usually in the form of training, operational safety procedure, operator limitations, operational restrictions, or less-than-optimum operations due to poor ergonomic or operator designs. Therefore, human factors must be an integral part of the safety hazard analysis and safety risk assessment process.

Safety Assessments

Safety Assessments are not paperwork exercises; they must be factual and accurate. Many assessment reports will be required on large projects or complex systems. A Final Safety Assessment is required at the SSC prior to system operation. Since it is important to identify all safety procedures of the hardware, software, and system design and to identify procedural hazards that may be present, the following areas need to be included in Safety Assessment Reports:

- Safety criteria and methodology used to classify and rank hazards.

- Analyses and tests performed to identify hazards inherent in the system.

- Those hazards that still have a residual risk and the actions that have been or will be taken to reduce those risks to acceptable levels.

- Results of the safety program effort, including a summary of all significant hazards along with specific recommendations, precautions, and mitigation methods required to ensure safety of personnel and property.

- A list of any hazardous materials generated by or used by the system. Identify material type, quantity, potential hazards, precautions, and safety procedures for use, storage, transportation and disposal. Also include copies of Material Safety Data Sheet (OSHA Form 20).

- Conclude with a signed statement that all risks have been resolved or controlled to acceptable levels. If there are any hazardous conditions or areas of risk that are unacceptable or questionable, these shall be clearly documented in the report and a deviation/waiver or variance shall be presented to the managing agency with sound justification and rationale for the lack of resolution of the hazard. As a general rule throughout industry, high risks must always be suppressed to a lower level. Any medium risk is subject to a written, time-limited waiver endorsed by management before operations are allowed.

SUMMARY

The ultimate goal of a comprehensive System Safety program is to minimize the potential for mishaps and thereby maximize operational effectiveness. Inherently safe designs and prudent safeguards and controls can help ensure systems availability and operational safety. To ensure that risks are minimized, many safety tasks must be performed. Safe systems must be planned. Systems must be designed with safety in mind. Shortcuts in safety and human engineering designs, invitations to disaster, must be avoided. Since human error is a contributing cause of the majority of mishaps, human factors must always be considered during design and during selection and evaluation of control and mitigation methods for severe hazards.

Hazard prevention, elimination, and control, and the four System Safety areas of hazard identification, risk assessment, hazard control, and risk management are all interrelated. System Safety hazard analyses and risk management processes, when coupled with Human Factors Engineering concepts, are proven methods and techniques that influence the design and operation of safe systems. Those methods and techniques are currently being implemented at the SSC.

SHIELDING CALCULATIONS FOR SSC

A. Van Ginneken

Fermi National Accelerator Laboratory
P. O. Box 500, Batavia, IL 60510

Abstract

Monte Carlo calculations of hadron and muon shielding for SSC are reviewed with emphasis on their application to radiation safety and environmental protection. Models and algorithms for simulation of hadronic and electromagnetic showers, and for production and transport of muons in the TeV regime are briefly discussed. Capabilities and limitations of these calculations are described and illustrated with a few examples.

1 Introduction

Along with a few empirical rules of limited applicability almost all high energy accelerator shielding calculations rely on the Monte Carlo method. Combining all the information necessary to simulate particle production and transport from TeV down to MeV results in large code systems. From the shielding point of view, four basic types of particles (problems) are usually distinguished: (1) *hadrons*, mostly nucleons and pions, which in a proton accelerator are the primary group of particles, but treating separately: (2) *low energy neutrons*, produced in abundance when the higher energy particles interact with nuclei, (3) *electrons and photons,* derived mostly from the decay of π^0 which are produced in hadronic collisions, and (4) *muons* resulting mainly from $\pi^\pm$ or D meson decay, but also produced in significant numbers by a host of other processes.

The main function of these codes is to trace the cascades, which typically develop when a high energy particle interacts in bulk matter, through some idealized version of the beam/target geometry and magnetic fields present in some apparatus or facility (real or fancied). The program we're most familiar with is CASIM [1] which was written at Fermilab and continues to see extensive use there. Others are, e.g., FLUKA [2] from CERN and MARS [3] written at Serpukhov. There exist also a number of codes to simulate detector response to high energy particle interactions, which is indispensible in the design and analysis of high energy experiments. Although they have a great deal in common, the needs of the two types of programs are quite different: shielding calculations can assume that a large number of (beam) particles is involved, correlations among outgoing particles can be ignored, a 'common sense' approach to questions of geometry and fields is generally sufficient. By contrast detector response codes consider individual events, fluctuations and correlations are extremely important and since a large detector operates in one—or at most a few—configurations one can take the time to represent the geometry, etc., in painstaking detail. Once patience will be rewarded by a smaller error in correcting for efficiencies, etc. For the above reasons there is only limited cross application of the two. Shielding type calculations might be useful in the preliminary stages of detector design but cannot simulate individual events. Detector codes can in principle be applied to shielding problems but are too slow and cumbersome for routine use. Their CPU time demands increase linearly with the accelerator energy vis-a-vis logarithmic for CASIM.

In recent years, shielding calculations have seen extensive use in problems concerned with protection of essential or expensive equipment rather than personnel or the environment, e.g., problems of avoiding quenching of the superconducting magnets cite and of radiation damage to silicon detectors. Because most of the action takes place within the accelerator structure, these problems tend to pose a stiffer computational challenge vis-a-vis the more mundane health physics applications to which this review is addressed. Another important application of these codes which is omitted here is that of target heating, radiation induced shock waves, etc. This is important in the design of beam dumps and thereby has, at least tangentially, some environmental impact. Also omitted here are the paralegal aspects of this business—the laws and the recommendations on dose limits, what are the assumed routine and accidental beam loss conditions, etc.,—which ultimately determine how much shielding must be provided. The remainder of this brief review is heavily biased towards CASIM, with apologies to the other codes and their authors as well as to the reader, for such lack of evenhandedness. It is also biased towards the high energy end, i.e., towards what is different at the SSC.

2 Hadrons

Hadron shielding deals with both the prompt ('beam on') radiation and with radioactivity produced in the environment around the accelerator. A key parameter for an accelerator or experimental area is the amount of lateral shielding required to reduce the hadron dose to an acceptable level. Especially around beam dumps there is the additional question of protecting the groundwater against induced radioactivity. It should be remembered that all such hadron shielding considerations boil down to shielding against low energy neutrons and an SSC calculation differs from one at Fermilab typically only with what happens in the first generation.

In any realistic shielding situation one always averages over a large number of particles and one therefore gains enormously using *weighted* calculations for shielding (in contrast with, most notably, detector simulation). This results in significant economies in modeling—both conceptually and in the amount of coding. The chief advantages of the weighting are (1) hadron production is represented by inclusive distributions, (2) it allows one to concentrate on the particles which contribute most to the problem, and (3) by exponentially reducing the weight of a particle as it traverses the geometry it is well suited to study problems of deep penetration which are the essence of shielding problems. More detail on weighted vs analog calculations may be found in refs. [4].

The particle production model used in the original version of CASIM [1] is the Hagedorn-Ranft model to which is added a high p_T component and a low energy nucleon component. [5]. In terms of physics, the model is quite outdated but has a large number of loose parameters which have been tied to experiment so that it generally works well in the sub-TeV region. However it does not predict the multiplicities and other gross features observed, e.g., in $\bar{p}p$ collisions at CERN and Fermilab, very well. To remedy this the hadro-production model in CASIM has recently been updated [6] to reflect better what is now known from the hadron colliders and from theory. However evaluation of practical hadron shielding requirements is not very sensitive to what is basically only the first generation of the cascade chain [7] when care is taken to conserve energy and momentum and provided none of the other important gross features (multiplicity, limited p_T, π^0 inelasticity, etc.) are flagrantly misrepresented. Some general calculations [8] performed with the old model during the early SSC design stages are still expected to be a useful guide. At the low energy end, hadrons are followed only down to some cut-off (50 MeV typically) which allows for considerable simplification of the code.

Hadron transport in the TeV regime differs from the lower energies in that direct electron pair production and bremsstrahlung become increasingly important mechanisms of energy loss. [9] This again affects only the first few generations of the cascade and is only of limited importance in hadron shielding calculations. While both processes have very broad energy loss distributions, they are adequately represented by an average dE/dx in these calculations. For TeV pions bremsstrahlung becomes an important source of angular diffusion.

In a typical hadron shielding calculation one keeps track of the nuclear interactions (≥ 50 MeV), occurring as a result of the cascade development, as a function of location throughout the target. Charged particles below this cut-off quickly slow down (though π^- in heavier targets will still interact in a nucleus). Neutrons below 50 MeV can migrate considerably, though generally with a shorter interaction length and with limited multiplicative power. In heavy targets the largest source of neutrons is the 'evaporation' process of excited nuclei which have a typical energy of a few MeV. Particularly when hydrogen is present in the medium they lose energy fairly quickly and get absorbed. One envisions then a situation in which an equilibrium is established in the outer regions of the shield (generally soil or concrete) so that the dose may be predicted from the nuclear interaction (or 'star') density of the energetic ($\geq 50 MeV$). A conversion factor [10] is applied which is based on a low energy neutron spectrum calculated deep inside a soil shield. [11] Such spectra are rather insensitive of location and of their detailed origin. For limited radial shielding there may be considerable contribution from high energy hadrons and from electromagnetic showers, so that a simple conversion factor does not apply.

A typical hadron shielding calculation might start from ref. [8] plus some rules of thumb for a rough estimate of the requirements. One then proceeds to model (an over-shielded version of) the particular beam/target geometry of the facility in question. This last part varies from one calculation to the next and must be supplied by the user. In principle the only limit to how realistic a model to furnish, is one's patience. But in practice the point of diminishing returns is usually reached much earlier and the real skill is to distinguish what might be important from all the rest and to simplify as much as possible what remains. This is even more important in the design stages when what's being modeled is not yet firm. The star density is computed throughout the entire geometry. These are readily converted to dose and one can then decide on the correct shield dimensions.

The Monte Carlo is done in three dimensions but, where possible, the analysis (binning of stars in this example) is done assuming cylindrical symmetry. This is mostly for reasons of statistical accuracy since there can be fewer, larger bins (and one can leave the binning procedures of CASIM alone). Where this cannot be reasonably maintained one retreats (gingerly) to some lesser symmetry, perhaps four-fold: left/right and up/down symmetric, or cylindrical but with as few azimuthal bins as possible. Another reason for keeping down the number of bins is to make the information generated more presentable (and digestible). For a simple geometry one can resort to contour plots such as shown below but this may not be suitable for more complicated geometries. Fig. 1 shows an example of such a calculation from the collection of ref. [8]. The star density is also easily converted into a dose from residual radioactivity. A CASIM derived dose can serve as input for calculations of transmission through labyrinths, etc. Ref. [12] cites some of the comparisons of CASIM with observation at Fermilab. On the whole such comparisons have been gratifying.

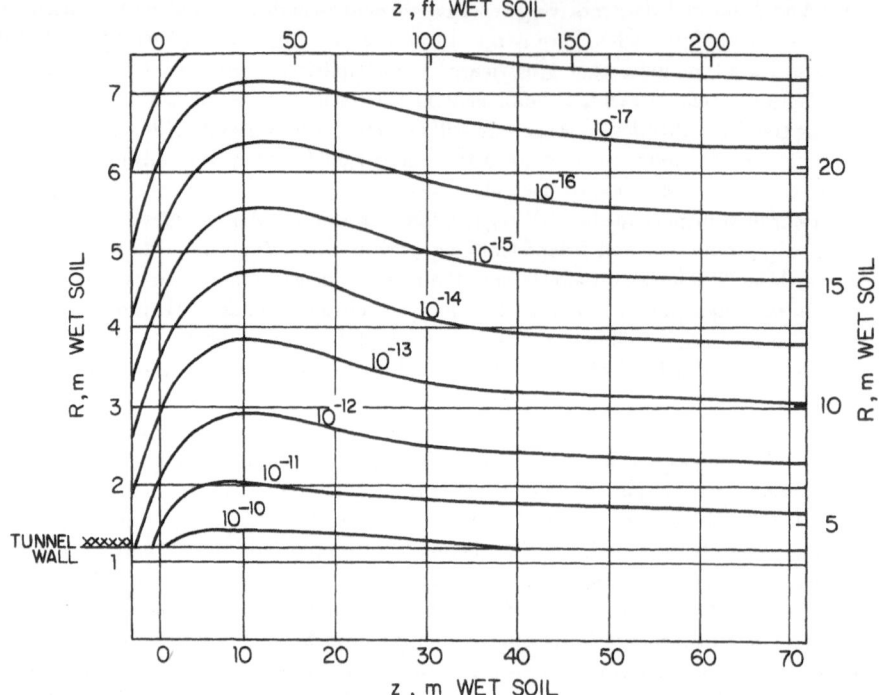

Figure 1. Iso-dose contours (rem/interacting proton) in soil around tunnel when 20 TeV protons are lost on the side of a bare beampipe.

3   Electrons and Photons

As mentioned above, γ and $e^{\pm}$ originate mainly from the decay of π^0 produced in hadron cascades. Although other sources have been identified they are negligible by comparison. The π^0 decays into 2γ which converts into $e^{\pm}$ pairs when traveling through bulk matter and the e^- or e^+ in turn emit γs by bremsstrahlung. This multiplicative behavior leads to an electromagnetic shower. In the typical hadron shield such a shower get absorbed long before the neutrons are, and they can therefore be neglected in most dose calculations. This is not the case for radiation problems with accelerator components or beam dumps and for this purpose AEGIS [13], a companion to CASIM, has been written also using the weighted Monte Carlo approach. The physics of the underlying processes has been known for at least fifty years and although some new effects are predicted to occur at multi-TeV energies these will not yet have any practical significance at the SSC. Production of hadrons by the electromagnetic shower, while of some consequence around electron accelerators, is negligible compared to that by the hadrons themselves. Muon production by these showers has some significance. There are a variety of mechanisms (see below), but the most copious one is photon conversion into $\mu^{\pm}$.

4   Muons

As far as hadrons, electrons, and photons are concerned shielding at the SSC is just a bit 'more of the same'. The extra shielding required is measured in feet. Muon shielding however introduces a real qualitative difference in scale. If, for example, we take muons of one tenth of the incident energy as being produced in significant numbers: it takes about 200m of soil to stop a muon of 0.1 GeV (at Fermilab) but about 1600m to stop a 2 TeV muon (SSC). Moreover the 2 TeV muon shows much more straggling and one still expects to see some at 2500m. For these reasons site criteria, land acquisition, etc. for the SSC are largely a direct result of muon shielding consideration. In contrast to hadron shielding calculations one cannot resort to armwaving about conservation laws, etc., to make things easier. The muons one must shield against come typically from the first few generations of the combined hadron and electromagnetic cascade and on average they carry off only a tiny fraction of the energy of the beam particles. Muon transport is for the most part based on 'old' physics but a shift in emphasis among the different processes responsible for muon energy loss and angular diffusion takes place as the muon reaches the TeV regime. [9] As a result *fluctuations* become much more important, leading to the aforementioned increase in straggling.

Muon production is usually split into (1) muons from π and K decay and (2) 'prompt' muons (which includes decays of short lived particles). This division has practical consequences: the π and K decay times are long enough to make this component very sensitive to the geometry and composition of the shield. In bulk matter its magnitude will increase (almost linearly) with the nuclear interaction length. Moreover, around an accelerator one typically finds large regions of negligible density (beam aperture, tunnels, etc.) where πs and Ks have the opportunity to decay without removal by nuclear interactions. As their name suggests prompt muons are, at least for our purpose, produced instantaneously and therefore do not exhibit this extra sensitivity to geometry and composition. *Hadronically* produced prompt muons which are presently included in CASIM calculations derive from (a) D meson decay, (b) vector mesons (ρ, ω, ϕ, J/ψ) decaying into $\mu^+\mu^-$, and (c) dimuon continuum both from Drell-Yan and 'soft annihilation' mechanisms. *Electrons* and *photons* are assumed to produce muons via (d) pair creation by photons, (e) decay of verctor mesons produced by photons on nuclear targets (coherent, incoherent, and inelastic production), and (f) annihilation of e^+ with atomic electrons. There are still large uncertainties involved in prompt muon production particularly from soft annihilation in the TeV regime. The representations in CASIM are based mostly on experiment and the models are used just as convenient tools of extrapolation.

The electromagnetic processes have typically much lower cross sections than the hadronic ones yet they are important since (1) with increasing primary energy (20 TeV at the SSC) an increasing fraction of this energy is spent on producing electromagnetic showers at the expense of the hadronic part, and (2) in mechanisms (d-f) essentially all the energy of the γ or e^+ converts into energy for the $\mu^+\mu^-$ pair (the *inelastic* part of (e) is the sole exception). The hadronic processes are much less energy efficient in this regard.

At 'low' energies ($E_\mu \leq 300 GeV$, say) the muons lose energy predominantly via collision with atomic electrons and change their direction mostly by multiple Coulomb scattering. Slowing down and angular diffusion can be to good approximation handled on a continuous basis. Above a few TeV, on the other hand, one finds that energy loss by direct (e^+e^- pair production, bremmstrahlung, and deep inelastic scattering (in that order) all exceed atomic collisions as a source of energy loss. As for angular spreading, bremsstrahlung becomes the most important process while deep inelastic scattering becomes competitive with multiple Coulomb scattering and pair production is still negligible. The above statements refer to *averages* only but to get the straggling right most of muon transport must be simulated on an event-by-event basis.

Muons from the initial collision, e.g., of a 20 TeV proton with an iron nucleus or of 20 on 20 TeV protons, usually form a significant part of a muon shielding problem. But to get the full picture one must include muons from all generations of the hadronic and electromagnetic cascade. In CASIM both prompt muons are created in each generation while decay muons are produced in weighted fashion, depending on the length of the meson trajectory. Because muons below a few GeV are absorbed by a typical hadron shield one can run with a higher energy cut-off. The muons are then transported using a continuous approximation only for atomic collisions and for multiple Coulomb scattering. The other processes listed above, plus energetic collisions with electrons (δ rays), are considered as individual events. For each such process a variable energy loss is first randomly selected from the appropriate distribution. For a given a particular energy loss an rms angular deflection is determined in the manner of ref. [9] whence a particular angular deflection is randomly chosen from a Gaussian.

The strong dependence of the π/K decay component on the geometry makes life a little more difficult in that one has to rely even more (relative to hadron shielding) on Monte Carlo and less on rules of thumb. This is all the more true when magnetic fields are present since, e.g., muons can follow the beam aperture for long distances being alternatively swept from the beampipe onto the return field of the magnet and thence bumped back into the aperture. While a CASIM calculation will include such effects, these facts make generalizations difficult including questions of what is a 'worst case' beam loss accident and what is its associated dose at a given location. Fig. 2 shows an example of a muon dose calculation for the contemplated Large Hadron Collider at CERN.

5   Special CASIM Calculations

The foregoing summary describes what is available to the more or less casual user. By programing for a particular geometry the results of such a calculation allow the dose to be determined as a function of location throughout the entire geometry. The biasing techniques, etc., applied in CASIM aim to do a good job for such general purpose calculations. While there are a number of options available in the user supplied data (energy cut-off, collision length biasing, etc.) one may occasionally make changes in the program itself to efficiently perform more specialized demands. A user may wish to know, e.g., the energy spectrum of positive muons at one or more specific locations. An example from real life is that of antiproton production and transport, to help with the design of the Fermilab Antiproton Source, [14] which required some extra routines to deal with the $\bar{p}$s. A recent addition to CASIM is that of *smoothing* the final results. [15] This is useful in the many cases where there are still statistical difficulties in regions where one must know the dose. Presently one gets around this by averaging over a larger region but it is more convenient and probably less biased to automate such a procedure. Fig. 3 shows an example of such smoothing. There has not been much experience with it yet.

6   Concluding Remarks

Shielding calculations, whether by CASIM or other such programs [2, 3], appear in good shape to meet the SSC's demands of radiation and environmental protection. There are some uncertainties, e.g., in muon production, but we know from cosmic ray studies that there will be no big surprises. For fixed target SSC we know this in even more detail from CERN and Fermilab collider results. But it will still be very ineteresting to compare predictions with results at SSC, particlularly for simple geometries, such as a dump. Since the demands for radiation safety appear to be growing with time (lower dose limits, higer quality factors, etc.) there are increased demands on the speed with which a particular geometry can be analyzed: faster computers, faster algorithms, and faster ways to interpret the results.

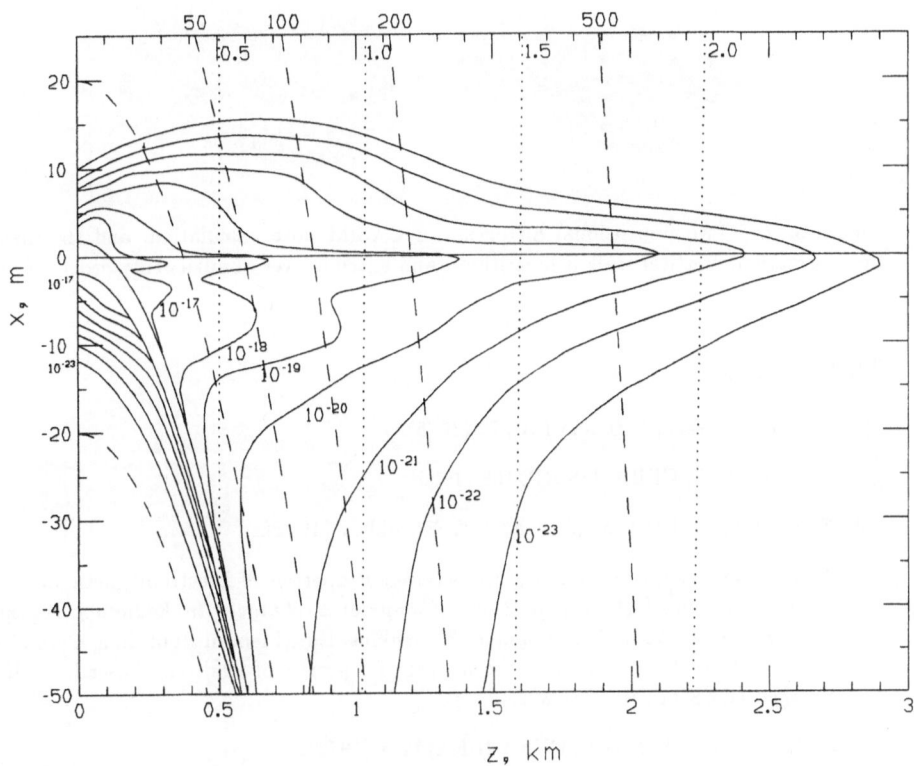

Figure 2. Iso-dose contours (sievert/interacting proton) in median plane in soil around tunnel when 8 TeV protons are lost on the out-side of the beampipe in a continuous dipole.

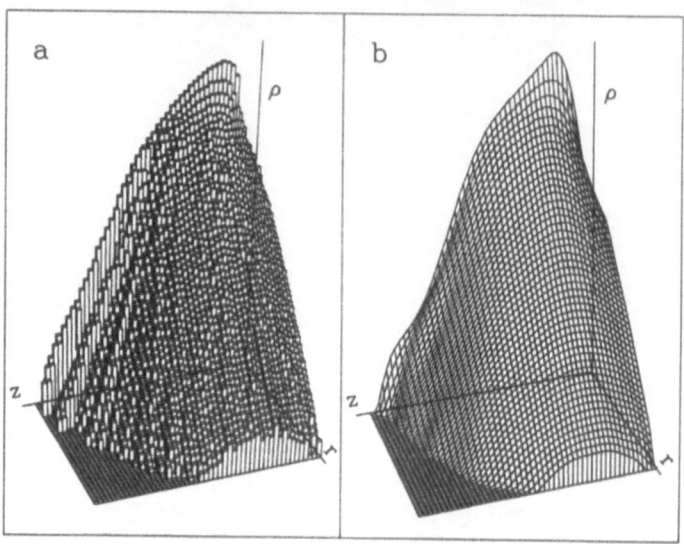

Figure 3. (a) Two dimensional histogram of CASIM dose calculation, and (b) smooth approximation. Vertical scale is logarithmic and extends over about twenty orders of magnitude.

References

[1] A. Van Ginneken, Fermilab-FN-272 (1975).

[2] P. Arnio et al., CERN TIS-RP/168 (1986).

[3] N. V. Mokhov, IHEP preprint 82-168, Serpukhov (1982).

[4] A. Van Ginneken, Calculation of the Average Properties of Electromagnetic Cascades at High Energies (AEGIS), p. 211 in *Computer Techniques in Radiation Transport and Dosimetry*, W. R. Nelson and T. M. Jenkins, Eds.,Plenum Publishing Corp., New York (1980) and Calculation of the Average Properties of Hadronic Cascades at High Energies (CASIM), ibid., p. 323.

[5] J. Ranft and J. T. Routti, CERN-II-RA/72-8 (1972).

[6] S. Qian and A. Van Ginneken, Fermilab FN-514 (1989).

[7] For problems involving colliding beams the production model used for pp collisions is a simple empirical model based on extending observations from CERN and Fermilab colliders and on calculated results [J. Ranft, DTUJET, p. 67 in *Radiation Levels in SSC Interaction Regions*, D. E. Groom, Ed., SSC Central Design Group (1988)]. For more detail see ref. [6].

[8] A. Van Ginneken et al., Fermilab FN-447 (1987).

[9] A. Van Ginneken, Nucl. Inst. Meth., *A251*, 21 (1986).

[10] A. Van Ginneken and M. Awschalom, High Energy Particle Interactions in Large Targets, Fermilab, Batavia, IL (1975).

[11] T. A. Gabriel and R. T. Santoro, ORNL-TM-3262 (1970).

[12] M. Awschalom et al., Nucl. Inst. Meth., *131*, 235 (1975); M. Awschalom et al., ibid., *138*, 521 (1976); J. D. Cossairt et al., ibid., *197*, 465 (1982); J. D. Cossairt et al., ibid., *A238*, 504 (1985); N. V. Mokhov and J. D. Cossairt, ibid., *A244*, 349 (1986).

[13] A. Van Ginneken, Fermilan FN-272 (1975).

[14] C. Hojvat and A. Van Ginneken, Nucl. Inst. Meth., *206*, 67 (1983).

[15] A. Van Ginneken, Smoothing Algorithm for Histograms of One or More Dimensions, Fermilab Pub-*90/197* (1990), to be published in Nucl. Inst. Meth.

16. Magnets III

DESIGN DEVELOPMENT FOR THE 50MM SUPERCONDUCTING

SUPER COLLIDER DIPOLE CRYOSTAT

Thomas H. Nicol

Fermi National Accelerator Laboratory
P.O. Box 500
Batavia, IL 60510
USA

ABSTRACT

The cryostat of a Superconducting Super Collider (SSC) dipole magnet consists of all magnet components except the magnet assembly itself. It serves to support the magnet accurately and reliably within the vacuum vessel, provide all required cryogenic piping, and to insulate the cold mass from heat radiated and conducted from the environment. It must function reliably during storage, shipping and handling, normal magnet operation, quenches, and seismic excitations, and must be manufacturable at low cost.

The major components of the cryostat are the vacuum vessel, thermal shields, multilayer insulation system, cryogenic piping, interconnections, and suspension system. The overall design of a cryostat for superconducting accelerator magnets requires consideration of fluid flow, proper selection of materials for their thermal and structural performance at both ambient and operating temperature, and knowledge of the environment to which the magnets will be subjected over the course their expected operating life.

This paper describes the design of the current 50mm SSC collider dipole cryostat and includes discussions on the structural and thermal considerations involved in the development of each of the major systems. Where appropriate, comparisons will be made with the 40mm cryostat.

INTRODUCTION

Fermilab has been involved in the design and production of cryostats for all prototype Superconducting Super Collider (SSC) dipole magnets built since 1986. That work is very well documented in the literature[2-7]. Late in 1989 a fundamental change in the dipole coil design; increasing the physical aperture from 40mm to 50mm, resulted in the need to redesign the dipole cryostat. The larger aperture gave rise to an increase in diameter of the overall cold mass assembly of approximately 25% and a weight increase of approximately 50% both of which impact the cryostat design significantly.

Cryostat designs for superconducting magnets are largely driven by thermal and structural considerations. Designers must continually be cognizant of the heat load to the helium system and of the structural loads imposed on the cryostat systems from static weight, shipping and handling, quench loads, and ambient ground motion. These two considerations are generally at odds with one another. Low heat load implies a minimum of structural material conducting heat from the environment. Sound structural design implies material with sufficient strength to resist both static and dynamic forces.

This paper attempts to summarize the results of the design effort to date on the collider dipole cryostat for the 50mm magnet. Thermal and structural aspects of the design of each of the major cryostat systems will be described in detail. The 40mm dipole cryostat purposely served as the starting point for this design work. Wherever possible, development work on that and other similar cryostats was borrowed in order to take advantage of proven technology. References to the 40mm cryostat will be made for comparison wherever appropriate. Each of the cryostat systems will be addressed in turn; vacuum vessel, 80K and 20K thermal radiation shields, multilayer insulation (MLI), cryogenic piping, suspension system, and magnet interconnect.

In order to begin the design process, one needs a good handle on the pertinent allowable heat loads to each thermal station as well as a good definition of the structural environment to which the magnet will be subjected. Table 1 summarizes the thermal and structural design criteria for the 50mm dipole which affect the cryostat design[1]. Figure 1 illustrates a cross section through the 50mm cryostat.

VACUUM VESSEL

The vacuum vessel is the outermost cryostat component and, as such, serves to contain the insulating vacuum. In addition, it functions as the major structural element to which all other systems are ultimately attached to the accelerator tunnel floor. Furthermore, it serves as a pressure containment vessel in the event of a failure in an internal cryogen line. The vessel is 685.5 mm OD, 7.92 mm wall with an overall length of 14922.5 mm. Its 40mm counterpart was 609.6 mm OD, 6.35 mm wall and had an overall length of 16306.8 mm.

The vacuum vessel has five penetrations at which the internal cold mass supports are attached. The internal supports are positioned to minimize the deflection of the cold assembly due to its own weight and are spaced on 3179.76 mm centers starting at the center of the magnet. Two external supports serve to tie the vessel to the tunnel floor. The external supports are spaced to minimize the deflection of the vacuum vessel due to its own weight and to the weight of the internal components. They are located 4154.17 mm on either side of center.

Table 1. Thermal and Structural Design Criteria

	4.5K	20K	80K
Static Heat Loads			
Infrared	0.053 W	2.335 W	19.1 W
Support conduction	0.160 W	2.400 W	15.8 W
Interconnect	0.150 W	0.320 W	2.1 W
Total static	0.363 W	5.055 W	37.0 W
Dynamic Heat Loads			
Synchrotron rad	2.169 W		
Splice IR heating	0.140 W		
Beam microwave	0.195 W		
Beam gas	0.136 W		
Total dynamic	2.640 W		
Total Dipole	3.003 W	5.055 W	37.0 W

Structural load summary
Cold mass weight:	11,360 kg	
Shipping and handling:	2.0g	vertical
	1.5g	axial
	1.0g	lateral

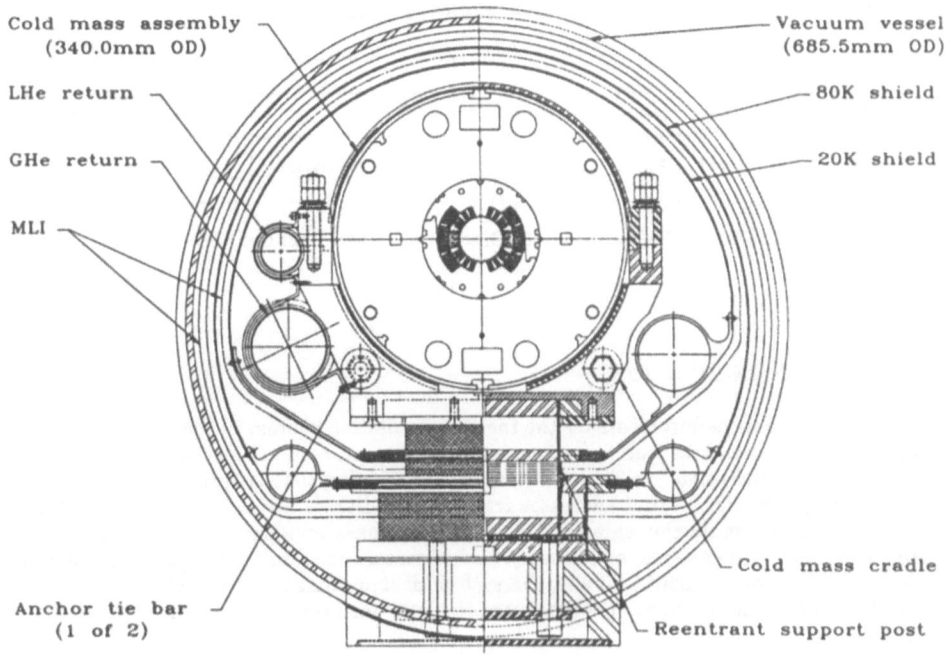

Cold mass assembly
(340.0mm OD)

LHe return

GHe return

MLI

Anchor tie bar
(1 of 2)

Vacuum vessel
(685.5mm OD)

80K shield

20K shield

Cold mass cradle

Reentrant support post

Figure 1. SSC 50mm Collider Dipole Cryostat - Cross Section

The 7.92 mm wall of the vessel is not sufficient to transfer suspension system loads to ground immediately around the internal support locations. At these positions a 19.05 mm thick, 559 mm long reinforcement transfers internally generated loads, primarily from shipping and handling, from the magnet support system to the vacuum vessel. Figure 2 illustrates the major features of the vacuum vessel and notes applicable dimensions. The vacuum vessel material is low carbon steel. Ordinarily one would like to use a material with greater fracture toughness in the event of a major cryogen spill inside the cryostat, however, cost mandated a more readily available material. In addition, failure analysis indicated that a high fracture toughness material was not required.

THERMAL RADIATION SHIELDS

Like its 40mm counterpart, the 50mm dipole cryostat uses two thermal radiation shields. The outermost shield is cooled by LN_2 and operates at 80K. The innermost shield is cooled by helium gas returning to the refrigerator and operates nominally at 20K.

Each shield is cooled by discrete pipes connected to formed shells which totally encompass the internal cold assembly. The 20K shield uses a single extrusion attached to the shell. The 80K shield uses two; one serving as the shield supply, the other as the return. This scheme requires that the shells have high thermal conductivity to minimize thermal gradients around their circumference. Copper and aluminum are the materials of choice for this application. Thermal analyses on both shields were performed using both materials. The results from these analyses indicate there is no significant difference in the thermal performance of either shield between these two materials. Material cost dictated the ultimate material choice. Copper and aluminum are approximately the same cost per unit weight, however, with a density over three times that of aluminum, the use of copper as the shield material would result in an assembly cost three times higher than an aluminum shield. The selected material for both 80K and 20K shields is 6061-T6 aluminum, 1.59 mm thick. The shields are segmented along their length to minimize the effect of bowing during cooldown caused by the asymmetry of the cooling tubes with respect to the shield shells.

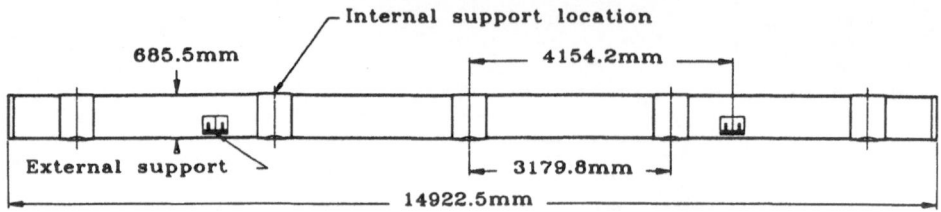

Figure 2. SSC 50mm Collider Dipole Cryostat - Vacuum Vessel Layout

MULTILAYER INSULATION

The design requirements for the thermal insulation system in the SSC dipole cryostat dictate that heat load from thermal radiation and residual gas conduction be limited to the values listed in Table 1. Essential to meeting these requirements is an insulation system design that addresses transient conditions through high layer density for improved gas conduction shielding, has sufficient mass and heat capacity to reduce the effects of thermal transients, and is comprised of materials suitable for extended use in a high radiation environment. Finally, the system design must be such that fabrication and installation techniques guarantee consistent thermal performance throughout the entire accelerator.

The insulation system must have a mean apparent thermal conductivity of 0.76×10^{-6} W/cm-K in order to meet the design heat load budget. This is achieved by using a multilayer insulation (MLI) system comprised of reflective layers of aluminized polyester separated by layers of spunbonded polyester spacer. The reflective layers consist of flat polyester film aluminized on both sides to a nominal thickness not less than 350 angstroms. The spacer layers consist of randomly-oriented spunbonded polyester fiber mats. The mean apparent thermal conductivity of an MLI blanket comprised of these materials has been measured to be 0.52×10^{-6} W/cm-K[6,7]. The MLI system for the SSC 50mm collider dipole cryostat will consist of full cryostat-length assemblies of MLI fabricated and installed as blankets on the 80K and 20K shields.

The MLI system for the 80K thermal shield consists of two 32-reflective-layer blanket assemblies, for a total of 64 reflective layers. The stack height of each 32-layer blanket is 8.86 mm, with a mean layer density of 3.61 layers per mm. The blanket design incorporates 32 reflective layers of double-aluminized polyethylene terepthalate (PET) film. The reflective layers are separated by single spacer layers of 0.10 mm spunbonded PET material. Single layers of 0.23 mm spunbonded PET cover the blanket top and bottom and serve to position the polyester hook and loop fasteners at the blanket edges. The fasteners are affixed to the cover layers by sewing. A third 0.23 mm PET layer is located midway through the blanket assembly and separates the upper and lower 16 reflective layers of MLI. The multiple blanket layers are sewn together as an assembly along both edges of the blanket. Non-lubricated polyester thread is used in all sewing operations.

The MLI system for the 20K thermal shield consists of ten reflective layers of PET film aluminized with a nominal coating of 600 angstroms per side. The reflective layers are separated by three spacer layers of 0.1 mm spunbonded PET material. Single layers of 0.23 mm spunbonded PET cover the blanket top and bottom and serve to position the polyester hook and loop fasteners at the blanket edges. The fasteners are attached to the cover layers by sewing. The multiple blanket layers are sewn together as an assembly along both edges of the blanket.

At each blanket edge, the upper MLI layers are sewn together from the upper cover layer through to the middle three layers with thread terminated in the three layers. The seam location is then incremented 7.62 cm laterally along the midlayer and the lower MLI layers are sewn together from the middle three layers through to the lower cover layer.

CRYOGENIC PIPING

In addition to providing the necessary structural support and thermal insulation for the cold mass assembly, the cryostat serves to contain the piping for all of the cryogenic services required for magnet and magnet system operation. Table 2 provides a summary of the pressure and flow parameters for each of these services. Table 3 lists the pipe sizes for each of the cryostat pipes. Each pipe is anchored at the center of the magnet assembly, and is free to slide axially at the remaining support points to allow for thermal contraction during cooldown.

SUSPENSION SYSTEM

The suspension system in any superconducting magnet serves as the structural attachment for all cryostat systems to the vacuum vessel which in turn anchors them to the accelerator tunnel floor. Conventional suspension systems were effectively reinvented during early SSC magnet development [2,3,5]. The intention in development of the 50mm dipole cryostat was to take advantage of that earlier work tailoring it only for the revised thermal and structural design parameters. Schedule requirements for the redesign work did not allow continued suspension R&D and, in fact, the thermal and structural performance of the 40mm design did not indicate it to be necessary.

Using the thermal and structural parameters outlined in Table 1 the suspension system for the new design was extrapolated from its 40mm counterpart. The emphasis was on meeting the allowed suspension system conduction heat load, satisfying the structural requirements, and maximizing the lateral suspension stiffness. This latter constraint is not explicitly defined in the design criteria but arose out of concerns during testing of 40mm prototypes that the lateral natural frequency was too closely coupled with resonances found in an over the road shipping environment.

As shown in figure 3, the suspension system consists of two major components; reentrant style support posts and anchor tie bars. There are five support posts located along the length of an SSC cold mass assembly spaced 3179.76 mm apart centered about the middle of the cold mass. Five being the number which limits the cold mass sag due to self weight to the allowed 0.25 mm. The support posts resist vertical and lateral loads imposed during shipping and handling. The cold mass is allowed to slide axially with respect to all but the center post to allow for thermal contraction of the cold assembly during cooldown. This implies that, given no other restraint, the center support would need to

Table 2. Cryogenic System Pressure and Flow Parameters

System	Pmax (atm)	Fluid	Poper (atm)	v (cm^3/g)	T (K)	Flow (g/s)	ΔP (psi)
20K shield	18	He	2.5	163.4	20	100	0.016
80K shield	18	LN$_2$	5.0	1.286	84	750	0.012
Single phase	18	He	4.0	7.255	4.25	100	0.010
LHe return	18	He	3.0	7.609	4.425	85.6	0.010
GHe return	9	He	0.9	73.07	4.3	26.8	0.00053
Vacuum vessel	2	He	0.0	na	300	na	na

Table 3. Cryostat Pipe Sizes

System	Pipe ID (mm)	Pipe OD (mm)	Flow Area (mm^2)	Material
20K shield	82.55	88.90	5352	Alum
80K shield	57.15	63.50	2565 x 2	Alum
LHe return	45.14	47.62	1600	Stn stl
GHe return	86.41	88.90	5864	Stn stl

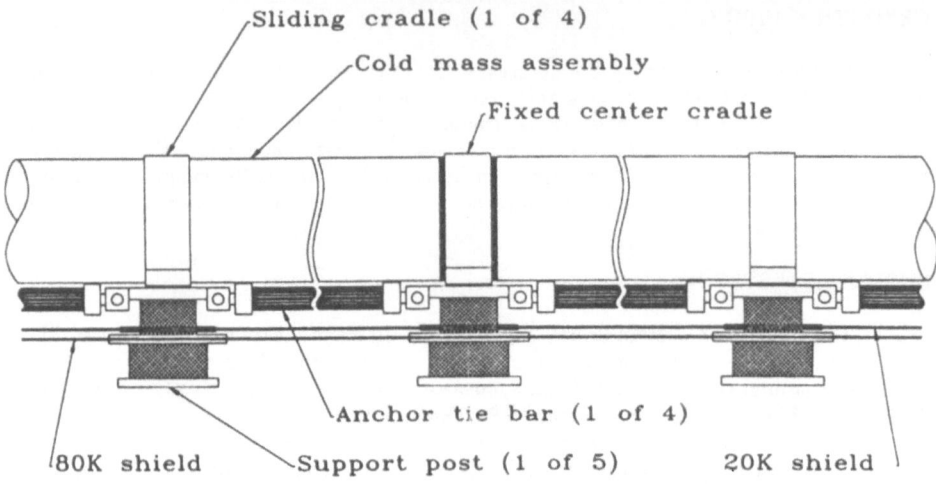

Figure 3. SSC 50mm Collider Dipole Cryostat - Suspension System Components

resist axial loads. A single support is not capable of resisting this potential 1.5g load. Using a special support at the center which could handle this load would impose inordinately high heat loads on the refrigeration system. Rather, a means was developed to distribute axial loads to all five supports without impacting the conduction heat load. Axial tie bars are used to connect the top of each support post to its neighbor(s). In this scheme, an axial load acts first at the center support and then is distributed to all five supports.

The tie bars must be dimensionally stable when cooled from their assembly temperature of 300K to their operating temperature of 4.5K, otherwise their contraction would impose high bending loads on the support posts. Uniaxial graphite fibers in an epoxy matrix provides the solution. This material is extremely stiff and yet exhibits virtually no change in length during cooldown. An elastic modulus of 82.5 GPa can be achieved in such a composite even with relatively low modulus fibers.

The configuration of the support posts for the 50mm cryostat is shown in figure 4. The outer composite tube is 2.16 mm thick and is filament wound using S-glass in an epoxy matrix. The inner tube is 3.18 mm thick and is filament wound using graphite fibers in an epoxy matrix. The ability of this design to use two different materials allows each material to be used in its optimum temperature range. Table 4 provides a summary of the critical design and performance parameters of the 50mm cryostat supports.

Table 4. Reentrant Support Post Design Parameters

Design lateral load:	32960 N (7410 lbf)
Design vertical load:	22240 N (5000 lbf)
Overall support height:	177.80 mm
Outer composite tube:	241.30 mm OD, 2.16 mm wall, 90.55 mm long, glass/epoxy
Inner composite tube:	177.80 mm OD, 3.18 mm wall, 150.88 mm long, graphite/epoxy
Maximum outer tube design bending stress:	132 MPa (19100 psi)
Maximum inner tube design bending stress:	154 MPa (22300 psi)
Theoretical lateral natural frequency:	9.41 Hz
Heat load to 80K:	3.133 W
Heat load to 20K:	0.492 W
Heat load to 4.5K:	0.033 W

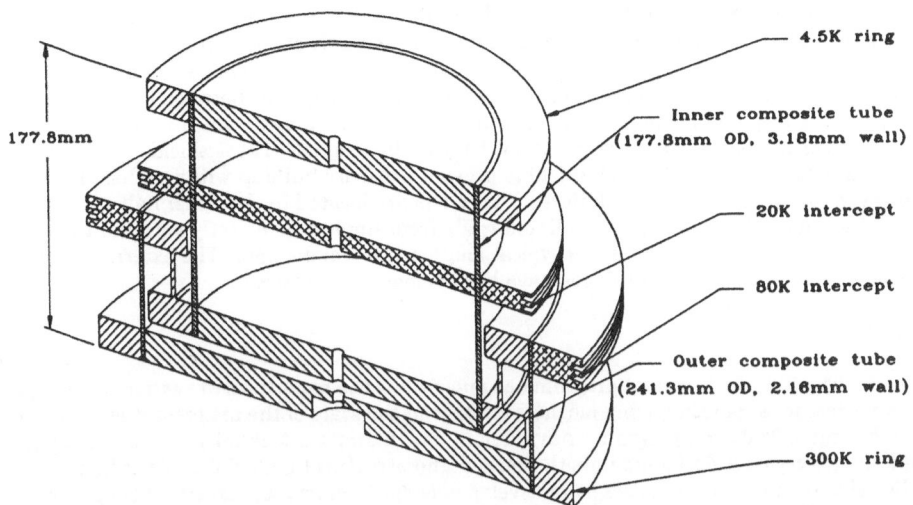

4.5K ring

Inner composite tube
(177.8mm OD, 3.18mm wall)

20K intercept

80K intercept

Outer composite tube
(241.3mm OD, 2.16mm wall)

300K ring

177.8mm

Figure 4. SSC 50mm Collider Dipole Cryostat - Reentrant Style Support Post

INTERCONNECT

As its name implies the magnet interconnect serves as the connection area between magnets at which each of the cold mass and cryostat pipes between magnets are connected. Each pipe is anchored axially at the center of the magnet which means they contract between approximately 25 mm and 32 mm depending on whether they are stainless steel or aluminum. The relative contraction at the interconnect is then 50 mm to 64 mm, i.e. twice the single magnet value. Bellows are required on each cold mass, shield, and cryogenic pipe to allow for this expansion. All bellows are hydroformed stainless steel. For the aluminum extrusions on the 80K and 20K shield pipes an aluminum to stainless steel transition joint is required for the bellows connection. Joints using diffusion bonding or brazing between these two materials have been successfully employed in 40mm prototypes. Lateral instability is a concern for interconnect bellows and requires the use of internal squirm protectors on each bellows assembly.

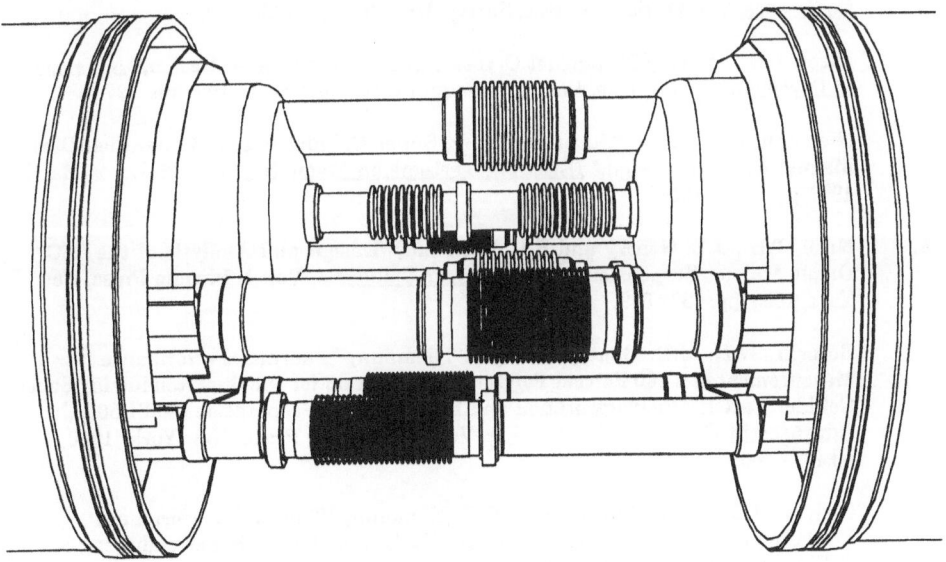

Figure 5. SSC 50mm Collider Dipole Cryostat - Magnet Interconnect

Radiative heat transfer must also be minimized in the interconnect area. This is accomplished by shield bridges which span the 80K and 20K shield gap between adjacent magnets. These bridges are little more than extensions to the magnet shields, modified to contain their respective bellow OD's if necessary. A sliding joint between bridge sections on adjacent magnets accommodates contraction during cooldown. Each is covered with the same MLI scheme used throughout the body of the magnet. These shield bridges also contain pressure reliefs for each shield to prevent pressure buildup within either shield in the event of an internal piping failure. The reliefs are located in the upper half of the shield sections in order to prevent liquid spills from impinging directly onto the vacuum vessel wall. Figure 5 illustrates a typical magnet interconnection. The external vacuum bellows and thermal shield bridges have been removed for clarity.

SUMMARY

The SSC development program has afforded us the opportunity to extend the design of cryostats for superconducting magnets far beyond the state of the art present at the end of the Fermilab Tevatron program. Advances in new materials technology have opened up options for cryostat designers in both thermal and structural materials. Strict limits on allowable heat load have forced us to develop new mechanisms for structural support and thermal shielding. The end result is a cryostat design which meets the demands of the SSC and which will serve as the starting point for the development of other magnet systems far into the future.

ACKNOWLEDGEMENTS

The author would like to recognize the contributions of D. Arnold, P. Heger, R. Richardson, and W. Boroski, without whose extreme dedication this work would not have been possible.

REFERENCES

1. Superconducting Super Collider Laboratory, "15 Meter Collider Dipole Magnet Prime Development Specification," March 6, 1990.

2. Nicol, T.H., et al., "A Suspension System for Superconducting Super Collider Magnets," Proceedings of the Eleventh International Cryogenic Engineering Conference, Vol. 11, Butterworths, Surrey, UK, 1986, pp. 533 -536.

3. Nicol, T.H., et al., "SSC Magnet Cryostat Suspension System Design," Advances in Cryogenic Engineering, Vol. 33, Plenum Press, New York, 1987 pp. 227 -234.

4. Niemann, R.C., et al., "Superconducting Super Collider Second Generation Dipole Magnet Cryostat Design," IEEE Transactions on Magnetics, Vol. 25, No. 2, March 1989, pp. 1615 - 1619.

5. Nicol, T.H., J.D. Gonczy and R.C. Niemann, "Design and Analysis of the SSC Dipole Magnet Suspension System," Super Collider 1, Vol. 1, Plenum Press, New York, 1989, pp. 637 - 649.

6. Boroski, W.N., J.D. Gonczy and R.C. Niemann, "Thermal Performance Measurement of a 100 Percent Polyester MLI System for the Superconducting Super Collider Part I: Instrumentation and Experimental Preparation (300K-80K)," Advances in Cryogenic Engineering, Vol. 35, Plenum Press, New York, 1990, pp.487 - 496.

7. Gonczy, J.D., W.N. Boroski, and R.C. Niemann, "Thermal Performance Measurement of a 100 Percent Polyester MLI System for the Superconducting Super Collider Part II: Laboratory Results (300K-80K)," Advances in Cryogenic Engineering, Vol. 35, Plenum Press, New York, 1990, pp. 497 - 506.

THE MECHANICAL RESPONSE OF THE SSC DIPOLE MAGNET TO

GROUND MOTION

A. R. Jalloh, R. Viola, and E. Daly

Magnet Division
Superconducting Super Collider Laboratory
2550 Beckleymeade Avenue
Dallas, Texas 75237

Abstract: One of the requirements for successful performance of the Superconducting Super Collider (SSC) particle beam accelerator is minimization of the effects of ground motion on the beam path. It is therefore important to be able to quantify this effect. To achieve this objective, a finite element dynamic model of the 40 mm dipole magnet was developed and validated with experimental data. The natural frequencies and mode shapes predicted by the model were in good agreement with experimental data. The model was used to predict beam path displacement resulting from ground motion. The magnitude and frequency of the ground motion had been previously measured at the SSC site in Ellis County, Texas.

INTRODUCTION

The dipole magnet is one of the major components of the Superconducting Super Collider (SSC) particle beam accelerator. Its main functions is to bend the path of the beam as it circulates in the collider ring.[1] This path must be maintained in precise orbits for optimum accelerator performance. One important factor that affects beam path stability is ground motion.[2] As such, it is important to quantify its effects in terms of the displacement of the beam path.

Several studies on ground motion effects on accelerator design in general, and the SSC accelerator in particular, have been carried out. A good overview of the effects of ground motion in designing accelerators was developed by Fischer.[2] Ng and Peterson[3] studied the effects of ground motion on the SSC using an ideal and uniform model of the SSC site. Tolerable ground motions for the SSC were investigated by Fischer and Morton.[4] Goss[5] characterized ground vibrations at the SSC site.

A dynamic model of the dipole magnet was developed to study seismic and transportation loads.[6] For that model the cold mass and vacuum vessel were modeled with beam elements.

There are several sources of ground motion to be considered in the design of the SSC magnets. The principal sources are ambient ground motion, trains and highway traffic crossing the main ring, and quarry blasting in nearby Midlothian, Texas.[1,7]

In this paper, a finite element dynamic model of the 40 mm dipole magnet was developed and was validated with experimental data. This was achieved by tuning the model until a good agreement was achieved between experimental and analytical natural frequencies and mode shapes. Ground motion test data taken from the SSC site in Ellis County, Texas,[7] was used as input data for the model to determine the displacement of the beam tube. The model development and analysis were carried out using ANSYS, a general-purpose, commercial finite element program.

TESTING EQUIPMENT AND PROCEDURE

In order to verify the finite element representation of the magnet, its vibrational characteristics were determined experimentally. The parameters of interest were the magnet's natural frequencies, mode shapes, and associated damping. This experimental program was carried out on a full-length 40 mm dipole magnet designated "DSHIP." DSHIP was assembled at Fermilab and was specially instrumented there for vibration studies. It has no heat shields or super insulation, but is otherwise identical to other SSC 40 mm dipole magnets.

The experimental procedure involved taking a series of frequency response functions (FRF) along the length of the vacuum vessel and the cold mass. A frequency response function is a complex function defined as the Fourier transform of a system's response (acceleration in this case) divided by the Fourier transform of the input (Figure 1). Vibrational input to the magnet was supplied by an electromechanical shaker driven by a random noise signal and was measured by a quartz force transducer. Output was measured by an array of piezoelectric accelerometers. These measurements were converted into FRFs by a Hewlett Packard 3567A dynamic signal analyzer. Finally, the FRFs were converted to the desired modal parameter using a PC-based structural analysis system.

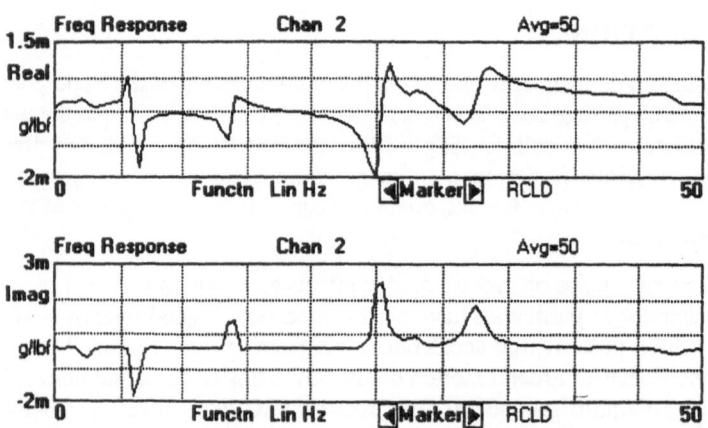

Figure 1.   A Typical Frequency Response Spectrum for the Cold Mass.

The specific technique used to determine the modal parameters was to fit a polynomial function, in rational fraction form, to the FRFs. Because the magnet's modes of vibration are in general lightly damped (less than 2% critical) and widely spaced, there was very little coupling between modes. However, a multi-degree-of-freedom version of the polynomial method was used in an effort to obtain the most accurate results possible.

The magnet's test configuration was developed to make the correlation with numerical results as straightforward as possible. Great care was taken in mounting the magnet to the floor in order to approximate a truly "constrained" (zero displacement/zero rotation) boundary condition. This involved pouring 5-ft- thick, steel-reinforced concrete footings to which the magnet was bolted using 1.25-in. all-thread rod. Measurements were made at 3-ft intervals down the entire length of the magnet in order to give the experimentally determined mode shapes the best resolution possible.

Results of the experimental program were consistent with earlier modal tests carried out on similar superconducting magnets.[8] Measurements showed a high degree of coherence and repeatability with very little noise. The quality of the mode shape resolution was generally excellent. This leads to confidence regarding the accuracy of the modal parameters determined.

MODEL DEVELOPMENT AND ANALYSIS

The major components of the 40 mm dipole magnet are the vacuum vessel, five support posts, two external support feet, a cold mass, and its support structure. These components were included in the dynamic model. Other components, namely the heat shields, cryogenic fluid pipes, and thermal insulation materials, were not considered significant contributors to the dynamic characteristics of the magnet and as such were not included in the model.

A thin shell ANSYS element, STIF-63, was used to model the vacuum vessel, support post tubes, and the cold mass skin. A thick plate element, STIF-43, was used to model the support post discs, external support feet, and cold mass support structure. The laminations contained within the cold mass skin were also modeled with STIF-43 elements. These laminations include the beam tube, inner and outer coils, and the iron yoke. While they do not contribute much to the structural strength of the cold mass, their contribution to the dynamic characteristics of the magnet has to be accounted for because of their enormous weight. It should be pointed out that the objective here is not to account for every detail of the cold mass but to include enough detail to give a good characterization of its dynamic behavior. The total weight of these laminations was distributed among 13 laminations. The locations of these laminations were selected to give an even distribution of the weight. One lamination was located at each end of the cold mass, and one was located above the center of each support post, for a total of seven. Six more laminations were added, one between every two that have already been defined. The model input values for these laminations were a pseudo-thickness and pseudo-density. These were determined such that they would result in the total weight of the actual laminations being represented.

Two separate models were developed and analyzed for this study. A model of the cold mass was developed and analyzed to ensure that its dynamic behavior was understood. This was then incorporated into a larger model of the entire magnet. Advantage was taken of symmetry to model only half of the magnet.

Model development was carried out in two phases. In the first phase the model was tuned to get a good match between natural frequencies and mode shapes predicted by the model and those obtained experimentally. Tuning was done by modifying the modulus of elasticity of the cold mass laminations until the natural frequencies predicted by the model matched those obtained from testing as closely as possible. In the second phase, the model was enhanced for seismic analysis so it could be used to predict beam tube displacements due to ground motions. The response spectrum method of seismic analysis was used. The modes were combined with the SRSS method. Also, Guyan reduction was used for both types of analyses to minimize problem size and to reduce computation time and space requirements.

A vertical displacement spectrum of 0.5 mμ magnitude and 2.7 Hz frequency was input into the model to represent ground motion. These values were extracted from ground vibration test data taken at the SSC site.[7] A 2% structural damping, which was determined experimentally, was used in the analysis. The maximum vertical displacement of the cold mass—and hence of the beam tube—resulting from this ground motion was determined. Model development and analysis were performed using ANSYS (4.4A) on a Sun SPARCstation 1+.

RESULTS AND DISCUSSIONS

The natural frequencies and mode shapes for the cold mass determined both experimentally and by the finite element analysis (FEA) model are in good agreement as shown in Table 1 and Figures 2 to 5. This indicates that all the dynamic characteristics are included in the cold mass model. The best resolution was obtained for the higher mode shapes. As such, the seventh and eight modes, which had excellent resolution, are presented here and are compared with the corresponding modes determined analytically. It should be noted that the boundary conditions used during the characterization of the cold mass were not meant to approximate the suspension system of the cryostat. Rather, they were chosen to facilitate a simple comparison of the experimental results with those of the FEA.

Table 1. Experimental vs. FEA Frequencies/Cold Mass Only.

Mode #	Freq (Expt.) (Hz)	Freq (FEA) (Hz)
1	2.50	2.50
2	3.50	2.72
3	5.95	6.37
4	13.71	15.75
5	25.25	26.12
6	27.98	28.11
7	32.43	35.27
8	47.44	47.75

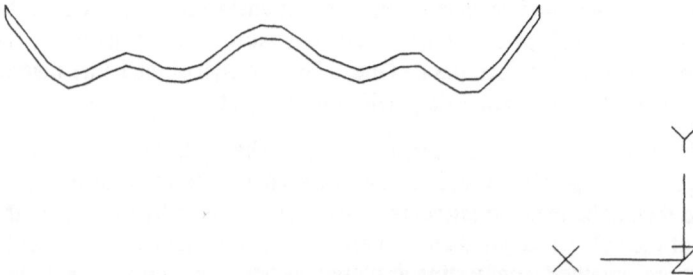

Figure 2. Seventh Mode Shape/Cold Mass/40 mm Dipole Magnet/Experimental.

Figure 3.   Seventh Mode Shape/Cold Mass/40 mm Dipole Magnet/FEA Analysis.

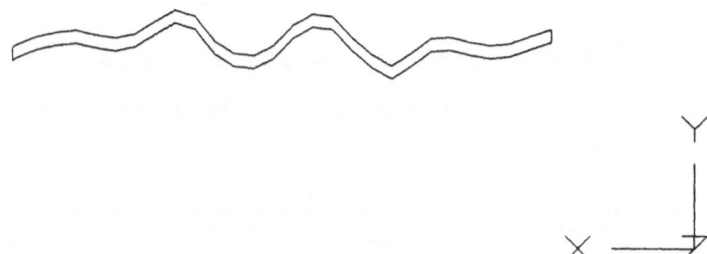

Figure 4.   Eighth Mode Shape/Cold Mass/40 mm Dipole Magnet/Experimental.

Figure 5.   Eighth Mode Shape/Cold Mass/40 mm Dipole Magnet/FEA Analysis.

Table 2.   Experimental vs. FEA Frequencies/Complete Magnet.

Mode #	Freq (Expt) (Hz)	Freq (FEA) (Hz)
1	6.5	5.7
2	7.5	6.6
3	15.5	15.6
4	16.5	16.4
5	-	17.8
6	20.5	25.8
7	-	29.6
8	37.5	39.4
9	39.0	41.6
10	44.0	45.1

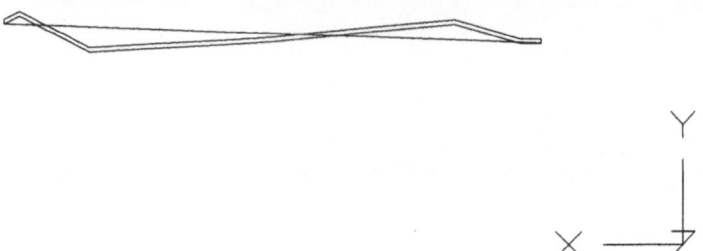

Figure 6.   Second Mode Shape/40 mm Dipole Magnet/Experimental.

Figure 7. Second Mode Shape/40 mm Dipole Magnet/FEA Analysis .

Figure 8.   Third Mode Shape/40 mm Dipole Magnet/Experimental.

Figure 9.   Third Mode Shape/40 mm Dipole Magnet/FEA Analysis.

While the agreement between experimental and FEA model results for the complete magnet (Table 2) is not as good as that for the cold mass, there is still good agreement. The second and third mode shapes for the complete magnet model obtained both experimentally and with the FEA model are shown in Figures 6 to 9. It would have been desirable to have experimental data to validate all of the individual component models before putting together the complete magnet model. However, the present model is a good representation of the magnet's dynamics characteristics.

Having validated the finite element model of the 40 mm dipole magnet, there are many applications which can draw upon this model for answers to important questions about the magnet's behavior in various dynamic environments. Of particular interest are the magnet's stresses and displacements due to ground motion and shipping loads. For this study, the maximum beam tube displacement for a vertical ground motion displacement of 0.5 mμ with a frequency of 2.7 Hz was determined by the model as 0.765 mμ. This displacement spectrum was derived from the ground motion test data taken at the SSC site in Ellis County, Texas. It should be cautioned that this represents the magnet's response to only one particular forcing function and does not represent the worst case

CONCLUSIONS

The following conclusions were made about the study presented here:

1. A dynamic model of the 40 mm dipole magnet was developed and was validated with experimental data.

2. Using seismic data taken at the SSC site in Ellis County, Texas, the dynamic model was used to predict displacements of the beam tube caused by ground motion using seismic data taken at the SSC site in Ellis County, Texas.

FUTURE WORK

1. The present model accounts only for vertical motions of the magnet. It should be enhanced to account for horizontal and twisting motions as well.

2. This work will be extended to include the 50 mm dipole and the quadrupole magnets.

3. These magnets will undoubtedly be transported by tractor-trailer to the construction site in Ellis County, Texas. They will experience nonlinear transient shock loads as well as continuous vibratory loadings. These shipping loads may adversely affect the mechanical integrity of the magnets. The model developed will be used to predict the stresses resulting from these shipping conditions.

4. The magnet will not sit directly on the floor when it is installed in the tunnel. A pair of magnets will be mounted, one above the other, on support stands fastened to the floor. The upper magnet's dynamic behavior will be better investigated with a model of this configuration.

ACKNOWLEDGEMENTS

This work was supported by the Director, Office of Energy Research, Office of High Energy and Nuclear Physics, High Energy Physics Division, U.S. Department of Energy, under Contract No. DE-AC02-89ER40486.

REFERENCES

1. J. R. Sanford and D. M. Matthews, *Site-Specific Design of the Superconducting Super Collider*, SSCL-SR 1056, July 1990.

2. G. E. Fischer, *Ground Motion and its Effects in Accelerator Design*, SLAC-3392 Rev., July 1985.

3. K-Y. Ng and J. Peterson, *Ground Motion Effects on the SSC*, SSC-212 and FN-511-Rev., December 1989.

4. G. E. Fischer and P. Morton, *Ground Motion Tolerances for the SSC*, SSC-55, January 1986.

5. D. Goss, *Some Characteristics of Ground Vibrations at the SSC Site*, SSC-N-635, May 1989.

6. _____, *SSC Dipole Magnet System: Stress Analysis for Seismic and Transportation Loadings*, EQE Engineering, January 1991.

7. K. Hennon and D. Hennon, *Field Measurements and Analysis of Underground Vibrations at the Superconducting Super Collider (SSC) Site in Waxahachie, Texas*, Earth Technology Corporation, Report No. SSC-SR-1043, December 1989.

8. W. Vogen, *SSC Magnet 0014 Vibration Test Report*, Vibration Engineering Consultants, October 1988.

9. G. J. DeSalvo and R. W. Gorman, *ANSYS Engineering Analysis User's Manual*, Swanson Analysis Systems, Inc., May 1989

DESIGN OF A THERMAL EXPANSION JOINT

FOR THE SSC MAGNET LEADS

W. Clay and J. Cox

Magnet Division
Superconducting Super Collider Laboratory
2550 Beckleymeade Avenue
Dallas, TX 75237

R. Thome

Massachusetts Institute of Technology
Cambridge, MA 02139

Abstract: In this paper we present the design of a thermal expansion joint
for the main power leads of the magnets used in the Superconducting Super
Collider (SSC) rings. This design ensures cryogenic stability of the leads in
an environment of low flow rate, single phase helium, survives high-energy
neutron and gamma radiation, and satisfies the connection scheme to reduce
dump resistor voltage. The finite element analysis used to model the leads for
both thermal and mechanical responses is discussed. The radiation resistance
of candidate materials for use as structural and electrical components at 4 K is
discussed. The possible connection schemes for the magnets in the collider
ring are detailed and the current design are presented.

INTRODUCTION

Because of the large thermal contractions of the magnet during cooldown, a thermal
expansion loop is required for the main leads. It is essential that the expansion loop not

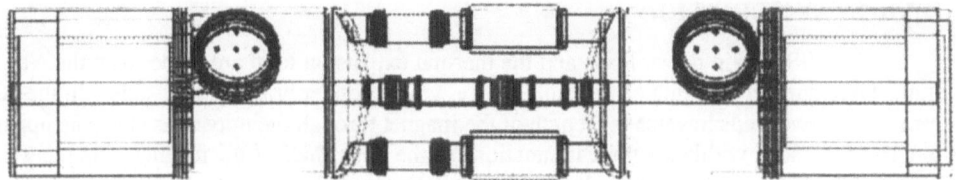

Figure 1. 50 mm Collider Dipole Interconnect Region.

*Operated by the Universities Research Association, Inc., for the U.S. Department of Energy under Contract
No. DE-AC02-89ER40486.

increase the overall magnet length. To accomplish this, we have placed the main lead thermal expansion loop within an annular cavity, bounded by the iron yoke laminations, the outer shell, the end clamp, and the magnet end plate.[1] See Figure 1.

In addition to operating in a unique cryogenic environment, the main leads must be spliced with minimal heat generation at operating current. The total heat generation budget for the splices associated with one collider dipole is 0.14 W. This specification requires that each splice joint be limited to about 20 mW. Preliminary joint resistivity measurements indicate that this budget limit can be met. A 10 cm overlap length of the superconducting joint resulted in an average resistance per splice of 5.16×10^{-10} ohms. This results in an average joint heat loss of about 20 mW at 6.5 kA, the operating current of the collider dipole.

The integral of the magnetic field produced by the bus and the length of the bus must be small in relation to the integrated field of the magnet. For this reason, the two leads of the magnet must be kept close together to minimize the field produced in the magnet bore. However, putting the leads close together increases their mutual repulsion force and the support structure of both the bus and expansion loop must be designed to accommodate the increased force.

To ensure reliable operation of the bus, the criterion of a pseudo-stabilized bus was established. The goal of this criterion was to ensure that a bus could not cause quenches in a magnet system. Because the bus and expansion loop are located in single phase helium with low flow rates, it was recognized that methods of cryogenic stabilization based on pool boiling were not applicable. Therefore, methods based on the stabilization of extended surface conductors were investigated by the design team at MIT. The design team used a thermal-hydraulics code to model the heating and subsequent flow of the helium fluid in the bus assembly.

The mechanical stresses produced in an expansion loop operating in the SSCL environment were also considered. An ANSYS model of an early design for an expansion loop is shown in Figure 8. The magnitudes of the Von Mises stresses in the early design are shown in Figure 9.

An additional complication to the design of the bus was a requirement to have "generic" magnets, which could be located in any magnet slot in the ring and still provide an electrical connection scheme which would reduce the voltage across the dump resistors and diodes to acceptable values. This requirement has been deleted for the present series of tests and the dipole magnets are now pre-wired for either a clockwise or counter-clockwise location in the collider. The wiring scheme for a typical half-cell is shown in Figure 6. This scheme considerably reduces the complexity of the magnet connections.

DESIGN DESCRIPTION

The SSC main power leads and the thermal expansion loop are made from the NbTi inner-layer superconducting cable flanked by a woven copper braid on both sides. In the 50 mm dipole, two leads traverse the length of the magnet through the upper bus slot. The upper bus slot is located within the yoke laminations of the upper half of the magnet. The bus slot serves as a path to support the bus leads. The main bus leads are supported within the bus slot by a composite pultrusion, shown in Figure 2. This composite pultrusion insulates the leads from each other and ground, and it orients the leads in a manner which is favorable to field-quality harmonics. Figure 3 shows the cross-section of the pultrusion for the thermal expansion loop.

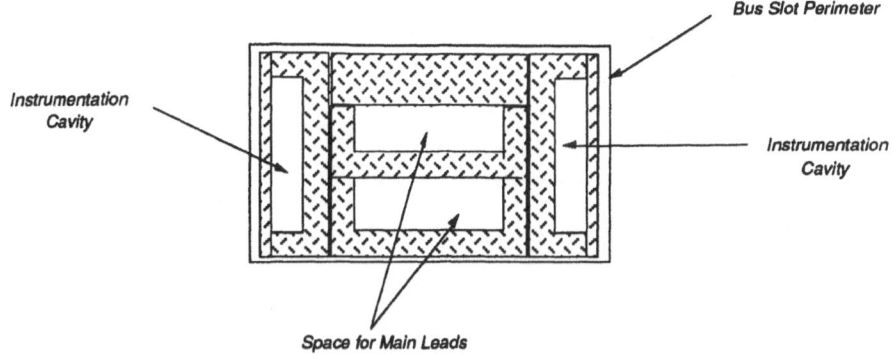

Figure 2. Composite Pultrusion Cross Section in Upper Bus Slot.

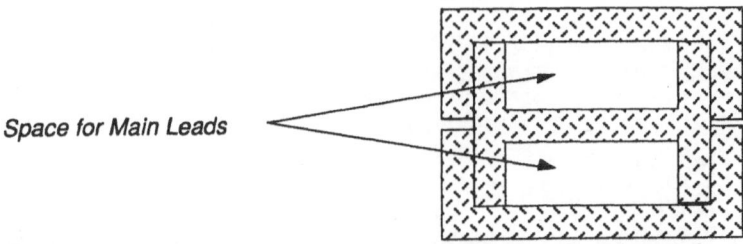

Figure 3. Composite Pultrusion Cross-Section for Expansion Loop.

The superconducting cable used in the expansion loop is the same type of cable used in the inner windings of the dipole coils. This cable has a minimum critical current (I_c) of 9,990 amps @ 4.2K, 7T.

The copper woven braid that flanks both sides of the superconductor is a standard "off the shelf" industrial copper braid that plays a substantial role in the performance of the main bus. Because the main bus is exposed to discrete energy inputs, it must be able to automatically recover from certain finite energy inputs in order to operate with some degree of reliability. Frictional heat input has the most destabilizing effect on the bus. This disturbance can be attributed to the relative motion of two surfaces due to differences in thermal expansion coefficients. The copper braid acts as a barrier, and shields the superconductor from local energy deposits. Because steady-state heat transfer to stagnant helium (approximately 50 mW/cm^2) is very poor relative to two-phase boiling helium, an increase in surface area is required to achieve the desired degree of stability. The copper braid enhances the surface area of the superconductor and provides a trapped volume of helium which encourages "pressure wave" cooling.

The copper braid is soldered to both sides of the superconductor with a pre-measured amount of solder (Figure 4). The solder is a 96/4 tin-silver solder foil and is placed between the superconductor and the copper braid during assembly of the bus and is then heated to approximately 600 K. It is important to note that when soldering the copper braid to the superconductor, care must be taken to insure that the copper braid is not saturated with solder (see Figure 5). If the void of the copper braid is filled, it will reduce the effectiveness of the additional surface area enhancement provided by the copper braid. The solder foil used in soldering the cables together is 0.05 mm thick and 15.2 mm wide. This pre-measured amount of solder ensures that the void in the copper braid is not filled.

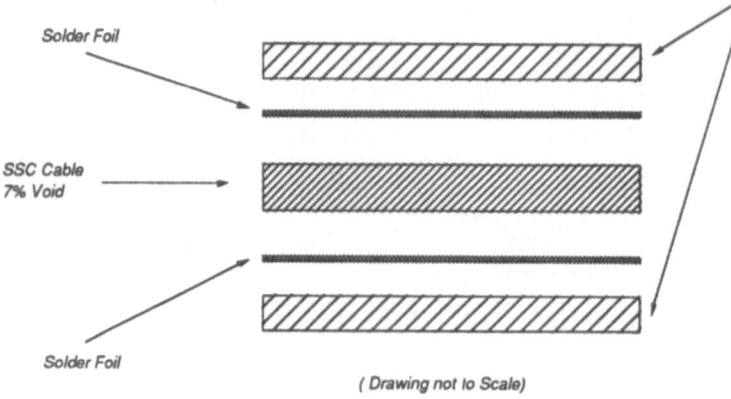

Woven Copper Braid–50% Void
Surface Area = 20 cm/cm² length

Solder Foil

SSC Cable
7% Void

Solder Foil

(Drawing not to Scale)

Figure 4. 50 mm Bus Components.

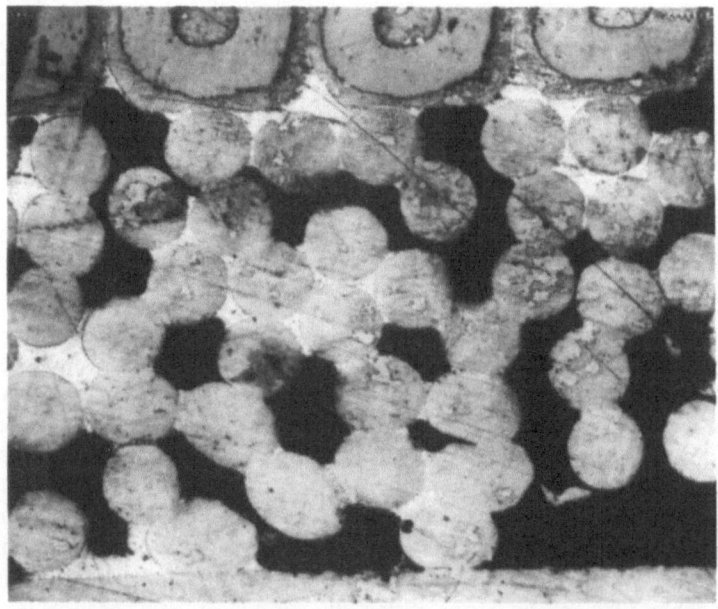

Figure 5. Superconductor/Copper Braid Joint Micrograph.

PSUEDO-CRYOGENIC STABILIZATION

To show the effect of the increased surface area provided by the copper braid, an analysis was done to compare the stability margin of the superconductor with varying surface area enhancements. The analysis of the cable-plus-braid assembly was performed by the MIT design team using a code they developed specifically for this type of analysis. Table 1 shows the effects of this surface area enhancement.

Table 1. Energy Margin for Recovery.

SSC Dipole Cable Only	3.4 mJ/cm
Cable + Copper Bars	12 mJ/cm
Cable + Braid + Copper Bars	300–800 mJ/cm

Energy margin is limited by cooling surface, not copper cross-section.

Range on energy margin with braid depends primarily on braid compaction.

HALF CELL POWER CONNECTION SCHEME

The power leads that provide current to the magnets are defined as a clockwise or counter-clockwise bus. The magnets in a half-cell are connected as shown in Figure 6. This connection scheme assures the protection of the magnets in the event of a quench. An active quench protection system is proposed for the SSC. This active quench protection system consists of warm bypass diodes connected to the magnets via safety leads at every quadrupole location and heaters to propagate the quench in the magnets on one bus of one half-cell.

The magnets connect to the appropriate bus at each interconnect region. The magnet is connected in series with either of the two busses depending upon its location in the half-cell.There are four bus splices in the interconnect region. Two of these splices are made at the lead end of the magnet on the face of the endplate. The other two splices are made in the upper single phase pipe between the dome ends of the magnet. Three splices are made within the magnet.The half-cell is split into two groups: the first, a group of three dipoles, and the second, two dipoles and a quadrupole. See Figure 6.

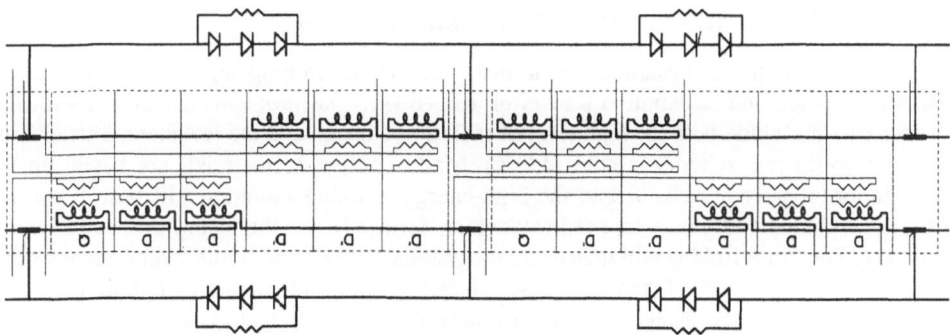

Figure 6. SSC Magnet Connection Scheme.

Up-Down Orientation   Vertical Orientation   Horizontal Orientation

Figure 7. Possible Configuration of Leads in Bus Slot.

MAGNETIC FIELD QUALITY CALCULATIONS

Field quality calculations for the three possible arrangements of the conductor in the bus slot have been completed. These calculations were published by Ramesh C. Gupta in SSC Note SSC-MD-232. Gupta considered the placement of the leads in the bus conduit in the vertical, horizontal, and up-down configuration (see Figure 7). Gupta used the computer codes POISSON and MDP to analyze these configurations and was able to show that the up-down configuration produced the least change in field harmonics.

STRESS CALCULATIONS FOR EARLY EXPANSION LOOP

The initial stress calculations for the expansion loop were done for a design where the superconductor was soldered to a solid copper bar of 1.1 cm^2 cross-section. To keep the stresses in the copper bar low and thus increase fatigue life of the copper bar, the loop was to be compressed 25 mm and then expanded 50 mm. This would keep the stress in the copper below a value of 138 MPa (20 Ksi) while still providing the expansion required. The limit of 138 MPa was set to maximize the fatigue life of the copper and was based on experience at Brookhaven National Laboratory. The ANSYS model used to calculate the stresses is shown in Figure 8. The magnitudes of the Von Mises stresses are shown in Figure 9. The design currently proposed by the MIT design team involves the use of a composite "H" beam and two "C" beams formed into the loop for the support of the leads in the expansion loop. A stress analysis of that configuration has yet to be done, but the design has been cycled several times at room temperature and appears to give very satisfactory results.

RADIATION DAMAGE OF SUPPORT MATERIALS

The composite materials used to make the expansion loop support sections must be resistant to gamma radiation at a level of 10E+6 Gy. The problems associated with this requirement include both the high level of radiation damage and the fact that the irradiation occurs while the material is cold (4.0 K). Many composite materials are based on the formation of polymer chains and the high-energy ionizing radiation will both break the polymer chains and create free radicals which further break the chains and cause cross-linking. The cross-linking will then cause the polymer to become brittle and/or change other properties. The amount of damage can be reduced by using filler materials (additives) which do not form polymer chains. Examples of these are inorganic materials such as aluminum oxide, magnesium oxide, mica, and quartz. The amount of damage to the composite can be further reduced by selecting polymers which form ring structures or complex hierarchical structures. The study of radiation damage for the SSC will be an continuing program for the materials group of the Magnet Systems Division.

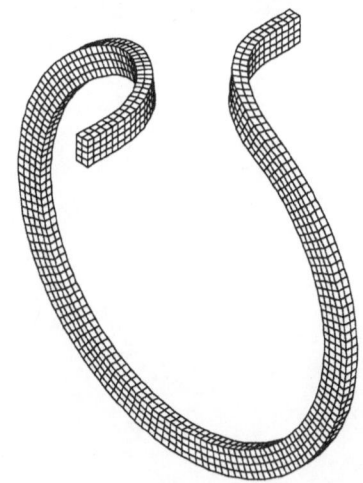

Figure 8. ANSYS Model of Copper Expansion Loop.

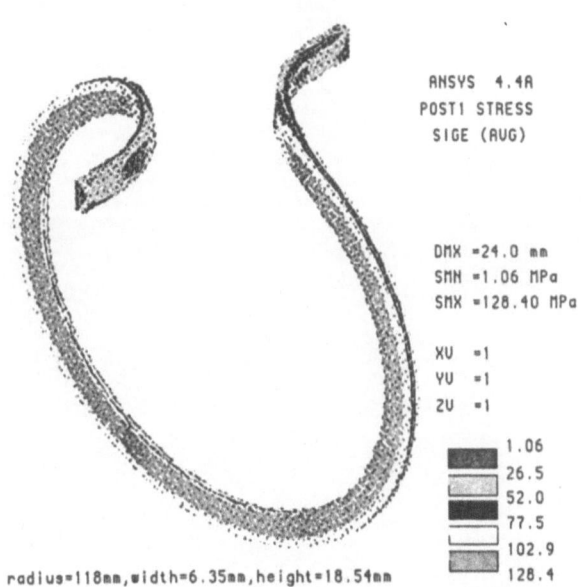

```
ANSYS  4.4A
POST1 STRESS
 SIGE (AVG)


DMX =24.0 mm
SMN =1.06 MPa
SMX =128.40 MPa

XU  =1
YU  =1
ZU  =1

            1.06
            26.5
            52.0
            77.5
            102.9
            128.4
```

radius=118mm,width=6.35mm,height=18.54mm

Figure 9. Von Mises Stresses in Copper Expansion Loop.

REFERENCE

1. *Requirements and Specifications for 50mm Collider Dipole Magnet,* Superconducting Super Collider Laboratory, October 10, 1990.

SSC DIPOLE MAGNET

ACCELERATED LIFE TEST PROGRAM

G. W. Albert, A. Kytasty, and R. E. Bailey

General Dynamics, Space Systems Division
PO Box 85990, MZ DC-9540
San Diego, Ca 92186-5990

ABSTRACT

An accelerated life test program is proposed to verify reliable operation of the SSC dipole magnet over its designed useful lifetime. A shortened series of thermal, magnetic, quench, vibration and environmental exposure cycles will be applied at above rated stress levels to demonstrate the integrity of the collider dipole magnet design. A statistical model is presented as the basis of our approach to reducing the span-time and cost of this testing. Details of the test plans including automation of the cyclic stimuli and data acquisition are addressed.

INTRODUCTION

One of the most profound test tasks required by the SSC collider dipole magnet (CDM) statement of work is the accelerated life test (ALT) program. It requires the CDM contractor to verify by test that the magnet (design) is capable of providing a minimum of 25 years of useful life in the collider with a probability of failure of 4.0×10^{-8} per hour of operation. The operational life of the magnet is further defined to include:

- 20,000 magnetic cycles
- 100 quenches
- 50 thermal cycles
- radiation exposure
- vibration and shock loads
- environments (temperature, dust, humidity, etc.)

To verify these life requirements with a 50% confidence level, testing at nominal stress levels would require many years of operational cycles. By contrast, the rapid pace of our CDM program schedule offers less than 15 months between fabrication of the first Prototype magnets and start of the Low Rate Initial Production phase. To support the CDM design and development process, the ALT program should ideally consume no more than three or four months. In response to the need for timely ALT results, we propose performing a shortened series of operational test cycles applied at above-nominal stress levels. An overload-stress reliability model was developed by co-author Andrew Kytasty to determine the necessary overstress levels and numbers of magnetic, quench and thermal test cycles.

APPROACH

The ALT operational reliability requirements are derived from the specified CDM useful life and failure rate requirements as follows:

The allowable probability of failure (PF) per CDM over its specified lifetime is:

$$PF = 1 - e^{-(25 \text{ years} \times 8,760 \text{ hrs/yr} \times 4 \times 10^{-8} \text{ failures/hr})} \qquad (1)$$
$$= .009$$

Baseline PF requirements for each of the three planned operational test modes are established by allocating the CDM PF equally among them,

$$.009/3 = .003/20,000 = 1.5 \times 10^{-7} \text{ failures per magnetic cycle} \qquad (2)$$
$$.009/3 = .003/50 = 6.0 \times 10^{-5} \text{ failures per thermal cycle} \qquad (3)$$
$$.009/3 = .003/100 = 3.0 \times 10^{-5} \text{ failures per quench cycle} \qquad (4)$$

With these requirements in mind, we shall review the ALT model and the relationship between overstress and number of test cycles. Figure 1 depicts the concept of distribution of stresses. The left distribution represents the nominal level stresses, while the right distribution can be thought of as the "allowable ultimate stresses". With deterministic analysis the ratio of the arithmetic means of these distributions would represent the design margin or safety factor. The following probabilistic analysis, however, models the inherent variability of processes and materials and the uncertainty in loadings and analysis. The result is a distribution of stress, instead of merely a discrete value.

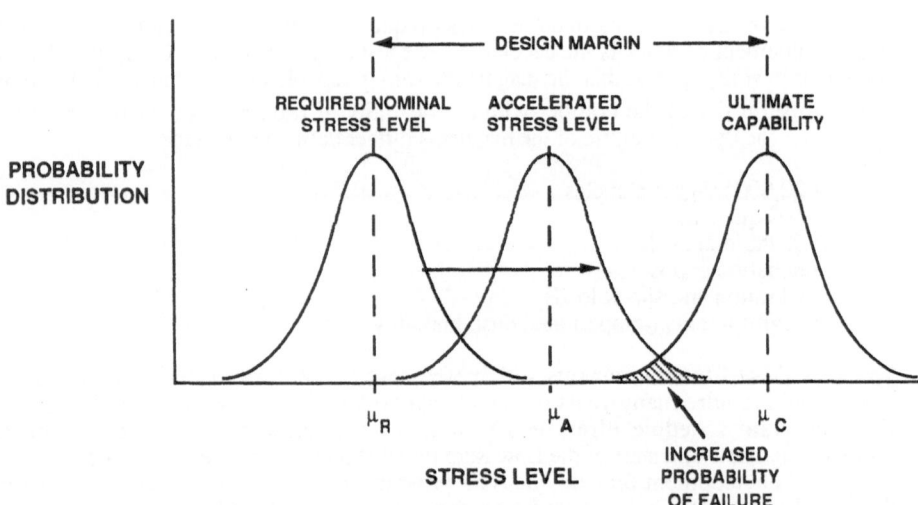

Figure 1.   Testing at above nominal stress levels reduces the required number of test cycles by increasing the probability of failure.

The problem arises then, how far apart are the nominal and allowable levels (i.e. how far apart are the two distributions) such that the probabilities of failure (eq. 2,3,4) are valid. We approach the problem by conducting a series of trials (i.e. tests). Each trial will represent a chance condition whereby an operational stress may exceed the ultimate capability, resulting in failure. By running a number of successful tests, we can prove that the two distributions are separated by a specific minimum margin. Elimination of fatigue or degradation problems allows our ultimate capability distribution to remain fixed, so that the probability of failure does not increase with time or tests.

Our ALT methodology increases the stress levels during test by a factor (B) above nominal. We are thus attempting to shift to the right the stress levels that govern the same failure modes as at the nominal level. One can see in Figure 1 that as a shift to the right occurs, the margin decreases, and so the probability of failure goes up. We seek to capitalize on this increased probability of failure, by a corresponding decrease in the number of required tests. For example, we can express our magnetic cycle PF requirement as follows:

$$PF(\text{magnetic cycle}) = P(\text{nominal} > \text{capability}) = 1.5 \times 10^{-7}$$
$$= f\{Z(\text{nominal})\}$$

Since we have assumed the distributions to be normally distributed, the resulting difference is also normally distributed with:

$$Z(\text{nominal}) = \frac{\mu_r - \mu_c}{\sqrt{\sigma_r^2 + \sigma_c^2}} \tag{5}$$

where:
    μ_r and μ_c are arithmetic means of the requirement and capability distributions respectively,

    σ_r and σ_c are the standard deviations of the requirement and capability distributions respectively,
    and Z is the standard normal variate used to compute probability.

The equation for the accelerated PF is,

$$Z(\text{accelerated}) = \frac{\mu_a - \mu_c}{\sqrt{\sigma_a^2 + \sigma_c^2}} \tag{6}$$

where:
    μ_a and σ_a are the arithmetic mean and standard deviation of the accelerated probability distribution.

Assuming standard deviations of 0.1μ, we can solve equations (5) and (6) for accelerated test levels of 10, 20 and 30% above nominal. The resulting curves, shown in Figure 2, relate the number of tests required to demonstrate the desired probability of failure requirement with a 50% confidence level using a "0" failure binomial testing scheme. Overstress factor B is defined as the ratio of accelerated stress level, μ_a, to the nominal stress requirement, μ_r. The PF requirements calculated in equations (2),(3) and (4) are used to determine the following proposed test scenario as shown in the chart of Figure 2.

- 4,000 magnetic cycles at 30% overstress
- 100 thermal cycles at 25% overstress
- 200 quench cycles at 25% overstress

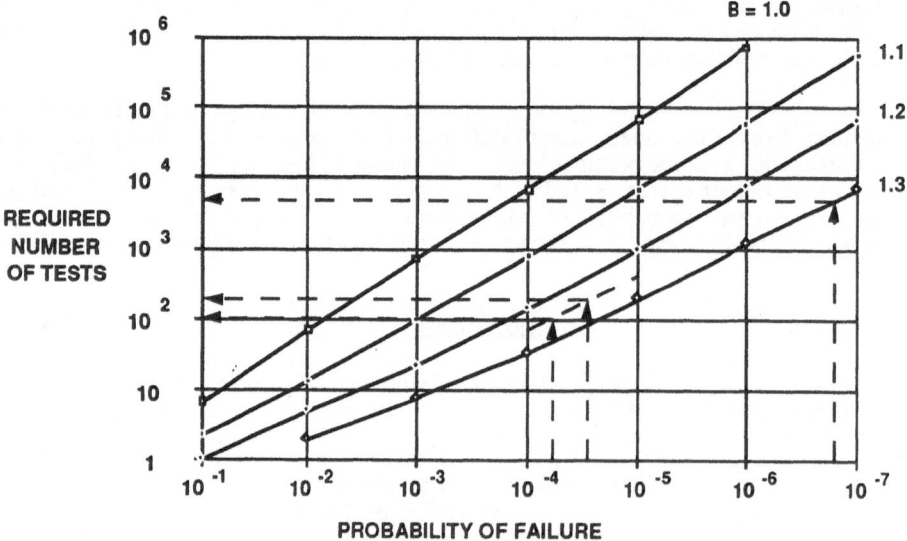

Figure 2. By increasing magnetic stress to 30% above the nominal requirement, the required number of tests is reduced by more than three orders of magnitude.

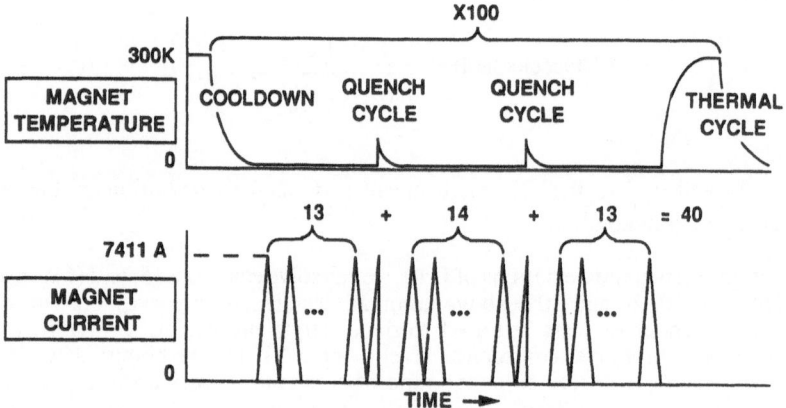

Figure 3. The estimated total time required for the proposed operational cycles is less than two months assuming a three shift, seven day per week test operation.

Figure 3 shows our proposed sequence of operational test cycles. Periodically during this sequence, the test cycles will be halted to verify the health of the test specimen and measurements of key magnet quality parameters will be taken. Any observed changes in these parameters will be used to support prediction of the magnet's useful life. We plan to test two magnets, applying half of the total test cycles to each. While one magnet is undergoing cyclic operational tests, the other will be in environmental exposure testing and vise versa.

Overstress may be applied to the magnet system in a variety of ways depending on the probable failure modes. Overstress must be applied gradually, and carefully limited to prevent failure modes atypical of collider use. Magnetic overstress may be controlled by applying increased current to increase Lorentz loads on the coils, collars and yokes. Current ramp-rate can be increased to 50 a/s to reduce test time. Stress induced by thermal cycles can be increased by using faster warmup and cooldown rates. For the quench tests, a higher resistance dump resistor may be used to increase the voltage stress on the coil insulation systems. Quench protection systems can be delayed to increase coil hot-spot temperatures. Quench pressures in the cold mass and cooling lines may also be increased by constricting flow of escaping gasses.

Our cold test stands will provide computer control of vacuum pumps, leak detection, cryogenic, power and quench protection systems for all modes of testing. Acquisition of quench voltage measurements will be supported by a CAMAC system (IEEE 583) of transient recorders and memory. Automatic monitoring and control of test stimuli will provide consistency in the application of the test cycles. This approach will minimize test time and manpower requirements as well.

The ALT plan will be based on fault tree and failure modes and effects analysis to identify failure mechanisms of magnet components and integrated systems[1]. Our proposed test methodology, applicable to determining the rate of constant magnet failures, is shown in Figure 4.

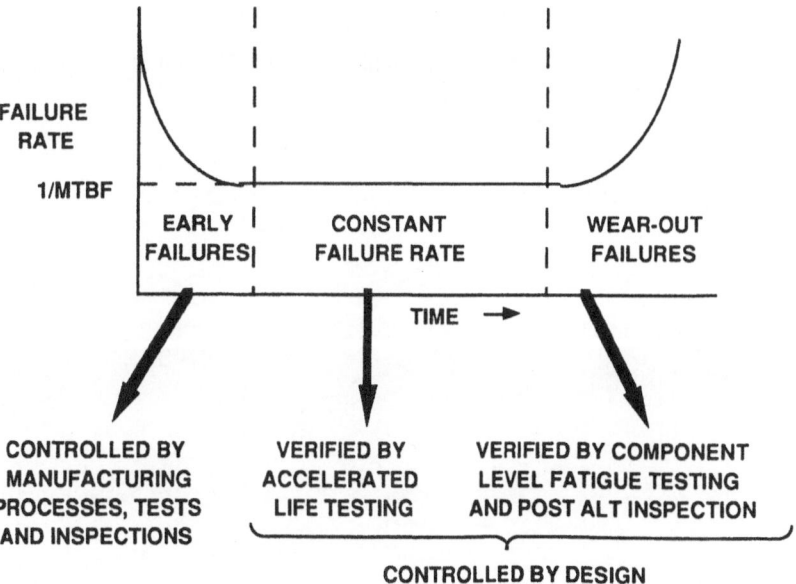

Figure 4. The magnet lifetime is characterized by this failure-rate curve which is typical of most manufactured items.

The rate of early magnet failures is largely controlled by the manufacturing process. Our manufacturing facility will have a system of statistical process control with in-process tests and inspection to maintain magnet quality. Producibility design also contributes to improving the robustness of the assembly processes and alleviating defectives. Warm and cold CDM acceptance tests will facilitate final screening of possible early failures. The life testing program will also include a series of fatigue and wear-out tests at the component level to verify the right-hand portion of the failure-rate curve (Figure 4) as these are difficult to perform at the magnet assembly level. Other types of overstress methodologies may be applied to accelerate this phase of testing[2]. A post-test tear-down and inspection of the ALT magnets will also be performed to help identify possible areas of magnet wear-out or degradation.

CONCLUSION

The model which we have constructed for the ALT program provides a test approach which is timely, economic and suited to model the types of failure modes and operational conditions of the CDM. Actual time required for the tests is dependent on many variables including cryogenic system capacity. Our approach is success oriented, but one should not neglect the significant technology base upon which the CDM design and development program is based. If necessary, the ALT program can be adapted to a reliability growth test scenario similar to that used for rapid development of military equipment. At this point in the program, we are confident that we will achieve the CDM test objectives as defined by the SSC laboratory.

The work described herein shall be accomplished under contract to the Universities Research Association in support of the Superconducting Super Collider project for the U.S. Department of Energy.

REFERENCES

[1] A. Kytasty, "CDM Reliability, Essential to Achieving an Operable Superconducting Super Collider", (these procedings).

[2] K. M. Shewfelt, et. al., "Accelerated Reliability Growth Testing of Complex Mechanical Systems", Proceedings- Inst. of Env. Sci..

STABILITY OF BELLOWS USED AS EXPANSION JOINTS BETWEEN SUPER-
CONDUCTING MAGNETS IN ACCELERATORS

R.P. Shutt and M.L. Rehak

Accelerator Development Department
Brookhaven National Laboratory
Upton, New York 11973

ABSTRACT

For superconducting magnets, one needs many bellows for connection of various helium cooling transfer lines. There could be approximately 20,000 magnet interconnection bellows in the SSC exposed to an internal pressure. When axially compressed, internally pressurized, or insufficiently supported at their ends, bellows can become unstable, leading to gross distortion or complete failure. If several bellows are contained in a magnet assembly, failure modes might interact.

If designed properly large bellows can be used to connect the large tubular shells that support the magnet iron yokes and superconducting coils and contain supercritical helium for magnet cooling. We investigate here bellows design features and end supports to insure that instabilities will not occur in the bellows pressure operating region, including some margin. A model of three superconducting accelerator magnets connected by two large bellows is analyzed in order to ascertain that support requirements are satisfied and in order to study interaction effects between the two bellows. Specific details of large and small bellows design and reliability for our application will be addressed.

INTRODUCTION

Bellows can be compressed, extended, bent, internally or externally pressurized, supported more or less rigidly at the ends by magnets which act as lateral and torsional springs. (The torsion, or rotation, is in the magnet plane, not about the bellows axis). Instabilities are due to various reasons: axial compression, internal pressure, support system, manufacturing and installation inaccuracies. External pressure is not a cause for instability unless the bellows buckles, or is "crushed", due to excessive pressure.

Early magnetic field test set-ups at Fermilab and Brookhaven National Laboratory resulted in two apparent large bellows failures. These were caused by the design of the bellows end supports in the test facility. At one end the bellows was mounted on a magnet but on the other end it was supported by a rod with a much higher

---

* This work has been supported by U.S. Department of Energy.

slenderness ratio (effective length/radius of gyration of the cross–section) than that of the magnet. Historically, it is the investigation of these incidents that led to the work reported in SSC Technical Note No.89 (SSCL–N–742) and in a shortened version (BNL–44548). These reports establish design guidelines for bellows, understanding of failure modes including those due to the support system, and interaction between bellows in a magnet–bellows assembly configuration. Some of the basic results derived in these reports are presented here again in a simpler, if somewhat less general, formulation which is sufficient to illustrate the theory's salient features. Specifically: force exerted by internal pressure on bellows, Euler buckling mode, bellows axial stiffness, critical pressure, critical precompression. Other aspects of the work which do not lend themselves to concise treatment are simply reiterated. These are: detailed bellows parameter design, stresses and deflection of bellows both unpressurized and internally pressurized, model of a bellows (equivalent to a bar) with offset and initial shape mounted on lateral and torsional springs, and assembly of three magnets and two bellows.

The current note draws on our previously developed theory to address the specific details of bellows design for the SSC. Two types of bellows are analyzed: the first type is a "large" bellows with diameter somewhat larger than the magnet shell, the second type is a "small bellows", specifically the small bellows used in the current Fermilab interconnect design for the SSC. This work shows that large bellows can be reliable if properly designed and supported. Small bellows require the use of "squirm–cans" which are tubes lining (internally or externally) the bellows to prevent them from distorting.

BASIC RELATIONS

Force Exerted by Internal Pressure on Bellows

It was shown in the above mentioned report that the walls of an internally pressurized (pressure p) bellows of average diameter D are subjected to the force increment

$$\Delta f = \frac{\pi}{4} D^2 p \frac{\Delta x}{\rho} = \frac{\pi}{4} D^2 p y'' \Delta x \tag{1}$$

where the curvature $\frac{1}{\rho}$ can increase with the internal pressure p (fig.A1). This can lead to instability at some critical pressure p_{cr}. Bending calculations show that the effect of internal pressure can be replaced by a force

$$F_x = pA \tag{2}$$

acting axially at the bellows ends ($A = \pi D^2/4$ is the average bellows cross–section).

Euler Buckling Mode

When pressurized, a bellows initially unstressed (fig.A2.a) of axial stiffness K with unsupported ends would extend by $\delta = F_x/K$ with F_x given above (fig.A2.b). However the bellows whose free length is l is mounted between magnets which act as fixed ends. Thus due to the lack of room to expand, the bellows is effectively compressed by the length d (fig.A2.c). The effect of internal pressure on the bellows is identical to a compressive force F_x on a bar. The lowest buckling mode for a bar simply supported at both ends ("hinged") is

$$F_{cr} = \frac{\pi^2 EI}{l^2} \tag{3}$$

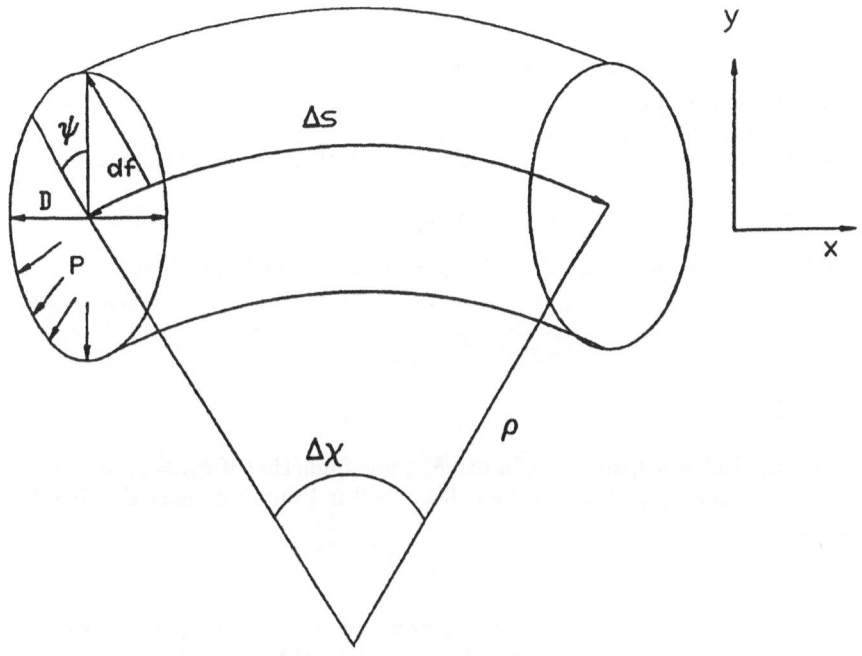

Fig.A1

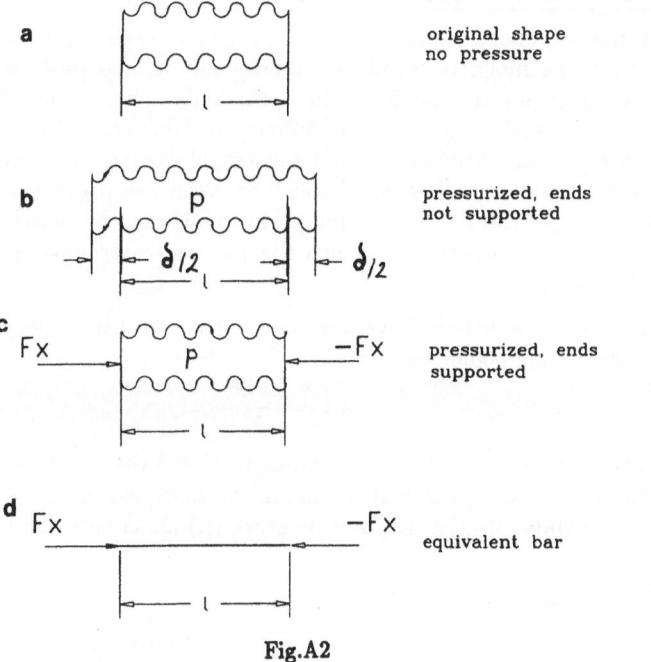

a		original shape no pressure
b	p	pressurized, ends not supported
c	Fx p −Fx	pressurized, ends supported
d	Fx −Fx	equivalent bar

Fig.A2

(fig.A2.d), with E=elastic modulus (or equivalent elastic modulus for bellows) and I=moment of inertia of the bellows cross–section.

Bellows Axial Stiffness

The bellows axial stiffness is obtained from the following basic equations:

- stress–strain formula : $\sigma = E\epsilon$
- definition of strain : $\epsilon = \delta/l$
- relation between stress and force : $F = \sigma S$ (S=average bellows wall cross–section)
- relation between axial force and axial stiffness: $F = K\delta$

these relations give the "equivalent" Young's modulus of the bellows:

$$E = \frac{Kl}{S} \tag{4}$$

Critical Pressure

Using the Euler critical load of a bar F_{cr}, one finds that if $F_{cr} = p_{cr}A = \frac{\pi^2 EI}{l^2}$ then the critical pressure p_{cr} at which the bellows will fail under compression is related to the axial stiffness by

$$p_{cr} = \frac{\pi K}{2l} \tag{5}$$

where K is assumed to be given. It is noted that this expression is independent of the average bellows diameter D which is equal to the inner diameter D_i plus the convolution height d. The moment of inertia $I = \frac{\pi D^3 t}{8}$ and cross–section $S = \pi Dt$ have been used for the bellows where t (much smaller than D) is the bellows wall thickness.

Precompression Length and Beam Tube Bellows

At installation, magnets are warm and bellows are precompressed by an amount λ_c. When cold, the magnets shrink, stretching the bellows past the free length and extending them by an amount λ_e. The total bellows travel usually must be split between extension and compression as follows: 3/4 for compression and 1/4 for extension in order to minimize bellows wall stresses, including considerations concerning material fatigue. This distribution of travel between compression and extension applies at room temperature. At lower temperatures it may be possible to increase the extension by the same fraction by which the material's strength is increased (about 20% to 40% at 4K).

From the relations listed above, one finds that the critical length by which one can precompress the magnets is

$$\lambda_{cr} = \frac{\pi^2 D^2}{8l} \tag{6}$$

The design pressure of the magnet containment is 300 psi (=20 atm=2 MPa). Using the usual margin of 50 %, the bellows should be designed for $p_{cr} = 450$ psi. (In the general theory axial deflection (λ) and pressure (p) effects are treated together.)

REVIEW OF PREVIOUS WORK

The previously mentioned report contains an approximate theory which is applicable if the bellows radial width of a convolution (d) is much smaller than the bellows diameter (D).

Bellows Design

The report gives the dependence of K on the wall thickness (t), convolution valley radius (r_i), other geometrical parameters, bending stresses (which are very close to yielding and thus require fatigue considerations), lateral spring constant, etc... The theory permits adjustment of bellows parameters to fit requirements and to compare with manufacturers' recommendations. Axial deflections and pressure effects are considered.

Spring – Supported Bellows Model

A model (fig.A3) of a bar with an initial offset and initial (prebent) shape mounted on lateral and torsional springs is used to gain insight into bellows behavior, in particular instabilities resulting from a weak support system. In the limit this analytical model which is a generalization of the conventional calculation for buckling of bars reduces to the classical cases (hinged, fixed, guided etc..) corresponding to various boundary conditions.

Magnet – Bellows Assembly

Finally a numerical model of any applicable combination of three magnets (dipoles or quadrupoles) and two bellows was also presented (fig.A4). It illustrates the effect of magnet prebend, magnet support and support location, and bellows to bellows interaction. In particular the question as to whether complete failure ("catastrophe") of one bellows causes failure of adjacent bellows is addressed. 18 bending equations, one for each span between supports or nodes, force and momentum equilibrium, boundary conditions resulted in a system of 51 equations. The solution of this system yields the deflection of the magnet and bellows. Of particular interest is the expression

$$\Delta l_{max} = |\lambda| + \frac{|y''|Dl}{2} \tag{7}$$

which is a measure of the behavior of the bellows. This expression combines the axial compression or extension λ of a bellows with the maximum local elongation or compression of the convolutions due to bending of the bellows. ($|y''| = \frac{Bending Moment}{EI}$). Δl_{max}, the maximum bellows local elongation, should not be confused with the strain in the bellows wall material. When plotted versus internal pressure, Δl_{max} shows peaks at instabilities and must be compared with the allowable elongations suggested by manufacturers. These in turn are based on maximum usable bellows wall bending stress and fatigue considerations.

NUMERICAL RESULTS

If the maximum operating pressure is $p_{op} = 300$ psi, a safety factor of 1.5 results in a required critical pressure of $p_{cr} = 450$ psi. Stability of large and small bellows at this critical pressure and effects of supports are presented next.

Large Bellows

A possible design for the interconnect region, originally considered but currently abandoned in favor of small bellows, made use of large bellows connecting the shells of adjacent magnets. Requesting that the bellows fail in compression at 450 psi for a free length of 9.3 in, inner diameter D_i=14.3 in, radial convolution height d=0.5 in, and precompression $\lambda = 1$ in, gives a computed axial stiffness K=2760 lbs/in. (These numbers are tabulated in forthcoming tables). The first concern, whether the support system is adequate, is addressed by examining the spring–supported bellows model

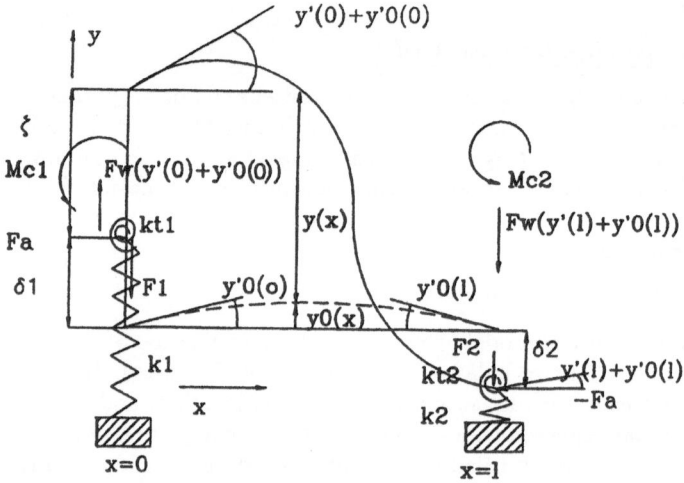

Fig.A3

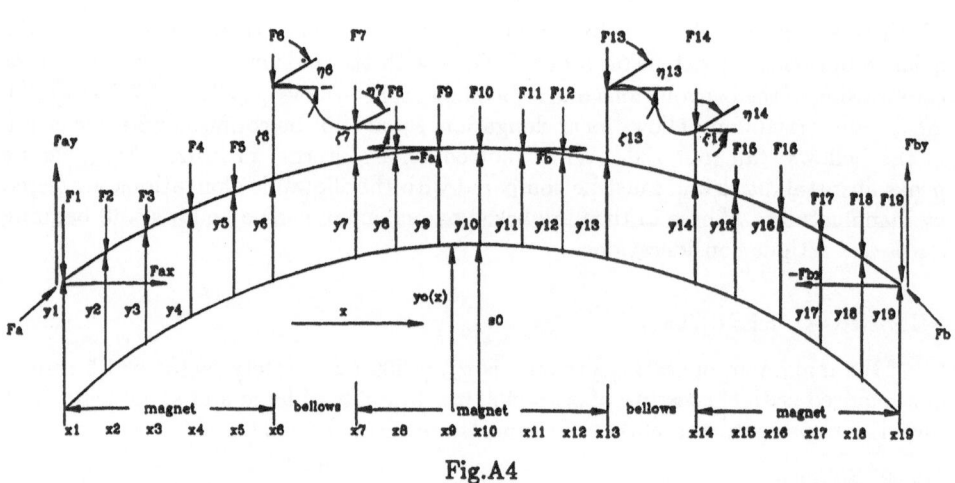

Fig.A4

with lateral stiffness k_1=7000 lb/in and torsional stiffness k_{t1}=3e7 lbs in/radian. (Left and right supports are assumed to have identical springs).

Fig.B1 shows instabilities due to bellows end supports only (the instability due to the Euler buckling mode at 450 psi here reduces to an indeterminacy and does not appear as a peak in the spring–supported model but is quite visible in the magnet–bellows assembly). Δl_{max}, in inches, is plotted versus the bellows internal pressure in psi. The bellows is assumed to be mounted with an initial offset of $\zeta = 0.03$ in and to be perfectly straight as $d_1 = d_2 = 0$ indicate on the legend. The value of Δl_{max} at the operating pressure of 300 psi is of interest since it must remain below manufacturer's recommendations. This value is quite small for the present case (no precompression was assumed, the effect of precompression would be to shift the curve upwards by a corresponding amount).

Since it is unlikely that the bellows is perfectly straight, one can assume that it has an initial sine shape $d_1 \sin(\pi x \; l)$ of maximum amplitude $d_1 = 0.02$ in and and initial cosine shape $d_2 (1 - \cos(2\pi x/l))$ with maximum amplitude $d_2 = 0.02$ in. Comparison between fig.B2 and fig.B1 reveals a large increase of Δl_{max} at p_{op}. (All curves are truncated at 5 in for clearer display of details.)

Should the supports be weak, instabilities due to support might well occur at lower values than $p_{cr} = 450$ psi for which the bellows axial stiffness was chosen. Such an occurrence is illustrated in fig.B3 with $k_{t1} = 1e4$ lbs in/radian where an unstability appears at 425 psi. Fortunately, this is far from being a realistic situation since torsional magnet stiffness is estimated to be 3e7 lbs in/radian.

Proceeding now to the model described in fig.A4, two of the bellows are inserted between three dipoles in fig.C1. Both bellows are assumed to be perfectly straight (no sine or cosine shape) and ideally mounted (no lateral or angular offsets). The bellows are designed to fail at 450 psi where the mesh of points is denser on the curve. The graph shows two curves, one for each of the bellows. Since both bellows are identical, the two curves overlap. The first support peak which appears above 600 psi is well above 450 psi. No peak is visible at 450 psi because of the zero value of the mounting offset, but it is anticipated that the peak will appear in the next figure for non–zero offsets.

Fig.C2 shows the case where the first bellows only has a mounting offset $\zeta_6 = 0.03$ in and an initial sine and cosine shape of maximum amplitudes 0.02 in each. The Euler peak is now visible for the misaligned bellows (curve made of triangles). The support peak is much larger: in general, misalignments, offsets, eccentricities do not cause instabilities but they increase the rate of rise to peak. In the limiting case where these parameters are zero as in fig.C1 the peak is no longer visible on the curve but a potential instability is always there in the form of an indeterminacy. Companion tables appear in figs.C2a and C2b. Table 1 shows the input to the bellows design part of the computations, table 2 gives parameters which are the result of the input of table 1. Table 3 shows the corresponding wall stresses and deflections, the total maximum stress is the sum of σ_{max} and σ_{pw2}. Table 4 lists the misalignments of the two bellows and table 5 lists the maximum amplitude of the intial sine and cosine shapes. Tables 6, 7 give some of the geometrical details such as length, overhang, moment of inertia of the dipoles and the quadrupoles. Table 8 gives the maximum bellows elongations at the operating pressure of 300 psi. In table 9 are listed nodes' coordinates, support stiffnesses, and force at operating pressure. One can distinguish magnet nodes from bellows nodes by noting that external support stiffnesses for bellows ends are zero.

Spring–supported Bellows model:max elong vs pres

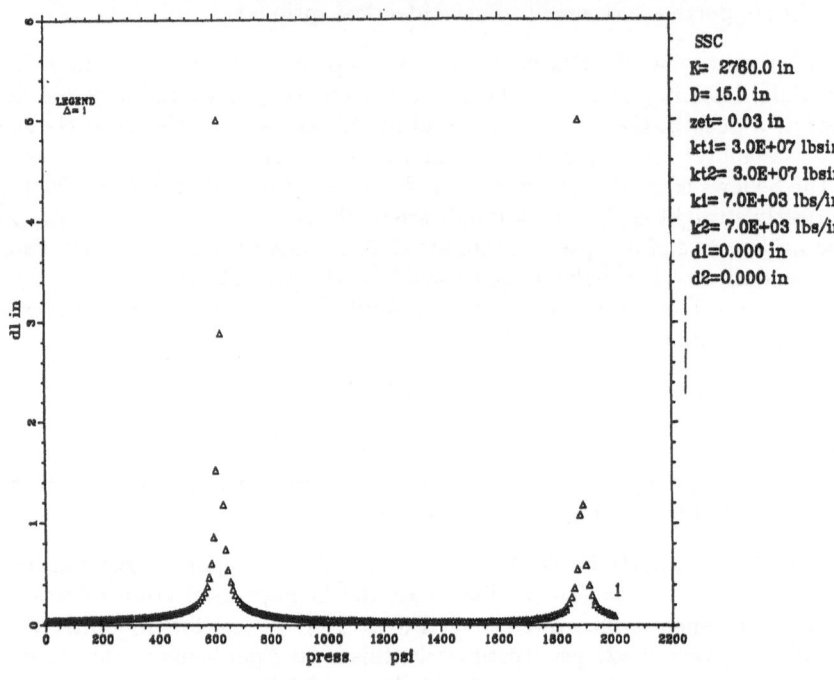

Fig.B1

Spring–supported Bellows model:max elong vs pres

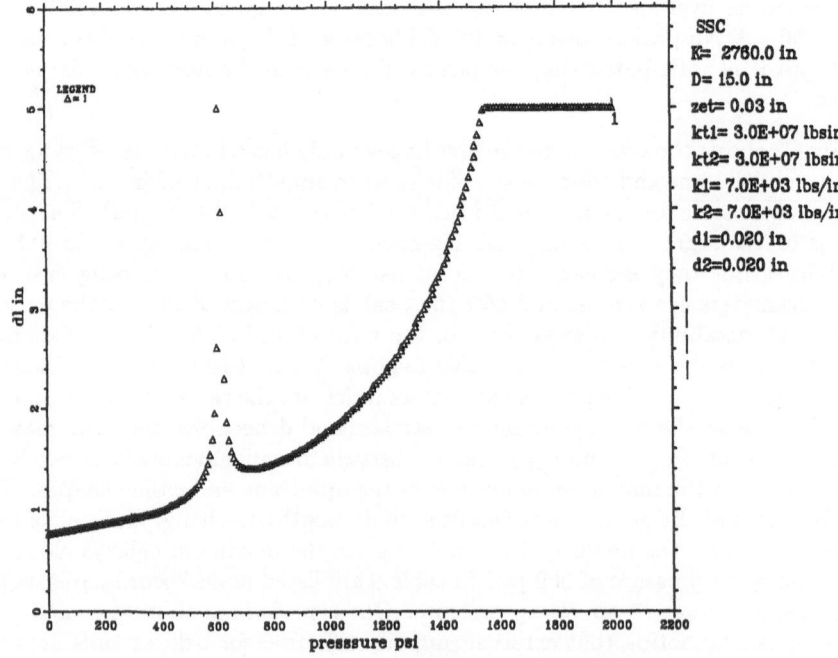

Fig.B2

Spring–supported Bellows model:max elong vs pres

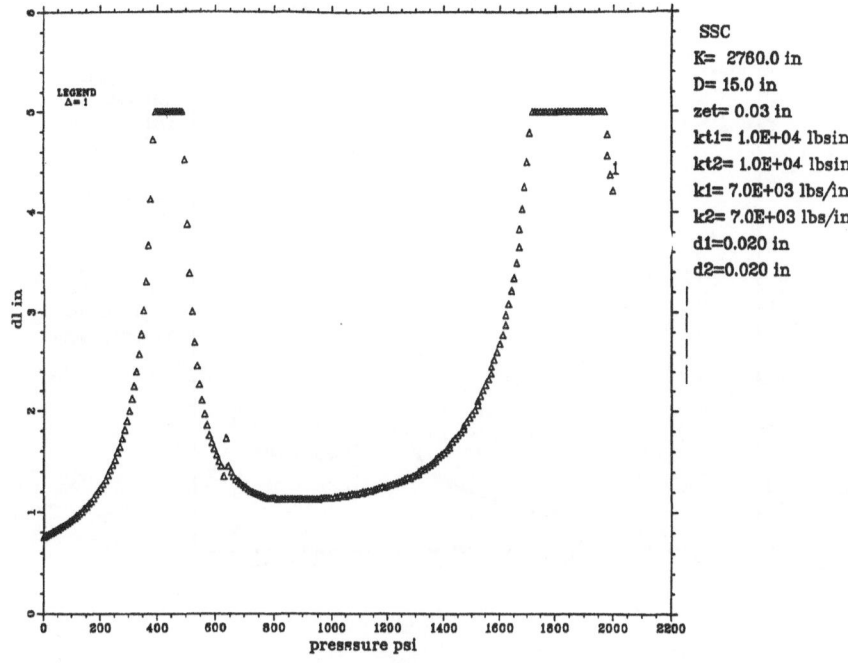

SSC
K= 2760.0 in
D= 15.0 in
zet= 0.03 in
kt1= 1.0E+04 lbsin
kt2= 1.0E+04 lbsin
k1= 7.0E+03 lbs/in
k2= 7.0E+03 lbs/in
d1=0.020 in
d2=0.020 in

Fig.B3

MAX BELLOWS LOCAL ELONGATION

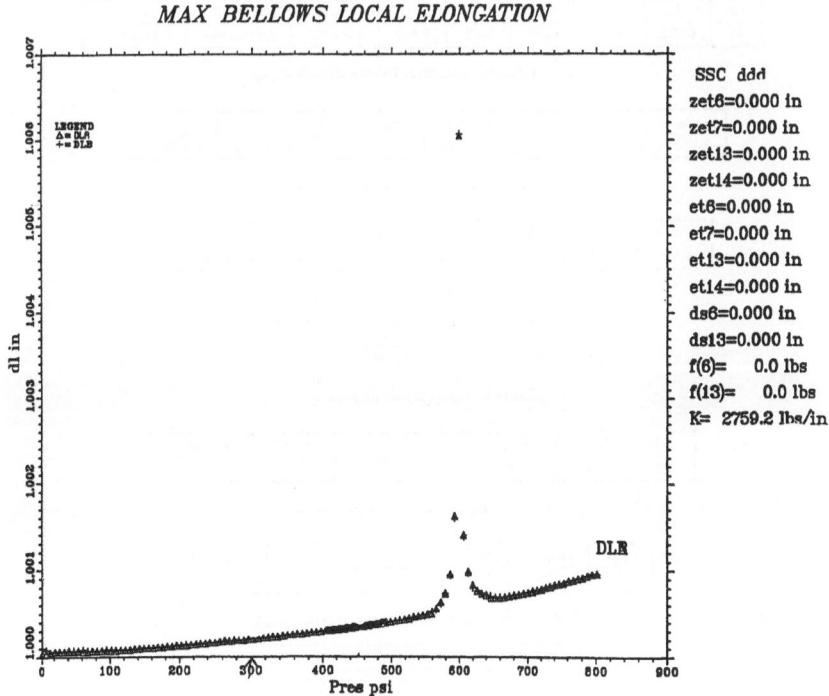

SSC ddd
zet6=0.000 in
zet7=0.000 in
zet13=0.000 in
zet14=0.000 in
et6=0.000 in
et7=0.000 in
et13=0.000 in
et14=0.000 in
ds6=0.000 in
ds13=0.000 in
f(6)=   0.0 lbs
f(13)=   0.0 lbs
K= 2759.2 lbs/in

Fig.C1

MAX BELLOWS LOCAL ELONGATION

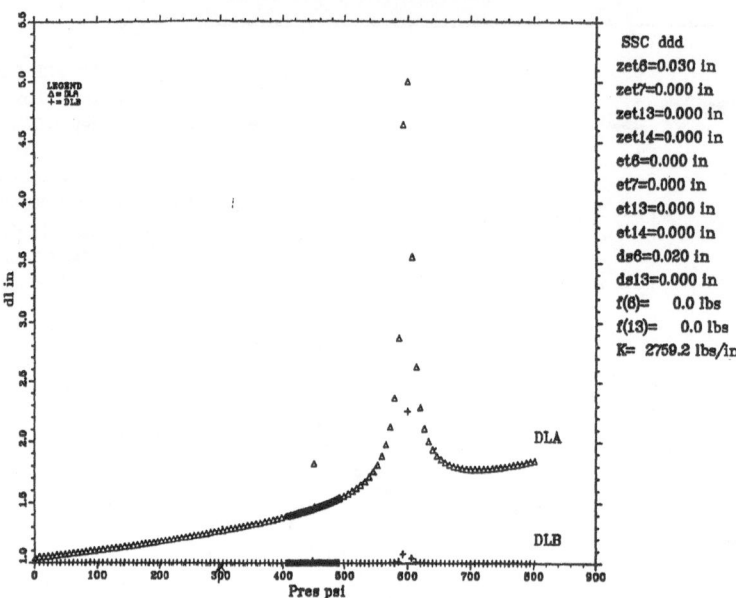

Fig.C2

Table 1: Input: bellows design parameters

r_i in	D_i in	d in	β	l in	pcr psi	λ in	Eb psi
0.060	14.300	0.500	0.094	9.300	450.000	1.000	3.297E+07

Table 2: Results: bellows properties

Nb	r in	D in	A in	t in	bc in	K in lbs/in	Kl in lbs/in	I in⁴
26	0.072	14.800	0.356	0.025	0.342	2759.212	1.048E+04	3.151E+01

Table 3: Results: maximum bellows stresses at p_{op}

σ_{max} psi	σ_h psi	σ_{yw1} psi	σ_{yw2} psi	σ_{yo}	y_{max} in
1.523E+05	2.778E+04	6.640E+04	6.782E+04	2.967E+04	1.472E-03

Table 4: Input: bellows misalignments

$\zeta(6)$ in	$\zeta(7)$ in	$\zeta(13)$ in	$\zeta(14)$ in	$\eta(6)$	$\eta(7)$	$\eta(13)$	$\eta(14)$
0.030	0.000	0.000	0.000	0.000	0.000	0.000	0.000

Table 5: Input: bellows initial shape

d_s6 in	d_s13 in	d_c6 in	d_c13 in
0.020	0.000	0.020	0.000

Table 6: Input: dipole properties

s in	L_D in	ovh_D in	D_m in	t_m in	Im in⁴	Em psi
0.200	670.	63.6	10.5	0.188	111.	3.000E+07

Table 7: Input: quadrupole properties

L_q in	ovh_Q in	D_q in	t_q in	Iq in⁴	Eq psi
210.	52.5	10.5	0.188	111.	3.000E+07

Table 8: Results: bellows maximum elongation at p_{op}

p_{op} psi	dl6 in	dl13 in
3.000E+02	1.262E+00	1.000E+00

Fig.C2.a

Table 9: Results: node location, stiffness, force at p_{op}

node	x(i) in	k(i) in lbs/in	F(i) lbs
1	0.00	25000.00	87.54
2	135.69	25000.00	35.69
3	271.38	25000.00	24.72
4	407.07	25000.00	17.59
5	542.76	25000.00	-12.71
6	606.38	0.00	0.00
7	615.68	0.00	0.00
8	679.30	25000.00	115.28
9	814.99	25000.00	-13.68
10	950.68	25000.00	18.37
11	1086.37	25000.00	26.95
12	1222.06	25000.00	27.80
13	1285.68	0.00	0.00
14	1294.98	0.00	0.00
15	1358.60	25000.00	27.45
16	1494.29	25000.00	24.13
17	1629.98	25000.00	22.63
18	1765.67	25000.00	35.16
19	1901.36	25000.00	87.70

Fig.C2.b

MAX BELLOWS LOCAL ELONGATION

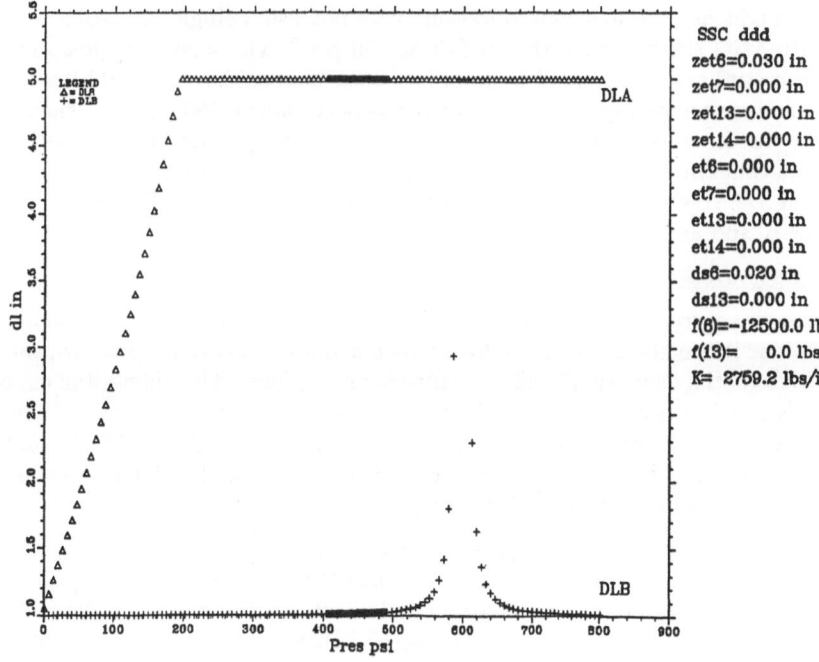

SSC ddd
zet6=0.030 in
zet7=0.000 in
zet13=0.000 in
zet14=0.000 in
et6=0.000 in
et7=0.000 in
et13=0.000 in
et14=0.000 in
ds6=0.020 in
ds13=0.000 in
f(6)=-12500.0 lbs
f(13)=   0.0 lbs
K= 2759.2 lbs/in

Fig.C3

Fig.C3 investigates a "catastrophe" where the first bellows was made to fail by applying a couple of forces of plus/minus 12500 lbs at the bellows ends. Complete failure of one bellows does not seem to affect the behavior of the adjacent bellows. The interaction, if any, between bellows is negligible in the operating region.

In fig.C4 the bellows are connected by two dipoles and a quadrupole. There is an interaction effect which is moderate at operating pressure and is due to the fact that a quadrupole has only 3 supports instead of the five for a dipole. The interaction can be minimized by increasing magnet support stiffnesses or moving the supports closer to the bellows to reduce the overhang as much as possible.

Fig.C5 shows the deflected shape of the assembly of three dipoles and two bellows. The 0.03 in offset is the dominant feature of the first bellows deflected shape, the second bellows shows little distortion.

Small Bellows

The current Fermilab design for the SSC interconnect uses small bellows made of 2 ply material that have a very low axial stiffness of $K = 161$ lbs/in in order to accommodate the required travel of 2 in. These bellows are mounted on pipes and not directly on the magnets themselves as were the large bellows. Fig.D1 is an enlarged detail of fig.D2 and serves to illustrate the fact that these small bellows can become unstable already at 35 psi: they must thus be used with liners (called "squirm supports"). Bellows design parameters are: inner diameter $D_i = 3.875$, radial convolution height $d = 0.33$, free length $l = 5.75$, precompression $\lambda_c = 0.83$. The first bellows only is misaligned and has an initial bent shape. Fig.D2 shows the location of support peaks which occur well below 450 psi for bellows with such a low axial stiffness and mounted on pipes. The stresses of this two ply bellows are estimated by applying half of the internal pressure on one ply only. By demanding a critical pressure at $35/2=17.5$ psi, the stiffness of a bellows with same inner diameter and length is about half of the 161 lbs/in and the stresses are 113 ksi (sum σ_{max} with σ_{pw2} of table 3 fig.D2.a) which is high but seems to be in the range used by manufacturers.

One might ask the following question: why not use a single ply bellows without liner with axial stiffness such that it fail at 450 psi ? The answer appears in fig.D3 where the stiffness of the small bellows has been increased to K=2100 lbs/in. This bellows is mounted on pipes and support peaks occur above 450 psi. It is shown again in fig.D4 but mounted between magnets instead of pipes. The problem with such a bellows is that the bending stresses are greater than 370 ksi (sum σ_{max} with σ_{pw2} of table 3 fig.D4.a) which is unacceptable and explains why one needs a two ply bellows with low K and anti–squirm protection.

Beam Tube Bellows

There are currently two identical bellows in the beam tube region of the interconnection: one is the "beam tube" bellows which is not internally pressurized, the other is the "differential expansion" bellows which is pressurized. The differential expansion bellows is put in place to accommodate sudden contraction of the ends of the beam tube in case of too rapid cooling. Using Eq. (6) with $D_i = 2$, $d = 0.336$, $l = 5.25$, one finds $\lambda_{cr} \simeq 0.93 \ in$. The total motion that both bellows must be able to accommodate due to the thermal contraction of the magnets is 2 in.

The critical pressure (Eq. (5)) which was derived in our previous report is not affected by the presence of an internal tube inside the bellows. This expression was obtained by considering pressure acting on elements on the bellows walls and consequently did not depend on what was inside the bellows. In the case of the differential

MAX BELLOWS LOCAL ELONGATION

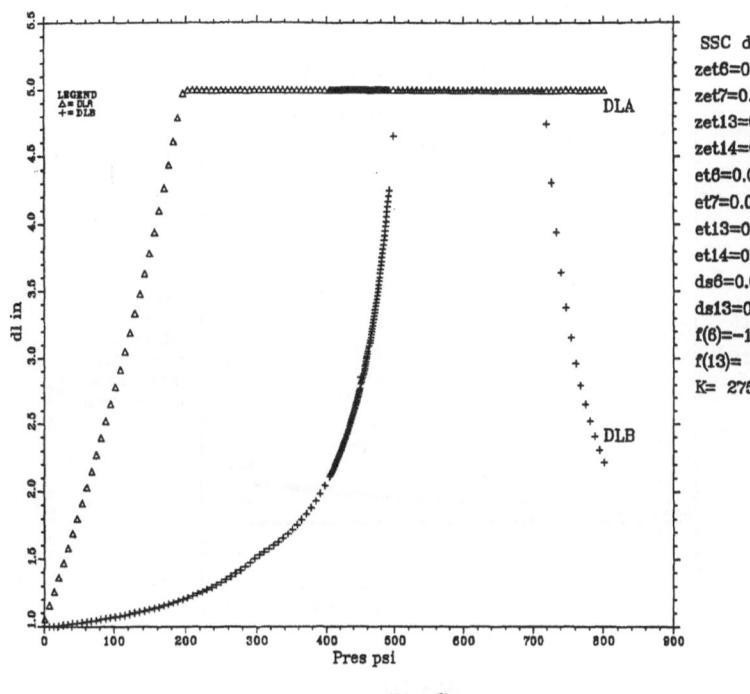

SSC dqd
zet6=0.030 in
zet7=0.000 in
zet13=0.000 in
zet14=0.000 in
et6=0.000 in
et7=0.000 in
eti13=0.000 in
et14=0.000 in
ds6=0.020 in
ds13=0.000 in
f(6)=-12500.0 lbs
f(13)=   0.0 lbs
K= 2759.2 lbs/in

Fig.C4

ASSEMBLY DEFLECTIONS AT 303.5 psi

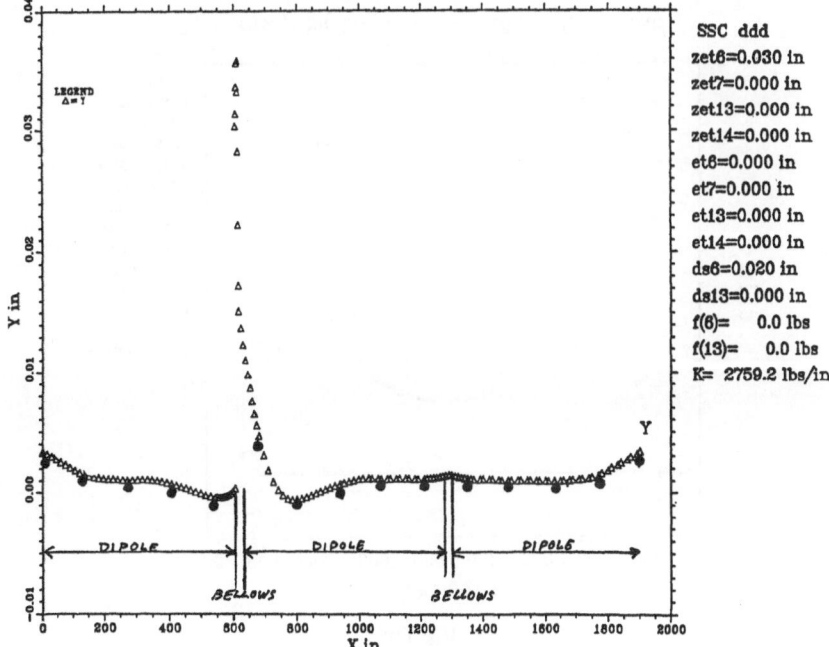

SSC ddd
zet6=0.030 in
zet7=0.000 in
zet13=0.000 in
zet14=0.000 in
et6=0.000 in
et7=0.000 in
eti13=0.000 in
et14=0.000 in
ds6=0.020 in
ds13=0.000 in
f(6)=   0.0 lbs
f(13)=   0.0 lbs
K= 2759.2 lbs/in

Fig.C5

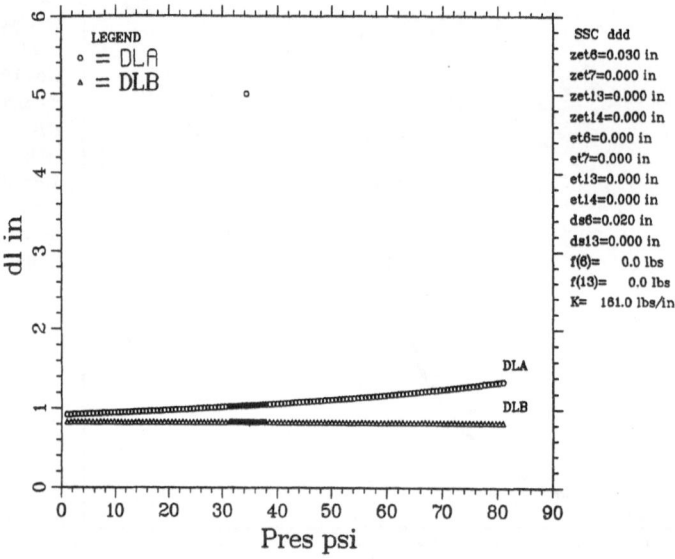

Fig.D1

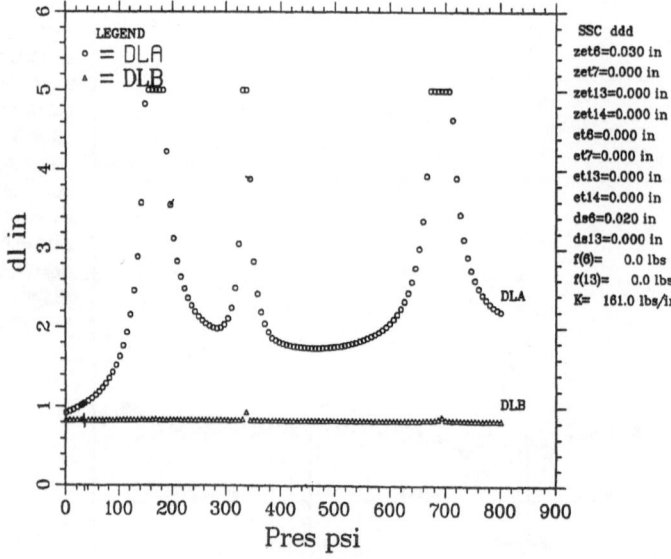

Fig.D2

Table 1: Input: bellows design parameters

r_i in	D_i in	d in	β	l in	pcr psi	λ in	Eb psi
0.060	3.875	0.330	0.094	5.750	17.500	0.830	3.297E+07

Table 2: Results: bellows properties

Nb	r in	D in	A in	t in	bc in	K lbs/in	Kl lbs/in	I in^4
19	0.065	4.205	0.297	0.010	0.203	82.000	6.578E+01	2.787E-01

Table 3: Results: maximum bellows stresses at p_{op}

σ_{max} psi	σ_h psi	σ_{pw1} psi	σ_{pw2} psi	σ_{ps}	y_{max} in
1.041E+05	9.557E+02	8.680E+03	8.779E+03	3.625E+03	2.573E-04

Table 4: Input: bellows misalignments

$\zeta(6)$ in	$\zeta(7)$ in	$\zeta(13)$ in	$\zeta(14)$ in	$\eta(6)$	$\eta(7)$	$\eta(13)$	$\eta(14)$
0.030	0.000	0.000	0.000	0.000	0.000	0.000	0.000

Table 5: Input: bellows initial shape

d_s6 in	d_s13 in	d_c6 in	d_c13 in
0.020	0.000	0.020	0.000

Table 6: Input: dipole properties

s in	L_D in	ovh_D in	D_m in	t_m in	Im in^4	Em psi
0.200	670.	63.6	10.5	0.188	111.	3.000E+07

Table 7: Input: quadrupole properties

L_q in	ovh_Q in	D_q in	t_q in	Iq in^4	Eq psi
210.	52.5	10.5	0.188	111.	3.000E+07

Table 8: Results: bellows maximum elongation at p_{op}

p_{op} psi	dl6 in	dl13 in
1.167E+01	1.884E+00	8.306E-01

Fig.D2.a

Table 9: Results: node location, stiffness, force at p_{op}

node	x(i) in	k(i) lbs/in	F(i) lbs
1	0.000E+00	1.000E+08	1.597E+00
2	3.000E+00	1.000E+08	-1.338E+01
3	6.000E+00	1.000E+08	5.743E+01
4	9.000E+00	1.000E+08	-2.268E+02
5	1.200E+01	1.000E+08	2.058E+02
6	2.290E+01	0.000E+00	0.000E+00
7	2.865E+01	0.000E+00	0.000E+00
8	3.393E+01	1.000E+08	-7.317E+01
9	3.693E+01	1.000E+08	6.076E+01
10	3.993E+01	1.000E+08	-1.506E+01
11	4.293E+01	1.000E+08	2.705E+00
12	4.593E+01	1.000E+08	3.931E-01
13	5.683E+01	0.000E+00	0.000E+00
14	6.258E+01	0.000E+00	0.000E+00
15	6.786E+01	1.000E+08	2.306E-01
16	7.086E+01	1.000E+08	1.108E-01
17	7.386E+01	1.000E+08	-3.695E-01
18	7.686E+01	1.000E+08	4.103E-01
19	7.986E+01	1.000E+08	-6.577E-01

Fig.D2.b

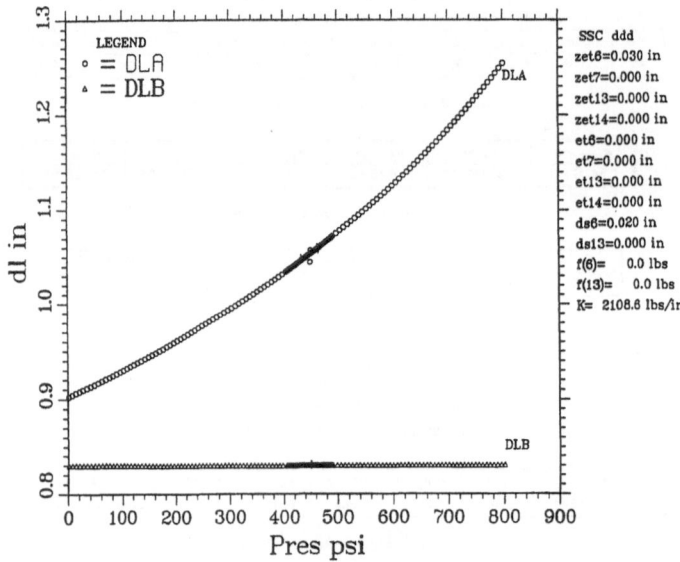

Fig.D3

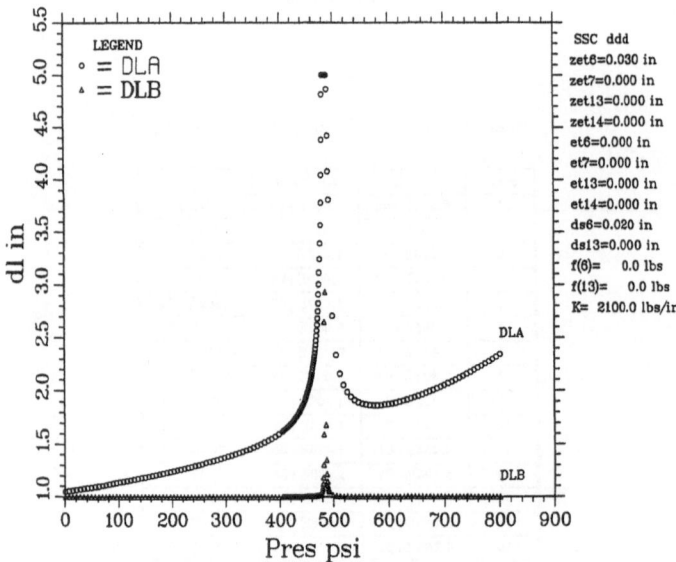

Fig.D4

Table 1: Input: bellows design parameters

r_i in	D_i in	d in	β	l in	pcr psi	λ in	Eb psi
0.060	3.875	0.330	0.094	5.750	450.000	0.830	3.297E+07

Table 2: Results: bellows properties

Nb	r in	D in	A in	t in	bc in	K lbs/in	Kl lbs/in	I in^4
17	0.072	4.205	0.321	0.024	0.176	2108.561	1.692E+03	6.996E-01

Table 3: Results: maximum bellows stresses at p_{op}

σ_{max} psi	σ_h psi	σ_{pw1} psi	σ_{pw2} psi	σ_{pz}	y_{max} in
3.310E+05	1.106E+04	3.504E+04	3.596E+04	1.400E+04	4.165E-04

Table 4: Input: bellows misalignments

$\zeta(6)$ in	$\zeta(7)$ in	$\zeta(13)$ in	$\zeta(14)$ in	$\eta(6)$	$\eta(7)$	$\eta(13)$	$\eta(14)$
0.030	0.000	0.000	0.000	0.000	0.000	0.000	0.000

Table 5: Input: bellows initial shape

d_s6 in	d_s13 in	d_c6 in	d_c13 in
0.020	0.000	0.020	0.000

Table 6: Input: dipole properties

s in	L_D in	ovh_D in	D_m in	t_m in	Im in^4	Em psi
0.200	670.	63.6	10.5	0.188	111.	3.000E+07

Table 7: Input: quadrupole properties

L_q in	ovh_Q in	D_q in	t_q in	Iq in^4	Eq psi
210.	52.5	10.5	0.188	111.	3.000E+07

Table 8: Results: bellows maximum elongation at p_{op}

p_{op} psi	dl6 in	dl13 in
3.000E+02	9.929E-01	8.301E-01

Fig.D4.a

Table 9: Results: node location, stiffness, force at p_{op}

node	x(i) in	k(i) lbs/in	F(i) lbs
1	0.000E+00	1.000E+08	7.835E-01
2	3.000E+00	1.000E+08	-4.369E+00
3	6.000E+00	1.000E+08	1.908E+01
4	9.000E+00	1.000E+08	-7.498E+01
5	1.200E+01	1.000E+08	4.596E+01
6	2.290E+01	0.000E+00	0.000E+00
7	2.865E+01	0.000E+00	0.000E+00
8	3.393E+01	1.000E+08	2.032E+02
9	3.693E+01	1.000E+08	-2.371E+02
10	3.993E+01	1.000E+08	6.036E+01
11	4.293E+01	1.000E+08	-1.503E+01
12	4.593E+01	1.000E+08	3.367E+00
13	5.683E+01	0.000E+00	0.000E+00
14	6.258E+01	0.000E+00	0.000E+00
15	6.786E+01	1.000E+08	4.946E-01
16	7.086E+01	1.000E+08	-2.362E-01
17	7.386E+01	1.000E+08	6.159E-02
18	7.686E+01	1.000E+08	9.065E-02
19	7.986E+01	1.000E+08	7.356E-02

Fig.D4.b

Table 1. small bellows properties

bellows type	D_i in	d in	l in	K lbs/in	λ_{cr} in	p_{cr} psi
single phase He	3.875	0.33	5.75	161	3.8	35
beam tube	2	0.336	5.25	91	0.93	no int pres
diff. expansion	2	0.336	5.25	91	0.93	27.2

Table 2. small vs large single phase bellows

bellows type	small	large
safety factor at 450 psi	0.12	1.5
squirm can	friction+guided	none
He interconnect volume	reduced	ok
He flow impedance	increased	ok
beam tube vacuum to vacuum	yes	need double bellows
access to beam tube liners	yes	more difficult

expansion bellows which contains the beam tube inside, one finds that the critical pressure is $p_{cr} = 27.2$ psi.

CONCLUSIONS

Large bellows which can travel the required two inches do not require liners since their axial stiffness is quite high and they can be designed to be stable at operating pressure with a safety factor of 1.5. Attention should be payed to proper design of the supports, in particular to their torsional stiffness.

Properties of small bellows and beam tube bellows used in the current interconnect design are summarized in Table 1 where eqs.(2.5) and (2.6) were used to compute the critical pressures and compression lengths.

In order to accommodate the required travel without overstressing the bellows walls, the axial stiffness must be so low that instabilities occur well below the design pressure and liners must be used for these small bellows.

Table 2 lists the respective advantages and disadvantages of large versus small bellows.

INVESTIGATION OF FACTORS AFFECTING THE CALIBRATION OF STRAIN GAGE

BASED TRANSDUCERS ('GOODZEIT GAGES') FOR SSC MAGNETS

M. Davidson, A. Gilbertson, and M. Dougherty

Fermi National Accelerator Laboratory*
P.O. Box 500 - M.S. 333
Batavia, Illinois 60510

ABSTRACT

These transducers are designed to measure stresses on SSC collared coils. They are individually calibrated with a bonded ten-stack of SSC inner coil cable by applying a known load and reading corresponding output from the gages. The transducer is supported by a notched 'backing plate' that allows for bending of the gage beam during calibration or in use with an actual coil. Several factors affecting the calibration and use of the transducers are: the number of times a 'backing plate' is used, the similarities or differences between bonded ten-stacks, and the differences between the ten-stacks and the coil they represent. The latter is probably the most important because a calibration curve is a model of how a transducer should react within a coil. If the model is wrong, the calibration curve is wrong.

Information will be presented regarding differences in calibrations between Brookhaven National Labs (also calibrating these transducers) and Fermilab - what caused these differences, the investigation into the differences between coils and ten-stacks, and how they relate to transducer calibration, and some suggestions for future calibrations.

INTRODUCTION

A brief description of how and why calibrations are done provides some background. The calibrations of beam transducers began at Fermilab in late 1989. Initially there were large differences in calibration data between Fermilab and Brookhaven. These differences persisted until our calibration system came fully on line during the summer of 1990. Basically, the two calibration setups are the same equipment. They consist of a loading press (manually controlled at Fermilab), a load cell to read loading forces, an HP3457A multimeter to read load cell and strain gage input/output, and a PC to control the meters and convert the gage output into microstrain. Transducers are calibrated two at a time by mounting them at the top and bottom of a ten-stack of SSC inner coil cable. A known load is transmitted to the beam. The beam is allowed to bend because of the presence of a notched 'backing plate' that supports only the ends of the beam. The gage is mounted to the beam and stretches and changes resistance to correspond with the applied load.[1] The transducers are calibrated up to a load of 18,000 psi. Data is collected continuously during the loading cycle at Fermilab. BNL collects data at ten discrete points along the loading curve. After the data is collected, the resistance measurements are

---

*Operated by Universities Research Association under contract with the Department of Energy

converted to microstrain. A cubic equation is fitted to the microstrain versus coil psi data. After the calibration, the transducers are installed into notched collar sections of a magnet. They are used to measure pre-stress during collaring and later during the actual tests of the magnets. The calibration curves at the two Labs initially varied markedly as far as the gage output was concerned. See Figure 1 and 2. The gages calibrated at BNL had far less output than the ones at Fermilab. The following is an investigation into what caused these differences and how they were overcome.

TRANSDUCER CALIBRATION

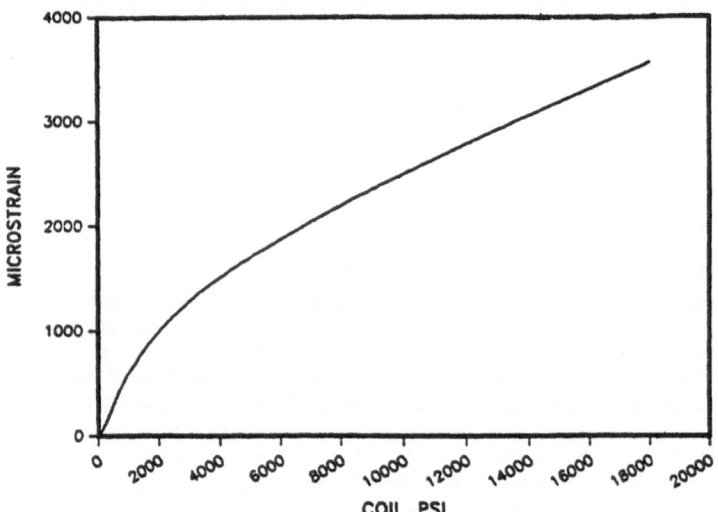

Fig. 1 Initial calibration done at Fermilab.

TRANSDUCER CALIBRATION

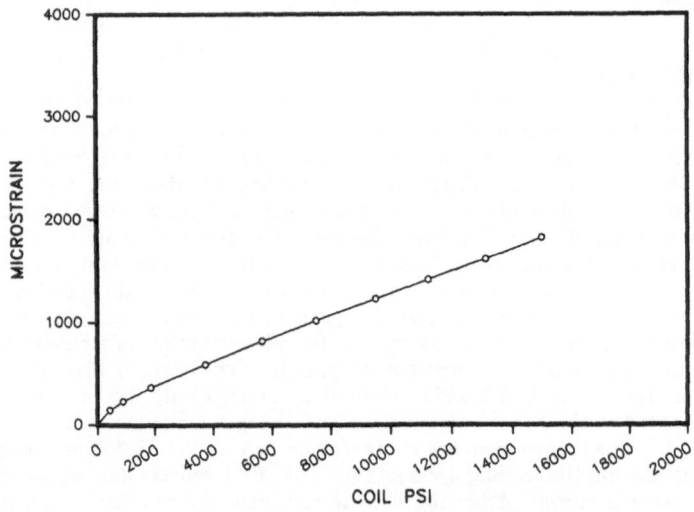

Fig. 2 Calibration of a transducer done at BNL.

INVESTIGATION

Major Differences in Calibrations

Initially, the backing plate was reused for every calibration. The backing plates were inspected and found to be bowed by approximately 0.0025" in the loading direction. This was causing the beam to load unevenly resulting in inaccurate calibrations. This can be seen in the calibration curve in Figure 3, where the calibration was done with an overused backing plate. It can be seen that when the plate was replaced, the curve became much steeper and corresponded with other calibrations. See Figure 4. As a result, new backing plates were used for each calibration.

TRANSDUCER CALIBRATION

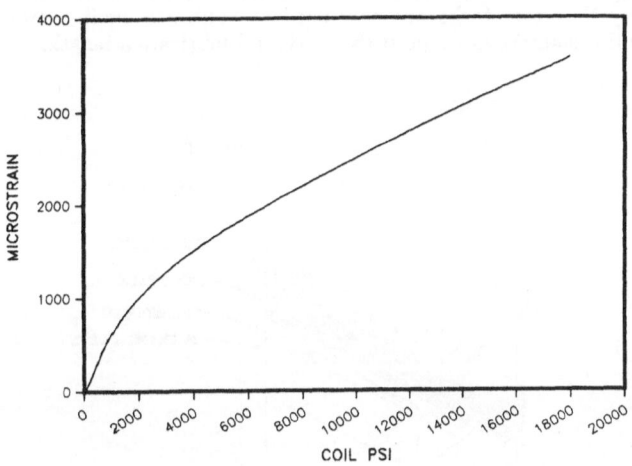

Fig. 3 Calibration of a transducer with a bent backing plate.

TRANSDUCER CALIBRATION

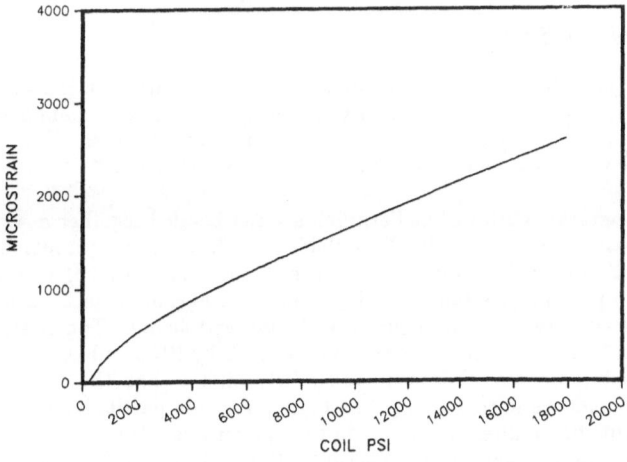

Fig. 4 Calibration of transducer done with new backing plate.

During the initial calibrations of the beams, several variations on the calibration setup were used. The ten-stacks were initially cut to the length of the calibration fixture. When a ten-stack comes out of the fabrication press, it is approximately 4" in length. The thought was then to cut the ten-stack down to the length of the fixture, 1.5", so the stack would fit with little clearance. See Figure 5. The length of the ten-stack itself did not contribute to inaccurate calibrations. The damage during the cutting process caused the calibration curves to vary. Compare Figure 6 and 7. It is of interest to note that coils made at Fermilab have some of these same qualities. They are not an example of poor coils or ten-stacks, but are a reflections of the epoxy content of the fiberglass tape used to wrap the coil. The comparison can be made between cutting a BNL ten-stack and a Fermilab ten-stack. The calibration curves indicate that the transducer calibrated with a cut BNL ten-stack and an uncut BNL ten-stack are relatively the same. See Figure 8. This is due to the compact nature of BNL's ten-stack. The cutting had very little effect on the bonds between cables. The more loosely bonded was the ten-stack, the more susceptible it was to damage during calibrations. Their reuse led to further damage of the bonds between cables. This caused the stack to become looser and calibrations worse as the condition of the stack deteriorated. See Figure 9. New ten-stacks are now used to calibrate against two gages for each calibration, and the stacks are kept at their original fabrication length.

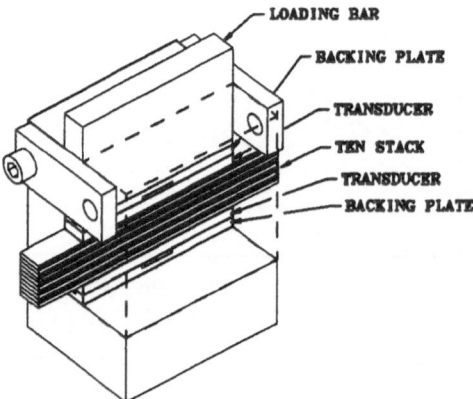

Fig. 5 Calibration fixture and set-up.

Investigation of Ten-Stacks

The major difference between calibrations is currently the ten-stack. Ten-stacks fabricated at BNL are very compact and have an appearance of a solid block. The cables are wrapped in an epoxy tape with an epoxy content of 23%. The ten-stack is heated during bonding to cure the epoxy, and the stacks are approximately 0.639."

The ten-stacks fabricated at Fermilab are not bonded together as tightly as BNL's, and are approximately 0.650" tall. Fermilab's tape has an epoxy content of about 18%. Upon examination of the two ten-stacks stress/strain curves, the BNL curve has a much lower change in percentage strain per coil psi than the Fermilab stack. Load deflection measurements were taken of an actual coil during coil sizing. The elastic modulus was calculated from this and compared to ten-stacks made by BNL and Fermilab. See Table 1. The elastic modulus of the BNL stack is slightly higher than the Fermilab stack. The measured coil modulus during sizing was higher for this particular coil. It is apparent that the modulus has a direct relationship to gage output. The greater the modulus, the lower the gage output. Compare a BNL calibration to that of a Fermilab calibration. See Figures 2 and 10. There are, however, other coil characteristics that merit attention.

TRANSDUCER CALIBRATION

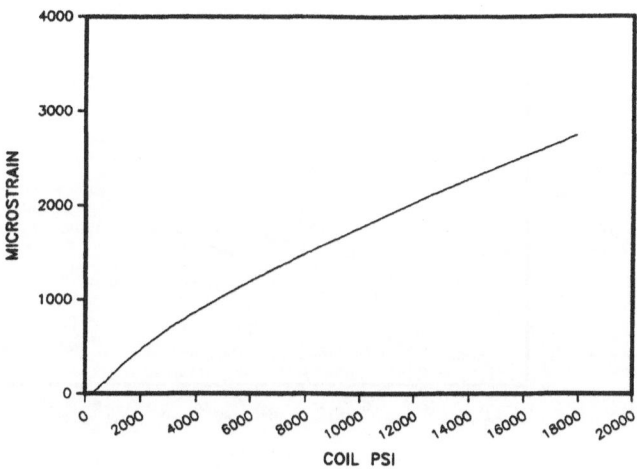

Fig. 6 Calibration of transducer with an uncut Fermilab ten-stack.

TRANSDUCER CALIBRATION

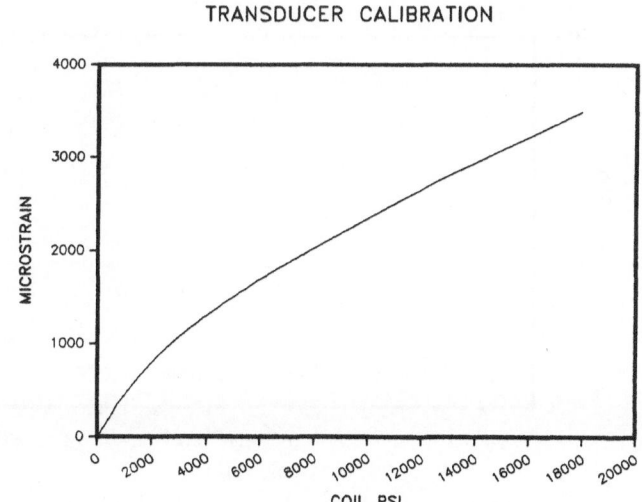

Fig. 7 Calibration of transducer with Fermilab cut ten-stack.

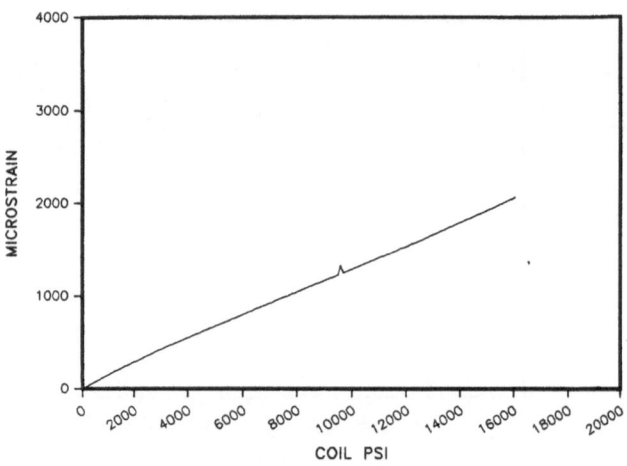

Fig. 8   Calibration of a transducer with BNL uncut ten-stack.

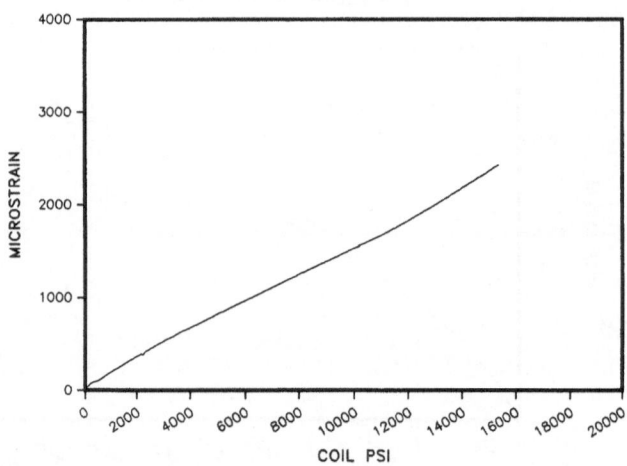

Fig. 9   BNL ten-stack cut to just fit in a calibration fixture.

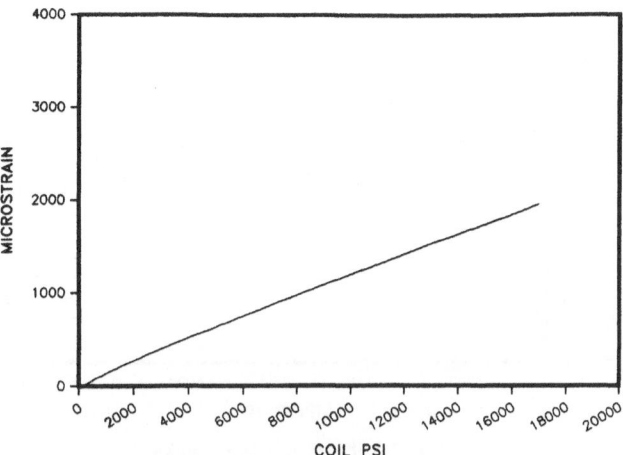

Fig. 10 Current calibration curve of a transducer.

The characteristic curve of a ten-stack should closely resemble the characteristic curve of an actual coil. This difference between the curves is believed to be influencing the low end of the calibration curve. This influence is evident in calibration curves resulting from stiff ten-stacks as opposed to looser ten-stacks. See Figures 2 and 10. The calibration curve of the looser stack does not become linear until higher loads. The opposite case is true with the stiffer stack. This implies that the non linear modulus of the stack causes the non-linearity of the calibration curve at the initial loading. With this fact in mind, characteristic curves were made of deflection versus load of an actual coil and compared to ten-stacks. The analysis of the curves shows that the actual coil became linear at approximately 3000 psi, the BNL stack at approximately 2000 psi, and the Fermilab stack at approximately 4000 psi. See Figures 11, 12, and 13.

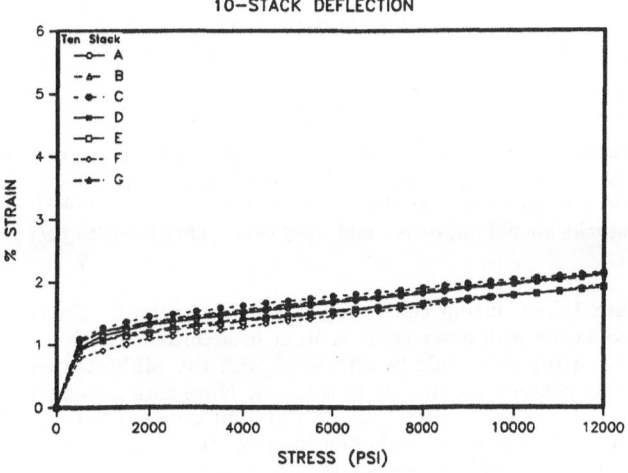

Fig. 11 BNL ten-stacks.

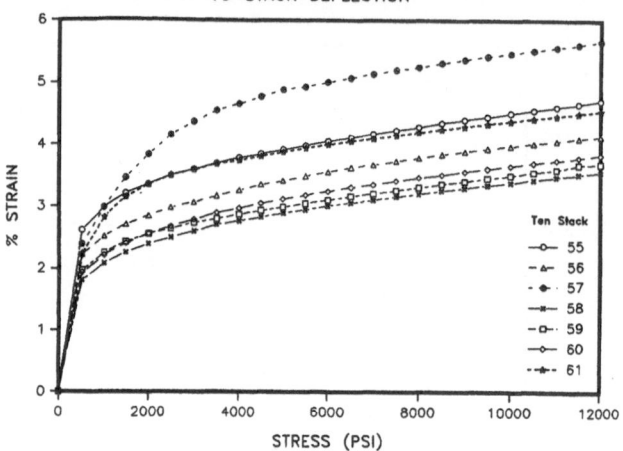

Fig. 12 Fermilab ten-stacks.

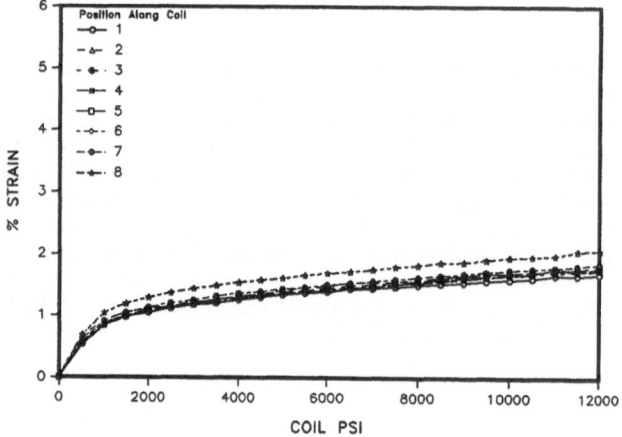

Fig. 13 Coil sizing.

CONCLUSIONS

As a result of these observations, new backing plates must be used for each calibration to avoid deforming the plate. This also implies that a backing plate should be kept paired with a transducer even after calibration up to and including installation in the magnet. Ten-stacks should not be cut and used over. This leads to poor and unrepeatable calibrations.

The data taken during coil sizing is very important. Currently, Fermilab is constructing SSC coils with lower epoxy content in accordance with the current Fermilab ten-stacks. Coil sizing data could be correlated with the calibrated transducers that are going to be used to measure the stresses on that coil. More data of this sort needs to be taken to determine how to model a ten-stack after a coil. The data gathered from coil sizing could be used to establish an allowable window for the properties of a ten-stack. The stacks could be tested and would have to fit this window to be acceptable.

Table 1

Coil Modulus*

BNL Stack	10-Stack E x 10e-6	FNAL Stack	10-Stack E x 10e-6	FNAL Position	Inner Coil E x 10e-6
A	1.30	55	0.950	1	1.96
B	1.40	56	1.20	2	1.76
C	1.50	57	0.98	3	1.75
D	1.59	58	1.30	4	1.66
E	1.50	59	1.02	5	1.70
F	1.23	60	1.07	6	1.65
G	1.43	61	1.15	7	1.67
				8	1.54

Mean = 1.42   Mean = 1.1   Mean = 1.71

* = Units of psi

REFERENCE

1.   C. L. Goodzeit, M.D. Anerella, and G.L. Ganetis, "Measurement Of Internal Forces In Superconducting Accelerator Magnets With Strain Gauge Transducers," SSC-MAG-R-7312, (1989).

17. Project Management and Systems Engineering

SYSTEMS ENGINEERING AT THE SUPERCONDUCTING SUPER COLLIDER (ONE YEAR LATER)

John A. Nonte

Superconducting Super Collider Laboratory*
2550 Beckleymeade Ave.
Dallas, Texas 75237

Abstract: After one year of systems engineering at the Superconducting Super Collider (SSC), the project baseline of costs, schedule milestones, and top-level (point design) physics parameters has been accepted by the Department of Energy (DOE). This paper describes the role of systems engineering in developing the baseline and in establishing requirements specifications, change control, and methods of tracking to a baseline. The differences between the Department of Defense and DOE—specifically at the SSC Laboratory (SSCL)—in application of systems engineering disciplines and tools are discussed. The aim of the paper is to inform participating industries of the anticipated requirements format and of the emphasis that will be placed on physics requirements as opposed to procedures. Industry subcontractors should have a better understanding of the systems engineering expected by the SSCL.

INTRODUCTION

The paper begins with a definition of systems engineering, develops the theme of baselines through the systems engineering tools used, and closes with conclusions and a statement of what subcontractors can expect when bidding to support the SSC.

SYSTEMS ENGINEERING

Systems engineering was defined at the 1990 IISSC as "the management function which controls the total system development effort for the purpose of achieving an optimum balance of all system elements. It is a process which transforms an operational need into a description of system parameters and integrates those parameters to optimize the overall system effectiveness." This definition is adequate in a classical systems engineering environment such as an aerospace company, but it is impractical in the environment of the Superconducting Super

---

*Operated by the Universities Research Association, Inc., for the U.S. Department of Energy under Contract No. DE-AC02-89ER40486.

Collider Laboratory (SSCL). Two phrases make it impractical: "controls the total systems development effort" and "a process which transforms an operational need into...." First, let there be no doubt that systems engineering does not control the system development effort. The scientists, specifically the physicists, control the technical management and the technical systems, and thus control the technical systems, and thus control the development. Second, the physicists describe the system parameters and integrate those parameters. Systems engineering is a tool to help bring some discipline to the process, to document and track the process, and to support development of "procedures" (as understood at the SSCL), including the process of availability (reliability and maintainability), logistics, safety, materials, etc. This author offers a different definition of systems engineering, one proposed at a Lockheed Corporation systems engineering task force meeting: "Whatever it takes to get what you want." This simple definition says it all. But note that it does not say, "Whatever it takes to get the job done." It is possible, even probable, that one can get the job done without getting what he wanted. With this definition as a guide, we can proceed to develop baselines (including specifications and interfaces), configuration management, and specialty engineering procedures.

BASELINES AND SYSTEMS ENGINEERING TOOLS

The Department of Defense (DOD) uses three kinds of technical baselines: functional, allocated, and product. The functional baseline is a high-level baseline usually formed by the system specification. The allocated baseline is formed by the development specifications flowing out of the system specification. This includes the "design to" requirements for the subsystems and components. The product baseline is set by "build to" specifications for subsystems or components which consist of a complete set of drawings and specifications which have been validated through testing of prototypes. This product baseline allows mass production of subsystems and components. At the Department of Energy (DOE), Order 4700.1, Chapter III, defines functional, technical requirements, design requirements, and final baselines. One way to think of these requirements, though imprecise, is the association with 30, 60, and 90 percent complete. DOE alludes to Title I design, which is associated more appropriately with construction than with technical systems. In contrast, DOD technical baselines apply more to technical systems than to construction. The SSC deals heavily with both technical systems and construction, and demands a total integration of the two. This sometimes confuses the definition and application of traditional baselines.

The definitions of technical baselines are not as important as the fact that there is a baseline, that baseline is established by *approved* specifications, and those specifications allow one to "design to" or "build to". At the SSCL, the functional baseline is the Site Specific Conceptual Design Report (SCDR), SSCL-SR-1056, July 1990. The lower level baselines are developed to create the detail which supports the functional baseline. The baseline is not formed at one time; it is formed as the detailed specifications are approved.

Once established, the baseline is controlled through configuration or change management; i.e., no changes are made to the baseline unless they are justified, approved, and documented. Anyone can originate a change and prepare a change package. The change package includes the requested change (as compared to the baseline), reason for the change, and the technical, cost, and schedule impact of making or not making the change. The Configuration Management Plan depicts the levels of change authority for the project, from the DOE through the SSCL divisions. Changes may be approved by the chairman of the change control board at the appropriate level of authority. Change packages are discussed with the members of the board but there is no vote. The change control board is chaired by the Project Manager, who has sole authority to approve or disapprove changes at the project level. Configuration management then tracks the changes to ensure that they are implemented and that a trail of the machine configuration is documented and maintained.

Systems engineering is a tool for project management to establish and maintain baselines. At the SSCL, systems engineering personnel reside within the divisions to assist in establishing and tracking lower level baselines. This discussion continues with emphasis on the technical baseline and assumes that the reader understands that cost and schedule baselines are also established and tracked. The tools for systems engineering to use are discussed in the following paragraphs.

Plans

Several plans, e.g., Project Management, Systems Engineering Management, Configuration Management, Reliability, and Software Development, are key to the structure for establishing and controlling the technical baselines. Note that major subcontractors will be expected to establish and control baselines on their product(s) and to produce similar plans as stand-alone or as supplements to be compatible with the SSCL plans. Subcontractors will receive copies of the SSCL plans if they are requested to provide such plans for the items to be produced. The Project Management Plan, completed January 28, 1991, SSCL control number P40-000021, is an agreement between the SSCL and the DOE on how to conduct business. Section 7, Configuration Management, and Section 11, Systems Engineering, set the stage for conduct of systems engineering at the SSCL. For brevity, the various other plans will not be discussed in this paper.

Specifications

DOD specifications usually follow the format of Mil-Std-490A, a time-tested, recognized method of formatting specifications. However, at the SSCL, the Mil-Std-490A format is believed to be too cumbersome, at least at the highest levels of specifications. These include the system and segments at the machine level: Collider, Linear Accelerator, Low Energy Booster, Medium Energy Booster, and High Energy Booster (levels 1, 2, and 3A in Figure 1). This philosophy entails depiction of physics parameters in tables such that they are readily accessible, without unnecessary verbiage, to physicists and engineers at the SSCL. This format will be also be used at the next level, 3B (elements for each of the above referenced machines). All of these levels of specifications will stay within the SSCL and do not, per se, result in procurements. Thus the process of brief, tabular specifications will be used for all specifications that are perceived as for use only to inform those at the SSCL of the physics requirements. The top level procedures—reliability, safety, logistics, etc.—will be documented separately in SSCL standards or procedures. These procedures will be prepared by the engineering standards group, in conjunction with the engineering groups, and will be published for the entire SSCL. These procedures will also be available for subcontractors, if required.

The engineering requirements normally found in specifications are discussed in the lower level specifications, level 4 and below, on the specification tree (Figure 1). These lower level specifications will be used to procure items; consequently, they require more detail for engineering requirements and will be in a format that industry is accustomed to seeing. That does not mean that the format will always be Mil-Std-490A. We have already seen a combination of formats at level 4. The collider dipole magnet specification is in Mil-Std-490A format, and the subcontractors have also been requested to prepare the flow down specifications in Mil-Std-490A format. The specification for the refrigerator plant was in another format which set out the requirements but assumed a continuing dialogue between the contractor and the SSCL to complete the requirements and the subsequent design. This particular specification was criticized for lack of "procedures," specifically reliability and safety considerations. However, subcontractors may see this type of specification again, particularly a small or disadvantaged subcontractor or one who is known not to have participated in large government contracts. This does not diminish the need for the subcontractor to provide a quality product, but it does relieve the burden that is sometimes perceived with a Military Standard. The point is that specifications will be in any format acceptable to both the SSCL and the subcontractor as long as the result is what the SSCL needs. Content and ability to determine the requirements to produce a quality product are the drivers; format is not.

Interface Control Working Groups

The interface definitions will be developed by the Interface Control Working Group (ICWG) under Dr. Don Edwards, head of the Accelerator Design and Operations Division. It is proposed that the ICWG oversee development of interface requirements, assign the individuals to do the work, direct the development, monitor the progress, and approve the interface document. The ICWG would be made up of the division heads or deputies, or at a minimum, the technical group leaders. Each required interface document would be assigned to an interface control working panel (ICWP) made up of members assigned by the ICWG. That is why such high-level membership is required on the ICWG. In the absence of an approved process, the ICWP is operating under this influence but with individual charters. At the present time there are two interface control working groups: one between the cold magnets and the spool (corrector magnets), and one required for the Accelerator System String Test (ASST). Systems engineering is an integral part of both of these interface groups. Systems engineering personnel bring the engineering parties together to work out the interfaces, follow up on the action items, and document the interface requirements.

Configuration Management

Configuration Management is more than just change control as depicted in Figure 2. It actually deals with knowing what the SSCL has and with ensuring that everyone is working on the correct set of documentation. The baseline and configuration management meticulously protects the baseline. When changes are approved, configuration management ensures that the change is properly distributed through document control. Document control is responsible for releasing a document or a change. The document is available in a central library and anyone can obtain a current version. This guards against the possibility of someone obtaining a copy of an out-of-date version from another person. Configuration management keeps careful records so that the status of approved changes is known, whether they are pending or implemented, and how many units (if applicable) are affected.

Subcontractor/vendors are required to manage their configurations in a similar fashion so that the product delivered to the SSCL is well-documented and its makeup can be verified. As a final item, configuration management participates in reviews and audits. Reviews may result in a change to the baseline; audits verify that the product meets specifications.

Design Reviews

DOE Order 4700.1 lists the reviews that the SSCL must conduct to manage the project from the DOE perspective. These reviews are also required by the Project Management Plan. Figure 3 depicts the reviews on a time line relative to the design phase, along with the types of specifications that accompany the reviews. These reviews include the Systems Requirements Review, Preliminary Design Requirements Review, Preliminary Design Review, and Critical Design Review, sometimes called the Final Design Review.

The Systems Requirements Review, not shown on the chart, is complete. The results are contained in the SCDR and in the minutes of the June 1990 review.

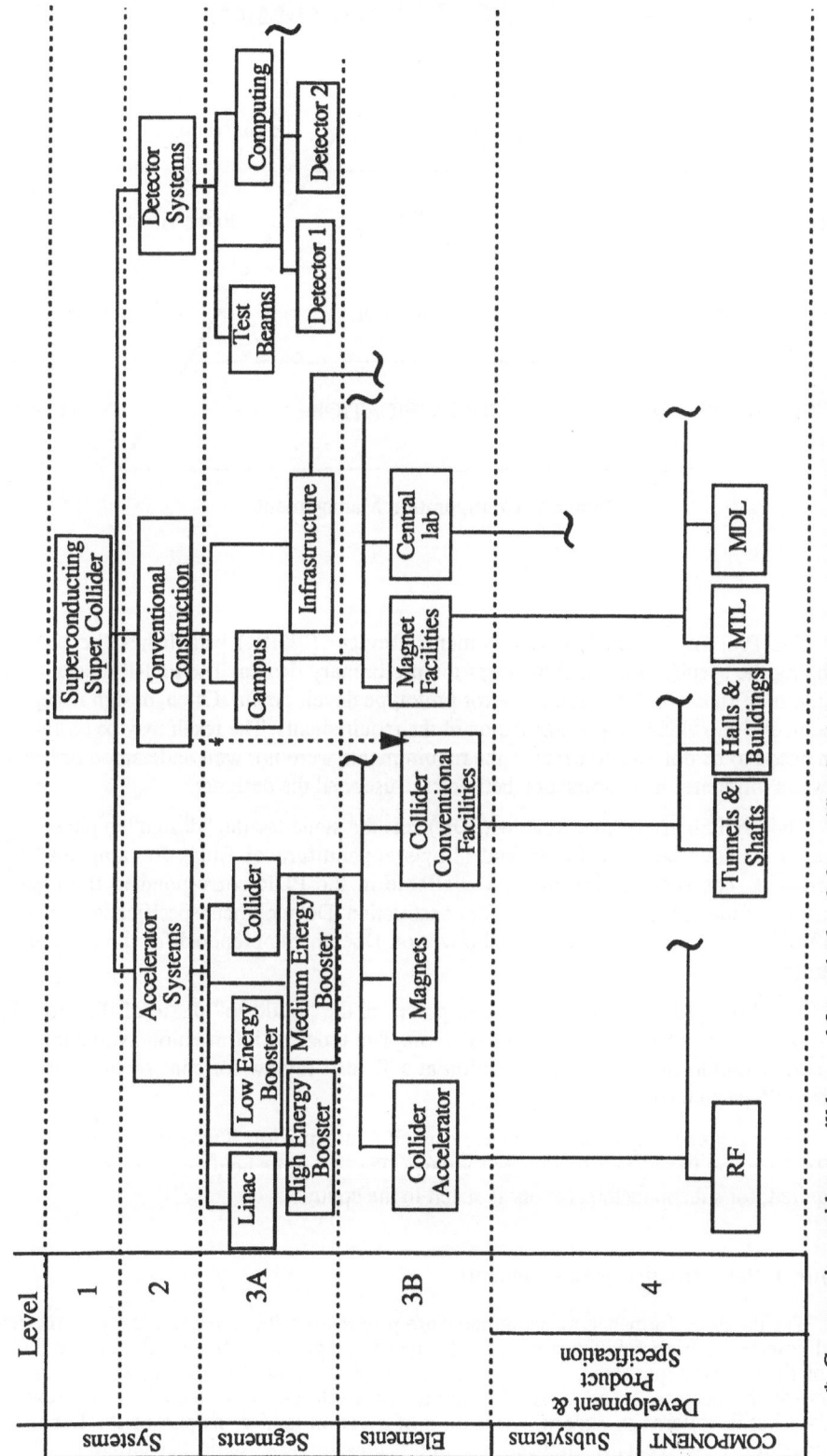

Figure 1. Example of Specification Tree.

* Construction guidance will be used from the level 2 specification but technical guidance will be from the technical system supported.

1093

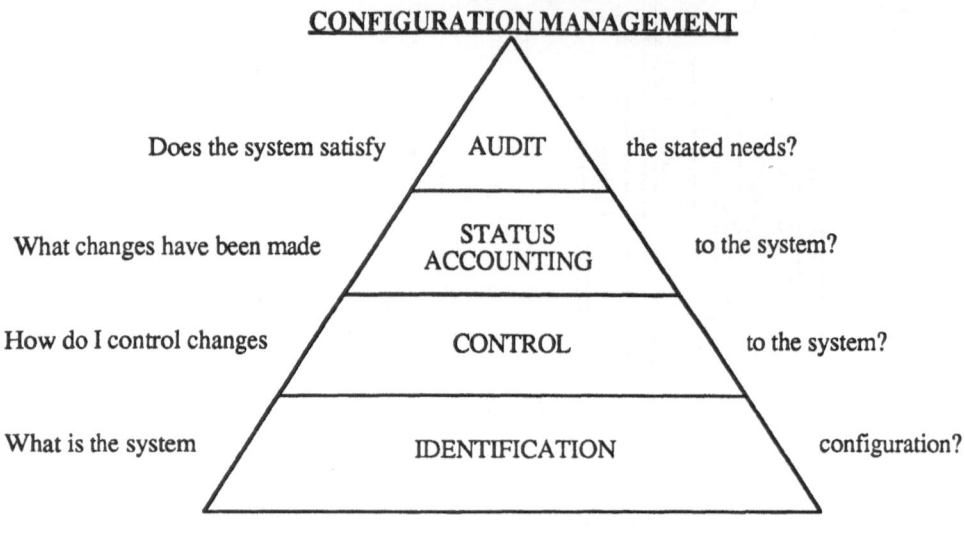

Figure 2. Configuration Management.

The Preliminary Design Requirements Review (PDRR) establishes the basis for developing the specifications and launches the preliminary design. The PDRR also approves expenditure of funds for long lead items for prototype development. Often, design is begun in the absence of knowledge or documentation of the requirements. The result may be costly if the design needs to be done again because the requirements were not well understood or because there was a difference in interpretation between the user and the designer.

The Preliminary Design Review (PDR) sets the stage for the "design" to phase of the project. It initiates detailed design and allows expenditure of funds for long lead item procurement. The types of information available in the PDR correspond to the types of information required for a Title I review for construction. Development specifications available in the PDR correspond to the Work Authorization Documents prepared for the construction contractor.

The Critical Design Review (CDR) precedes the "build to" phase of the project. It allows expenditure of funds to begin production. The types of information available for the CDR correspond to the types of information at a Title II review for construction, albeit, 60 percent or 90 percent complete.

Other reviews may be held to convey specific information. Two such reviews are Test Plan Reviews and Production Readiness Reviews. These will be held on an as-required basis. As required, for subcontractors, means as stated in the contract.

Technical Performance Measurements

Technical performance measurements are progress or status reports/analyses on items of high interest or high risk. Currently, the Project Manager has a list of 12 items, including each of the injector machines, which are reviewed once each month. The reviewers, task leaders for the project, spend about 30 minutes on each, covering technical and schedule aspects. Cost is covered in a monthly meeting of division leaders as part of the Laboratory management review. Systems engineering supports these reviews indirectly through studies or analysis.

Spec Level	Approval	Review
1	OSSC	ESAAB
2	SSCPO	OSSC
3A	SSCL Project Management	SSCPO
3B	ADOD	SSCL Project Management
4	Technical Divisions* MSD, ASD, CCD	ADOD

Cost Change Approval Authority**

WBS Level	SSCPO	SSCL/PM	ADOD	SSCL Div
0	Any Change in TPC***	None	None	None
1	Less Than $50M or More Than $10M	Less Than $10M or 5%	None	None
2	$5M or More	Less Than 10% or $5M	None	None
3	N/A	More Than 5%	Less Than 5%	None
4	N/A	Receive Report For More Than $1M	More Than 5%****	Less Than 5%
5	N/A	NA	Receive Report For More Than $1M	All****

Schedule Change Approval Authority

Level	Approval Authority
1	OSSC, based on contractual agreement
2	SSCPO, supports achievement of Level 0 Milestones
3A	SSCL Project Management, used to control the project
3B	ADOD
4	Division Head
5	Group Leader or Subcontractor

Figure 3. Reviews.

Trade Studies/Decision Papers

Trade studies will be an important part of the systems engineering process. However, the systems engineering currently emphasizes requirements, definitions, documentation, and specifications. Once specifications are completed and design begins, major emphasis will shift to interfaces and trade studies, which are already being conducted. Most of the studies are related to the machine parameters and are conducted primarily by physicists. These are called "optimization" studies, and they include the machine lattice and specific design parameters. Other major studies relate to the construction aspects, such as building sites and building functions.

Decision papers, signed by the SSCL General Manager, Project Manager, Technical Director, or the Deputy Project Manager, are processed through document control, where they are given a control number and distributed as a released document. In this way decisions can be tracked and people desiring information on a particular decision know where to find it. An example of a decision paper residing in document control is the definition of system availability.

CONCLUSIONS

Clearly, the application of systems engineering is more difficult at the SSCL than in a DOD environment, where the concept of systems engineering is widely accepted. It is not clear, however, that systems engineering is well understood, even at the DOD. If we accept the definition of systems engineering stated earlier in this paper, then there is no clear demarcation of systems engineering principles except for discipline in the process of designing, building, and operating a system. At the SSCL, systems engineers must be creative in applying systems engineering. The environment will not accept strict discipline nor a "cookbook" approach. One recent example comes to mind: An engineering group leader was concerned about controlling engineering drawings, 30 of which were located at the SSCL and the rest at a national laboratory. The group leader was unsure about changes to these drawings or whether he could control the changes. He approached the systems engineering group for help, but made it clear that he did not want configuration management. The systems engineering leader for that division agreed to help with "drawing control," thus avoiding the term configuration control. Maybe the lesson here is that all of the standard configuration management functions are not needed; in this case, only drawing control was called for. Still, the lasting impression is that the systems engineering function, specifically configuration management in this example, is not understood.

Without much elaboration, the lessons learned can be summarized as follows:

1. Cultural differences are real, and they represent a real test. Subcontractors dealing with the SSCL should understand the cultural differences and work very hard on definitions of terms. A misunderstanding of terms could lead to dissatisfaction by both parties.

2. Reaching the proper people is still difficult and time-consuming. A recent layoff at a major aerospace company provided a potential source of people. By job title and resume, many satisfied the criteria for systems engineers and configuration management. However, interviews revealed a very narrow application of systems engineering, and not one had the required ability to interface with various disciplines and to prepare specifications and interface control documents.

3. The new definition of systems engineering now includes attitude. A systems engineer must be positive and have a direction in mind. His education and experience provide the tools to accomplish the mission. His attitude provides the motivation and courage to persevere toward the goal.

4. The most important lesson learned is that systems engineering can be drastically tailored and still be effective. It can be used, as opposed to building plans and specifications that sit on the shelf. A statement attributed to Benjamin Franklin puts it in perspective: "I wrote you a long letter because I didn't have time to write a short one." When members of the systems engineering staff at the SSCL complete their jobs, they will be true systems engineers because they had to be creative in the application of systems engineering.

WHAT SUBCONTRACTORS CAN EXPECT

In summary, subcontractors can expect the following:

- A different culture from DOD contracting and a different understanding of terms.

- Conduct of plans and project management similar to that of DOD contracts due to the on-site presence and influence of DOE.

- Specification formats of Mil-Std-490A for major procurements and other formats for lesser procurements.

A PRACTICAL SYSTEMS ENGINEERING PROCESS FOR INTEGRATING SSC COLLIDER RING COMPONENTS

Jonathon Y. Simmons

Magnet Division
Superconducting Super Collider Laboratory*
2550 Beckleymeade Avenue
Dallas, Texas 75237

Abstract: Successful development and integration of the Superconducting Super Collider (SSC) collider rings requires coordination of unique design implementations formulated by the SSC Laboratory organizations, and later by individual superconducting magnet and spool piece manufacturers. Application of the Interface Control Working Group/Interface Control Document process provides a systematic review and concurrence of these designs, while minimizing constraints on the design process by avoiding excessively detailed specification. Design details necessary for integration of the separate components are documented and reviewed, with only inconsistencies and exceptions being addressed by management direction and/or specification revision.

INTRODUCTION

The ultimate goal of the interface control process is to coordinate the specific design implementations of development organizations so that they are compatible where they interface with systems or features developed by a different organization. The process should provide maximum flexibility of design by dictating the absolute minimum of design details. Where design details are defined, the process must allow changes with a minimum of effort, while ensuring that the changes are reviewed and accepted by all affected parties. In cases where such acceptance is not obtained, the process must provide a review procedure to reconcile the discrepancy. Effort should also be made to keep other interested parties informed of the current configuration as much as practical.

One specific feature of the Superconducting Super Collider (SSC) interface control process represents a unique problem. The superconducting magnet cross section is very

---

*Operated by the Universities Research Association, Inc., for the U.S. Department of Energy under Contract No. DE-AC02-89ER40486.

constrained: little variation of this design is permissible without impacting critical system performance parameters. This cross section in turn drives many interface details which impact adjacent components. Many of these components, including the magnets and spool pieces, have aggressive requirements which call for substantial technical ingenuity. Such ingenuity is best served by a systems engineering environment that places a minimum number of constraints on the designers.

INTERFACE CONTROL

The Interface Control Working Group (ICWG) and its associated Interface Control Working Panels (ICWPs) have primary responsibility for interface control. These groups are used respectively to organize the interface control process and to reconcile discrepancies, and to bring the appropriate personnel together to prepare and review Interface Control Documents (ICDs). The SSC utilizes two classes of ICDs: system-level ICDs, which detail the interfaces between components, and ICDs which communicate the specific interface details to a supplier charged with developing a component for the SSC.

The Standard Arc Half-Cell ICD, associated with the Collider Superconducting Magnet Systems Specification in Figure 1, defines interfaces which are required to be common to, or between, the standard collider string components. The collider string components are defined as those magnets, corrector elements, and spool pieces which are serially connected to complete the arcs of the collider, and which may be located in other elements in the collider lattice. The interfaces which are controlled by this document include the magnet to magnet, magnet to spool, corrector element to magnet, string component to tunnel (including stands and exterior alignment features), and the quench protection system (including voltage taps and dynamic electrical loads).

The task of controlling the interfaces to more than 12,000 collider string components is manageable by maximizing commonality. The collider is predominantly composed of recurring strings of eight components known as half-cells (Figure 1). In turn, the interfaces between components within the half-cell have maximum commonality. Each of the interfaces between components in Figure 1 (i.e., dipole magnet to spool, or dipole magnet to quadrupole magnet) is identical or very similar.

The half-cell is the smallest functional system within the collider string. The superconducting bus power, insulating vacuum, thermal insulation, and quench protection systems function on the half-cell level. For example, the insulating vacuum and thermal insulation systems are continuous through the half-cell and behave differently if tested as subsystems of a single magnet or spool piece. Satisfaction of the requirements on these systems must be evaluated at the half-cell level, using the performance of the integrated components.

An ICD is used to communicate the interface details of components procured as development programs. The ICD associated with the Collider Dipole Magnet in Figure 2 is a prime example of such an ICD. These ICDs for the Arc Half-Cell components are essentially a subset of the Standard Arc Half-Cell ICD tailored to the particular requirements of the components. This limits the demands for participation by the supplier to those issues which are of direct interest, and provides a well-defined communication path. The collider spool pieces are to be a "build-to-print" type procurement. Therefore, there is no ICD required between the SSCL and the vendor(s).

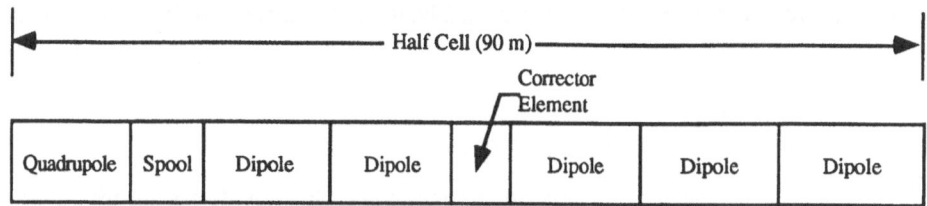

Figure 1. Half-Cell Schematic.

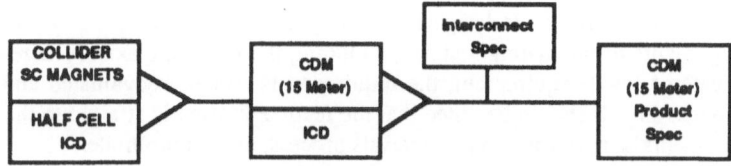

Figure 2. Specification Tree Excerpt.

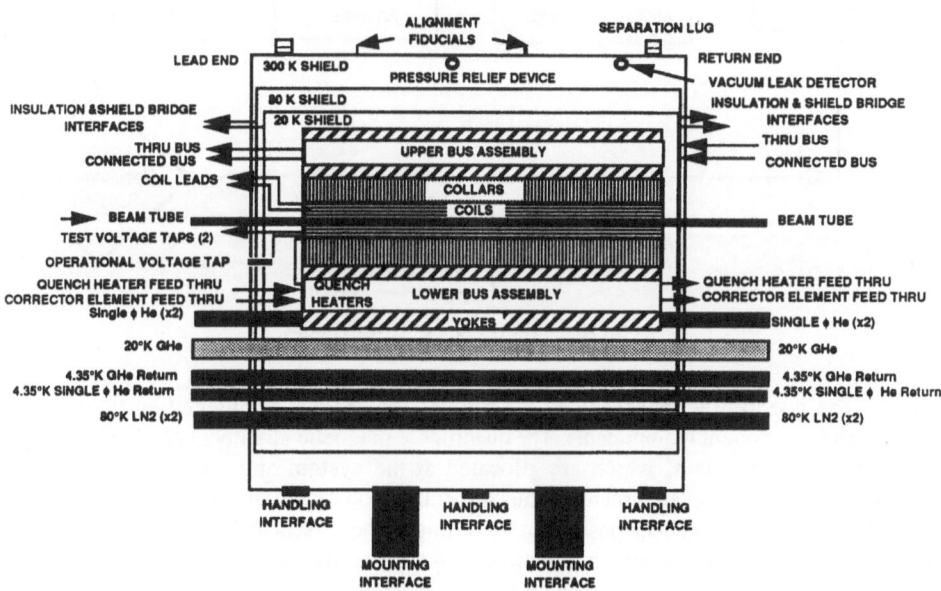

Figure 3. Collider Magnet Interfaces.

The ICDs at the half-cell and component level must address interfaces by fully describing their form, fit, and function. Figure 3 schematically represents these interfaces on a magnet "control volume." These interfaces all require that the physical geometry, including tolerances and misalignments, be described, as well as any additional stresses or conditions applied at the interface. Mechanical interfaces also require that the mechanical loads be defined. Cryogenic interfaces require that mechanical loads, flow rates, temperatures, and pressures be defined. The superconducting power interfaces require the definition of mechanical loads, currents, and any voltages that occur. Vacuum levels must be defined, as must any forces they exert at the interface. Finally, the beam image current and the associated longitudinal and transverse impedance are defined for the interface, including any geometrical effects due to misalignment.

ICDS AS A DESIGN VERIFICATION

The Half-Cell ICD also serves to partially verify that the designs of the individual components satisfy system-level requirements when the components are integrated as a system. Certain requirements such as the magnetic axis alignment of sequential magnets can be determined at the half-cell level, but not at the individual magnet level. Figure 4 represents this process. After the magnet is designed to its requirements, the Half-Cell ICD tabulates the relative alignments of the mating features, including the magnetic axis. If there is a conflict with the systems level requirement, the requirements can be reevaluated and the design iterated if necessary. It should be noted that the feedback often is straight to the component level requirements or to the design and analysis process, as is appropriate.

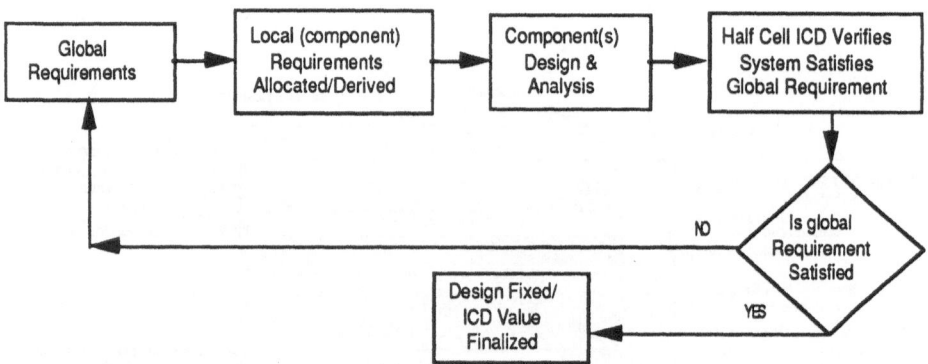

Figure 4. Interface Control Process.

The quadrupole magnet is an example of the process of integrating the ICD with system and component requirements. The quadrupole magnetic alignment and field quality are interrelated requirements which are allocated at the system and component levels. The magnetic field error and alignment tolerance specification on the quadrupole magnets applies to the magnet after it has been positioned according to the offsets and rotations defined during measurement of the magnetic axis relative to the fiducials on the magnet. The tolerances associated with positioning the magnet in the tunnel accumulate with these single magnet tolerances. The following expression gives the magnetic field in this magnet as measured with respect to the external fiducials, with the appropriate offsets and rotations compensated for:

$$iB_x + B_y = 10^{-4} \, g \, R_r \sum_{n=0}^{\infty} (ia_n + b_n)\left(\frac{x + iy}{R_r}\right)^n$$

where R_r is the reference radius, defined as 0.01 m, and g is the magnet gradient in Teslas per meter. The skew and normal values of the multipole coefficients (field quality) are designated by a_n and b_n, respectively. The specification on the random dipole field of 200 units is equivalent to a magnetic centering specification of 0.2 mm, and the specification on the random skew quadrupole of 5 units is equivalent to a rotational error specification of 0.5 milliradian. Thus, the positional accuracy of aligning the quadrupole in the tunnel changes the apparent field quality (field harmonics as experienced by the particle beam) of the installed magnet.

The quadrupole alignment is determined by integrating the tunnel survey, locational accuracy of the magnet, internal magnet alignment, and physical location of adjacent magnets and any mechanical loads they transmit across the interface. Individual magnet performance alone cannot be used to determine whether the specified field quality is adequate, the alignment is satisfactory, or the corrector elements are of sufficient strength. The Half-Cell ICD integrates the alignment process to determine the actual quadrupole locational tolerance. These data are then fed back into the allocation process to verify the system requirements (see Figure 4).

CRITICAL DESIGN DRIVERS

The concept that only the design implementation at the interface is controlled is vital to achieving interface control without overly constraining design. The common temptation is to describe many details which are known to work but are not required to interface the components. In meeting the ICD, the designer is then locked into many design details, which tend to cascade through his system. A seemingly innocuous detail may impact many internal component features which are not interfaces. However, this approach requires that the interface be completely defined in the ICD, including stressors such as mechanical loads or fluid pressures as well as the details of fit and location.

The Collider Dipole Magnet (Figure 5) is the critical design driver for the half-cell. This magnet is the most common component in the collider lattice (approximately 8000 units) and is designed to very severe engineering constraints which take available technologies to their limits. Design margins for critical dipole magnetic parameters are on the order 8 to 12%, rather than the traditional 20 to 200% or more allocated for minimum risk engineering development. The extremely low allowable heat leak from the tunnel to the coldmass and parameters such as the magnetic field alignment, strength, and quality are unique and severe constraints on these magnets. Other components without these unique constraints, such as the spool pieces, could take a radically different and more appropriate cross section than the magnets. However the need to interface with the magnets and provide continuity of the beam tube, superconducting buses, thermal shield, and vacuums restricts the non-magnet components to a magnet-like cross section. Without these constraints, the design tradeoffs on these components would emphasize other factors.

The location of the clockwise and counterclockwise superconducting buses is a prime example of this approach. These buses must be installed within the coldmass, yet they are joined at the interconnect regions between adjacent magnets and spools to form a single system. Magnetic field quality requires that they be kept in an exact location in the coldmass cross section within the magnets, with no allowance for change in this position should interfacing components require it. Yet the spool will probably require different bus location(s) in the coldmass cross section. Also these buses must be allowed to thermally contract at a different rate than the magnets and spools in the string, yet be rigidly attached to each component in the string. The current solution requires that the bus be rigidly attached at one known location in each component and have thermal contraction compensation for the distance between that location and the next one. The Half-Cell ICD must furnish the location and amount of thermal contraction compensation provided between the rigid ties of the bus to the component, and the details of the superconducting joints. However, the methods of tying the bus to the magnet and of compensating for the differential thermal contractions are left to the design organizations.

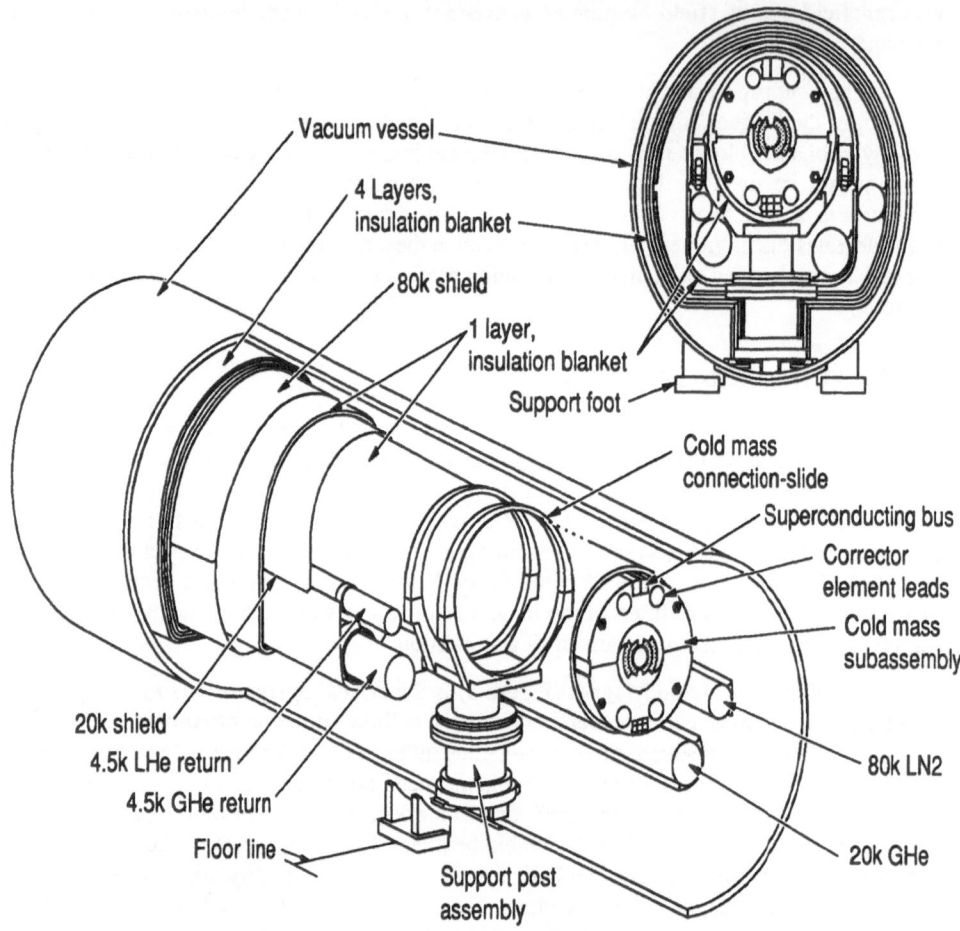

Figure 5. Dipole End Cutaway.

The dominant local alignment problem is the potentially large misalignment of interfaces at the interconnect regions. This situation is a result of the accumulation of tolerances both within the magnets or spools themselves, and in series with the alignment tolerances of the component installation in the collider tunnel. Reference fiducials on the magnet exterior are used to accurately position the component in the tunnel. The magnetic axis (coldmass centroid) is carefully surveyed relative to these same fiducials, and the beam tube is located relative to the magnetic axis. All other internal interfaces are then located relative to the beam tube. This long tolerance chain is dictated by the need to have fiducials representing the magnetic axis accessible when the components are installed, and by the mechanical design of the magnets necessary to minimize the heat leak from the 4° K coldmass to the exterior of the magnet. Figure 6 schematically represents this tolerance accumulation, as well as the tolerance accumulation calculations for the beam tube misalignment.

The beam tube rf joint is a critical alignment driver and an excellent example of the interface control process. The impedance of the beam tube system is a measure of the work required to attach an imaging current from the proton beam to the beam tube (transverse impedance) and the work required to circulate the current down the length of the beam tube (longitudinal impedance). This imaging current is effectively the electrical reciprocal of the proton beam, with the work required to overcome the impedance being lost energy from the beam. The beam tube rf joint longitudinal impedance (Z/n) is allowed a maximum of $10.39 \times 10^{-6}\Omega$, and the transverse impedance is allowed a maximum of $888.1 \times 10^{-6}\Omega$.

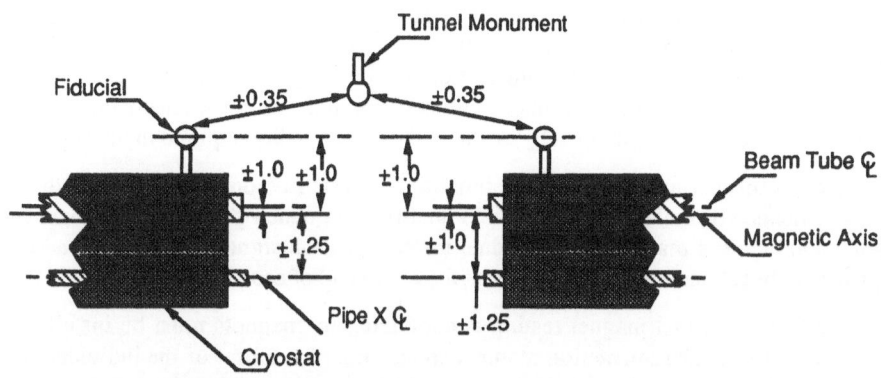

Tolerance between opposing Beam Tubes = 1.0 + 1.414 + 0.35 + 0.35 + 1.0 + 1.414
$$= 4.8 \text{ mm } (0.2 \text{ in.})$$

RMS Tolerance between opposing Beam Tubes $X = \sqrt{1.0^2 + 1.414^2 + 0.35^2 + 0.35^2 + 1.0^2 + 1.414^2}$
$$= 2.4 \text{ mm } (0.1 \text{ in.})$$

Figure 6. Interconnect Alignment Tolerance Schematic.

These numbers are derived from a budget of 0.117Ω longitudinal impedance and 10.0Ω transverse impedance for the total collider, distributed over 11,260 beam tube rf joints. The beam tube is lined with very pure copper to meet these stringent requirements. However, the beam tubes of sequential components must be joined at the interconnects, and these electrical joints are very sensitive to axial misalignment. Occurring at an interface between components, the actual axial offset situation is described in the Half-Cell ICD. The interconnect specification defines the offset requirement to be satisfied and the magnet ICDs define the attachment of the joint to the magnet assemblies. As the analysis in Figure 6 illustrates, the misalignment in these joints is somewhere between 2.4 and 4.8 mm, depending on analysis methodology. The inability of existing technology rf joints to accommodate these misalignments defined in the ICD is driving a redefinition of the alignment requirements at the installation level and within the magnets themselves.

ASST CONSIDERATIONS

The Accelerator Systems String Test (ASST) is a demonstration/validation exercise of industrially assembled dipole magnets in a half-cell type configuration. While not exactly duplicating the collider system, most collider type interfaces are present. This demonstration is configured to provide data on many operational parameters which can only be measured at a system (half-cell) level. Therefore, there is much additional instrumentation which would not be required in a functional accelerator. Figure 7 presents the systems to be integrated and their interrelationships.

The magnet to spool (and magnet to magnet) interface is a discrete interface, not interrelated to the other systems, except by providing continuity from the other ASST systems through the spool pieces to the magnets. Therefore, the magnet to spool interface is treated as a separate interface with its own ICD. A system-level ICD encompasses the other interfaces. The spool being a common element to both of these ICDs ensures integration of the magnets into the full system. The ASST ICD addresses a limited set of interfaces dedicated to operating a string of about a half-dozen components, but there are time constraints. The interface control process is at best concurrent with design, and is quite often a check of existing design. Magnetic alignment is not critical since there will be no particle beam in the ASST, minimizing alignment requirements. The use of an above-ground facility minimizes space (packaging) constraints, allowing less optimized designs than would be suitable for use in the collider. Additionally, the limited number of components allows many interfacing parts to be custom fabricated to fit , thereby avoiding the commonality required in the collider.

Each magnet and spool will be individually tested before installation in the ASST. Measurements will include magnetic field quality and quench performance. These magnet signals total approximately 147 individual instrumentation signal lines at the lead end and approximately 138 individual signal lines at the return end of each magnet.

After individual magnet testing is complete, the magnets must be installed in the ASST. The ASST instrumentation requires connection of a subset of the individual magnet instrumentation to a bus system which runs through the string. This instrumentation consists primarily of voltage taps and temperature and pressure transducers which are required to be mounted within the coldmass and routed to buses which are also within the coldmass. These buses are transferred from the coldmasses to the tunnel instrumentation systems at the spool pieces. The coldmass must be opened to connect the buses and terminate the unused instrumentation. Additional instrumentation, including accelerometers and pressure and temperature transducers, is installed in the string. This instrumentation is installed in the interconnect regions between magnets, with the signal lines routed within the cryostats but outside the coldmass, so there is little magnet impact. However, the spools will also be required to transfer these lines from the cryostat to the tunnel systems. These ASST signals total approximately 32 individual instrumentation signal lines per magnet and approximately 70 bus channels through the string.

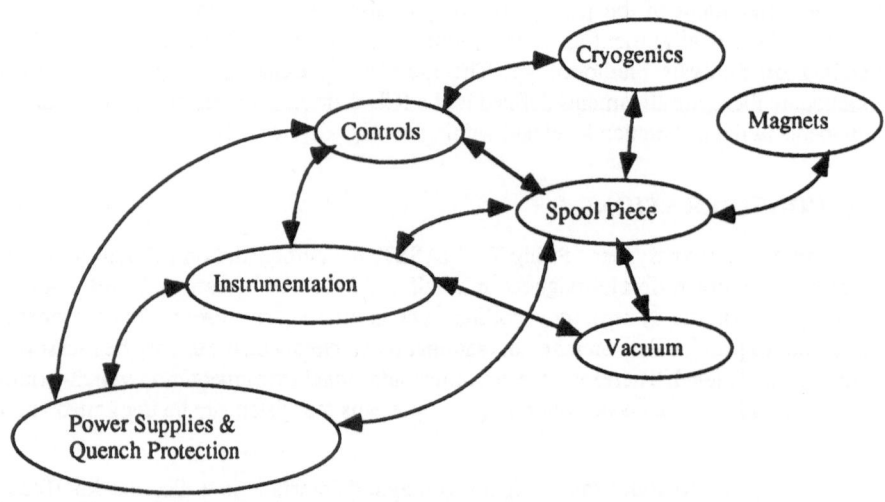

Figure 7. ASST Functional System Interrelationships.

CONFIGURATION MANAGEMENT

Specific interface control documents are created and maintained by ICWPs staffed by individuals representing the technical disciplines and organizational interests charged with implementing the interfaces. Overall authority for the Collider String ICWP resides with the Accelerator Design and Operations Division. Systems Engineering chairs this ICWP, with representation on the working panel designated by each division head. Membership on this panel changes, however, as necessary to address the different types of interfaces.

The magnet-specific ICDs are created and maintained by the Magnet Systems Division (MSD). This division is responsible for ensuring that these ICDs conform to the Half-Cell ICD and/or to other appropriate interface systems. If deemed necessary by the MSD, manufacturers under contract to develop collider string items may also represent their specific design interests on the Collider String ICWP.

The Configuration Management process is an integral part of interface control. Configuration management ensures that all impacted parties concur with the proposed changes, or that the issues are elevated to the appropriate authority for resolution. The ICDs are released as their technical content is required as a baseline for design. As the technical content is revised or amplified to impact this baseline, the ICD is revised, with an Engineering Change Request initiating the Configuration Management process, thus ensuring that the impacted parties concur with the proposed ICD revision. Document Control maintains the original released version and all revised versions of the ICD for access by all affected or interested parties.

Ultimate authority for the Half-Cell ICD and associated drawings rests with the ICWG under the Director of the Accelerator Design and Operations Division (responsible for system level requirements and integration and operation of the SSC). Acting as chair of the Configuration Control Board, the Project Manager approves this ICD, making it part of the Technical Baseline.

Authority for the magnet-specific ICDs rests with the Associate Director in charge of the Magnet Systems Division. The Configuration Management process ensures that these documents are reviewed by the appropriate MSD functions, including Quality Assurance, Production, Systems Engineering, Engineering, and Test. Any conflicts between MSD functions are ultimately resolved by the Associate Director (or designee). Conflicts between MSD and the vendor(s) are resolved by the applicable MSD Product Manager, or by the Associate Director (or designee). If the Half-Cell ICD is impacted, the conflict must be elevated to the ICWG, with the Project Configuration Control Board responsible for the final decision.

CONCLUSION

The use of a system level Interface Control Document (ICD) for interface control at the half-cell level allows maximum flexibility of design, while ensuring that the rigid design constraints of the dipole and quadrupole magnets are satisfied. This ICD also has the benefit of providing a system-level analysis and check of parameters which are not observable on a single magnet.

The Accelerator System String Test (ASST) is a demonstration/validation exercise requiring an ASST half-cell level ICD, as well as a system-level ICD to interface the string of magnets and spools to the control, cryogenic, vacuum, instrumentation, and power supply and quench protection systems.

Interface Control Working Groups and Panels are used to respectively organize the interface control process and to reconcile discrepancies, and to bring the appropriate personnel together to prepare and review ICDs. Authority for the Half-Cell and ASST ICDs resides with the SSCL Technical Director. Component-level ICDs are controlled by the appropriate division Technical Director. Configuration management is used to ensure review and approval of the ICDs by affected parties, with periodic re-releases as the technical content changes. Document Control ensures that current release version of the ICDs are available for affected and interested parties.

SSC ACCELERATOR AVAILABILITY ALLOCATION

Kirby T. Dixon and Jim Franciscovich

Superconducting Super Collider Laboratory*
2550 Beckleymeade Avenue
Dallas, Texas 75237

Abstract: Superconducting Super Collider (SSC) operational availability is an area of major concern, judged by the Central Design Group to present such risk that use of modern engineering tools would be essential to program success. Experience has shown that as accelerator beam availability falls below about 80%, efficiency of physics experiments degrades rapidly due to inability to maintain adequate coincident accelerator and detector operation. For this reason, the SSC availability goal has been set at 80%, even though the Fermi National Accelerator Laboratory accelerator, with a fraction of the SSC's complexity, has only recently approached that level. This paper describes the allocation of the top-level goal to part-level reliability and maintainability requirements, and it gives the results of parameter sensitivity studies designed to help identify the best approach to achieve the needed system availability within funding and schedule constraints.

OVERVIEW

A systematic approach to achievement of 80% availability of the Superconducting Super Collider (SSC) accelerator requires equitable apportionment of the allowable downtime among the accelerator subsystems, followed by subdivision of the subsystem allowances to the level at which design alternatives can be selected to arrive at a design choice satisfying the allocation. The approach used to allocate the system-level availability requirement to design-level reliability and maintainability requirements is summarized.

Using current requirements and available information, preliminary availability allocation to the segment level has been completed. The collider segment allocation has been allocated to the component level, the Magnet System has been allocated to the magnet and interconnect level, and the interconnect requirements have been allocated to the part level. The current allocation allows up to six Magnet System failure intervals per year, satisfying a perceived need to accommodate three collider dipole magnet failures per year, and requires no

---

*Operated by the Universities Research Association, Inc., for the U.S. Department of Energy under Contract No. DE-AC02-89ER40486.

improvement of the Injector Complex availability over Fermi National Accelerator Laboratory (FNAL) experience. When fewer than 3.5 Magnet System failures per year are allocated, the number of short-downtime failures allowed causes excessive time loss in ring fill and tune operations. The remainder of the segments need further analysis to understand the degree of redundancy and any special support provisions required to meet the allocations.

At this stage of development, many assumptions are necessary for completion of the analysis. As the program progresses through detailed design, development, testing, commissioning, and early operations, the assumptions will be refined and then replaced with test/operational data to maintain visibility of availability potential, actual achievement, growth progress, and problem areas needing management attention.

MATHEMATICS, MISSION PROFILE AND BEAM CYCLE

SSC reliability and maintainability requirements are derived from the availability goal stated in the 1986 SSC Conceptual Design Report (CDR):

$$\text{Availability} = \frac{\text{(Scheduled uptime) - (Unscheduled downtime)}}{\text{(Scheduled up-time)}} \geq 0.8.$$

Substitution of

$$\text{Unscheduled downtime} = \frac{\text{Scheduled up-time}}{\text{MTBF}} * \text{MDT}$$

into the Availability expression allows calculation of the mean time between failures (MTBF) requirement, provided the mean downtime (MDT) per failure is known from experience or can be estimated from the steps required. When unscheduled downtime is caused by multiple (i through n) subsystems or items,

$$\text{Unscheduled downtime}_{\text{total}} = \sum_{i=1}^{n} \text{Unscheduled downtime}_i$$

and

$$\text{Availability} = 1 - \sum_{i=1}^{n} (1\text{-Availability})_i$$

Accelerator usage is assumed in the Mission Profile[1] to be continuous, 24 hours per day, seven days per week, nine months per year, with an average of one day per week allowed for scheduled maintenance (to be scheduled only on an as-needed basis) followed by three months of scheduled downtime in which maintenance may be accomplished without affecting system availability. Figure 1 summarizes the average annual Mission Profile time allocations, showing 939 hours allowed for scheduled maintenance and 1126 hours allowed for unscheduled maintenance, leaving a total of 4505 hours available for beam operations. Up to 4 hours per day (751 hours/year) are allowed for operations associated with loading and accelerating the beams, leaving 3754 hours of stable beam time.

Figure 2 shows the assumed standard operating cycle, including the time required for collider ring fill [operating time required of the Injector Complex [LINAC, Low Energy Booster (LEB), Medium Energy Booster (MEB), and High Energy Booster (HEB)], the time required for accelerating/decelerating the beams between 2 and 20 TEV, and the number of cycles required/possible during available operating time. The time requirements are consistent with the Site-Specific Conceptual Design Report dated July 1990. No time allocation was included for special beam tube vacuum controls or other factors which may be necessary for operation beyond the baseline requirements.

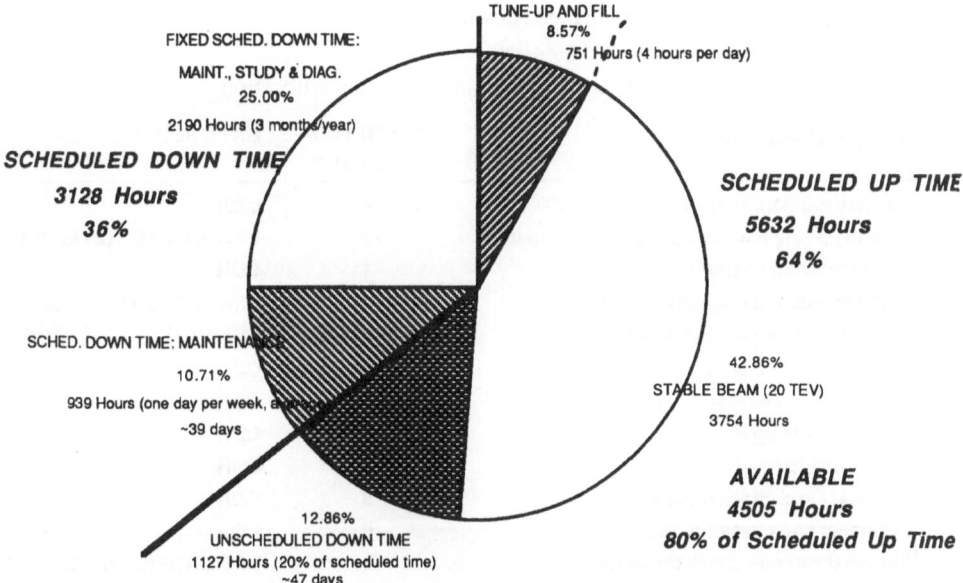

Figure 1. Standard SSC Time Allocation.

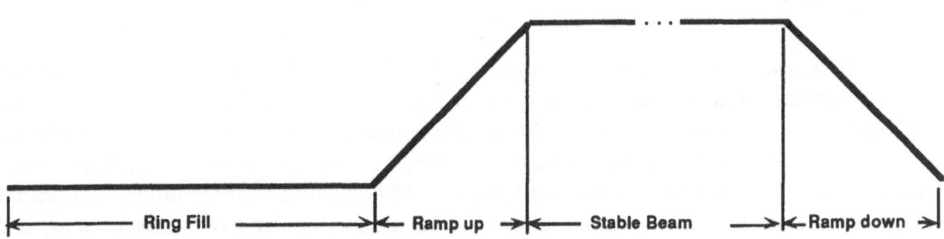

Ring Fill = 64 minutes. All SSC Accelerator subsystems must operate.

Ramp up = 25 minutes. Failures in Injector Complex do not affect beams.

Stable Beam = 20 hours, nominal. Failures in Injector Complex do not affect beams.

Ramp down = 25 minutes. Failures in Injector Complex and other subsystems have no effect.

Figure 2. Standard Accelerator Beam Cycle.

ANALYSIS APPROACH

The driving factor in SSC accelerator availability is expected to be the superconducting magnet rings (two Collider and one HEB). For purposes of the allocation, the Magnet System was defined to include all items in the superconducting magnet rings whose maintenance will require warm-up and cool-down of portions of the rings (includes all superconducting magnets, spools, and their interconnection regions). Downtime per failure of these items is estimated to be on the order of 10.5 days, as summarized in Table 1, and there is no feasible possibility of design redundancy. About eight failures (0.032%) per year would consume all the available scheduled and unscheduled downtime.

Table 1. SC Magnet Maintainability Allocation.

MAINTENANCE TASK	DOWN TIME (Hours)	DATA SOURCE
Isolate problem to the faulty magnet	16	CDR
Warm up ring section containing bad magnet	72	SYSTEM SPEC (24 in CDR)
Spoil relevant vacuum	1	CDR
Disconnect bad magnet mechanically	4	FNAL REPORT (4 total
Disconnect bad magnet electrically	4	ESTIMATE in CDR)
Remove bad magnet	2	CDR
Prepare for new magnet, as required	4	CDR
Place new magnet	2	CDR
Align magnet	4	CDR
Connect and check electrically	6	CDR
Connect and check mechanically	6	CDR
Pump down and check for leaks	48	ESTIMATE (36 in CDR)
Hi-pot test	2	CDR
Cooldown	72	SYSTEM SPEC (48 in CDR)
Pre-start checkout	8	CDR
TOTAL Hours	251	CDR + 88 Hours
TOTAL Days	10.5	

It is assumed that most (807 hours of the 939 hours) of the one-day-per-week-as-needed scheduled maintenance will be used for Magnet System replacements. The remaining 132 hours of scheduled maintenance allowance are reserved for possible priority needs such as restoration of critical protection system redundancies. Assuming a Magnet System failure is detected during scheduled operating time, the first 16 hours of downtime time will be used to diagnose and isolate the faulty item and to verify the need to replace it. Once the need has been determined, about ten days of downtime will be scheduled to complete the maintenance. Should the scheduled downtime total reach the 939 hour limit, *all* subsequent downtime for maintenance would be included as unscheduled downtime.

The second area of concern is the Injector Complex, which includes the LINAC, the LEB, the MEB, and the HEB. These subsystems are assumed equivalent, at least from an availability perspective, to the FNAL LINAC, Booster, Main Ring, and Tevatron, respectively. These subsystems are also non-redundant and historically have demonstrated availability lower than the 80% required of the entire SSC Accelerator. Fortunately, these subsystems are required to operate during less than 5% of the standard Accelerator operating cycle. Furthermore, failures during Collider operations generally do not degrade Collider operation, most failures are repaired in less than one hour, and any needed scheduled maintenance may be accomplished during stable Collider operation.

Table 2 shows FNAL statistics for the most recent collider mode operating period. The availability assigned to each Injector Complex subsystem was obtained by adding the unavailabilities associated with the FNAL Utilities, Controls and Miscellaneous "subsystems" to the four primary subsystems in proportion to the unavailability of the primary subsystem.

Table 2. Fermilab Downtime/Availability Summary SSC Availability Comparison.

	# ACTIONS	TOTAL HOURS	MEAN DOWN TIME	AVAIL **	SSC AVAIL	SSC SUBSYSTEM	
LINAC	581	110	0.19	0.987	*0.984*	LINAC *	
BOOSTER	322	66	0.21	0.992	*0.990*	LEB *	
MAIN RING	454	448	0.99	0.946	*0.935*	MEB *	
UTILITIES	47	118	2.52	0.986	0.991	COLL. UTIL.	
CONTROLS	158	75	0.47	0.991	*0.995*	GAS	
MISC	45	64	1.42	0.992	0.995	COLL. SAF./INTLK.	
TEV INJECTOR TOTAL	1607	881	0.55	0.894			
TEVATRON TOTAL	517	1121	2.17	0.865	*0.891*	HEB *	
CORRECTORS	46	62	1.35	0.993	0.996	CORRECTOR P.S.	
CRYOGENICS	47	75	1.59	0.989	0.994	CRYOGENICS ***	
INJECTION	10	16	1.61	0.998	0.999	INJECTION/ABORT	
MAGNETS	8	333	41.61	0.960	0.876	MAGNET SYSTEM	
MISCELLANEOUS	84	99	1.18	0.988	0.993	MISCELLANEOUS	
POWER SUPPLIES	74	134	1.81	0.984	0.990	RING P.S. & REG.	
QPM	43	60	1.40	0.993	0.996	QUENCH PROT. SYS.	
QUENCH	113	195	1.73	0.976	0.996	MAG. SYS. QUENCHES	
RF	47	61	1.31	0.993	0.996	RF SYSTEM	
VACUUM	9	30	3.31	0.996	0.998	VACUUM SYSTEM	
CEN HE LIQ	4	13	3.19	0.998			
CONTROLS	32	43	1.35	0.995			
TEV/INJ TOTAL	2124	2002	0.94	0.759	0.986	INJECTOR COMPLEX	
COLLIDER TOTAL					0.816	COLLIDER	
SSC TOTAL					0.800		

*   Includes part of Utilities, Controls, and Misc. in proportion to subsystem unavailability
**   1988/89 Collider run scheduled operating time =   8304.00   hours.
***   Includes Central Helium Liquifier

The remainder of the SSC Accelerator consists of the Collider ring support elements and the Global Accelerator System. Where cost effective, fault tolerance may be designed into these subsystems to achieve high availability. Because these subsystems are analogous to Tevatron support subsystems, availability data from FNAL Tevatron operations were used as a basis for the downtime per failure (one hour additional downtime per failure was assumed necessary and sufficient to account for the larger Collider geographic scale), and as the basis for the relative availability of the Collider subsystems other than the Magnet System. A summary of the pertinent data compared with the SSC Accelerator allocations is included in Table 2.

AVAILABILITY MODEL AND ALLOCATION RESULTS

An allocation model incorporating the above assumptions was developed to calculate the number of failures allowed each Collider subsystem for a given number of Magnet System failures. Each Accelerator failure (other than Injector Complex failures) was assumed to require Collider ring refill plus one full 2-20 TEV magnetic cycle. Failures during ring fill were assumed to require restart of the fill process, averaging a loss of one-half injection cycle per failure. Figure 3 shows the allocated maintenance and fill/tune times, and Table 3 shows these times plus failure counts, availability, and cycle reliability (defined as the probability of operating through a complete Accelerator operating cycle without a failure).

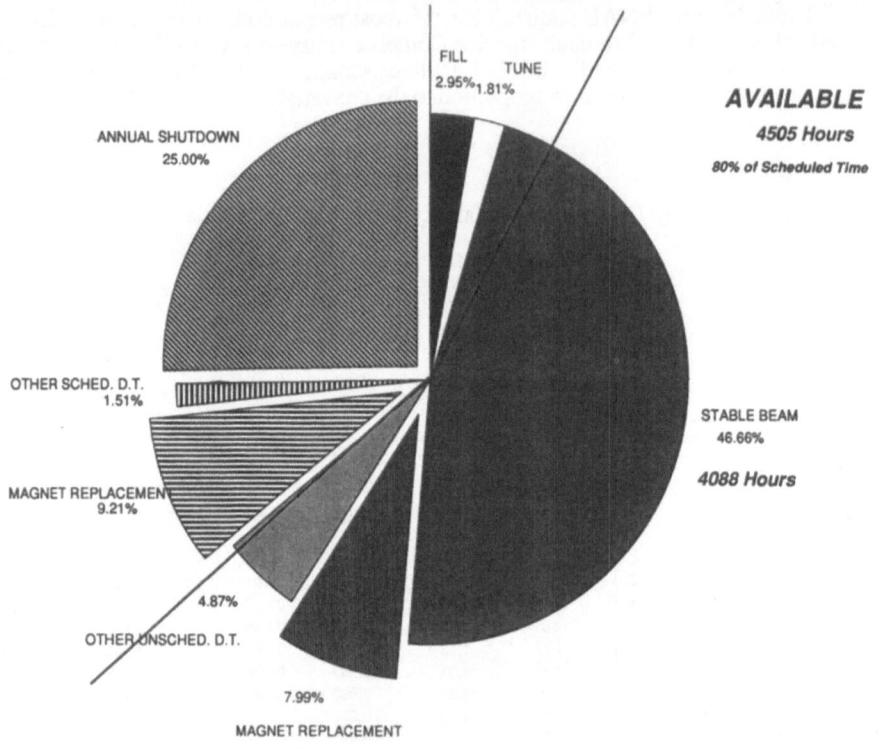

Figure 3. SSC Accelerator Annual Time Allocation.

Figures 4 through 8 show how the above figures of merit vary with the number of magnet failures, with downtime per magnet failure, with the downtime delta due to Collider ring scale, with Injector Complex reliability, and with beam luminosity lifetime. Figures 9.1 through 9.6 show the sensitivity of key figures of merit to the parameters varied in Figures 4 through 8 (reference the appropriate figure for applicable horizontal scale). Because Accelerator availability is a constant for all calculations, non-magnet failure allocations increase rapidly as magnet failure count or downtime per failure decreases. However, as failure count allocations for the short-downtime-per-failure segments increase, cycle reliability falls and time required for ring fill and beam acceleration operations rises. The current goal of no more than four hours per day for beam fill and tune operations limits the allowed number of failures per year to about 450 (consistent with about 3.5 Magnet System failures per year), which is less than the experience data for the Tevatron support subsystems (Table 2). Other goals, like 20 hour nominal stable beam time per beam cycle, place further restrictions on the number of failures allowed (Figure 9.5).

It is anticipated that improved basic reliability, coupled with selective redundancy and scheduled maintenance, will promote achievement of the goals. Figure 10 shows that most of the Tevatron downtime in the reference period was caused by only a few events, hence the major emphasis placed on improved reliability of superconducting magnets. On the short downtime end of the scale are momentary interruptions with a multitude of causes (basically built-in-test errors), and readily isolated hardware failures near the control center where maintenance delay times were minimal. The short-delay types of faults, which would have a greater effect on SSC beam time due to the longer beam restoration process and the wider distribution of the hardware, will be reduced by improved hardware design and quality, improved software techniques, and redundancy where essential.

Table 3. SSC Availability Allocation.

| | TUNE/FILL (Hours) | | DOWN TIME (Hours) | | UNSCHED | # | AVAIL. | CYCLE |
	FILL	RAMP	UNSCHED.	SCHED.	DT/FAIL (Hours)	FAILS		REL.
Magnet System R&R	6	5	700	806	16.0	6	.876	.977
Injector Complex	56		63		1	56	.989	.729
*	52	41	17		1	15	.997	.926
Correction P.S.	11	9	25	24	2.3	11	.996	.960
Ring P.S. and Regulators	20	16	54		2.8	19	.990	.930
Quench Protection System	11	8	24	24	2.4	10	.996	.962
Cryogenics	14	11	35	24	2.7	13	.994	.952
Vacuum	3	2	12		4.3	3	.998	.989
Control and Instr. (GAS)	22	17	30	60	1.5	20	.995	.925
RF	11	9	25		2.3	11	.996	.960
Injection/abort	3	2	6		2.6	2	.999	.991
Utilities	14	11	48		3.5	14	.992	.950
Safety and Interlocks	11	9	26		2.4	11	.995	.960
Magnet System Quenches	5	4	22		5.0	4	.996	.983
Miscellaneous	19	15	40		2.2	18	.993	.933
								.582
ACCELERATOR SYSTEM	259	159	1126	939	5.3	213	.800	.393

* FILL and RAMP times for normal cycles without failures.

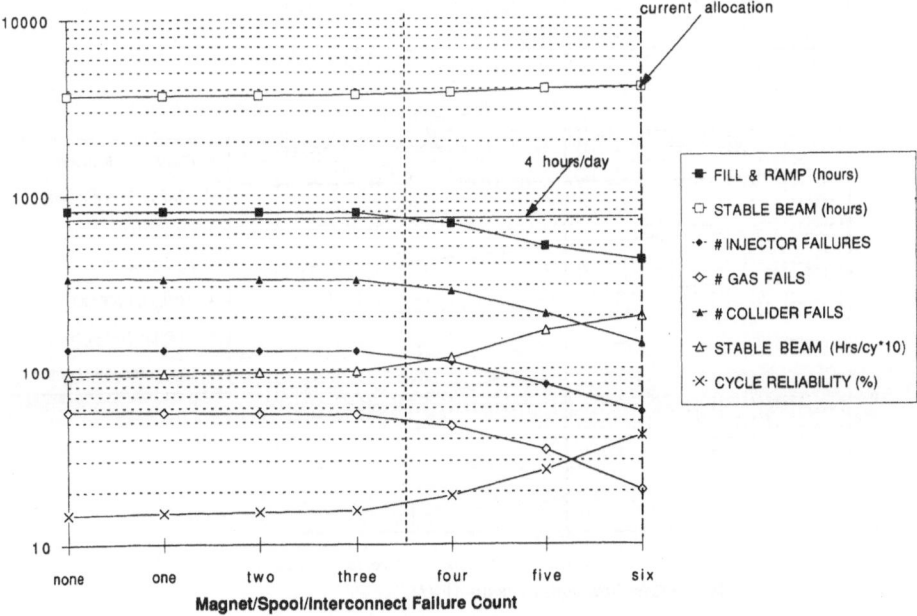

Figure 4. Parameter Sensitivity to Magnet System Failures.

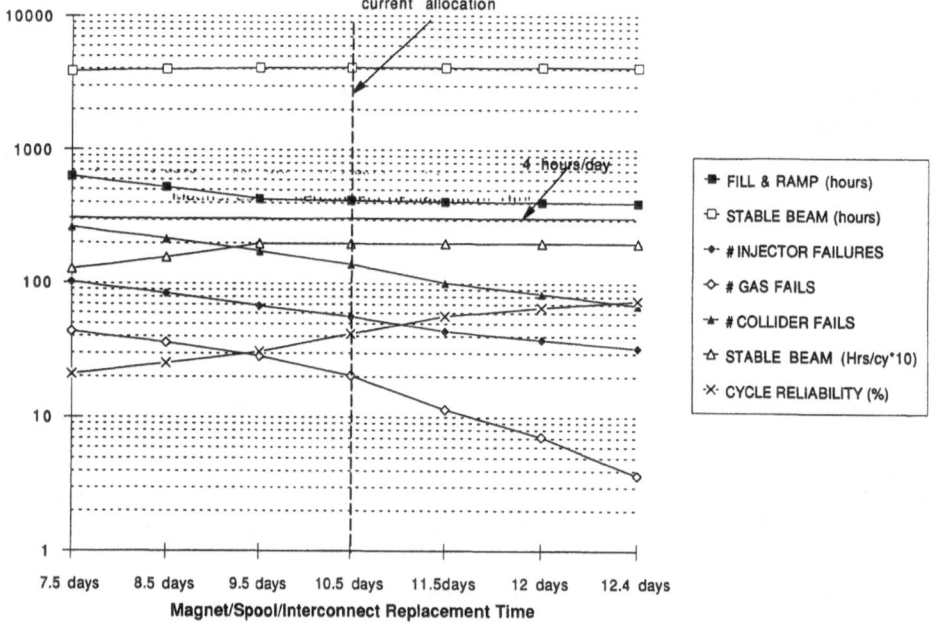

Figure 5. Parameter Sensitivity to Magnet System Failures.

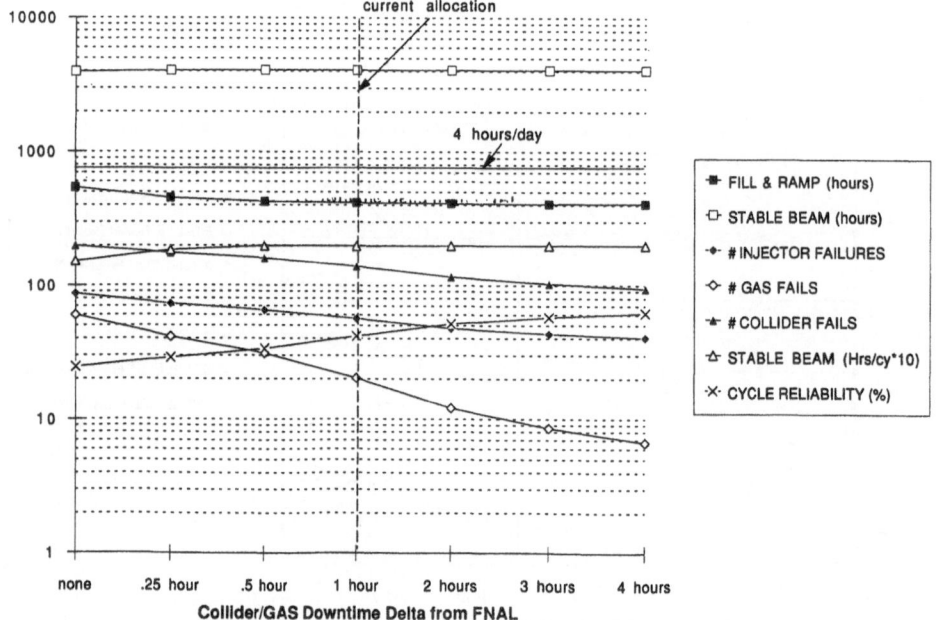

Figure 6. Parameter Sensitivity to Collider/Gas Downtime.

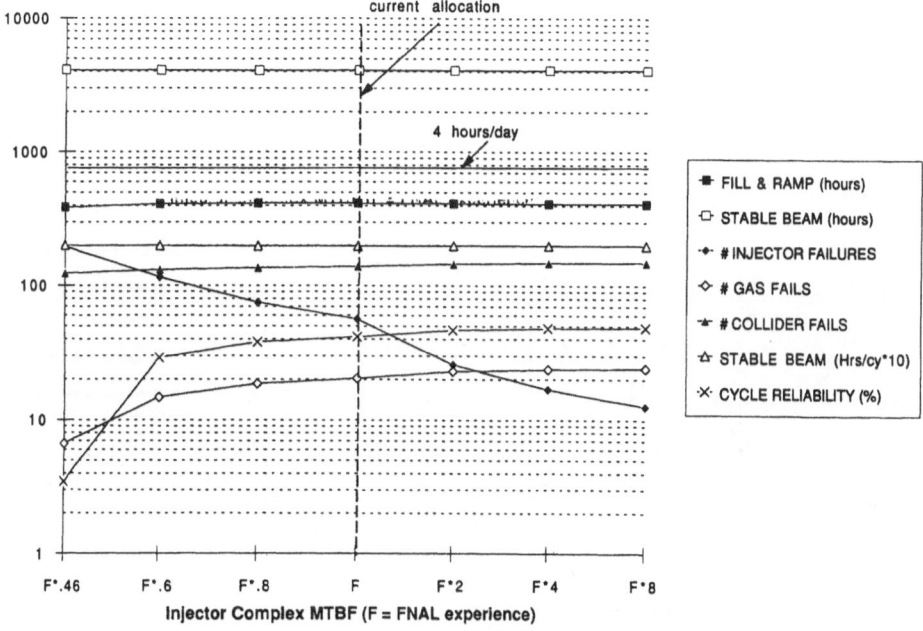

Figure 7. Parameter Sensitivity to Injector Complex Reliability (MTBF).

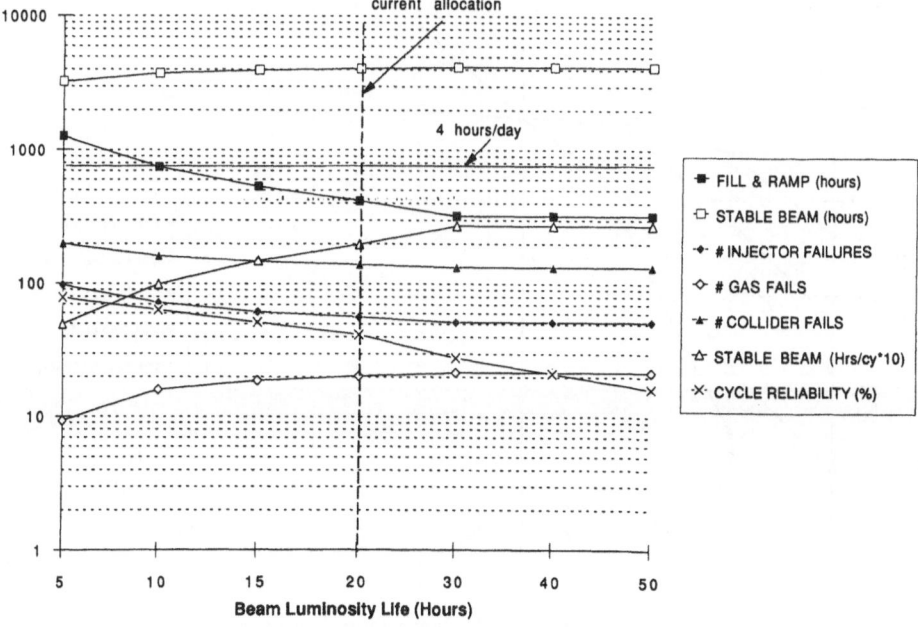

Figure 8. Parameter Sensitivity TO Beam Luminosity Life.

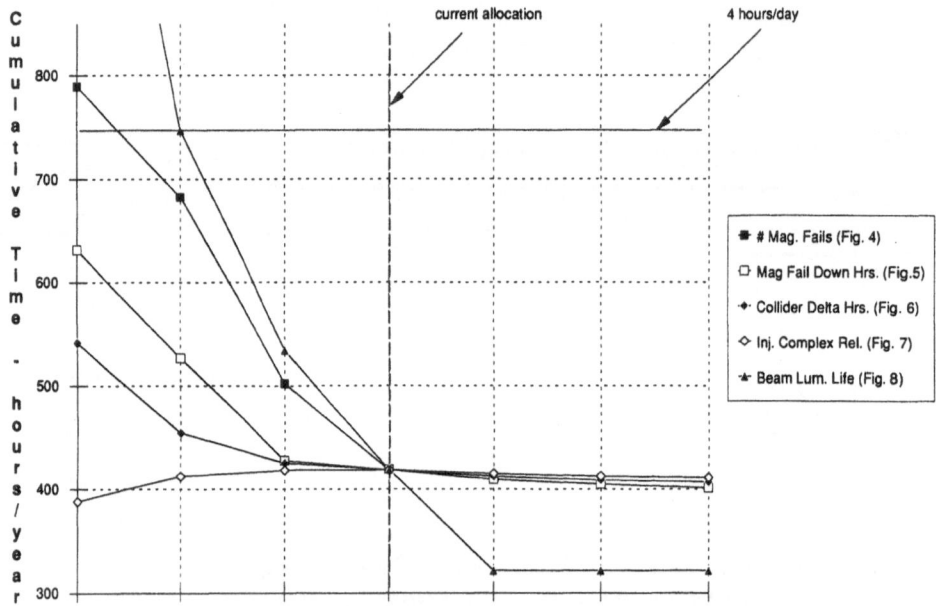

Figure 9.1. Beam Fill and Tune Time Sensitivity.

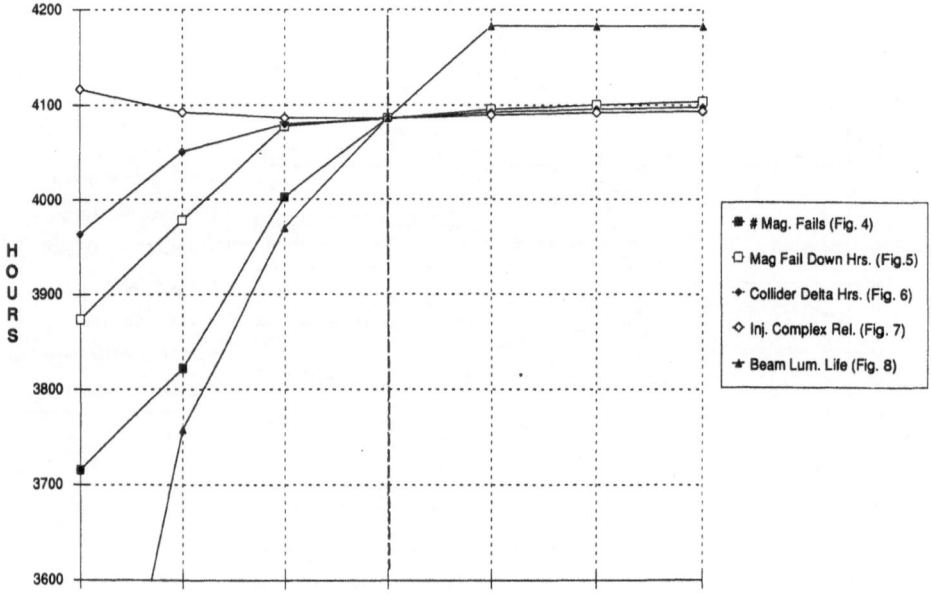

Figure 9.2. Stable Beam Time Per Year.

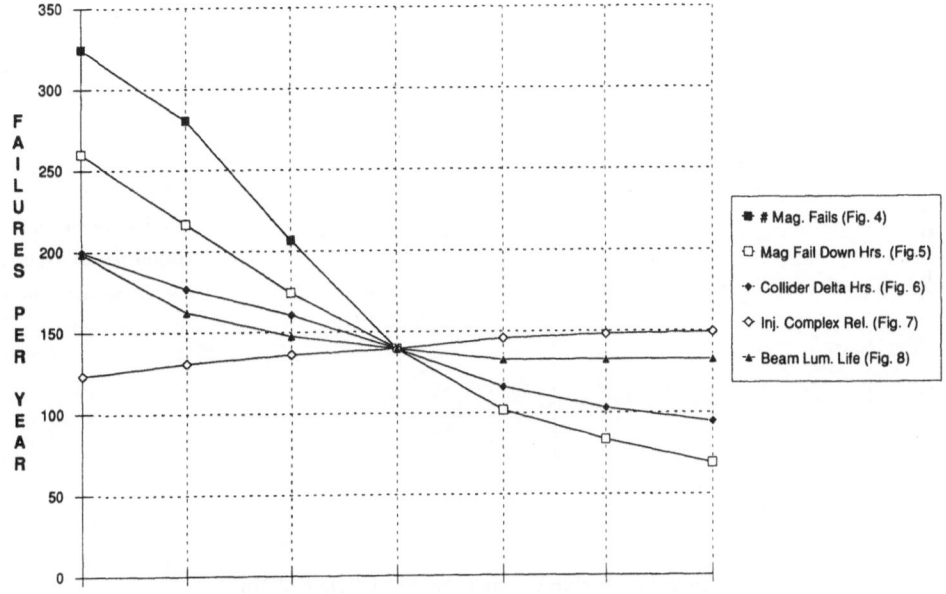

Figure 9.3. Collider Failure Count Sensitivity.

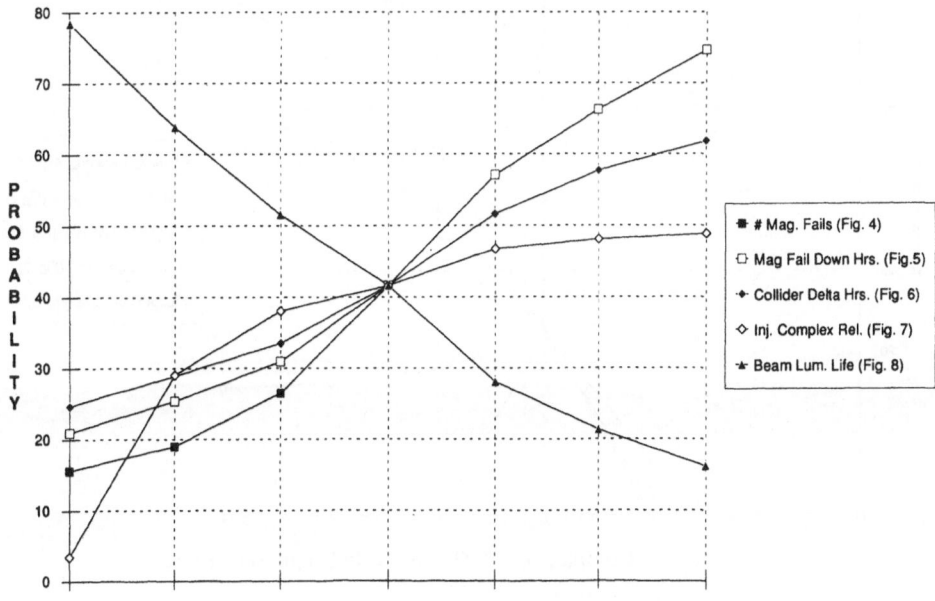

Figure 9.4. Collider/Gas Cycle Reliability Sensitivity.

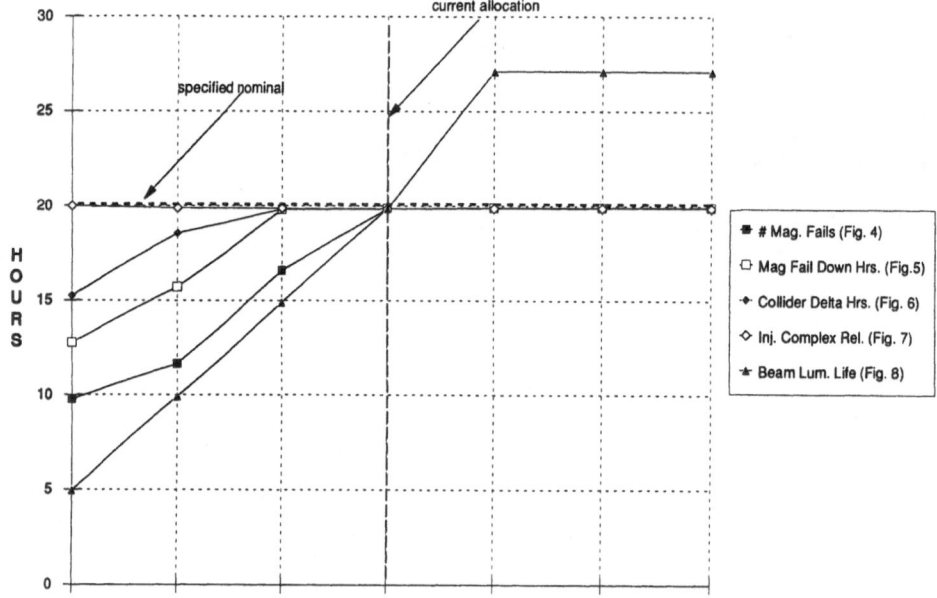

Figure 9.5. Stable Beam Time Per Cycle Reliability.

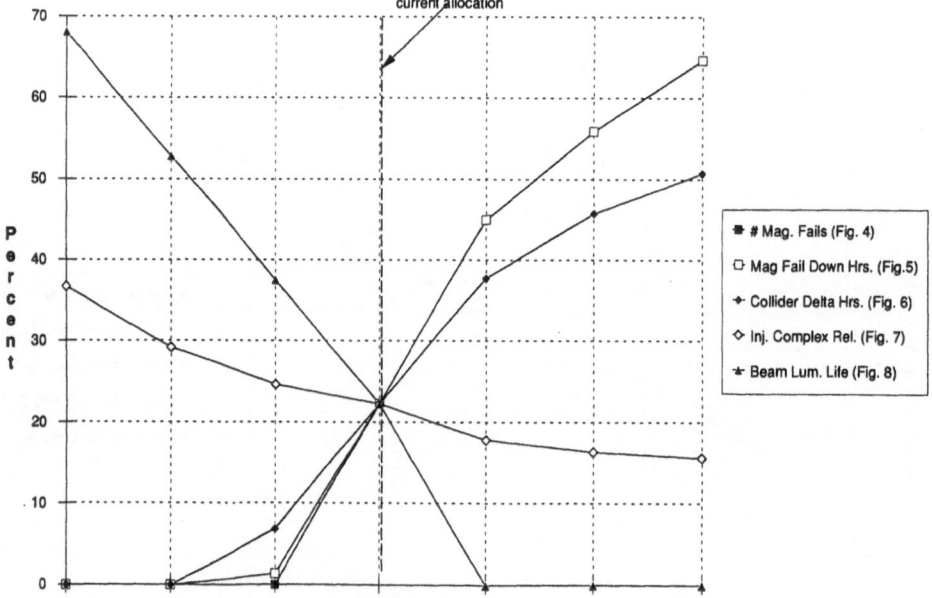

Figure 9.6. Complete Cycle/Total Cycle Count Sensitivity.

CONCLUSIONS AND RECOMMENDATIONS

Figure 11 shows the allocated values at the SSC Accelerator segment specification level. Figure 12 shows the allocation down to the major component level for the Collider Accelerator and Collider Magnet Systems elements. Table 4 shows the magnet level reliability allocations compatible with the Collider Magnet System and HEB Magnet System availability allocation. Allocation of the other segment values to lower levels is in progress.

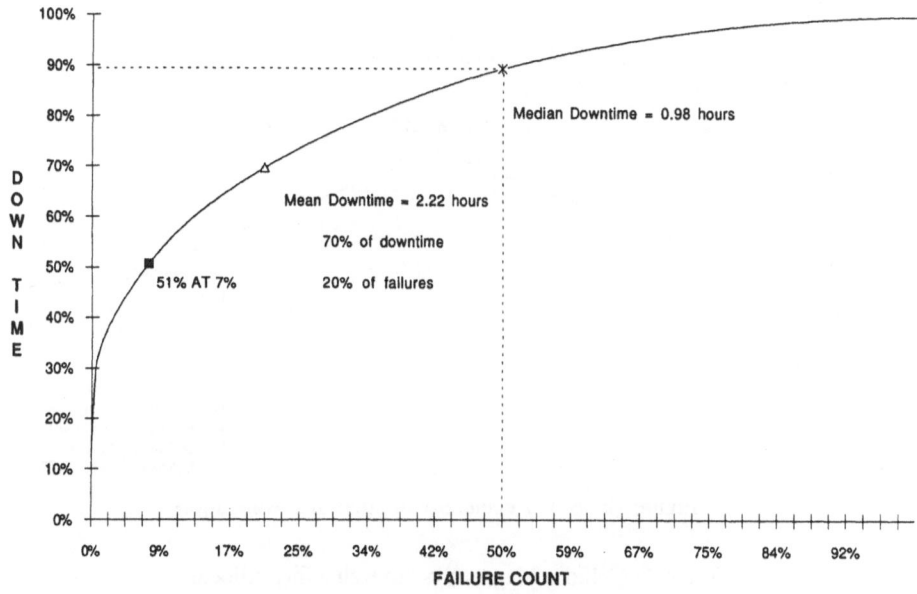

Figure 10. Tevatron Downtime Pareto Analysis.

Comparison of the allocated number of failures with the number of failures occurring for similar subsystems at FNAL show that significant reliability improvements must be made in all subsystems outside the Injector Complex. Not only are the required failure counts smaller, the hardware complexity is greater, and the hardware in many cases is spread around a 54-mile tunnel instead of a 3.9-mile tunnel, making fault isolation and repair time potentially hours longer.

Subsystem and magnet-level reliability requirements allocated from the SSC Accelerator availability requirement will not be achieved without significant, systematic attention to the design details that ultimately drive the reliability achievement. A significant part of the availability achievement program will be the expansion, detailed development, and maintenance of the model initiated with this analysis. Proper model development and maintenance will promote availability growth tracking and visibility of the design features most likely to prevent availability achievement. Ranking of problem areas will guide incorporation of the most effective change actions, leading to availability achievement at minimum cost.

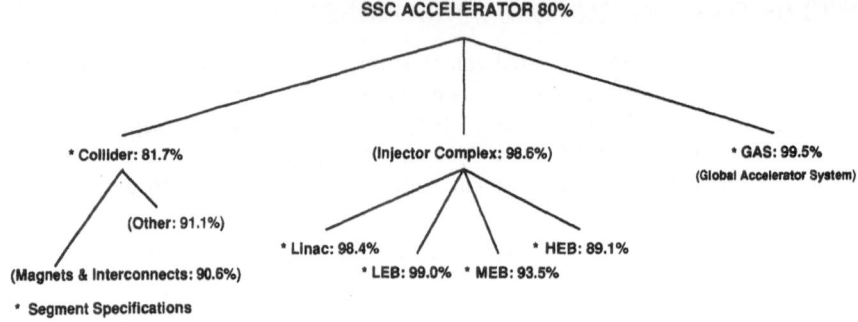

Figure 11. SSC Accelerator Availability Allocation.

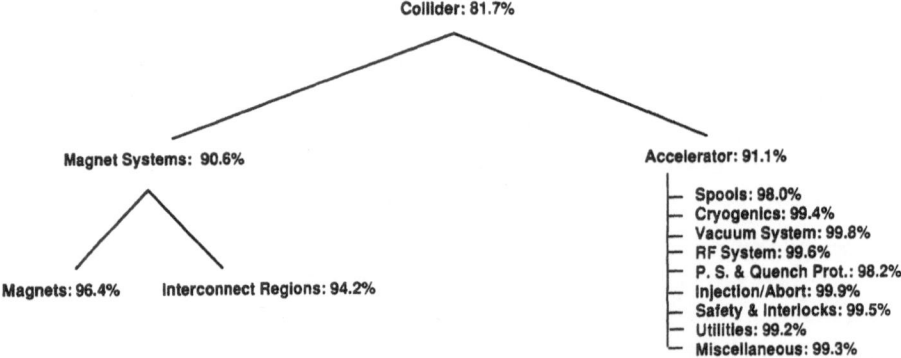

Figure 12. SSC Accelerator Availability Allocation.

Table 4. Collider Magnet System Reliability Allocation.

	# Types	QTY	UNIT AVAIL.	Unit MTBF (Years)	Unit F. R. (Failures/ Billion Hrs.)	25 Year # Unit Failures	25 Year # Failures Incl. I.R.	Annual # Failures Incl. I.R.	F. R. (Failures/ Billion Hrs.) Incl. I.R.	ANNUAL AVAIL.
Dipoles		8596								
CDM	2	8460	0.9999956	6,501	17.56	32.53	79.44	3.18	42.88	0.9730
Vertical, BV1	2	64	0.9999911	3,250	35.12	0.49	0.85	0.03	60.44	0.9996
Vertical, BV2	2	72	0.9999823	1,625	70.24	1.11	1.51	0.06	95.56	0.9991
Quadrupoles		1960								
CQM	1	1664	0.9999956	6,501	17.56	6.40	15.62	0.62	42.88	0.9947
Dispersion suppressor, QS	4	120	0.9999911	3,250	35.12	0.92	1.59	0.06	60.44	0.9992
M-1, QV	1	64	0.9999823	1,625	70.24	0.98	1.34	0.05	95.56	0.9992
Low beta focusing, QL	6	40	0.9999911	3,250	35.12	0.31	0.53	0.02	60.44	0.9997
Medium beta focusing, QM	6	40	0.9999911	3,250	35.12	0.31	0.53	0.02	60.44	0.9997
Utility region, QU	4	32	0.9999911	3,250	35.12	0.25	0.42	0.02	60.44	0.9998
COLLIDER MAGNET TOTAL		10556	0.9501731	0.58	18.73	43.30	101.82	4.07	44.05	0.9641
INTERCONNECT REGIONS (IR) *		12568	0.9999936	4,509	25.32	69.68	69.68	2.79	25.32	0.9423
			0.9198228	0.36						
SPOOLS										
SP	1	900	0.9999867	2,167	52.68	10.38	15.37	0.61	78.00	0.9914
SPR	4	978	0.9999845	1,857	61.46	13.16	18.59	0.74	86.78	0.9891
SPC	3	1309	0.9999991	32,504	3.51	1.01	1.73	0.07	6.04	0.9992
SPC (Empty)(20%)	1	327	0.9999999	325,043	0.35	0.03	0.04	0.00	0.60	0.99998
SPOOL TOTAL		3514	0.9717177	1.02		24.58	35.73	1.43	46.43	0.9796
HEB										
Dipoles		432	0.9999934	4,334	26.34	2.49	4.89	0.20	51.66	0.9979
Quads		234	0.9999947	5,417	21.07	1.08	2.38	0.10	46.39	0.9991
Special quads		44	0.9999889	2,600	43.90	0.42	0.67	0.03	69.22	0.9996
Spools		250	0.9999856	2,000	57.07	3.12	4.51	0.18	82.39	0.9974
Interconnects		960	0.9999936	4,509	25.32	5.32	5.32	0.21	25.32	0.9956
HEB TOTAL		1920	0.9856836	2.01		12.44	12.44	0.50	29.59	0.9897
COLLIDER TOTAL			0.842	0.18		137.56	137.56	5.50		0.8860
TOTAL (Including HEB)			0.827	0.17		150.00	150.00	6.00		0.8757

* Assumes no bellows required for SPC spool IR; empty SPC IR C. F. = 10% of normal IR, other SPC IR C. F. = 50% of normal IR.

REFERENCE

1. H. Edwards and R. Stefanski, "Availability—Mission Profile—Collider Operations," memorandum, September 21, 1990.

TECHNOLOGY TRANSFER CONSIDERATIONS
FOR THE COLLIDER DIPOLE MAGNET

C. Goodzeit and R. Fischer

Magnet Division
Superconducting Super Collider Laboratory*
2550 Beckleymeade Avenue
Dallas, Texas 75237

Abstract: The R&D program at the national laboratories has resulted in significant advances in design and fabrication methods for the Collider Dipole Magnets. The status of the transfer of the technology developed by the laboratories is reviewed. The continuation of the technology transfer program is discussed with a description of: (1) the relation of technology transfer activities to collider dipole product development; (2) content of the program relating to key magnet performance issues; and (3) methods to implement the program.

THE ROLE OF TECHNOLOGY TRANSFER
IN THE DIPOLE ACQUISITION PLAN

The acquisition strategy for the Collider Dipole Magnets (CDM) outlines some of the aspects of the role of the technology transfer in the dipole magnet procurement program. The leader subcontractor is responsible for the development of the CDM to meet system performance requirements as specified in the Prime Item Development Specification. As he is expected to provide significant intellectual content, he will not be required to accept as "proven" any technology or design provided. The subcontractor will be held accountable for all material provided and will have to defend the validity of his design approach to the CDM. The lone exception is the development of the superconducting cable for the magnets, which is the responsibility of the Superconducting Super Collider Laboratory (SSCL). The subcontractor will be free to question and alter the present baseline of the magnet that was developed in the R&D programs at the laboratories, if such changes lead to a more reliable, less costly design.

The national labs, particularly Brookhaven (BNL) and Fermilab (FNAL), that have been involved in the R&D program for the CDM will participate in the transfer of technology to the subcontractor and assist the SSCL in providing scientific and technical support to the

---

*Operated by the Universities Research Association, Inc., for the U.S. Department of Energy under Contract No. DE-AC02-89ER40486.

subcontractor. Existing contracts between the participating labs and the SSCL include provisions for supporting industrial activities on behalf of the SSCL. Contracts for Fiscal Years (FY) 1991 and 1992 will include increasing emphasis on these activities. Although there will no longer be SSC magnet production activities at the supporting national laboratories beyond FY 92, contracts for continued industrial support are expected to be awarded.

It is expected that laboratory personnel will make important contributions to the magnet development program and that subcontractor personnel will visit the labs. However, all interactions involving the subcontractor will be closely coordinated through the SSCL.

When the subcontractor requests support in a specific area, the SSCL will evaluate the request and determine the most beneficial method for providing the required assistance. If it is decided that lab support is needed, the SSCL will make the necessary arrangements with the lab to ensure that the desired results are achieved. As the program advances it is anticipated that the expertise being accrued by the SSCL will lead to less laboratory involvement being requested by the subcontractor.

STATUS OF THE R&D PROGRAM TECHNOLOGY TRANSFER

Previous Technology Transfer Activities

A formal program at BNL in January 1989 under the auspices of the Central Design Group acquainted potential magnet fabrication subcontractors with the technology for the 40-mm-aperture dipole magnets developed there. There were three activities associated with this program. One was a series of lectures on magnet technology, including performance requirements, design procedures, magnet test results, and quality assurance issues pertaining to the cold mass design developed at BNL. In addition, copies of specifications of all processes and materials as well as detailed drawings of magnet components were presented. The program was completed with work station demonstrations and displays of all major magnet fabrication steps.

A meeting held at FNAL in February 1989 emphasized work on the cryostat that had been done there and included discussion of the design and testing of demonstration cold masses and work station demonstrations. In October of that year another Phase I Development Program was held at the SSCL in conjunction with a meeting for potential bidders for CDM development and production. This program also included technology transfer data in the form of presentations and, in addition to the update of the BNL cold mass technology, presentations by FNAL personnel on the design approach for the cold mass and associated tooling in the version of the 40-mm-aperture magnet being developed there.

Objectives of the Present Program

Initially, the technology transfer program appeared to be aimed at instructing prospective vendors in the technique of building the versions of the dipole magnets being developed at the laboratories. However, the magnet acquisition strategy, as explained above, calls for the development of the magnet design by the subcontractor. Thus, the technology transfer program has now been geared to that approach. The present objective of the program is not to tell the subcontractor how to design and build the magnets but rather to present him with the knowledge and expertise gained in the R&D programs at the laboratories so that he can best determine the design approach to be taken for the magnet design that he will develop and defend according to the acquisition plan.

Technology transfer is organized to recognize that the subcontractor is responsible for designing the magnet to meet performance requirements:

- Quench performance requirements.

- Field quality specifications.

- Electrical integrity requirements.

- Heat leak and magnet alignment requirements.

- Requirements for operational environment conditions.

In addition, two other key items are regarded as necessary to verify magnet performance:

- Cryogenic testing and measuring technology.

- Superconducting wire and cable technology.

The program will provide information on why and how the laboratories resolved these key issues in a certain way, the test results obtained from that approach, and an assessment of the limit of knowledge to resolve key magnet performance issues.

IMPLEMENTATION PLANS FOR TECHNOLOGY TRANSFER

Working Groups

The formation of working groups to convey technical information to the subcontractor is considered to be a formal part of the program. These groups or committees are organized to address the key issues stated above. Each committee is chaired by an SSCL person; however, the committees are made up of people who have had hands-on experience in the particular technology to be conveyed by the group. Since the R&D work has been carried out mainly at the national laboratories, the groups would consist mostly of people from the laboratories who are willing to contribute their time and effort to the success of this program.

Initially, six working groups would be set up to address the key issues.in the following areas:

Cold Mass Design and Fabrication

Topics to include BNL experiences in the R&D program on 40-mm-aperture dipoles; FNAL variations of the 40-mm design and experiences; 50-mm-aperture demonstration magnets fabricated at BNL and FNAL; current status of the R&D program with conclusions and assessment of unresolved issues.

Cryostat Design and Fabrication

Provides an overview of designs used in various accelerator magnets; describes development of current design done at FNAL with test data relating to meeting requirements for heat leak and alignment; discusses fabrication and assembly techniques.

Cryogenic Testing

Warm and cold magnetic measurements. Covers methods of measuring quench performance and magnetic field quality developed in the R&D program; existing cold test facilities at the laboratories and SSCL test facility plans; special test instrumentation for measuring coil mechanical data, quench locations, thermometry, etc.; design experiences and operation of magnetic measurement probes; measurements during magnet fabrication and methods used in supporting documentation; status of the measurement technology and unresolved issues.

Superconducting Wire and Cable Technology

Historical evolution leading to present design of the cable; fabrication process variables; vendor qualification program for superconductor procurements; cabling technology; technology for measuring strand and cable superconducting properties; low field and persistent current effects and present status; quality assurance considerations; unresolved issues pertaining to time-dependent effects, manufacturing process control, and manufacturing process parameters.

Quench Protection and Other Systems

Electrical, cryogenic, and vacuum systems required for the collider ring; quench protection systems and design experience with test results; magnet cooldown analyses and measurements; current status of systems topics and unresolved issues.

Operational Environment Considerations

Dynamic loads due to transportation and ground motion; temperature, voltage stress, and radiation effects on the magnets; reliability and safety considerations.

Hands-On Experience

Requirements for the fabrication of the five dipoles to be used in the Major Milestone Systems Demonstration Test in September 1992 call for the magnet to be assembled by subcontractor personnel. Since these magnets are planned to be assembled at Fermilab, the technology transfer program includes provisions for the magnets to be fabricated in compliance with the major milestone requirement. Thus, the subcontractor personnel will use SSC developed tooling at Fermilab to perform the required assembly operations, thereby obtaining early hands-on experience in fabricating magnets. It is planned that the industrial personnel under their own supervision will gain experience in the following areas of magnet fabrication:

- Wind and cure coils
- Inspect and test coils.
- Apply insulation, quench heater strips, and primary voltage taps.
- Assemble collar packs to coils and collar coils.
- Splice coil end connections and install end clamps.
- Install electrical bus work and power lead expansion joints.
- Assemble yokes to collared coil and weld helium containment shell to enclose cold mass.
- Assemble end plates and interconnection parts to cold mass.
- Check for leaks and perform pressure checks of the cold mass.

TECHNOLOGY TRANSFER PROGRAM

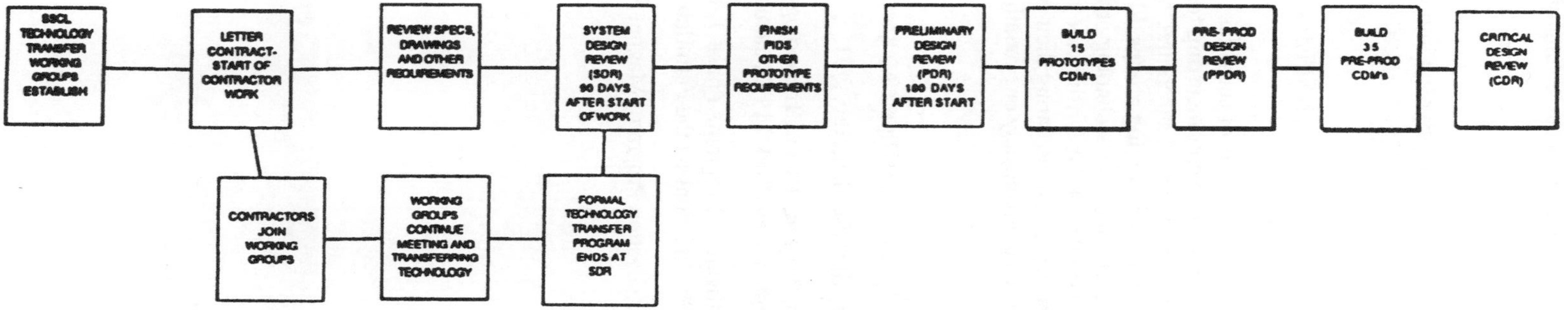

Figure 1.

- Perform electrical and warm magnetic testing of assembled cold mass.

- Install cold mass onto its support posts and assemble cryostat from its various sub-assemblies and components.

- Insert cold mass into cryostat and cryostat into vacuum vessel.

- Final mechanical and electrical tests.

SCHEDULE FOR TECHNOLOGY TRANSFER—INTERACTION WITH MAGNET DEVELOPMENT

Figure 1 shows the tentative schedule of activities related to technology transfer in the magnet acquisition program. As soon as cost incurring for technology transfer is permitted, the activities would begin with assignment of subcontractor personnel to working groups and scheduling of formal sessions with the working groups. Since the CDM program is paced by the various design reviews which follow the technology transfer program, it is important that these activities proceed on schedule. At this writing, the complete outline for content of the sessions is in place and the presenters are preparing their outlines and material. It is expected that the formal session can begin approximately one month after cost incurring begins.

REFERENCES

1. Acquisition Strategy for SSC Collider Dipole Magnets, Rev. 12/13/89.

2. Prime Item Development Specification, Rev. 2/91.

3. Magnet Industrialization Program, Phase I Technology Orientation Meeting, Brookhaven National Laboratory, Upton, New York 11973, January 23-27, 1989.

4. SSC Magnet Industrialization Program. Phase I Technology Orientation Meeting, February 20-24, 1989 (Presentations by the Fermilab Staff).

5. Phase I Magnet Industrialization Program, Magnet Systems Division Update, October 31, 1989.

A SYSTEM APPROACH TO THE DESIGN OF
SUPERCONDUCTING MAGNETS

A. Pennington

Magnet Systems Division
Superconducting Super Collider Laboratory*
2550 Beckleymeade Avenue
Dallas, Texas 75237

INTRODUCTION

The design of superconducting magnets poses numerous technical, engineering and physics challenges to ensure that the magnets, when installed in the operational environment, will satisfy performance requirements of the accelerator. The discussion presented here will address the systematic, methodical process, as applied by the magnet system design engineer, that defines magnet performance requirements, the allocation of requirements to magnet components, and the development of design solutions which will satisfy the requirements. The approach presented will show how the system engineer parses top-level performance requirements as established by the system specification ("shall" statements), develops design solutions, performs design trades to ensure selection of the best solution in terms of life-cycle cost and performance, verifies the design through analysis and computer modelling, and then validates the design by build-and-test. The process shows the methods for achieving traceability of design and correlation of design solutions to performance parameters.

Since the early 1950s, the development of large and sophisticated systems has provided the impetus for the application of a systems engineering approach to the development and implementation of design solutions. The difficulties experienced in evolving new systems have led to the development of specific tools and techniques within the systems engineering discipline which permit better control and insight into the development process, and to better documentation to show traceability to optimized design solutions.

Systems engineering is the application of scientific and engineering efforts to (a) transform an operational need into a description of system performance parameters and a system configuration established through the use of an iterative process of definition, synthesis, analysis, design, test, and evaluation; (b) integrate related technical parameters and ensure compatibility of all physical, functional, and program interfaces in a manner that optimizes the total system definition and design; and (c) integrate reliability, maintainability, safety, integrity, human, and other such factors into total engineering effort to meet cost, schedule, and technical performance objectives.

---

*Operated by the Universities Research Association, Inc., for the U.S. Department of Energy under Contract No. DE-AC02-89ER40486.

The systems engineering process is a systematic analytical and management methodology which provides both a technical and a management process leading to the successful completion of system development. The systems engineering process begins with the identification of needs, constraints, and capability requirements to satisfy mission objectives through the application of technology.

RESPONSIBLE AUTHORITY

Responsibility for the systems engineering effort leading to the design of superconducting magnets for application in the Superconducting Super Collider (SSC) lies with the associate director and division head of the Magnet Systems Division of the SSC Laboratory. A Task Responsibility Matrix, developed by the magnet product manager and, negotiated with the responsible functional area group leaders, identifies the responsibility of the assigned personnel for detailed system development disciplines. A close correlation with SSCL program management planning is maintained to keep SSC program and technical functions synchronized. The system design major milestones and product development charts form a common basis for both management functions to ensure proper tracking of activities.

TECHNICAL PLANNING AND CONTROL

In defining the elements of this process, a standard system of technical planning and control is applied. The first technical planning task is to analyze the SSC baseline documentation to establish a tailored research and development series of activities and a detailed program schedule based on a standard PERT technique. The results of these activities by the engineering group, in consonance with the product manager and systems engineering, forms the basis for planning and conducting of the engineering efforts and monitoring of magnet development progress. In the case of the SSC, the baseline was established by the initial Collider Design Group (CDG) and was published in the Conceptual Design Report (CDR) in October 1989. This report was subsequently updated to the Site-Specific Conceptual Design Report (SCDR), released in June 1990, to reflect the move to the 50-mm collider dipole magnet configuration. In parallel with the release of the SCDR, the Magnet Systems Division developed and released a Research and Development Plan in June 1990, with a reissue in December 1990. The SSC design baseline, as revised to the 50-mm configuration essential to satisfying field quality requirements, will be analyzed as the technical design baseline for the collider dipole magnet (CDM). The CDM serves as the basis for the development of all other superconducting magnets for the SSC.

Any credible systems engineering approach must, prior to implementation, establish a means to ensure traceability of design and measurement of progress in reaching the desired solution. This is accomplished through the implementation of specific technical program assurance methods. These methods include:

- Technical Performance Measurement (TPM). TPM is exercised throughout the system design process following the System Design Review (SDR). Key technical performance requirements and acceptable tolerances are identified and monitored during the follow-on system design. These performance requirements will be measures of effectiveness (MOE) type parameters such as system availability, quench performance, cool-down and warm-up times, and mean-time-to-replace. A Performance Compliance and Verification Matrix is used to document the results and status of all TPM activities. Design deficiencies identified through TPM will be investigated, and appropriate corrective actions will be recommended to the engineering group leader.

- Quality Assurance. Early in the program, i.e., during the "Requirements Definition and Operations Analysis" process, system test concepts will be developed and documented in a Master Integration and Test Plan (MITP). This plan defines the detailed quality assurance activities necessary to ensure that all system level requirements comply with developed acceptance criteria. System test plans and procedures based on this system test concept will be developed during the system engineering process. A Verification Cross Reference Index will be developed to identify test levels and methods for each system requirement.

- A Risk Management and Control Process. This process is based on a Risk Management Plan and is initiated during the "System Analysis and Design" phase of the system engineering process. The prime efforts of this process are to identify and analyze the dominant risk areas, develop an overall program risk profile, and recommend the most advantageous approaches to minimizing overall program risk.

MAJOR PHASES OF THE SSC LIFE CYCLE

The life cycle for the SSC development and acquisition is depicted by Figure 1. The life cycle is initiated by the determination of a need for a collider system that can provide a controlled acceleration of proton masses to the 20 TeV level and cause a collision of two such masses to occur, resulting in the combined energy of 40 TeV and the measurement of the event. The initial exploration of alternative approaches to satisfying the desired SSC system objectives results in preliminary design concepts developed by the SSCL, collaborating national laboratories, and selected subcontractors. These concepts are documented in conceptual design reports and form the basis for the issuance of requests for proposal to industry.

During Phase I of the industrialization program, technology transfer to industry provides a definition of the initial baseline design for demonstration and validation through analysis and test.

Phase II proceeds with full-scale engineering development by the selected subcontractor(s) and with preliminary design, including the design and build-and-test of prototype magnets to prove both the magnet design and design-specific tooling. Following successful completion of a prototype test and a design iteration to incorporate enhancements identified during test evaluation, a critical design review is conducted prior to the authorization to build production units.

Finally, Phase III carries the process to production, installation, and checkout in the tunnel, and then to commissioning for operation.

This systematic and analytic flow process is depicted in more detail in Figure 2.

PLANS AND SCHEDULES FOR DESIGN AND PROGRAM REVIEW

As shown by Figure 2, the systems engineering process documents all program reviews from start to finish and maintains design baseline control. Specifications and interface design documents are established during the design process to ensure that the evolving design will be based on the appropriate set of system requirements as established for the SSC by the Accelerator System Division (ASD) of the SSCL. The scheduling of program reviews and the establishment of design baselines are based on both the program management plans and the technical planning and control (TPC) plans and milestones.

MILESTONE		I		II		III

MISSION NEED DETERMINATION	CONCEPT EXPLORATION PHASE	DEMONSTRATION & VALIDATION PHASE	FULL SCALE DEVELOPMENT PHASE	PRODUCTION INSTALLATION PHASE
DOE	SSC LABORATORY STUDIES	PARALLEL EFFORT BY SSCL, NAT. LABS & SUBCONTRACTOR(S)	ONE (OR MORE) SUBCONTRACTORS	CONVENTIONAL CONSTRUCTION SUBCONTRACTOR

		RFP	RFP	RFP
• SSC MISSION ANALYSIS	• ALTERNATIVE STUDIES	• SUBCONTRACTOR SELECTION	• PDR DETAILED DESIGN	• INITIAL PRODUCTION
• STATEMENT OF NEED	• PRELIMINARY SYSTEM SPECIFICATION	• SYSTEM REQUIREMENTS REVIEW	• CDR	• PHYSICAL CONFIGURATION AUDIT
• OPERATIONS CONCEPT	• SOWs/RFPs	• TRADE STUDIES	• TEST & EVALUATION	• PRODUCTION
	• PREPARE CONCEPT DESIGN REPORT	• SYSTEM DEFINITION	• PRODUCT BASELINE	
		• ALLOCATE BASELINE		
		• SYSTEM DESIGN REVIEW		

Figure 1. Major Phases of the SSC Life Cycle.

Requirements Definition Methodology and Control

System requirements are classified and allocated to magnet system functions. Functions, in turn, are allocated to configuration items (CIs), and these CIs are allocated to prime items (hardware or software components). This process provides a consistent allocation of requirements over several design stages leading to the final magnet product. In addition, all requirements can be traced back from any design entity to the top-level requirement specified by the ASD. The specific magnet Prime Item Development Specification prepared and controlled by the MSD Engineering Group establishes and controls all magnet system requirements.

Subcontractor Control

The physical size of the SSC (approximately 87 km in circumference) and the magnitude of the total project have generated a need for superconducting magnets in quantities never before envisioned. Approximately 13,000 magnets of 27 differing configurations will be designed, manufactured, tested, and delivered by industry. Because of the large number of differing magnet configurations and the quantities in demand, several subcontracts will be let to multiple vendors. Full-scale engineering development procurements of superconducting magnets of this complexity and magnitude have imposed the need to establish practices for supplier evaluation, qualification, risk analysis, and source selection.

Subsequent to subcontract award, subcontractor progress requires monitoring through regularly scheduled in-process reviews and test witnessing. Where necessary, corrective actions are initiated through a Corrective Action Request process using a Subcontractor Product Assurance Rating system.

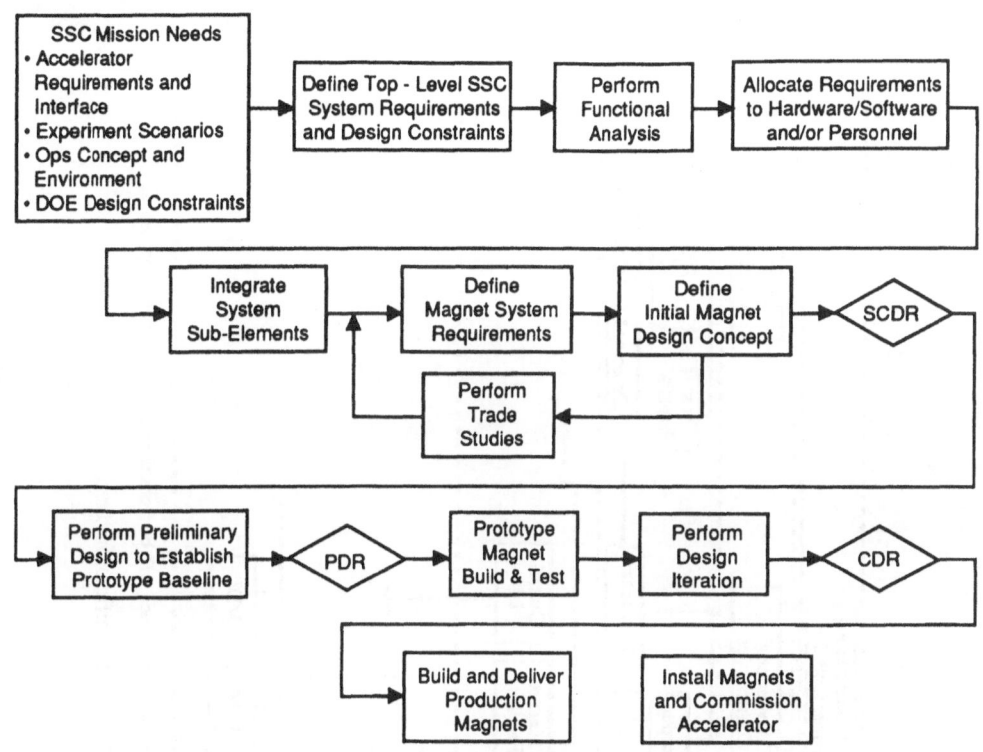

Figure 2. System Engineering Process.

Documentation Control

Design baselines have been or are in the process of being established for the three major families of superconducting magnets: the Collider Dipole (CDM), Collider Quadrupole (CQM), and the High Energy Booster (HEB) magnets. Documentation defining these design configurations including Prime Item Development Specifications (PIDS), drawings, and supporting analyses are provided to the magnet subcontractors as elements of the technology transfer program sponsored by the SSCL. Beginning with the first in-process review, the subcontractors system design review (SDR), the PIDS (B1 Specification) and configuration drawings are placed under internal subcontractor configuration control. Changes to this baseline documentation are tracked to ensure traceability of design through the development process. The PIDS is revised to a hardware product specification (C-Level) at the subcontractors' preliminary design review (PDR) and is placed under internal configuration control board (CCB) management. The subcontractor must notify the SSCL of any subsequent changes to the PDR baseline design. Subsequent changes to the baseline may be implemented only through formal (internal subcontractor) engineering change proposals (ECPs). The next milestone in the control process occurs at the subcontractors' formal critical design review (CDR) at which time the baseline is "frozen" and any subsequent changes require formal ECP submittal and review and approval by the SSCL prior to implementation. The SSC specification hierarchy is depicted in the Superconducting Super Collider Specification Tree (Figure 3).

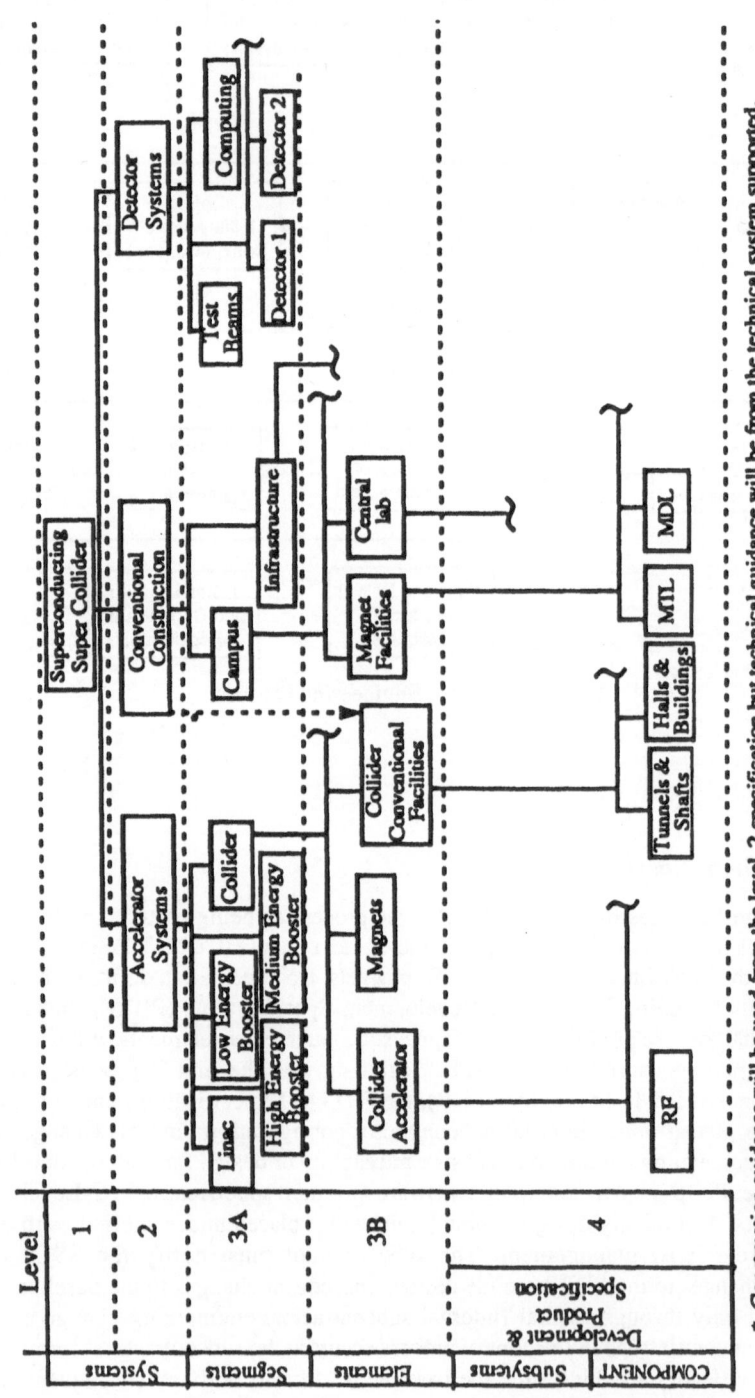

Figure 3. Superconducting Super Collider Specification Tree.

* Construction guidance will be used from the level 2 specification but technical guidance will be from the technical system supported.

1136

THE SYSTEMS ENGINEERING PROCESS

The systems engineering process is really quite simple and is defined as the methodical, analytical approach to defining top-level (the SSC) system performance and functional requirements, and decomposing these into lower-level (magnet and other system segments) performance and functional requirements. Once the magnet system performance and functional requirements have been adequately defined, these requirements can be allocated to each of the magnet families, i.e., CDM, CQM, HEB, and other specialty magnet designs. A further decomposition of magnet system functions will then lead to the identification of a complete set of magnet functions required to support SSC system-level performance within the operational environment. These functions are then analyzed to define magnet interfaces and the optimum allocation to magnet hardware and/or software, operational and support equipment, and personnel consistent with the operations concept for the SSC. The process flow shown by Figure 2, Systems Engineering Process, consists of the following distinct tasks.

Definition of Top-Level System Performance Requirements and Design Constraints

The first step in evaluating SSC concepts is to determine all factors which must be considered in satisfying SSC mission objectives. This mission analysis results in the translation of the accelerator functional requirements into design requirements. Needs expressed by the high-energy physics community—the ultimate users of the SSC—are formulated as a set of operational objectives and are quantified in broad terms and basic functions which satisfy operational SSC needs. In addition to these operational functional and performance requirements, certain design constraints are imposed by the Department of Energy and other government regulatory agencies. Mission analysis then proceeds to identify and express requirements in measurable parameters which state the needs in terms that the magnet technologists can use to develop magnet design concepts. The following inputs are typically considered in the establishment of the accelerator and accelerator components in arriving at performance requirements and design solutions.

- SSC Mission Needs

- Accelerator Requirements and Interfaces

- Experiment Scenarios

- Operations Concept

- Operational Environment (Site-Specific)

- DOE Design Constraints

Functional Analysis

Functional analysis is the method for analyzing performance requirements and decomposing them into discrete tasks or activities. This process involves the decomposition or parsing of the SSC functions into subfunctions at ever increasing levels of detail; it supports the mission analysis in defining functional areas, sequences, and interfaces. Functional analysis also provides an understanding and permits the development of the SSC accelerator configuration and its operational capability. Functional analysis is used by the engineering and physics specialties and support organizations to develop requirements for equipment, software, personnel, and operational procedures to complete implementation and commissioning of the SSC.

Flowdown and Allocation of Requirements

Completion of the functional analysis will result in a hierarchical tree structure which progressively divides and allocates requirements to the lowest level (component or sub-routine) that fulfills a definable requirement. This approach permits hierarchical modularity where independent modules are arranged in downward levels of increasingly specialized functions to promote management efficiency. Advantages in design, design verification through modelling, simulation, testing, and design iteration for enhancement of performance are realized.

The decomposition of system functional areas or segments, each of which satisfies an allocated portion of the basic system functions, permits the clear definition of interfaces and documents the interface requirements. Ideally, decomposition is a totally logical process, carried to completion without need for definition of a physical design concept or solution. Adherence to this process prevents the premature definition of design solutions with improper constraints and limitations imposed.

System Integration

Tradeoff Analysis

Trade studies are performed throughout development of the magnet configuration to ensure that the best solution is selected to satisfy accelerator requirements, including SSC performance objectives, cost, schedule, reliability/availability, and maintainability. During the concept design phase, trade studies are performed to establish the optimum magnet configuration. Once a subcontractor has been selected, trades studies serve as the tool for analysis of the detailed design for each magnet configuration. Trade studies are applied to the production phase for evaluation of production alternatives to ensure repeatability and producibility. Trade studies are developed and implemented as the design process progresses toward the definition of the magnet system requirements.

Magnet System Requirements Definition

Definition of the magnet system requirements typically focuses on the following specific critical issues of magnet design:

- Field Strength

- Field Quality

- Luminosity

- Quench Performance

- Availability

- Endurability

- Transportability

- Operational Safety

- Others as identified during development.

Design Concept Definition

Through the process of performing and documenting the engineering analyses described to this point in the systems engineering approach to the design of magnets, a design concept evolves.

Establish Baseline Design

Baseline designs are established at three distinct levels to define and document the magnet configuration for management and control:

Functional. The initial, approved design configuration documentation describing a specific magnet's functional characteristics and the method by which each characteristic is verified—by test, analysis, or inspection—to demonstrate the achievement of the specified characteristic or set of characteristics.

Allocated. The initially approved documentation describing a magnet's functional characteristics allocated from the set of accelerator requirements, interface requirements with other associated items, additional design constraints, and the method by which each characteristic is verified—by test, analysis, or inspection—to demonstrate the achievement of the specified characteristic or set of characteristics.

Product. The initially approved documentation describing all of the necessary functional and physical characteristics of the magnet configuration (including a comprehensive summary of the other interfacing configuration items, systems or equipment), and the selected functional and physical characteristics designated for production acceptance testing and tests necessary for support of the magnet.

Magnet System Build-and-Test

Magnet system build-and-test is normally performed at a minimum of two levels of design completion. The first level of testing is performed on a model magnet to prove the functional feasibility of the design to satisfy performance requirements. The second level is a full-size prototype of the preliminary design which is tested to verify that all magnet performance requirements are satisfied. Prototype testing also verifies production tooling and the ability to produce the magnet in production quantities. Prototype units are also tested under accelerated life conditions to verify the ability of the final design to satisfy system integrity requirements under operational environments.

Iterative Design Process

Results of the test activities are analyzed and are provided as inputs to the design engineer to permit the correction of design deficiencies or enhancements to the design which were identified as a result of magnet performance under test.

Finalize Design

Upon documentation of any design changes made as a result of the design iteration process, the design is finalized through the conduct of a critical or final design review (CDR). Additional model magnets may be built to incorporate the design iterations evolving from the test program. Successful completion of the CDR results in the authorization to proceed with magnet production.

The following phases leading to the completion of the SSC system design and implementation are the final steps in arriving at an operational Superconducting Super Collider.

- Build and Deliver Production Magnets
- Install Magnets
- Commission Accelerator
- Operate Accelerator.

TEAM-BASED ORGANIZATION FOR COLLIDER DIPOLE MAGNET (CDM)

DEVELOPMENT AND PRODUCTION

M.D. Packer and L.R. Page

General Dynamics, Space Systems Division
PO Box 85990, MZ DC-9540
San Diego, Ca 92186-5990

G.C. Winters

Systemics
San Diego, CA

ABSTRACT

The most influential factor in developing a magnet design and the manufacturing processes capable of mass producing Collider Dipole Magnets (CDMs) for the Superconducting Super Collider (SSC) is the work system or organization design. It is essential that design of the organization balances the demanding quality, schedule and cost aspects of the SSC program with the extraordinary technological challenges of the CDMs. The organizational approach taken by the General Dynamics team is based on high employee involvement. This approach entails more widely distributed access to information, coordination and control of work, decision-making and rewards for overall performance. Implementation of this approach will apply team-based organizational concepts and proven methodologies such as concurrent engineering, work teams, skill-based pay and gainsharing. This paper focuses on the structural facets of the General Dynamics organization design to accomplish the CDM Program. Why this management approach is being taken, how it was developed and tuned for the CDM Program and how it will be incorporated in personnel staffing will be described in this paper along with general operational characteristics. The issues of pay and gainsharing, while recognized as vital constituents of the overall design and effectiveness, are not discussed in this paper.

WHY A NON-TRADITIONAL APPROACH?

The CDM production program has many unique features that require all contractors develop the capability to upgrade their methods of operation to ensure that they have the ability to effectively deal with a dynamic and complex process. All programs have their own dimensions of complexity and their own set of environmental dynamics to which suppliers and contractors are subjected.

We have elected to reduce risk, cost and schedule by employing an organization with the capability and motivation to respond to changes in the economic environment as the program passes through successive stages.

The requirements that this organization must satisfy include:

- FLEXIBILITY to allow employees in the lower levels of the organization to establish strong alliances to strengthen their ability to increase performance within a well defined set of organizational values that are uniformly understood and well communicated throughout the organization.

- EFFECTIVENESS, without requiring extensive resources.

- SMALLER and relatively self-contained business units that can focus on the success of the enterprise because each member understands the operation.

- ADAPTABLE to permit modification of the basic business unit to enhance group performance.

- SKILLS DRIVEN because the organization itself thrives on the skills, knowledge and quality of worklife of its members.

The organizational focus we are developing is one which assures that all personnel are fully aware of the critical elements for organizational success and are motivated to perform on those critical items. This concentration of focus carries with it the management responsibility of ensuring excellent communications throughout the organization.

Employees must have enough information and be thoroughly familiar with the values of the organization to know what is important to their operation. By definition, employees who are more productive do not seek someone to tell them what is important. The more they know about the operation, the less requirement those individuals have for supporting members of management and the larger their span of control.

In summary, we chose a nontraditional organizational design for this program for several reasons. First, it offers the opportunity to select fresh approaches for increased productivity, given that it is a new program start. Second, emerging management methods suggest that employees take the opportunity for increased performance if they are involved in a participative management process. Third, as employees are included in continuous communications and thoroughly involved in operations, they are well informed to take effective action on a broad range of issues without waiting for management direction. This improves responsiveness to changing business conditions and reduces the need for extensive middle management. When considered collectively, all of these factors point clearly toward the desirability to establish a high performance, team based organization.

STEPS TAKEN TO DEVELOP THE TEAM BASED APPROACH

It is widely acknowledged by authors and consultants that the implementation of a high performance organization cannot be accomplished by copying another organization's successful implementation. While the general flow of establishing a high performance organization is fairly common, the effective implementation is so unique to the nature of the business and the firm that it cannot be easily replicated. In this case, our design was influenced by our past experience in General Dynamics Energy Programs operating as a small team for a focused effort. The reliance on teamwork was a necessity for this relatively small group of people covering a broad area of technology. The adoption of the high performance team is a logical extension of our experience. Nonetheless, we have found it necessary to follow a systematic development of this philosophy. The following steps have been taken in starting up this program and are recommended for other groups considering the adoption of a team-based organization.

1. CONSULTANT SELECTION - The selection of a consultant to assist us in the design and implementation of a high performance organization is key. Most groups attempting to implement high performance management will need the services of a carefully selected consultant to achieve success with

designing the actual implementation. The consultant adds value by offering alternatives and by objectively assessing actions taken to achieve the objective. In selecting our consultant, we chose to emphasize:

    a.   Their credibility both academically and pragmatically.
    b.   Previous experience with General Dynamics.
    c.   Publications produced by the consultant organization.

2. A PRELIMINARY WORKSHOP was conducted for senior management and the heads of functional departments of the division where they were provided an overview of the concept. In this workshop, they were given the opportunity to confirm their interest in attempting to implement an organization of this type. They confirmed this interest and endorsed a proposal to employ the high performance, team-based organization on the CDM program.

3. A CDM TEAM MEMBER WORKSHOP was conducted to assist the team in its development of a mission and values statement to be incorporated in the CDM program.

SELECTED ORGANIZATION DESIGN AND VALUES

Organizations operate on a foundation of values widely held by the members of the organization. The extent of congruency and consensus of these values establishes the overall effectiveness of the organization. Therefore, it is paramount that these core values and guiding principles be clearly stated and continually emphasized. The genesis values for the General Dynamics CDM Program are outlined below:

OUR MISSION

We will produce quality superconducting magnet systems that are uniform and reliable. We will be efficient and profitable. We are dedicated to continuous improvement of everything we do.

HOW WE WILL ACCOMPLISH THIS

- PEOPLE - We will provide equal employment opportunity. We seek competent and ethical people who are committed to achieving our mission in an open and trusting team environment.

- PLANT MANAGEMENT - We will develop high performance work teams where members will be actively involved in goal setting and decision making. We will continually assess plant performance against our goals and take appropriate steps to meet them. We will provide resources and support to achieve continuous improvement. We will ensure open communication and be responsive to employee ideas and suggestions. Like any plant, we have rules and regulations that must be observed in order to maintain a safe, effective work place.

- WORK ENVIRONMENT - We will maintain a physical environment that protects our health, safety and well being -- one that fosters pride. We will treat each other equally and remove barriers that rob us of dignity and respect.

- JOBS - The work will be performed by teams of multi-skilled employees who will play an active role in managing their work.

- COMPENSATION - We will reward for acquired skills and plant performance.

- TRAINING AND DEVELOPMENT - Training and development are critical to the success of our operation and will be an on-going activity in which we participate both as learners and teachers.

- CUSTOMERS - -We will form a partnership with our customer, based on teamwork and open communications. We will deliver magnet systems on time that meet or exceed performance specifications.

- SUPPLIERS - Quality suppliers are critical to our success. They are essential members of our team. Together we will build strong, long-term working relationships.

- FAMILY - We recognize that we are members of families that have needs. We will act in a manner that promotes pride in their relationship with us.

- COMMUNITY - We will be actively involved citizens and foster the well-being of the community.

- COMPANY - We will be a profitable, well managed business segment of General Dynamics Corporation. Our performance will enhance the image of the company.

It is the extraordinary challenge of the CDM Program that shaped the values articulated above. It is these values that shape the structural design of an organization that effectively responds to this challenge. The following paragraphs describe the organization design believed to be most responsive to CDM Program needs by the General Dynamics Team.

The organization structure divides labor into two high-level groupings: The management team and the work team. There are two stages of work teams: concurrent engineering and manufacturing production.

The management team subdivides to functionally encompass (1) the program manager and (2) the support staff, which includes quality assurance, engineering services, human resources, master scheduling, procurement, information resources, finance and accounting, and facilities services.

The work team is subdivided between product assembly elements of comparatively high process interdependencies. Those subdivisions emerge most clearly as: (a) coil manufacturing, (2) cold mass assembly, and (3) final assembly and checkout. Once in production, each of these primary work teams is further subdivided into five to six focused work centers. Since these work centers will operate collectively as a team, this paper maps work at the primary team level only. The teams themselves will further map internal responsibilities and tasks as part of their charter.

Responsibility and task mapping or work design will be described in this order: Program Manager, Work Teams and Support Staff.

The **Program Manager** has ultimate organizational responsibility for development and implementation of overall strategies, establishment and administration of policy, and organization performance with respect to contractual obligations.

The Program Manager is tasked to:

- Create, articulate and persistently drive to completion the visions for the organizational future.

- Evaluate the activities and capabilities of similar enterprises and take action to improve competitive capabilities.

- Inform organization members of business conditions, other environmental factors and plant-wide performance.

- Help organization members recognize the need and time for change.

- Ensure that all environmental conditions are adequately considered by organization members in all that they do.

- Sort out and crisply articulate to affected team leaders the directions for handling multiple and sometimes conflicting priorities, needs and interests.

- Develop leaders and productive citizens

- Develop, inform, and empower production teams and support staff members to responsibly make all decisions necessary to achieve organization goals.

- Acquire and commit resources needed by production teams and support staff members to achieve organization goals.

- Ensure all policy is administered consistently and fairly throughout the organization

- Exhibit community citizenship through actions.

The Work Teams are completely responsible for achievement of jointly-established organization goals for quality, schedule and cost of production efforts and support needs. The Work Team Leader has ultimate responsibility for team performance.

The Work Team Members are tasked to:

- Develop operational and functional responses to implement organization strategies or adapt to changes in the business conditions. These responses include:

    - Staffing
    - Training and development
    - Work Assignment
    - Quality control
    - Performance Measurement

- Accomplish all actual tasks and directly supporting tasks necessary to transform raw material and purchased components into delivered product.

- Measure, discuss and improve the team's quality, schedule and cost performance from a process perspective.

- Review, discuss and develop strategies to improve plant's quality, schedule and cost performance from a process perspective.

- Review, discuss and develop strategies to address the implications of business conditions.

- Exhibit community citizenship through actions.

The Work Team Leaders are additionally tasked to:

- Cause effective team building within and between work groups and individual contributors.

- Ensure that policy, rules and regulations are appropriately administered.

- Ensure that all team members are fairly treated with dignity and respect.

The Support Staff members are responsible for providing value-added technical and administrative support to the Work Teams and the Program Manager. This support generally involves information and knowledge.

The Support Staff members are tasked to:

- Articulate external factors and conditions that have internal implications.

- Articulate internal factors and conditions that have external implications

- Coordinate the flow of incoming and outgoing information.

- Monitor and facilitate internal information.

- Provide areas of expertise consulting and direct support (training and development) as needed by Work Teams to accomplish their work.

- Perform audits of production and plant-wide processes.

- Articulate and audit results and strategies for improvement.

- Exhibit community citizenship through actions.

Summarizing at the highest level:

- The Program Manager sets the organization strategy and integrates all efforts to ensure its implementation.

- Work Teams provide all touch labor and their directly supporting efforts to implement the organization's strategy.

- The Support Staff provides knowledge and information needed by the Production Teams to implement the organization strategy.

SELECTING THE RIGHT KIND OF PEOPLE

It is clear that the kind of organization described above is not for everyone. It takes individuals who:

- Speak up - actively contribute to team discussions and goal setting

- Commit themselves to consensus

- Can admit to being in error

- Take responsibility for their own behavior

- Dedicate themselves to lifelong learning, growing and developing as human beings.

A person's interpersonal, communication and problem solving skills must be unusually strong to exhibit the above characteristics and desires.

Selection of people for the General Dynamics CDM team is a critical process. While the primary objective of this process is to select high potential employees, an important secondary objective is to use the selection system as an expression of the facility's operating philosophy. In this way, successful applicants will have been exposed to the values and beliefs of the organization from the moment they submit a completed application. "Orientation" does not start after hire; it begins with the hiring process itself.

The process is rigorous and thorough. It provides the applicant six opportunities to demonstrate excellence and gives the organization five distinct levels of applicant analysis. It gives the final approval of who gets hired to the employees themselves in the sixth gate. It produces new employees who are highly motivated, know they are the best, know they are surrounded by excellence, and know they are responsible and accountable for creating the best organization possible. This process flow is shown in Figure 1, the Six gate process diagram.

ASSURING THE RIGHT TRAINING FOR TEAM MEMBERS

A high performance management (HPM) organization cannot be considered for long without becoming aware of the high emphasis on training and development. This has been consistent among all firms who achieved some degree of success with this concept in their operation. The experience of the Robert Bosch Corporation plant in Charleston, SC is that employees spend 20% of their time in training, eight hours out of every week. This recurring training takes place after these employees have already undergone a very thorough pre-hire and entry level training program.

Another common experience among firms implementing HPM is evaluation of the training process itself. With such a pervasive and high volume requirement for training, these firms tend to look for ways to improve the effectiveness of training. They tend to break up training into smaller modules that can be tailored for specific requirements. These smaller

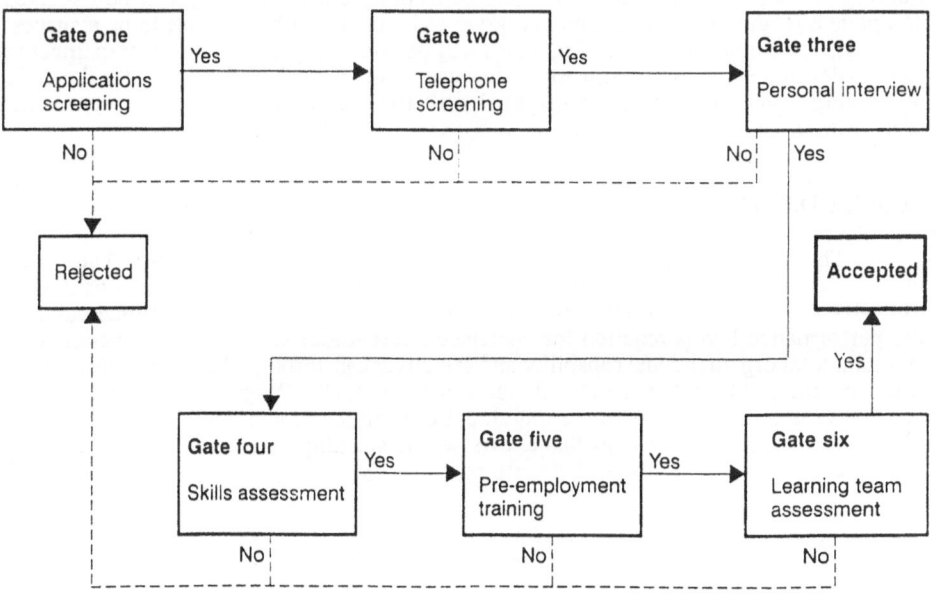

Fig. 1 The Six-gate process enhances selection confidence

units also make it possible to have team members teach much of what is required without relying exclusively on professional trainers.

A side benefit of using the smaller units of training, specifically tailored to the situation and delivered by team members, is that it tends to occur in the environments where it will be used. Delivering training at the point-of-need enhances the value of the provided training and ensures optimum sensitivity to the environment.

Start-up organizations require new skills as the organization evolves. Even if training is designed to meet every expected need during implementation, that will change. Implementing HPM requires a high degree of both organizational and personal learning. The organization must learn about itself and be constantly alert to the need for change. The focus on performance and the willingness to undertake change in the interest of performance implies that the basic training for HPM organizations consists of those skills that will allow team members to address and resolve problems as members of a team. Personal learning becomes important because individuals must assume responsibilities which extend beyond being just technically proficient in one's assignment. They will need to develop the skills to function well as the member of a team. Therefore the training they require will include:

- TEAM SKILLS to help them perform effectively as team members

- BUSINESS SKILLS to help all members correctly interpret business conditions and assess the economic impact of alternative courses of action being proposed to resolve problems.

- PROBLEM SOLVING SKILLS and methods to permit the team member to systematically approach and resolve diverse issues that impact the business at all levels.
- TECHNICAL SKILLS will be an on going need because the dynamic nature of the organization will require change for improved performance.

Companies implementing HPM are uniform in their assertion that a steady and consistent training program is required for successful implementation of HPM. What we will employ is likewise strongly rooted in training. Employees will receive 160 hours of pre-hire training and an average of over 80 hours of training annually. They will be expected to complete a lengthy pre-hire training program to be provided by the state to be a successful candidate for employment. Training topics will be wide-ranging and the team members will need to develop skills as an instructor so they can assume some responsibility for the delivery of training. This will become an essential skill for long term professional success in the organization.

CONCLUDING REMARKS

The thoughts expressed in this paper are really not new thoughts. They involve proven methods and operating philosophies that are carefully aligned and specifically tailored to satisfy the peculiar demands of the CDM Program, the desired organizational values, and the performance levels required for enhanced shareholder value. It is our belief that the resultant total organizational capability and effectiveness, through the contributions of every team member, will dictate the overall success of the CDM Program and, in turn, the SSC project. It is this linkage between individual contributions and SSC Project success that plants the seed of human energy that allows people working together to achieve the greatest successes.

REFERENCES

[1]   Lawler, E. E., "High Involvement Management," Jossey-Bass Publishers, San Francisco (1986)

[2]   Mohrman, S. A., and Cummings, T. G., "Self-Designing Organizations," Addison-Wesley, New York (1989)

ACKNOWLEDGEMENTS

The work described herein shall be accomplished under contract to the Universities Research Association in support of the Superconducting Super Collider project for the U.S. Department of Energy.

DESIGNING SUPERCONDUCTING MAGNETS

FOR RELIABILITY AND AVAILABILITY

Craig S. Arden

Magnet Division
Superconducting Super Collider Laboratory*
2550 Beckleymeade Avenue
Dallas, Texas 75237

Abstract: The successful operation of accelerator systems such as the Superconducting Super Collider (SSC) is limited by the ability of the system design engineer to design components, including the superconducting magnets, which are capable of extended operation under all encountered environmental conditions. The reliability of these critical magnet components can have a major impact on the availability of the SSC, and can mean the difference between success and failure of SSC mission objectives.

The availability requirement for the SSC has been specified at 80%; when allocated to reliability at the magnet level, this translates to a mean-time-between-failure (MTBF) greater than ever before achieved. Obviously, this presents a major challenge to the reliability engineer, who must ensure that appropriate measures are taken to design reliability into the magnet at all levels, starting from the beginning of the design phase. No amount of quality control inspection will provide the assurance necessary to achieve reliability figures of this order.

This paper presents a logical system approach to the design of superconducting magnets which, when adhered to, will result in magnets of adequate reliability to meet SSC availability requirements. The paper shows the steps to be taken to identify potential failure modes and the feedback process required to ensure that reliability, integrity, and availability are considered at all steps of the design process. The issues of magnet maintainability and the impact of maintainability on system availability are also discussed.

---

* Operated by the Universities Research Association, Inc., for the U.S. Department of Energy under Contract No. DE-AC02-89ER40486.

INTRODUCTION

The Superconducting Super Collider Laboratory (SSCL) Conceptual Design Report (March 1986) sets the availability goal for the accelerator at 80%. The apportionment of the availability goal to the accelerator subsystems, and from the subsystems down to the major sub-subsystems such as the Collider Dipole Magnets (CDMs), is performed in the Reliability Allocation Analysis.

From the present Reliability Allocation Analysis, the Collider has an 82.4% availability, which further allocates an 88.2% availability goal for the magnets and interconnects. It can be shown that each CDM will have an availability of 99.99954%, which correlates to a 6305 Years Mean-Time-Between-Failure (MTBF) goal. This MTBF is based on the 10.5 Days Mean-Time-To-Repair (MTTR), which is defined in the Availability Analysis. From the availability requirement, the scheduled up-time, and the unscheduled down-time, it follows that an average of 1.3 CDMs can fail per year during the scheduled up-time.

As a result of the availability analysis, it is evident that the reliability goal for the CDM is an enormous technical challenge. The available data from national laboratories (e.g., Fermilab, Brookhaven, and Lawrence Berkeley) and from international accelerators provide information on design features to be avoided. However, the SSCL CDM design is differs to an extent that little confidence is obtained from analysis of historical data. The conventional statistical approach to reliability achievement and verification is not economically feasible due to the cost and time that would be required and the program delay that would result if the test failed.

The product integrity-based approach was proven effective by the aerospace industry and provides a systematic approach to reliability assurance suitable for SSC magnets. The following discussion is fairly specific to the CDM, but the same methodology is presently planned for the Collider Quadrupole, High Energy Booster Magnets and other special superconducting magnets.

INTEGRITY

At the Magnet Systems Division of the SSCL, the reliability program is based on the Integrity Methodology, a systematic, integrated, closed loop of reliability tasks performed in concert with the design/development/production tasks to ensure that the integrity built into the final product is compatible with lifetime reliability requirements. Figure 1 shows the basic order in which the tasks are performed and how the tasks relate to one another. The integrity approach includes a Failure Modes Effects and Criticality Analysis (FMECA), a Durability Analysis, a Test Program, a Failure Reporting Analysis and Corrective Action System (FRACAS), and a Maintainability/Prognostics task.

The emphasis of the integrity program is on identifying potential areas of significant problems and devising a scheme to work on the issues in a prioritized manner. From the availability allocation it was evident that the CDM would provide the best return for the effort. Reasons include the large number of CDMs (8460 units), the minimal experience building these magnets, the high amount of cross application to the other magnets to be studied, and the 25-year life requirement with a total of 33.5 failures permitted.

Conventional reliability programs are suitable for equipment with a permissible number of failures similar to a reference system that is available for comparison. This approach is usually started with an allocation and prediction using "same as" equipment, "similar to" hardware, mechanical handbooks, and electrical handbooks. The conventional approach may not include a FMECA, but it would usually include a FRACAS initiated at the qualification test. Unfortunately, many conventional reliability programs end with the approval of the required data items and the approved qualification report, with no assurance that the reliability provisions are implemented in production.

FMECA

The primary intent of the FMECA is to identify critical items and place them in a prioritized list. The critical item list is the starting point for the durability analysis. The FMECA is designed to identify all possible failure modes, the actions required to eliminate or control the risk, the failure detection method for each failure mode, the compensating provisions, and the fault isolation provisions. To accurately obtain this goal, the FMECA requires a closed loop review process, as shown in Figure 2. Such a process is being used at the Magnet System Division of the SSCL. The document is reviewed not only by Magnet Design Engineering but by other departments as well, including Quality Assurance and Test and Manufacturing. Also, due to division overlaps it was important that the Accelerator and Physics Divisions reviewed the FMECA. The best way to get strong support is to have the division head support the review. Due to inter-department and inter-division effects, success of the effort is strongly dependent on division head support of the review and action item assignment. The review of the FMECA is a tedious and time-consuming process, but it is absolutely imperative in order to close out any open issues raised in the FMECA.

The FMECA Worksheet, Table 1, consists of 12 fields, each of which was specifically defined to permit consistency within each field and within the document. The fields include the item name, item number, failure mode, failure cause, item/component effect, magnet effect, collider end effect, failure detection method, failure rate, severity class, remarks, and failure prevention. Each field is self-explanatory from its title. Note that the failure effect was reviewed at the level of the component which failed, then at the magnet level, and finally at the collider level.

The CDM FMECA analyzed 73 items and identified 181 possible failure modes with 488 failure causes. Forty-seven failure modes were classified as critical (defined as having potential to cause severe injury, radiation exposure, or major collider damage which will result in extended system down-time). No catastrophic failure modes (involving loss of life or collider) have been identified by the FMECA. Again, all the FMECA goals must be met in order to increase the level of confidence for the CDM.

DURABILITY

Once the preliminary FMECA is in the initial review cycle, the Durability Analysis is the next effort to be performed. The Durability Analysis initially takes each critical component identified in the FMECA and examines all the steps required to go from a drawing to an end item operational in the collider tunnel.

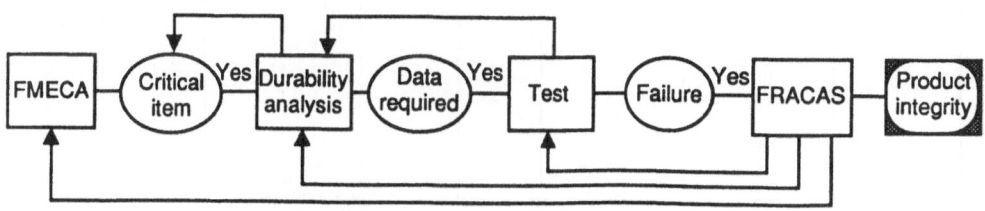

Figure 1. Integrity Reliability Tasks.

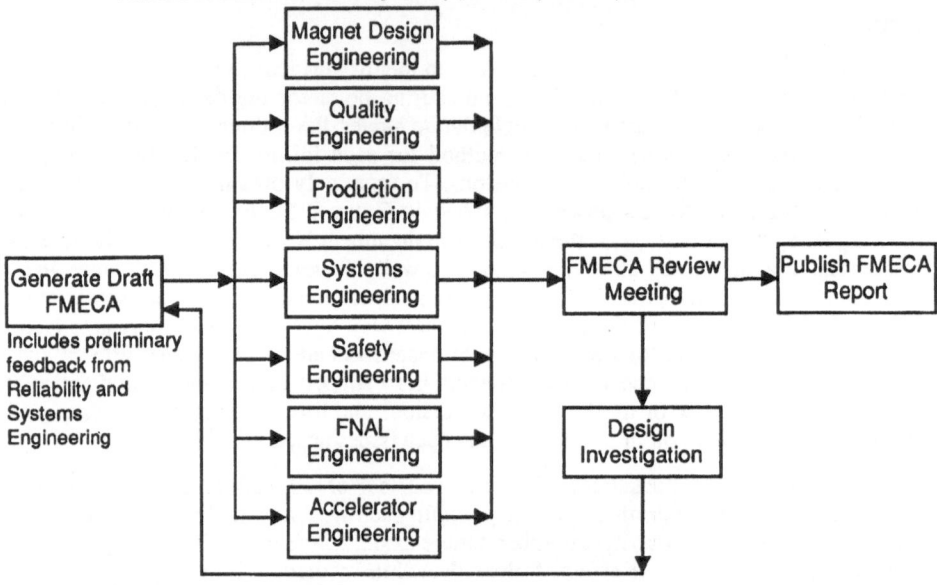

Figure 2. FMECA Review Process.

Table 1. CDM FMECA Worksheet.

ITEM NAME	ITEM NO.	FAILURE MODE	FAILURE CAUSES	ITEM EFFECT	MAGNET EFFECT	COLLIDER EFFECT	DETECTION METHOD	FAILURE RATE	SEVERITY CLASS	REMARKS	FAILURE PREVENTION

The intent of the analysis is to ensure that the design/development process has properly analyzed each part for life-determining stress levels, has selected materials with characteristics compatible with reliability and life requirements, and has developed manufacturing methods and controls consistent with material limitations and process variations. Once this effort is complete, quality control procedures can be expected to ensure production of reliable magnets. Table 2 shows the basic form used to document this analysis. The durability worksheet is composed of six fields: Environment, Item, Level, Specification, Method/Vendor Margin, and Method/SSCL Margin.

The Environment field includes subtopics such as manufacturing, test, packaging and handling, transportation, storage, installation, operation, and maintenance. The Item field includes topics such as temperature, humidity, stress, vibration, shock, magnetic cycles, thermal cycles, and quenches. The Level field allows a cursory look at the worksheet to quickly identify problem areas. The Level field may be 1) Acceptable, 2) Marginal, or 3) Unacceptable. Any Level 3 is highlighted on the worksheet for ease of identification. The Specification field is used to identify the requirement and where it is defined. The Method/Vendor Margin field provides room for a brief written explanation of the vendor's analysis, methodology, and margin. The Method/SSCL Margin field provides a place to state the SSCL's analysis, methodology, and margin for that specific item.

The manufacturing environments are unique for each part under analysis. To properly address the various manufacturing environments, the field permits sub-environments under the main manufacturing environment. Some examples of the manufacturing sub-environments include forming, pressing, welding. and inspections. The unique details studied during the manufacturing phase permit a strong understanding of the item and its characteristics to be obtained, deviations from requirements to be identified, and control processes to be specified, implemented and verified as effective.

Results of the Durability Analysis and the FMECA will help identify and concentrate efforts on the highest risk items first; lower-risk cases can be addressed later. Once the Durability Analysis is completed on a component/item, the level of confidence that integrity can be achieved and maintained over the life of the item is significantly increased. Information acquired on the item under analysis is reflected back into the FMECA for completeness and technical accuracy. If the Durability Analysis determines an unacceptable unknown, then a test may be defined to reduce the risk to an acceptable level.

TEST

Various tests are conducted simultaneous with the above analyses. Unlike conventional reliability programs, which do little or no tracking or failure analysis until the formal qualification process, the integrity approach uses as much of the development testing as possible to help find new modes and causes of failure and to take corrective actions. This approach allows the risk to be reduced as the program moves toward production by validating hardware performance and eliminating in a timely manner any potential design weaknesses discovered. Testing includes CDM component/item tests, CDM acceptance testing during assembly, performance testing, and development accelerated- life testing. The life test program, coupled with the product integrity design-for-reliability approach, will provide magnets that meet SSCL requirements.

Durability testing of completed magnets is planned for the immediate future. A full-length, 40 mm CDM is presently scheduled for extended cycle testing. The test goal is 28 quenches, 14 thermal cycles, and an equivalent of 7,400 magnetic cycles, as well as experience in life-testing a CDM. The test will make use of acceleration factors including operation 10% over the normal magnetic field and electrical power ramp rates up to 25 times the normal value of 4 A/sec. Life-test of a 50 mm CDM built to the Accelerator Surface String Test (ASST) configuration is planned.

Table 2. Durability Analysis Worksheet.

PART NUMBER DESCRIPTION	XX-XXXXXX ASSEMBLY		AVERAGE ENVIRONMENT 253-310 K 10-96% RH VIBRATION SHOCK 2g/1g/1g	LEVEL CODES 1-ACCEPTABLE 2-MARGINAL 3-UNACCEPTABLE INVESTIGATION REQUIRED	
ENVIRONMENT	ITEM	LEVEL	SPECIFICATION	METHOD/VENDOR MARGIN	METHOD/SSCL MARGIN
MAGNET					
PACKAGE/HAND	AMBIENT TEMPERATURE HUMIDITY SHOCK VIBRATION AIRBORNE PARTICULATES REQUIREMENTS PROCEDURES				
TRANSPORTATION	AMBIENT TEMPERATURE HUMIDITY SHOCK VIBRATION				
INSTALL	AMBIENT TEMPERATURE HUMIDITY AIRBORNE PARTICULATES SHOCK VIBRATION				
OPERATION	MAGNETIC CYCLES QUENCH CYCLES THERMAL CYCLES AMBIENT TEMPERATURE HUMIDITY SHOCK VIBRATION				
MAINTENANCE	AMBIENT TEMPERATURE HUMIDITY AIRBORNE PARTICULATES SHOCK VIBRATION				

Component testing plays an important role by allowing individual components to be evaluated prior to being included in the magnet design. Component testing provides early feedback during the design process and permits development magnets to include "proven reliable" items. Component testing to date includes 40 mm re-entrant posts, electrical bus expansion loops, and aluminum-to-steel cryogen transition joints. Component testing presently planned includes 50 mm re-entrant posts, cryogen bellows, electrical insulation, cable splices, and instrumentation connections. Any component determined from the durability or failure modes analysis to require individual testing is expected to be tested during the magnet development phase.

The lessons learned from these tests provide the insight needed for long-term magnet production. Failures during manufacturing also provide insight as to which present tests are inadequate and which tests if any need to be incorporated into the quality inspections. Also, by examining development assembly problems, the design engineering and production groups can iterate the design or obtain understanding of the design margin.

This explanation of how reliability engineering approaches testing while functioning under the integrity methodology makes it plain that it involves not only the design engineering group, but the quality assurance, production, and test engineering groups as well. Constant communication between these groups is imperative to making the system a positive entity.

FRACAS

To ensure that the test program results are fed back into the design, FRACAS is applied to all testing being performed. FRACAS is a closed-loop discrepancy reporting system that collects/analyzes data and ensures that corrective action is taken when discrepancies are reported on procured/fabricated items.

FRACAS covers discrepancies or non-conformance during the following tests or operations: 1) development test, 2) functional acceptance test, 3) reliability test, 4) systems and integration test, 5) verification test, 6) testing and/or repair associated with hardware returned from field usage, 7) tests performed at vendors, other labs, and the SSCL test lab facilities, 8) SSCL operations, 9) manufacturing non-conformance, and 10) other discrepancies identified in tooling, manufacturing procedures, or other areas directly impacting quality of the end item .

Figure 3 presents the basic steps of FRACAS. Notification of the failure by the receipt of the data report leads to convening of the Failure Review Board (FRB) "As Required" to discuss the failure and the proposed failure analysis plan or the proposed corrective action. The failure analysis plan is assigned to a cognizant engineer. Once the cause of the failure has been determined, corrective action is defined, agreed upon, implemented, and verified.

For Class 1 failures—defined as any affecting form, fit, function, reliability,or safety—initial failure notification is required within 24 hours of discovery by either the subcontractor, the national laboratories or the SSCL. Once notified, the FRACAS coordinator provides failure notification simultaneously to the product manager, Engineering, and Production and Quality Assurance. The product manager has the authority to initiate an investigation of the discrepancy and analysis of the data and to take any repair and/or corrective action required as a result of the discrepancy. This includes involvement of the Failure Review Board.

Within five working days of the initial failure notification, the responsible party is required to provide any additional data from the original notification and a plan to perform any necessary failure analysis. Within 30 days the reporting organization is required to provide the completed failure investigation data, with a determination of cause, suggested corrective action, and possible verification method. Note that the Failure Review Board has final approval of the failure report and corrective action proposal.

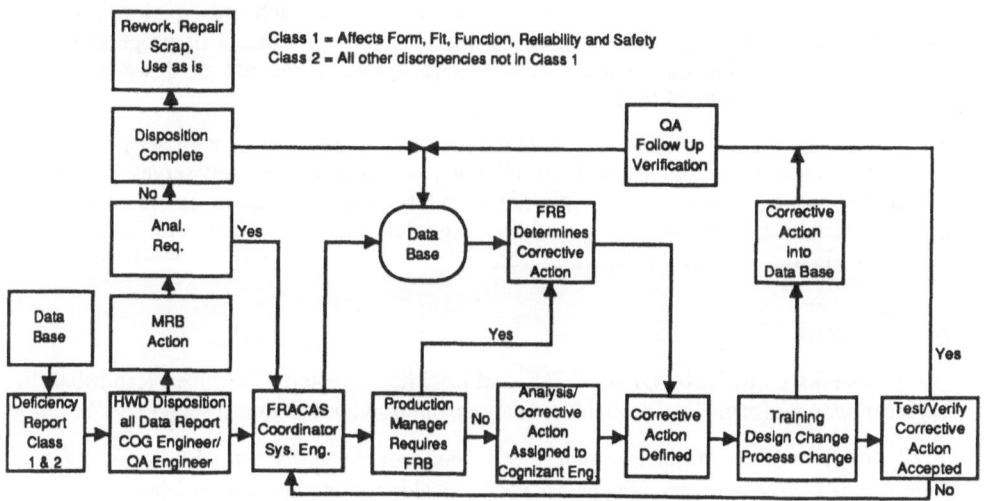

Figure 3. FRACAS Flowchart at SSCL.

Reliability Engineering is responsible for maintaining the FRACAS data base, which includes all the magnet Class 1 failure analysis data and can link to the Class 2 failure data and performance test data loaded into the MSD main computer. The FRACAS data base provides trend analysis capability, input for status meetings, and support for the monthly FRACAS Summary Report. The last two items are provided to and supported by the head of Magnet System Division. Reliability Engineering is also responsible for maintaining a record of data accumulated through FRACAS to update the FMECA.

MAINTAINABILITY

The program described above will provide the lowest number of magnet failures possible within the cost and schedule constraints of the overall SSCL program. However, it is not yet clear that "the lowest number possible" is less than the maximum number allowed by the availability allocation. Therefore, the use of prognostics may become essential to high availability achievement. Prognostics is defined as the ability to predict impending failures and to correct the situation before failure occurs. Prognostics depends on the ability to monitor the health of the magnet in operation and to track parameter trends toward failure conditions. Prognostics may be enhanced by the capability to relate failures and drifting parameters to production and/or operational variables such as material quality imperfections, acceptance test parameter values, or excessive-life consuming events (e.g., quenches and temperature cycles).

Such a capability could require an enormous database and widespread monitoring/recording system. Techniques to identify poor or weak magnets during operation must be addressed and investigated during the magnet development phase. Data obtained during the magnet testing constitute the initial step in determining the means and capability of a prognostic system. If accurate prognostics is feasible, it becomes possible to have more than the allocated number of failures occur.

SUMMARY

This report presents reasons why a logical system approach, the integrity methodology, was chosen over a classical approach to reliability achievement. The primary reason for this selection is the design of the CDM to be failure-free for its 25-year life. The paper reviewed each major reliability task of the integrity approach and explained how each task interfaces with other applicable tasks. Finally, the importance of prognostics was discused as a possible maintenance approach to optimize the use of down-time, thus increasing the availability of the SSC.

Reliable magnet achievement is in the hands of those who actually design, test, and deliver the magnets. Accomplishment of reliability program tasks merely serves as a tool for identifying issues—often interdisciplinary or inter-organizational in nature—that could preclude reliability achievement. Timely management action to resolve these issues is a fundamental requirement for program success.

ACKNOWLEDGMENTS

Thanks go to Kirby Dixon, SSCL and Lockheed Systems Engineer, lead reliability engineer, who performed the availability analysis and provided technical support.

Also to James Franciscovich, SSCL and Lockheed Systems Engineer, Systems Engineering Group Manager, who provided technical input and document review.

And to Jim Potter, SSCL and Lockheed Systems Engineer, lead reliability engineer, who provided technical support and document review.

COLLIDER DIPOLE MAGNET TEST PROGRAM FROM DEVELOPMENT THROUGH PRODUCTION

R. E. Bailey

General Dynamics, Space Systems Division
PO Box 85990, MZ DC-9540
San Diego, Ca 92186-5990

ABSTRACT

Verification of CDM performance, reliability, and magnet production processes will be accomplished during the development phase of the program. Key features of this program include thorough in process testing of magnet subassemblies, verification of the magnetic field quality, and demonstration of the CDM performance during the formal qualification program. Reliability demonstration of the CDM design includes component tests and an accelerated life test program. The Prototype magnet phase will address achievement of magnet performance goals through a program of fabrication, test, analysis, redesign as required and procurement of modified parts for a second fabrication run. This process would be repeated again if necessary, and would conclude with a final design for the production magnets. Production process validation will address the effects that key production processes have upon magnet performance, using the magnets produced during the Preproduction phase.

TEST REQUIREMENTS ARE TIED TO PERFORMANCE REQUIREMENTS

Test requirements and their relation to Collider requirements

During the design and development phase of the SSC program, the Prime Item Development Specification (PIDS) will become the basic requirements document for the collider dipole magnet (CDM). The establishment of valid test requirements, based on the reality of single magnet testing within a factory environment, and as driven by the PIDS performance requirements, will drive the development of all of the required test planning, test procedures and test equipment design. An example of this process, and the differences between the collider system requirements and the related CDM test requirements, concerns the specified single phase helium flow through a collider dipole magnet. The PIDS requirement for this flow during collider operation is:

Flow: 100 gm/sec.
Temperature of inlet helium: 4.15K
Temperature (maximum at inner coil): 4.35K

These requirements are based upon a full cell of magnets in series, as installed in the collider, with the inlet temperature for each cell at the 4.15K value, and the maximum allowable exit temperature for the last magnet in the cell at the higher value. The rise in temperature within the cell is made up of static heat input to the magnet cold mass from the cryostat and support system, and from beam heating due to collider operation. For single magnet testing, the nominal test conditions will be as follows:

Flow: 0 - 50 gm/sec. (Demonstrate 100 gm/sec. for the design)
Temperature: 4.3K (to achieve a coil temperature of 4.35K)

The rationale that supports these test conditions is that one should test at the more severe conditions (4.3K), and that one should not penalize the design of the cryogenic supply facility unnecessarily when adequate test conditions can be achieved with flow rates much lower than the collider will use. By way of comparison, the helium flow used in the 40mm magnet testing at the National Labs has been 30 - 40 gm/s. Other similar differences exist between other collider requirements for the CDM, and the related test requirements that have been developed for the single magnet acceptance test program. These include the CDM cool down and warm up rates,80K shield cooling flow, 80K coolant quality, 20K shield cooling flow, 20K shield temperature, and testing with or without simulated beam tube heating.

Test equipment design driven by test requirements

These derived CDM performance test requirements also drive the design of the test equipment and cryogenic test facility. We have developed the test requirements in conjunction with the SSCL to provide for an adequate test program at minimum cost. One key concern is the size (and cost) of the cryogenic refrigeration facility that is required to perform the cold magnet acceptance test. By accepting the lower test value for single phase helium flow (and other similar differences from the PIDS performance requirements) the test facility cryogenic refrigeration requirements are significantly reduced, resulting in a lower cost program which still addresses the important performance issues.

VERIFICATION OF CDM PERFORMANCE

In-process testing

A major concern in the testing of the production CDM is the establishment of a test program which will detect magnet assembly faults early enough in the assembly process to minimize the cost incurred by any such fault. To implement this program, we have planned an extensive test program from the insulation of the conductor through to the final CDM acceptance tests, both cold and warm. This process is shown in Figure 1-1. We intend to implement a resistance test of each magnet coil as it is being wound so that any turn to turn short can be detected and dealt with prior to completing the winding process. In like fashion, we intend to monitor the coil resistance and coil to ground resistance (high potential tests) during the coil curing operation.

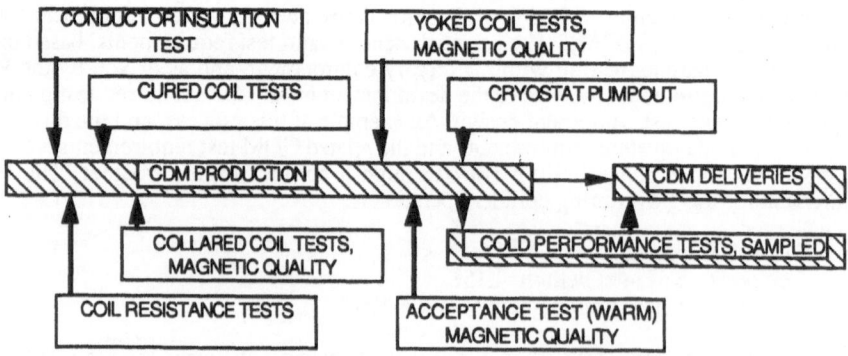

Figure 1-1 In-process and acceptance tests support production and deliveries

All of the in process and acceptance test stations will be computer based to provide automatic test equipment operation such as power supply current settings and voltmeter control and data acquisition, and also to provide the operator with process instructions. This computer interface will insure repeatable and predictable test results and minimize the opportunity for human error. Figure 1-2 illustrates the type of test stations (cells) that are planned. The acquired test data will be handed off to our shop floor control system and to the manufacturing resource planning (MRP II) system where the data will be available for archival storage, manufacturing process control, transmittal to the SSCL, and for Quality Assurance use.

Acceptance testing

Acceptance testing will provide a final check on the performance of a completed CDM. Warm acceptance testing will be accomplished at a typical test cell as illustrated in the preceding figure, and will include magnetic quality testing at low current levels. Cold acceptance testing will be accomplished at one of the two cold test stands located in the factory. The completed (and tested) magnets will then be prepared for shipment to the SSCL receiving facility. This preparation will have included pump down of the cryostat during acceptance testing to ensure that the majority of the absorbed contaminates have been removed. It will conclude with a purging of the cryostat and cold mass with dry gas and enclosing the magnet ends with hermetically sealed closures. These closures may include structural support to help minimize the transportation induced stresses applied to the magnet support system and cold mass.

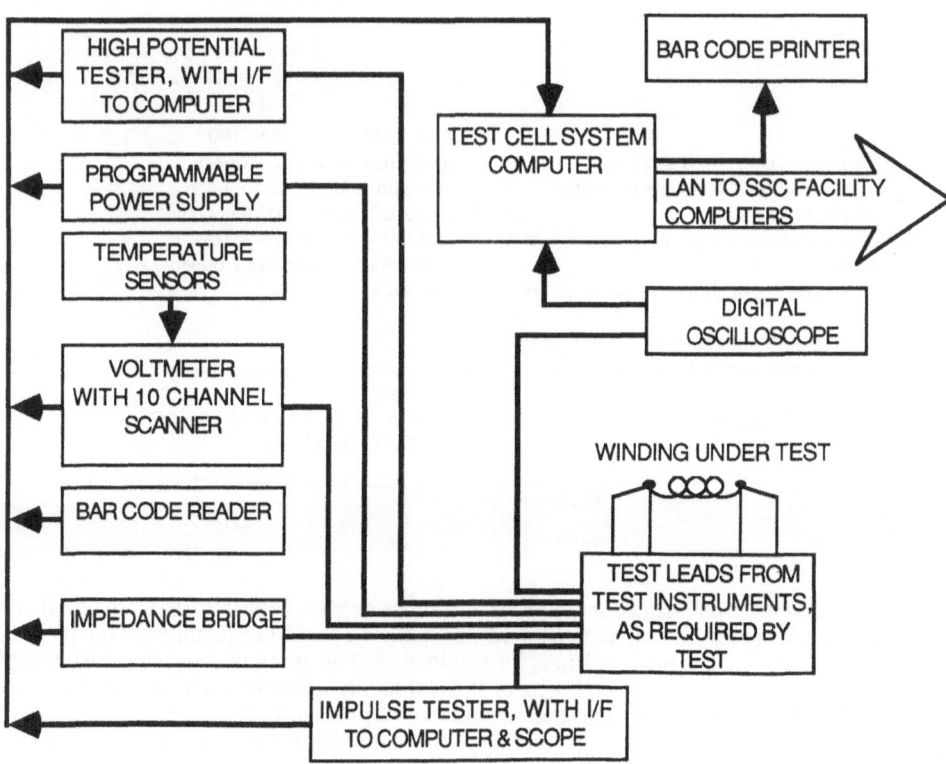

Figure 1-2 Generic Test Cell Arrangement

The cold testing of the completed CDM will be accomplished on each magnet in the Prototype and Production phases, and on a sampling basis during Rate Production. The primary aim of these tests is verification of CDM cold performance including magnetic field quality. The SSCL requirement is for a capacity to test a minimum of two CDMs per week. Our estimate of the test cycle time and test sequencing for two magnets being tested in two stands is given in Figure 1-3.

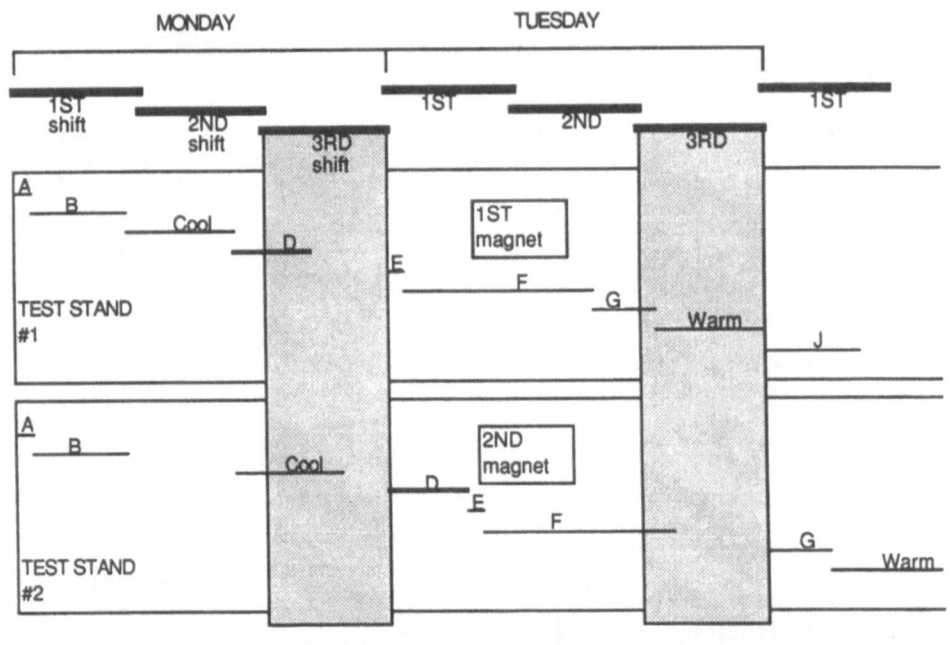

LEGEND:
A = room temp. elec. checks (1 hour.)
B = connect, purge & evacuate (6 hours)
C = cool down (7 hours)
D = cold elec. checks + low temp. cond. cycles (5 hours)
E = quench cycles at short sample current (1 hour)
F = cold magnetic measurements (12 hours)
G = di/dt quench, heater induced quenches (4 hour)
H = warm up (7 hours)
J = disconnect (6 hours)

Schedules based upon 100 g/sec. flow during cool down/ warm up, taking 7 hr. for each operation; no simultaneous cool down or warm up operations; 3rd shift when needed; 7 days per week.

TEST SCHEDULE FOR TWO TEST STANDS, PRODUCTION TEST CYCLE TIME OF 49 HOURS TOTAL

Figure 1-3 Cold test cycle time

Using these estimates (based on the test times expected for the 1000th magnet), our two test stands could test up to 5 magnets per 7 day week. Other features of the planned cold test facility include the provisions for eliminating all welding of cryogenic connections between the magnets and the end cans of the test facility, development of a high current splice design for the test stands that would not require soldering, and elimination of bellows for the test stand by incorporating flexible metal hose in a manner similar to the DESY cold test stand design.

Cold to warm test data correlation

Of significant concern to the CDM program is the correlation of the magnetic field quality data between cold and warm testing. To develop this data, the Prototype, Preproduction and Low Rate Initial Production cold testing (and associated warm testing) data will be carefully analyzed to insure that, during production, acceptance of a magnet based on the results of its warm magnetic test data is based on a sound foundation of prior experience.

Formal Qualification Testing

During the process of development of the CDM design, a formal qualification program will be instituted during the Preproduction fabrication and test phase. This program will required up to two magnets to verify that the CDM design can operate in an acceptable fashion prior to, during (when applicable) and following exposure to specific environments. Table 1-1 lists these performance concerns. The environmental effects of radiation from the proton beam will be assessed on specific materials, such as organic insulations. Of particular concern is the long term radiation resistance of epoxies and other non-metallic materials that may be used in the conductor insulation systems. We believe that the expected radiation levels are sufficiently low as to preclude any significant damage to metallics. Another concern is the effects of transportation and handling environments on the magnet performance and alignment. Available data suggests that (for the current CDM design) there is a degradation in accuracy of alignment of the CDM cold mass to the external mounting surfaces. Also at issue is protection of the CDM support system against unusual dynamic loads that may occur during transportation. This qualification test will verify that our design can withstand these environmental stresses and subsequently meet the SSCL performance requirements.

Table 1-1 Environmental/Functional Tests

1. Electrical integrity tests of magnet (room temperatures)
2. Magnetic field measurements (room temperatures)
3. Magnet performance measurements at operation temperatures (performed prior to and following various environmental exposures)
    a. Coil resistivity ratio
    b. Electrical integrity tests of magnet
    c. Quench performance
    d. Magnetic measurements, field purity, transfer function, magnetic length, field angle, persistent current effects
    e. Studies of manually induced quenchs
    f. Verification of cryostat performance, thermal and structural
4. Storage temperature and humidity exposure
5. Transportation and handling vibration and shock environments
6. Magnet alignment prior to and following transportation
7. Magnet alignment prior to and following thermal cycling

Development Testing

The developmental testing program is to be carried out during the early phases of the CDM engineering development program and will highlight design issues that need early resolution. These include verification of interconnect bellows design, both from a design adequacy for all operational loads and from a reliability standpoint. Other development tests will include conductor stack modulus tests, conductor insulation tests and material characterization of candidate materials for the CDM coil end pieces. These are currently being made of machined fiberglass (G-10 and G-11), but this method is expensive and there are candidate materials that would be suitable for injection molding for these parts providing their mechanical properties and radiation resistance is adequate.

Expected design iterations

Our current plans for the Prototype program of approximately 15 magnets will focus on early fabrication of several magnets and subsequent testing to concentrate on magnetic field quality, quench and electrical performance and structural performance. It is expected that this initial set of magnets (3 to 5) will require the cold mass design to be altered, based on their cold performance, to achieve acceptable magnetic performance. This analysis and design iteration will provide our suppliers with new component designs based on this initial test data so that they can make the necessary modifications to the procured parts (laminations, etc.). An alternate method of modifying the magnetic design is to rely on varying the dimensions of shims in order to effect the dimensional changes dictated by the results of this analysis. Following this design iteration, a second series of magnets will be fabricated and tested for performance verification. This second group of magnets will provide an additional opportunity for any necessary changes required to meet the magnetic field performance. This group of magnets will also provide test articles for initial qualification and accelerated life test programs. Following this second analysis and design alteration and resulting component modification, a third group of magnets will be fabricated and used for final magnetic performance verification. At this point, the program would concentrate on process verification during the fabrication of a series of 35 magnets for delivery to the SSCL. In addition, a Formal Qualification Test article would be designated from the first of this series of magnets. Following a successful qualification test completion and process verification, the program Critical Design Review would be held.

VERIFICATION OF CDM RELIABILITY

Reliability requirements

The detailed reliability requirements and a description of our approach to meeting these requirements is contained in another paper to be presented at this conference. In summary, a major driving requirement for the CDM design is the availability requirement for the entire collider of 80%. When partitioned between the various major collider systems, the dipole magnet system will require an availability of about 96%. To attain this availability the reliability of a single CDM must approach 0.999997. Achievement of this reliability is clearly a challenge. Our approach to achievement of this goal is illustrated in Figure 1-4.

Component testing for reliability demonstration

One of the methods for assuring this level of CDM reliability is to test, at a component level, various magnet subsystems or components to verify that the reliability of these items will support the CDM reliability requirement. For example, the various stainless steel bellows used in the interconnect region to connect a CDM to another magnet or spool piece will be subjected to various loads (cyclic and static) and temperatures excursions during magnet operation. We intend to perform life cycle testing to verify the acceptability of candidate bellow designs early in the development program to support our overall approach to CDM reliability.

Magnet reliability demonstration - Accelerated Life Test

On the magnet level, the approach to verifying reliability will be the Accelerated Life Test. This test program will subject two magnets to a series of environmental tests (transportation vibration, storage temperature, etc.) and then impose a program of thermal cycles (300K to 4K to 300K), power cycles (magnet current ramping from zero to a test current to zero), and quench cycles which includes both magnet voltage and cold mass operating pressure variations from nominal to some maximum. A major problem identified early in the planning for these tests was the realization that, to achieve test results that would verify that the next magnet operating cycle would be successful to a 50% confidence level, one would be forced to perform many years of testing if the operating stress levels (current, thermal stress, quench pressures) were kept at nominal values. To eliminate this

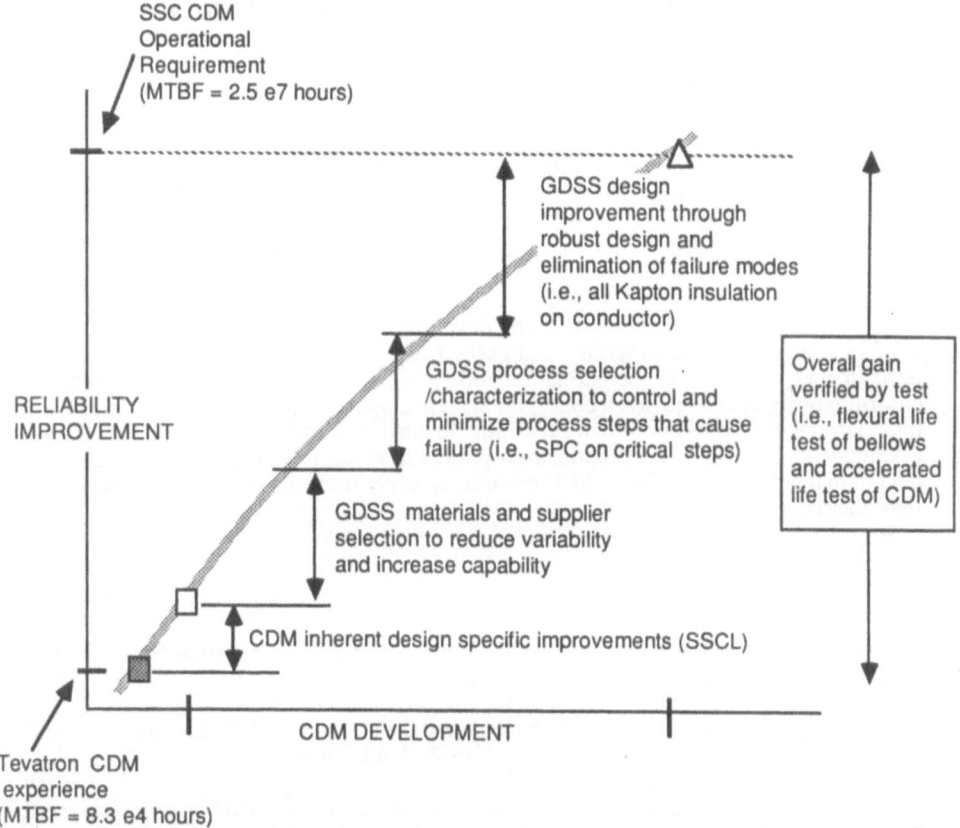

Figure 1-4 The required reliability improvement will be achieved through design, material selection, and process enhancements

programmatic issue, we elected to develop a program which increases these operating stresses during the various operating cycles to an extent that the testing time is reduced to several months. A detailed description of the planned accelerated life test program as given in references 1 and 2.

VERIFICATION OF CDM PRODUCTION PROCESSES

Magnet performance is determined by design of the production process and control of these processes

Following successful demonstration of a producible design, the next program phase will concentrate on validating the production processes and determining effects of process variation on magnet performance. In order to be able to consistently fabricate magnets which meet the required performance requirements, we must understand what effect the production processes have on magnet performance, and this becomes the first step in establishing a valid Statistical Process Control program for the magnet processes. In this effort, the various production processes will be analyzed, and tests performed to determine how the various process variations affect magnet performance. Typical processes include coil winding (winding tensions), coil curing press (curing pressures and temperatures), and collaring press (collared coil stress levels). Process parameter measurement will include measurement of collared coil prestress, collared coil magnetic field characteristics, collared coil magnetic verticality, yoked and skinned (cold mass shells installed) magnetic field characteristics and collared coil prestress levels. The magnet performance parameters to be checked will include performance at full current, quench performance, magnet training, and the quality of the magnetic field. Following successful process validation demonstration (and successful Formal Qualification Test), we will schedule the program Critical Design Review at which time the program Configuration Control responsibility will be transferred from General Dynamics to the SSCL.

CONCLUSIONS

This test program, currently under development at General Dynamics to support the very challenging CDM program, addresses fully the program's product development, performance verification, operational reliability and rate production test needs. It has been carefully designed to provide a cohesive, low cost, appropriate and integrated approach to the CDM testing concerns. When fully implemented this integrated approach will provide assurance that the CDM development program will result in a proven and reliable magnet design that will meet all of the SSCL performance requirements and can be produced at the maximum needed rate.

REFERENCES

1.G.W. Albert, A.H. Kytasty and R.E. Bailey: General Dynamics Space Systems Division, "SSC Dipole Magnet Accelerated Life Test Program,"

2.A. Kytasty: General Dynamics Space Systems Division, "CDM Reliability: Essential to Achieving An Operable Superconducting SUPER COLLIDER".

The work described herein shall be accomplished under contract to the Universities Research Association in support of the Superconducting Super Collider project for the U.S. Department of Energy.

DOCUMENT FORMAT CONSIDERATIONS FOR A

DOCUMENT TRACKING AND STORAGE SYSTEM*

Norman E. Wells, Ivan Chow, and Linn D. Johnson

Physics Research Division
Superconducting Super Collider Laboratory†
2550 Beckleymeade Ave, MS-2000
Dallas, TX 75237

Abstract: The design and development of the detectors for the Superconducting Super Collider Laboratory (SSCL) will be facilitated by a central system to track and store the design documents and drawings. Collaborators and SSCL personnel need a single system that they can access from a terminal on their desks and use to locate and view the latest version of a document. The SSCL Physics Research Division has developed a prototype system to track documents and drawings and to make them accessible via local and wide-area networks. The prototype is being used to refine the requirements and to develop the procedures for a larger system that will be acquired later. The format for storing documents is a major challenge because there is no interoperable standard. This paper discusses the system requirements and architecture, and presents results of research on standards for storing documents.

INTRODUCTION

The Physics Research Division of the Superconducting Super Collider Laboratory (SSCL) and the detector collaborations require a system to track and store the documents and drawings being used to design and develop the detectors. The system should also be usable for tracking and storing physics papers.

With the geographical separation of detector developers, it is important that a central system be provided so the latest version of a document or drawing is available for a collaborator working on that component or on the interface to another component. The goal is a system that can be accessed via the SSCL local or wide area network so that information is available to physicists and engineers at their desks.

*This work was supported in part by Dr. Tom Kirk and the Argonne National Laboratory for the U.S. Department of Energy under Contract No. W-31-109-ENG-48.

†Operated by the Universities Research Association, Inc., for the U.S. Department of Energy under Contract No. DE-AC02-89ER40486.

Current technology does not support the error-free exchange of information between the many computer-aided design (CAD) and word-processing systems in use at the institutions involved. A limited survey of members of the Solenoidal Detector Collaboration (SDC) found that 20 different CAD and 15 different word-processing systems were in use by the collaborators. The challenge is to develop a system that provides interoperability but does not require the replacement of hardware and software currently in use.

Operating procedures within the SSCL and in some of the collaborations are still being developed. A prototype Document Tracking and Storage System (DTASS) was developed in-house to ensure that a large and expensive system was not acquired until we know exactly how it will be used. The prototype also provides experience that will help specify the complete system to be developed or acquired later. The plan was to get a limited-capability system on line as soon as possible to work out the human/procedural issues and to refine the technical requirements for the larger system to come later.

A format for storing and using documents must be found that will allow system users at the SSCL and associated organizations to access and use the information without buying new hardware and software. Since current technology does not support universal format translations, the Physics Research Division is examining ways to tailor a system to meet its needs.

REQUIREMENTS

System requirements include the following:

- A database management system (DBMS) for tracking the documents and drawings.

- Processing and storage devices for hosting the DBMS and the documents/ drawings.

- Local-area and wide-area networks for communicating with the system.

- Document input and retrieval via menu selection and not by use of a database language.

- Access controls to keep unauthorized personnel out of the system. Controls are also required to allow certain groups to maintain private documents in the system that cannot be accessed by other groups.

- A system to control changes to documents so unauthorized personnel do not modify them without permission.

- The capability to scan and store A- and B-size documents and drawings.

- The capability to add configuration management to the initial system.

ARCHITECTURE

The system architecture is shown in Figure 1. Users log onto the system from a terminal at their desks using the same procedures as those used for accessing the SSCVX1 VAX.

Two Sun SparcStation 1+ workstations, each with 2.5 gigabytes (GB) of storage and tape capability for backups and for mailing large files, are on the network. The Sybase database management system is installed on one of the workstations and contains the index of all of the documents and drawings in the system. This database (similar to a card catalog in a library) can be searched for subjects, authors, key words, etc. When the desired document or drawing is found, it can be retrieved to the user's workstation.

An input-output workstation comprising a Macintosh IIfx, Fujitsu M3096E/F scanner and QMS PS-810 Turbo laser printer is part of the system. The scanner provides the capability to scan documents that are not received or generated electronically. Hand-written documents can also be entered into the system and stored as a graphic.

Intergraph workstations are connected to the system via the local area network and provide the capability to generate, translate, and store CAD drawings. Users will log onto the DTASS to search for the drawing desired. When a drawing is requested, the database system will retrieve it from the Intergraph server and send it to the requester's workstation via file transfer protocol (FTP).

The document and drawing archive system of the SSCL will be accessible via the local area network and will be able to store documents not in active use.

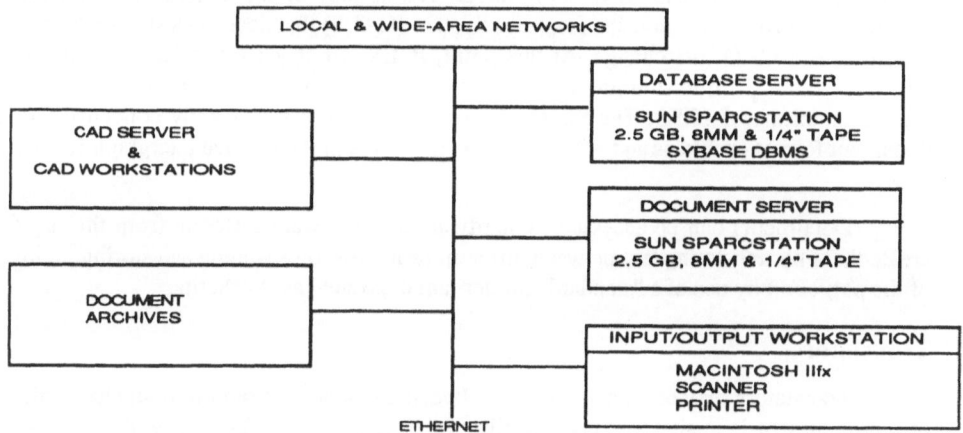

Figure 1. System Architecture.

SYSTEM USE

Users log onto the system via the local or wide-area networks. Choices at the opening menu are:

- *Find/Edit Document*: Users can search for a document by entering the author's name, title, key word, etc. Data similar to that on a library card catalog helps determine if this is the document needed. The current system capability allows the user to transfer a copy of the file to his/her local workstation via a file transfer procedure that is transparent to the user.

- *Enter Document*: Users enter data about the document by filling in the blanks on the screens. The document itself can be transferred into the system by completing additional screen inputs.

- *Utilities*: Users can change their password with this function. Standard reports will be available in the next upgrade of the system.

The system has demonstrated the capability to store and transfer word-processing, spreadsheet, scheduling program, and CAD files. The system hardware, database management system software, and network are all operating. The challenges are to determine the best format for the document files and to develop easy-to-use procedures.

DOCUMENT FORMAT ISSUES

There are many problems to be solved before a user will be able to view a document or drawing at his/her workstation without having to replace existing hardware and software. Current technology does not provide error-free translations between the many word processors and CAD systems in use by the physicists and engineers developing the collider and detectors. The ASCII (American Standard Code for Information Interchange) format for text documents and IGES (Initial Graphics Exchange Specification) format for drawings were initially selected as the standards for interoperability. However, ASCII text does not support storage of the many graphics in physics documents. A search for a better standard is underway.

The interchange of CAD drawings is a problem that may be solved by later versions of IGES. IGES (MIL-D-28000) is being developed as part of the Department of Defense Computer-Aided Acquisition and Logistics System (CALS) that deals with standards for text, graphics, and CAD interchange. At this point, IGES is not yet a standard because many vendors have product-unique extensions that do not convert to other vendors' implementations of IGES. The Physics Research Division is currently concentrating on document format problems and will store CAD drawings in their native Intergraph format for now.

Documents can be accessed by hardware and software different from that used to create them by translating text between different programs, by scanning and storing an image of the page, and by use of a "standard" printer language such as PostScript.

Translators

Translators may be able to convert between application programs if an electronic file of the document is available. This capability (also called a filter) comes with many word-processing programs and can also be purchased separately.

The Microsoft *Word* word-processing program running on the Macintosh was chosen as the initial format for documents because this combination was in wide use at the SSCL. It was believed that getting the system operating with some format was the first order of business; interoperability could come later. Since *Word* is also available for DOS machines and Unix workstations (running DOS for Unix), it was hoped that it could be the interoperable standard for the Physics Research Division. However, research indicated that a physics paper with graphics could not be translated from *Word* running on the Macintosh to *Word for Windows* running under DOS without losing the graphics. In at least this case, a particular word-processing program running on one type of hardware is not interoperable with a word processor with the same name running on another type machine or operating system.

Graphics and Greek letters/math symbols have proved to be major problems in translating between programs. Advertising indicates that translators are available to convert documents between word-processing programs. The authors were able to translate text between word processors and retain most of the formatting, but graphics and equations are a special case for which an easy solution has yet to be found.

Scanning

A hard copy of a document can be scanned and stored in several formats.

- *Raster Image*: A picture of the document can be stored after scanning. This can be done in a graphics format like TIFF (Tagged Image File Format) or PICT (PICTure of the document) used by applications such as *Canvas* or *Optix* for the Macintosh. (*Optix* is a program that controls a scanner and produces raster image files of a document of many pages.) A raster image converted to a graphics application like *Canvas* can be edited or pasted into a word-processing program like Microsoft *Word*. Research indicates that *Canvas* would not hold many pages of images before exceeding the 8 megabytes (MB) of random access memory (RAM) of the Macintosh IIfx. Individual page images can be stored and retrieved in *Canvas* but if the document contains many pages, it needs to be pasted one page at a time into an application that does not try to put the entire document into RAM, such as Microsoft *Word*. A 27-page viewgraph presentation was scanned using *Optix*, transferred to *Canvas*, and pasted together into a *Word* file. This file was 5.3 MB long and required that the RAM allocation to *Word* be increased to handle it. No attempt was made to compress these files since additional user software would be required to decompress them.

  The orientation of images must be considered when they are input. A page input in the portrait orientation with the graphics or text in the landscape orientation must be rotated so it can be viewed. Not all application software programs have a rotate capability.

  A raster image of a document can be stored and retrieved on a central server using a system such as that provided with an *Optix* server. Users with the *Optix* software on their Macintosh can call up and view pages stored in the server at their workstation via the local area network. Since raster image files are large, *Optix* compresses the file for storage and transmittal. The file is decompressed at the user workstation.

The authors have made no attempt to transfer raster images to computers other than the Macintosh.

- *ASCII Text:* Scanned text can be converted to a file of ASCII text by recognition programs such as *Omnipage*. The text output can be put into various word-processing or ASCII formats. However, text recognition is not perfect and many errors should be anticipated. The authors experienced 20-30 errors per page of text. Spell-checkers for text recognition programs are available that claim to clean up most recognition errors. (The authors did not test any of these programs.) Scanning at a higher resolution (300 dots per inch [DPI]) resulted in more errors per page than at 200 DPI. Files of recognized text are much smaller than raster image files: 2-3 kilobytes (KB) versus 50-200 KB. These text files can also be edited or input into other files as if they were created by the user.

  Text recognition programs may not handle graphics. In some programs like *Omnipage*, they need to be "cut" out of the text and handled by a graphics program such as *Canvas*. However, when graphics are pasted into a word-processing program, they may not be edited directly. Greek letters and mathematics symbols may introduce many errors in text recognition and require special handling.

PostScript

PostScript is a device-independent page description language that some printers use to prepare the instructions for printing a page. Files saved in PostScript may be used on various machines and thus may provide interoperability because the documents can be printed or viewed at workstations other than the type on which they were created. However, our initial investigations indicate that there are a number of versions of PostScript, and full interoperability has yet to be demonstrated by the authors. (Additional work is underway by the authors in this area.) Users can convert an existing file to PostScript by invoking the correct command for their computer. Instead of sending the new PostScript file to the printer, the computer stores it on the hard drive, from which it can be transferred to another user.

PostScript files tend to be very large because they define how to create individual letters and graphics elements on the page. For example, a seven-page physics paper (half text and half graphics) was scanned, run through a text-recognition program, and saved in Microsoft *Word*. The graphics were converted to *Canvas* and imbedded in the *Word* document. This *Word* file was 413 KB, but grew to 2.44 MB when converted to PostScript. An 8.5-by-11 in. page containing a CAD drawing of a detector was scanned at 200 DPI into a 449 KB TIFF file and saved as a 144 KB PICT file in *Optix*. The file grew to 287 KB when converted to *Canvas*, but expanded to 1.5 MB when translated to PostScript.

PostScript files can be compressed by about a factor of 10. The software to perform the compression is part of the Unix operating system. Additional work may be required to decompress files on a Macintosh or DOS computer that were compressed on a Unix workstation.

PostScript has the capability to rotate images to correct for storage in the wrong orientation. Panning and zooming capabilities are also provided.

Additional work needs to be done with PostScript to determine its true interoperability. PostScript offers promise for those who need to share files and can handle their large size.

SYSTEM CONSIDERATIONS

These document format issues have a large impact on the total system. Memory is a major challenge—both for storing a document and for accessing it over a network. The resolution of scanners, monitors, and printers may determine whether a document can be read. Manpower to translate or scan documents and input them into the system must be considered as well as the user's time required to actually see the document. Also, the total system should not require expensive licenses at every user's workstation.

Memory Considerations

Since graphics files are so large, a great deal of RAM is required to process them. Even scanning a page of text creates a large file. As previously stated, an 8.5-by-11 in. page scanned at 200 DPI creates a 449 KB file (scanning at 400 DPI results in a 1.7 MB file) in *Optix*. The authors found that the 8 MB of RAM in the Macintosh IIfx used for this research required them to quit applications before another could be opened. This was cumbersome because scanning a document, recognizing the text, processing the graphics, and creating a usable file required the use of *Optix, OmniPage, Canvas,* and *Word*. The following table indicates the application memory specified by the authors on a Macintosh IIfx for the processing of the documents used for this research. No attempt was made to find the minimum RAM that could be used since it would depend on the documents being processed.

We started with the standard allocation and, when it was insufficient, made arbitrary increases. The following were the final RAM allocations:

APPLICATION	SUGGESTED	USED
SYSTEM	924K	924K
FINDER	350K	350K
TELNET	594K	594K
WORD	512K	3000K
OPTIX	2048K	4096K
CANVAS	700K	3500K
OMNIPAGE	N/A	3000K

Files can be compressed by a factor of about 10. This helps in the storing and transmission of files but takes time for the computer to compress and decompress—at least 30 seconds should be expected. Additionally, the software used for decompression must be compatible with that used to compress the file. This potential threat to interoperability may be resolved because the standards used for fax compression (CCITT [Consultative Committee on International Telegraphy and Telephony] Group III and IV) are also being used for some image compression.

Because the files containing text and graphics are large, access time across the network (and the time is takes to display the file) takes longer than desired. The 27-page viewgraph presentation mentioned earlier was scanned and saved as *Canvas* images stored one after another in a 5.3 MB *Word* file. This file was stored in the Document Tracking and Storage System and retrieved across the SSCL local area network. Transferring a large file like this takes a long time--about two minutes for a computer hooked directly to the Ethernet (ten megabits per second). However, transferring via Appletalk (24.5 kilobits per second) takes about 20 minutes.

Paging through an electronic document to find desired material takes time in the absence of an index or titles on the graphics. Displaying the next page of graphics in the document requires a 10-second wait while it is drawn. A raster image of the document (text or graphics) is just a bit map picture that cannot be searched by a "find" utility in a word-processing program. An index to the document created by the author may be necessary to facilitate finding material if the document is stored as a raster image or PostScript file.

Resolution

The resolution of the various components of the system must be considered before the overall architecture can be determined. Scanning a document at 400 DPI is a waste of memory if the application is going to display it only at 72 DPI. Applications handle displays in two ways: (1) they show each pixel even if only a portion of the page can be viewed on the screen, or (2) they display the entire page on the screen by not showing all the pixels. If the application permits, all of the detail in a document scanned at 400 DPI could be seen by displaying part of the document and letting the user pan around the document to view it all. *Omnipage* displays the page (at least a portion of it, depending on the monitor) at 72 DPI, while *Canvas* and *Optix* show all the resolution available even though the entire graphic may not be seen at once. In *Word*, the user manually sizes a graphic (and loses resolution in the process) to fit on the page, where it is displayed at 72 DPI.

Resolution must also be considered when printing scanned documents. The authors could see no difference in the printed output of a graphic scanned at 200 DPI and at 300 DPI.

It would obviously make no sense to scan an A-Size document at 400 DPI if it were going to be printed at the same size on a 300 DPI laser printer.

The hardware used by the authors had the following resolutions:

SCANNER	200–400 DPI
LASER PRINTER	300 DPI
MONITORS	72 & 82 DPI
FAX MACHINE	200 DPI

Manpower Considerations

The DTASS was designed to allow users to input electronic files directly from their workstation with little effort or system knowledge. In fact, the database fields can be filled in and files transferred to the system in just a few minutes. Everything works correctly as long as the input file can be read by the user who retrieves it. Since there are incompatibilities, an operator may be needed to translate from the native version of the file to the interoperable standard or to scan the document and create a raster image. The system should provide the capability for the operator to input an entire document at one time and not be forced to process one page at a time.

An operator may also be necessary to ensure that the document stored in the system (the interoperable standard file format) is the latest document and matches that of the native/original version. Ideally, users would translate the native version to an interoperable standard, but they may not have the capability to do so. Time to retrieve and actually view the document on the user's workstation must also be considered.

License Considerations

A total system solution must not require all users to buy expensive licenses for their local computers in order to use the system. Public domain software or software included in operating systems such as Unix must be considered. It may also be appropriate to purchase a limited number of floating licenses running on a central computer that allow users to log onto the system, find and view the document they want, and retrieve it with no additional software needed on their machine.

User hardware will certainly need graphics capabilities so that graphics or raster images can be viewed. X Windows will probably also be required. The goal should be to design a system that takes advantage of the hardware that most users have (or are obtaining), and not try to accommodate the terminal with the lowest capability.

SYSTEM OPTIONS

Research to date indicates that the requirements of the Physics Research Division and its associated detector collaborations might be met by (1) a single central server running a word-processor which allows users to view and process documents, (2) a single central server where users can view raster images of documents, or (3) the storage of files in PostScript. Additional options may also be available. The following is only a listing of possible options and should not be considered as the acquisition path that the SSCL or the Physics Research Division is pursuing.

Central Word Processor

A multiuser server with a word processor could be connected to the DTASS and would be operated like the Sybase database management system is now. Users would continue to find documents as they do now, but they would also have a choice of viewing the document. In this case, the central word-processor would manage the viewing and be controlled by the user from his/her local workstation. This option should result in the lowest memory requirements because documents are stored in text format (with possible imbedded graphics).

If the central word processor were the same one as that used by the SSCL's Technical Information and Publications Department, documents already processed by the Lab would be in the same format. If not, procedures for converting from other formats could be developed to meet the needs of both organizations. Users should be able to convert from the central format to their own if that were necessary.

Central Image Viewer

A central image viewing system could be connected to the database management system to provide the capability for users to view an image of the documents they are interested in. A similar system is in use in the SSCL's Magnet Systems Division using Macintosh computers.

Memory requirements for such a system would be high, and access times over the wide area network would have to be demonstrated to be acceptable. Decompression software would probably be required at each user's workstation.

PostScript

Text and graphics files could be stored in PostScript format in the DTASS and accessed using the same procedures as currently in use. Documents could also be stored in their native format, but they would not necessarily be viewable by all users. Additional work is required to ensure that interoperability can be achieved by all types of workstations. Compatible compression/decompression software also must be demonstrated.

As with the central image viewer, memory requirements and access times would have to be acceptable. An indexing system may also be required to allow users to find desired material quickly.

Other

There may be other options unknown to the authors at this point. The SSCL and Physics Research Division are following the progress of the Defense Department's CALS to ensure that we take advantage of the latest technology in this field.

SUMMARY

The Physics Research Division of the SSCL has developed a prototype Document Tracking and Storage System (DTASS) to provide users with the capability to locate and retrieve documents and drawings. The system is operating and is being used to develop the requirements for a more capable system that will be required later.

A text, equations, and graphics format that can be used on all hardware and with any software has not yet been found. Text has been found to translate between different word-processing systems with little loss of format, but equations and graphics need special attention.

Files can be scanned and stored as raster images of the document. These files are very large. A scanned document can also have text recognized and converted to a text file in the format of several different word-processing programs.

The capability to convert files to the PostScript format allows printing or viewing on devices other than the one that created the file. PostScript files are also very large.

The large size of many of these files creates the most significant challenge to developing a usable system. Large memory requirements make document access and processing slow. The resolution capability of the hardware and software in the system must be matched to ensure that users can actually read the document selected. Manpower requirements to input and retrieve an interoperable document must also be considered in system development.

System options include a central server hooked to the database management system to enable the user to view/process the document on a common word processor or to view a raster image of the document. Conversion of files to PostScript for remote viewing and printing may also meet requirements.

Experience with the prototype DTASS is helping the Physics Research Division of the SSCL develop a usable system. Issues such as developing or locating an interoperable format for documents will be resolved before we acquire a larger system.

Attendees

James Abbee
U.S. Department of Energy

Ken Abbott
High Tech Services

Deborah L. Abrahamson
U.S. Department of Eneregy

Frank S. Adams
TSA

Pedro Aguirre
Aguirre Associates Inc.

Bulent Aksel
SSC Laboratory

Gary Albert
General Dynamics

T. M. Alexander Sr.
Alexander & Company

David Alger
Babcock & Wilcox

A. W. R. Allcock
British Embassy

Michael Allen
SSC Laboratory

Ed Altgilbers
Inland Steel

Bruce Andersen
H&J Tool & Die

Charles E. Anderson
Air Products & Chemicals Inc.

Kathy Anderson
SSC Laboratory

David Anderson
SSC Laboratory

Richard W. Anderson
Analogic Corp.

Owen W. Anglum
Armco

Thomas Ankermann
Koch Process Systems, Inc.

Narko Antti
University of Helsinki/SEFT

Thomas W. Appelquist
Yale University

William A. Appleton
CVI Inc.

Craig S. Arden
SSC Laboratory

Katsuhiko Asano
Hitachi, Ltd

Richard F. Atkinson
Silvex, Inc.

Gordon J. Aubrecht
Ohio State University

Stanislaw D. Augustynowicz
SSC Laboratory

Betzalel Avitzur
Metalforming Inc.

A. F. Axt
Westinghouse

Paul S. Ayres
Babcock & Wilcox

Neil V. Bagget
SSC Laboratory

Penny Baggett
SSC Laboratory

Scott Bagwell
SSC Laboratory

Steve Baier
Honeywell SRC

David B. Bailey
SSC Laboratory

Richard Bailey
General Dynamics

Jon Bakken
Martin Marietta

Robert W. Baldi
General Dynamics

Joseph Ballam
SSC/SLAC

Robert Balliett
NRC Inc.

Teresa Ballinger
EG&G Energy Measurements

Bailey S. Barash
CNN Science News

Oscar Barbalat
CERN

Marcel M. Barbier
Marcel M. Barbier, Inc.

Victoria Bardos
SSC Laboratory

Dan Baritchi
SSC Laboratory

Thomas A. Barracato
Air Products & Chemicals, Inc.

Rep. Joe Barton
(R)-Texas

Suzanne Bass
Office of Rep. Jim Chapman

Kurt E. Bassett
Martin Marietta

Irv Baston
AFA

Laura P. Bautz
Lawrence Berkeley Laboratory

Donald S. Beard

G. F. Beatty
LTV Steel Company

Didier Bederede
CEN Saclay

M. C. Begg
Tesla Engineering Ltd.

Mark R. Behan
Grumman Aerospace

Jay Benesch
U.S. Department of Energy

Barry R. Benroth
U.S. Department of Energy

Daryl Bever
General Dynamics

Jack P. Biegalski
Inland Steel Co.

William Birkholz
Brookhaven National Laboratory

Paul A. Bish
Lawrence Berkeley Laboratory

Charles D. Bizilj
Howden Compressors, Inc.

Alan Blakeley
Johnson Controls Inc.

George J. Blanar
LeCroy Corp.

R. J. Blanken
IBM

A. C. Blankenship
Trans-Tech Inc.

Miriam Bleadon
Fermilab

Michael C. Boivin
New England Electric Wire

John W. Bonn
CVI Inc.

Franz V. Boos
MPI International, Inc.

Regina Borchard
Martin Marietta

William Boroski
Fermilab

Rodger Bossert
Fermilab

M. Bruce Boswell
Westinghouse Electric Corp.

Clark Boyd
Pressure Systems Inc.

Kurt Brandenberger
Von Roll Isola

Frank L. Brandon
E-Systems Inc.

Richard J. Briggs
SSC Laboratory

Mark W. Broadley
Handy & Harman Tube

J. Broadus
Constr. Industry Institute

John E. Brogan
Danco Precision, Inc.

G. M. Brown
SSC Laboratory

Donald P. Brown
Brookhaven National Laboratory

J. Michael Brown
Minnesota Valley Engineering

Steve Brown
Long-Airdox Company

Kent Browning
Lockheed Missiles & Space Co.

John C. Bruno
J&L Specialty Products

Warren Buckles
Superconductivity

J. Fred Bucy
Texas Nat'l Research Lab.Commission

Mark Bugeaud
Digital Equipment

Lawrence D. Buhl
Copper & Brass Sales Inc.

Joe Bularzik
Inland Steel

Michael A. Burke
Westinghouse STC

Paul R. Burke
Kurt J. Lesker Co.

Thomas O. Bush
SSC Laboratory

J. M. Butler
Westinghouse

Robin Caldwell
IBM

Kateri Callahan
IASSC

Larry Camilli
Air Products & Chemicals Inc.

H. Glenn Campbell
Babcock & Wilcox

John Campbell
Cryogas International

Barry Cantor
Babcock & Wilcox

R. Michael Capers
Scientific Instruments

Donald W. Capone
SSC Laboratory

Claudio Caprio
IBM

Ruben Carcagno
SSC Laboratory

J. W. Carey
Triumf

Lee Carlson
Superconductor Technology Magazine

James R. Carney
U.S. Department of Energy

Annette Caudiano
Mitsui & Co. (USA), Inc.

Alberto Cerri
Swiss Metal

C. David Chaffee
AIS-Super Collider News

Rep. Jim Chapman
(D)-Texas

Anthony K. Chargin
Lawrence Livermore National Laboratory

Greg Chartrand
SSC Laboratory

Coby Chase
SSC Laboratory

Sherman Chen
SSC Laboratory

Dennis R. Cheney
INEL

Warren E. Chester
Crest Products Corp.

Denis Christopherson
SSC Laboratory

Robert E. Cieslukowski
Minnesota Valley Eng.

Joseph R. Cipriano
U.S. Department of Energy

Brad Clark
Digital Equipment Corp.

Alan Clark
McDonnell Douglas

Susan Claude
Sheldahl

Herschel Wayne Clay
SSC Laboratory

John L. Clendenin
Bellsouth Corp.

Tony Cochrane
Ziemer & Associates

Michael Coffey
Cryomagnetics Inc.

William S. Cogger
Allied Apical Co.

Ben Cole
SSC Laboratory

Mark Coles
SSC Laboratory

E.W. Collings
Battelle

Thomas C. Collins
University of Tennessee

J. A. (Buddy) Conner
Babcock & Wilcox

Maury L. Constien
IBM

Todd Cottrell
Norplex Oak

Larry Coulson
SSC Laboratory

William J. Courtney
IBM

Jeff Cox
SSC Laboratory

Dennis Cox
SSC Laboratory

Jimmy K. Cox
TU Electric

Robert Crane
IDM Corp.

Mark Crawford
New Technology Week

Sam Crivello
SSC Laboratory

Wayne Cross
c/o Steve Hicks IBM

Paul Cunningham
Welded Ring Products

Frank Cunningham
Swagelok Co.

Richard L. Curl
PB/MK Team

Roy Cutler
SSC Laboratory

Edward Daly
SSC Laboratory

Larry Darby
Babcock & Wilcox

Mark A. Davidson
Fermilab

Charles Davidson
Gwinnet Daily News

William J. Davison
Baltimore Spec. Steels

Dean De Benedet
Martin Marietta

Tom Deany
Nor-Cal

Artie G. Debeling
EG&G Energy Measurements

Earl Dedman
Aguirre Associates Inc.

J. G. Deley
Westinghouse

Bill Dennis
Dow Corning Corp.

Peter C. Dent
Lake Shore Cryotronics Inc.

Kurt Derr
EG&G Idaho Inc.

Arnaud Devred
SSC Laboratory

Nickolas Di Giacomo
Martin Marietta

Carl Dickey
SSC Laboratory

Robert E. Diebold
U. S. Department of Energy

Hollye C. Doane
National SSC Coalition

Stephen Doherty
Alcatel Vacuum Prod. Inc.

Thomas Dombeck
SSC Laboratory

Edward Donley
Air Products & Chemicals Inc.

Duncan Douglas
Douglas Eng. Co.

Lois Dowd
Digital Equipment Corp.

John R. Dowdle
C. S. Draper Laboratory. Inc.

Gilbert Drouet
CERN

Rens L. Dubbeldam
Holec Ridderkerk

Tom Dudley
Flexonics Inc.

Sandra P. Dunn
Texas Nat'l. Research Lab. Commission

William J. Dyess
Major Tool & Machine Inc.

John C. Dykstra
Minnesota Valley Engineering

James P. Eaton
Minnesota Valley Engineering

P. W. Eckels
G.E. Medical Systems

William R. Edwards
Lawrence Berkeley Laboratory

Don Edwards
SSC Laboratory

Kenneth R. Efferson
American Magnetics Inc.

Tor Ekenberg
Argonne National Laboratory

Anthony J. Elam
IBM Federal Sector

Mike Erdahl
Loral Corp.

Mark Erdmann
SSC Laboratory

Timo Erkolahti
Outokumpu Copper USA, Inc.

Roger A. Farrell
Intermagnetics General Corp.

Chris Farver
Allen-Bradley

Anthony Favale
Grumman Aerospace Corp.

Susan Feeney
Dallas Morning News

James H. Ferrell
SSC Laboratory

Michael L. Ferris
SAES Getters USA

J. A. Fickling
Westinghouse

William A. Fietz
SSC Laboratory

Ron Fincher
AT&T Federal Systems

Michele Fisher
IBM

Terry Fleener
Ball Aerospace

Adrian Flühmann
Werner Fluehmann Ag

David G. Fong
Energy, Mines & Resources, Canada

Martin Fortner
U.S. General Accounting Office

David Fortunato
SSC Laboratory

William B. Fowler
Fermilab

Ray Fox
SSC Laboratory

Jim Fraivillig
DuPont

Arthur W. Francis
Union Carbide Industrial Gases

Jim Franciscovich
SSC Laboratory

Keith Franck
Lawrence Berkeley Laboratory

Thomas K. Frasier
Gardner Cryogencis

Jim Freim
Furukawa Electric Technology

James French
SSC Laboratory

Ray Fricken
U. S. Department of Energy

Richard L. Frohlich
Westinghouse - Sunnyvale

David Frost
Supercon, Inc.

Touru Fukuhara
Mitsubishi Heavy Industries Ltd.

Gordon T. Fuller
Trans-Tech, Inc.

James Fusco
ITEN Industries

James E. Fye
IBM

Philip O. Gagnon
General Physics Corp.

Henry M. Gandy
Americans for the SSC

Rao Ganni
SSC Laboratory

Raj K. Gattu
SSC Laboratory

Marcel Gaudreau
MIT

Rod Gerig
SSC Laboratory

Robert J. Gibbs
Brookhaven National Laboratory

Garry W. Gibbs
U.S. Department of Energy

Paul H. Gilbert
PB/MK Team

Murdock Gilchriese
Lawrence Berkeley Laboratory

A. Van Ginneken
Fermilab

Giuseppe Scarfi
Ansaldo

L. David Godbey
HDR

Cathy Golden
City of Dallas

Leonard M. Goldman
Bechtel Corp.

Norman Gomm
Embassy of Australia

Jack Goodman
Lockheed Research Laboratorys

Carl L. Goodzeit
SSC Laboratory

Michael Gordon
KMI Services

Peter J. Gossens
General Dynamics

Stephen R. Gottesman
Grumman Corp. Research Center

James A. Gray
SSC Laboratory

Thomas H. Gray
Allegheny Ludlum Corp.

Michael I. Green
Lawrence Berkeley Laboratory

Michael A. Green
Lawrence Berkeley Laboratory

Arthur F. Greene
Brookhaven National Laboratory

S. Steven Greenfield
Parsons Brinckerhoff Inc.

Jacques Greetis
Flexonics Inc.

Eric Gregory
IGC Advanced Superconductor

Doug Grones
SSC Laboratory

Charles Grozis
Fermilab

John A. Gruver
Westinghouse Electric Corp.

Victor Guarino
Argonne National Laboratory

Jeff Guillot
Halgo Field Services, Inc.

George Gunn
Logicon

Ramesh Gupta
Brookhaven National Laboratory

David Gurd
SSC Laboratory

F. J. Gurney
Westinghouse

H. Gurol
General Dynamics

Ahmet Gursoy
PB/MK Team

Gregory Haas
U.S. Department of Energy

Bernhard Haberthuer
Sulzer USA Inc.

David Haenni
SSC Laboratory

John Eric Haggard
Fermilab

Phyllis Hale
SSC Laboratory

Ann Hardie
Atlanta Journal-Constitution

William E. Harrison
Brookhaven National Laboratory

Kenneth O. Hartley
PB/MK Team

Don Harwell
Cryenco Inc.

Nick Hassan
SSC Laboratory

William N. Hasselkus
DOE-OSSC

Dan Hatfield
SSC Laboratory

Bill Hawkins
Iron Worker

Robert Hedderick
SSC Laboratory

Tricia Heger
Fermilab

Warner Heilbrunn
SSC Laboratory

Ezra D. Heitowit
Universities Research Assn.

Keith Helart
Ball Aerospace

Roland Hellmer
Balzers

Barry L. Hendrix
SSC Laboratory

Steve L. Hensley
CVI Inc.

Frank Hervey Jr
Hyspan Precision Products Inc.

Wilmot N. Hess
U.S. Department of Energy

Susan Hickey
Lockheed Missiles & Space Co.

Steve Hicks
IBM

W. Kent Higgins
Koch Process Systems Inc.

Gale Hill
PB/MK Team

Ron Hill
KMI Services

John Hill
UTD Inc.

Gia Ky Hoang
Alsthom Intermagnetics

Dave Hodan
Digital Equipment Corp.

Matthew Hoff
Lawrence Berkeley Laboratory

Charles B. Hood
CVI Inc.

Richard P. Hora
General Dynamics

Ron Horan
KMI Services

Peter A. Horton
Taracorp Industries

Kenji Hosoyama
KEK

Michael B. Hoye
InterFET Corp

Robert E. Hughes
Associated Universities Inc.

Steve Hunt
SSC Laboratory

Dan Hutton
SSC Laboratory

Ralph Hutton
DuPont Co.

Masaru Ikeda
Furukawa Electric

J. Jeffrey Irons
Grumman

Hisashi Ishida
Kawasaki Steel America Inc.

Tutomu Isobe
Furukawa Electric

Hans J. Israel
Holec Ridderkerk

Jon R. Ives
SSC Laboratory

Robert Jacbosen
Directed Technologies Inc.

Abdul Jalloh
SSC Laboratory

Dan Janikowski
Trent Tube

Andrew J. Jarabak
Westinghouse Electric Corp.

R. Jayakumar
SSC Laboratory

Wendell W. Jesseman
New England Electric Wire

Don Johansen
Babcock & Wilcox

David H. Johnson
Howden Compressors, Inc.

David E. Johnson
SSC Laboratory

Curtis D. Johnson
University of Houston

Robert A. Johnson
General Dynamics

Proctor Jones
Energy/Water Dev., Appropriations

N. B. "Buck" Jordan
Waxahachie Chamber of Commerce

Donald Jordan
Spaulding Composites Co.

Tom A. Jouvanis
United Engineers & Constructors

Doug Juanarena
Pressure Systems

Ralph Kalkbrenner
Westinghouse PQC

Jill Kallsen
Teledyne SC

Hem Kanithi
IGC Advanced Superconductor

Raphael Kasper
SSC Laboratory

Osamu Kawamata
Hitachi Cable America Inc.

Eugene Kelly
Brookhaven National Laboratory

John M. Kennedy
Texas Nat'l Research Lab. Commission

John E. Kenny Jr.
Kenny Construction Co.

Willem Keyer
Koch Process Systems Inc.

Edmund Killian
Brookhaven National Laboratory

Glenn E. Kinard
Air Products & Chemicals Inc.

Nancy O. King
SSC Laboratory

Thomas B. W. Kirk
Argonne National Laboratory

P. A. Kitchin
Tesla Engineering

Sherrie "Sam" Kivlighn
SSC Laboratory

R. Kleinhaus
Oracle Federal Division

Mary S. Klorer
Allen-Bradley

Ted Kobel
Koch Process Systems Inc.

Ted Koizumi
Furukawa Electric Tech.

Thad T. Konopnicki
U.S. Department of Energy

David Koopman
Metallized Products

Arthur Z. Kovacs
Rochester Institute of Tech.

Michael Kramer
ACPT Inc.

David Kramer
McGraw-Hill Inside Energy

Gordon Kramer
Hughes Aircraft Co. - EDSG

Philip Kraushaar
SSC Laboratory

T. Scott Kreilick
Hudson International Conductors

John M. Kreisle
Team Industries

D. Krischel
Interatom GmbH

Thomas Krueger
Johnson Controls Inc.

Robert R. Kurth
MPI International Inc.

Jozef Kuzminski
SSC Laboratory

Tamio Kuzuhara
Kawasaki Steel America Inc.

Andrew Kytasty
General Dynamics

Aaron La Barge
SSC Laboratory

David S. La Fleur
MKS Instruments Inc.

Patrick La Master
Sci Tech

Horace N. Lander
Nippon Steel USA, Inc.

Dennis Lanning
AIT

F. C. Lanvie
PB/MK Team

Robert J. Lari
Vector Fields Inc.

Warren Larson
Supercon, Inc.

Charles Laverick
Consultant

George A. Leakey
AECL Research

Ken Lee
Dow Corning Corp.

Mitch Leff
IISSC

Daniel Lehman
U.S. Department of Energy

Phil Leibold
SSC Laboratory

Thomas A. Leiser
Trammell Crow Company

Louis J. Lestochi
Air Products & Chemicals Inc.

Kent Leung
SSC Laboratory

Eddie Leung
General Dynamics

Marc Levenson
CNN

Joyce Lewis
U. S. Department of Energy

Melvin Lindner
Brookhaven National Laboratory

Joseph Linteau
Climax Specialty Metals

Arie Lipski
Fermilab

Thomas Looser
MPI International Inc.

B. J. Lowe
Ebasco Services Inc.

Jill L. Ludwigsen
Nichols Research Corp.

Joseph A. Lujan
Mevatec Corp.

Roy Lusk
IGC Advanced Superconductor

Hylan Lyon Jr.
SSC II

Charles Lyraud
CEN Saclay

Jeff MacKinnon
Office of Rep. Joe Barton

John Mackay
SSC Laboratory

Craig Magruder
Nissei Sangyo America

David Main
NRC Inc.

Dub Maines
Office of Rep. Joe Barton

Milan Malec
Dour Metal - SCE Co.

Robert J. Malnar
SSC Laboratory

John Mangino
SSC Laboratory

Paul Mantsch
Fermilab

Gerhard Mara
Elin Energieanwendung

Bill Marancik
Oxford Superconducting Tech

John H. Marburger
SUNY, Stony Brook

Trent A. Marion
AT&T Federal Systems

Finley W. Markley
Fermilab

George Marks
Huitt-Zollars Inc.

Cal Markwood
Martin Marietta

Paul E. Marshall
Stone & Webster Engineering Corp.

Peter G. Marston
MIT/PFC

Lillian Mashburn
University of Tennessee

Lyle Mason
Martin Marietta

Raymond J. Massey
SSC Laboratory

Giovanni Masullo
Ansaldo Componenti

Roman Matherne
Martin Marietta

Hamel Maurizio
Laben SpA

Michael S. McAshan
SSC Laboratory

William McAuley
Koch Process Systems Inc.

David McBane
Alconex Specialty Products

Don McCredie
Von Roll Isola

Kim McDonald
Chronicle of Higher Education

J. R. McDonald
PB/MK Team

John McGill
SSC Laboratory

Neil McGladdery
Westinghouse Electric Corp.

Glen E. McIntosh
Cryogenic Tech. Services

Jim McKinnell
Teledyne Wah Chang

Del McLane
Ellis County

Clay McMullen
Crest Products

David McWilliams
Koch Process Systems Inc.

Thomas Mensah
Superconductor Technology

William Merz
SSC Laboratory

L. K. Mestha
SSC Laboratory

Robert B. Meuser
Lawrence Berkeley Laboratory

Frank Meyer
MTM Cryo Tech Laboratory

James R. Miller
Sheet Metal Workers

Scott Miller
SSC Laboratory

John Mitchell
Martin Marietta

Shinichi Miwa
Kawasaki Steel America Inc

Alex Miyatake
Furukawa Electric

James E. Monsees
PB/MK Team

Mike Montgomery
IDM Corp.

Suhas Mookerjee
Abb Technology Co.

Billy Moore
Office of Rep. Jim Chapman

Melissa Moore
Inco Alloys International Inc.

Linda M. Moore
APD Cryogenics

John J. Morena
ACE Inc.

Ira Lon Morgan
University of North Texas

Naoki Morimatsu
Sumitomo Corp. of America

Hiroaki Morita
Mitsubishi Heavy Industries Ltd.

Vicki Lynn Morris
S.C.I.

Paul Morris
Teledyne SC

Bill Mote
Best Southwest

Stu Motew
Koch Process Systems Inc.

George T. Mulholland
SSCL/Fermilab

Thirumaleshwar Muliya
SSC Laboratory

C. Richard Mullen
Intermagnetics General Corp.

Timothy K. Muller
Westinghouse Electric Corp.

Pamela J. Mundo
Midlothian Chamber of Commerce

Naseem A. Munshi
Composite Technology Development

Harry Murayama
Nippon Steel

Michael Murphy
Oxford Superconducting Tech

Michael D. Murphy
Edwards High Vacuum

Rep. John T. Myers
(R)-Indiana

Yasuchika Nagai
Hitachi Cable America Inc.

Masayuki Nagata
Sumitomo Electric Industries Ltd.

Kazuki Nakagawa
Hitachi, Ltd.

Koichi Naruse
Mitsui & Co. (USA), Inc.

Eric C. Nehrlich
Fermilab

Richard C. Neunzig
ITT Electron Technology Div.

Long Nguyen
General Dynamics

Renee Nichols
Ceramaseal

John Nicholson
U. S. Department of Energy

Gene Nicholson
Logicon Inc.

Tom Nicol
Fermilab

Ted Niehaus
SSC Laboratory

Fred Nobrega
SSC Laboratory

Kiyohiko Nohara
Kawasaki Steel America Inc.

J. A. Nonte
SSC Laboratory

Mary Ann Novak
Parsons Brinckerhoff Quade

Mark O' Mahony
Armco Advanced Materials Co.

James Ochsner
DuPont Co.

R. H. Oeler
Air Products & Chemicals Inc.

Fumio Ohkubo
Mitsubishi Heavy Industries Ltd

Tsuguo Ohkuma
Ishikawajima-Harima Heavy
Industries Co., Ltd.

Takao Ohmori
Ihi Research Inst.

Bob Oldefest
Advanced Plastics

Kenneth O. Olsen
Martin Marietta Corp.

Charles Orsak
Navarro College

John L. Ostrowski
Versa-tool Mfg. Inc.

Kevin D. Ott
Council On Superconductivity

C. E. Owen
Westinghouse

Satoshi Ozaki
Brookhaven National Laboratory

Peter Panasci
SSC Laboratory

John Papazian
S.A.I.C.

Chris Pappas
SSC Laboratory

Joseph Paradiso
C.S. Draper Laboratory

Harold C. Parish
CVI Inc.

Joseph Parish
DuPont Co.

David Parrish
Cray Research, Inc.

Jim Parrott
PPB Inc.

Libby Patnovic
Lockheed Missiles & Space Co.

Pamela Patrick
Alcoa

Lee R. Patterson
General Dynamics

Dennis Pavlik
Westinghouse STC

James B. Peeples
CVI Inc.

Clarence Peger
Hard Chrome Plating Consultants Inc.

Arlin M. Pennington
SSC Laboratory

Kurt Pennington
SSC Laboratory

John Peoples Jr.
Fermilab

Herb Peters
Koch Process Systems Inc.

James R. Pfefferle
General Physics Corp.

Steve Pidcoe
General Dynamics

James G. Pierce
CVI Inc.

Dan Pierson
Investigations & Oversight Subcommittee

A. Pinet
Thevenet-Clerjounie

Joe Pohlen
Martin Marietta

Paul Polakos
AT&T/Bell Laboratoryoratories

Ilene M. Pollack
U.S. General Accounting Office

Douglas A. Pollock
SSC Laboratory

William Pope
Lawrence Berkeley Laboratory

Nikolaus Porschek
P.E.C. Engineering

Ronald P. Pratt
Weldaloy Products Co.

Peter E. Price
Industrial Materials Tech.

Jean-Claude Puippe
Werner Fluehmann AG

Cindy Pulver
Honeywell SRC

Tom Purtill
Baylor Company

Hans H. Quack
Sulzer Bros. Ltd.

M. G. Rao
Aniga Enterprises

Deepak Raparia
SSC Laboratory

Shannon Ray
North Texas Commission

Paul J. Reardon
SSC Laboratory

Penny Redington
Ellis County Judge

R. P. Reed
Composite Technology Development

M. Rehak
Brookhaven National Laboratory

Anne Reifenberg
Dallas Morning News

Robert Remsbottom
SSC Laboratory

Keith Rensberger
Cryenco Inc.

R. L. Rerig
IGC

Richard L. Rhodenizer
Intermagnetics General Corp

Marc Rice
Associated Press

Philip R. Ridley
Silvex, Inc.

Hal J. Rietveld
General Dynamics Corp.

Butch Riffe
Long-Airdox Company

Edward C. Rinck
Edward C. Rinck Assoc., Inc

Verrill Rinehart
Cray Research, Inc.

Richard L. Riney
Martin Marietta

Ernst Ringle
Noell, Inc.

Michael Riordan
Universities Research Association

J. F. Roach
Westinghouse STC

William Roark
Nichols Research Corp.

George Robertson
SSC Laboratory

William Robinson
SSC Laboratory

William Robotham
Fermilab

Mike Rodemeyer
Committee On Science, Space & Technology

Jimmy D. Rogers
SSC Laboratory

Parke Rohrer
Brookhaven National Laboratory

Drew Rohrer
ABB Impell

Elisabeth Root
Gould Electronics

Carl H. Rosner
Intermagnetics General Corp.

Jim Rossi
Kyser Company

Mary Kay Rossman
Unistrut Corp.

William G. Rueb
MK-Ferguson Co.

Maurice M. Sabado
Science Applications International Corp.

Richard Sah
Maxwell Laboratories Inc.

Joan Sanchez
Associated Press

Jack Sandweiss
Yale University

Guy J. Scango
U.S. Department of Energy

Terrence J. Scanlan
AIRCO Industrial Gases

Samuel A. Scheer
Westinghouse

Earl A. Scheidemantle
SSC Laboratory

J. C. (Jack) Scheider
Lockheed TO-AO 310

Wolfgang Schellmann
Kloeckner Wilhelmsburger

D. W. Scherbarth
Westinghouse STC

L. Schieber
A.I.T.

W. E. Schiesser
SSC Laboratory

George C. Schmauch
Air Products & Chemicals Inc.

Larry Schneider
SSC Laboratory

Steve Schuermann
The Pb/mk Team

Thomas R. Schulte
Union Carbide

E. P. Schumacher
Westinghouse

William Schumacher
ARMCO

H. Donald Schwartz
Cabot Corporation

William Schwenterly
SSCL/ORNL

Roy F. Schwitters
SSC Laboratory

John Scudiere
Oxford Superconducting Tech

Jeff Seuntjens
SSC Laboratory

Howard W. Shaffer
Westinghouse Electric Corp.

Sol Shapiro
SSC Laboratory

Stephen Shapiro
SLAC

Ken Shearer
Martin Marietta

Kevin Sheehy
Von Roll Isola

Bob Sheldon
SSC Laboratory

Philip E. Shelley
SSC Laboratory

Mike Shrubsole
Murdock Inc.

Jack Shultz
Tempel Steel

Derek Shuman
Lawrence Berkeley Laboratory

Warren D. Siemens
Martin Marietta Energy Systems

Joe Simmons
Baylor Company

Jonathan Simmons
SSC Laboratory

Greg Simon
Chief Counsel, Subcommittee
 Investigations & Oversight

Charles Simpson
Morrison Knudsen Corp.

Richard E. Sims
Fermilab

Sharad K. Singh
Westinghouse STC

George Sintchak
Brookhaven National Laboratory

Edward J. Siskin
SSC Laboratory

John Skaritka
SSC Laboratory

Don Slanina
IBM

Dave Smathers
Teledyne Wah Chang Albany

Robert F. Smellie
SSC Laboratory

Lonnie Smith
SCI

Sally Anne Smith
SSC Laboratory

Christopher C. Smith
Dynapower Corp.

Gerald A. Smith
U. S. Department of Energy

John Smith
Inco Alloys International

Greg Snitchler
SSC Laboratory

John Sondericker
Brookhaven National Laboratory

Philip Spampinato
Grumman Space Systems

Charles L. Spangler
Handy & Harman Tube Co.

Frank R. Spinos
Grumman Corp.

Robert Sponsel
SSC Laboratory

Vic Stack
Cray Research, Inc.

Stuart Stampke
SSC Laboratory

Judy Stanford
AT&T Federal Systems

Michael J. Stanko
Westinghouse Electric Corp.

Allen F. Stansky
EMX Controls Inc.

Peter F. Stansky
EMX Controls Inc.

Elizabeth Stefanski
SSC Laboratory

Raymond J. Stefanski
SSC Laboratory

John Stekly
Intermagnetics General Corp.

Chris Stelter
AT&T

Richard Stephens
U.S. Department of Energy

R. Stiening
SSC Laboratory

William Stokes
Brookhaven National Laboratory

John (Jay) C. Stone
U.S. Department of Energy

E. Jack Story
SSC Laboratory

Richard Strader
Trammell Crow Company

Oscar Suarez
AT&T

Hidehiko Sumitomo
Nippon Steel Corp.

Wallace Sunderman
Westinghouse PQC

John Supancic
Von Roll Isola

Yoichi Suzuki
Hitachi Cable Ltd.

Vito Sylvester
GTE

Mike Syphers
SSC Laboratory

Greg Tanimoto
C. Itoh Pipe & Tube Inc.

Claude D. Tapley
KM-Kabelmetal America

David W. Taylor
ARMCO Inc.

Clyde E. Taylor
Lawrence Berkeley Laboratory

L. E. Temple Jr.
U.S. Department of Energy

Robert K. Tener
Sverdrup Corp.

Jukka Teras
Cerntech Ltd.

Camilo M. Tesone
N.E.I.T.D.O.

Roger E. Tetrault
Babcock & Wilcox

Juhani Teuho
Outokumpu Copper

Roberto Y. Than
SSC Laboratory

Dave Thomas
Babcock & Wilcox

Thomas J. Thompson
SSC Laboratory

John C. Thorne
SSC Laboratory

Tim Thurston
SSC Laboratory

Kathleen Tobin
CNN

Yoshi Tokunaga
Nippon Steel

John S. Toll
Universities Research Assn.

Yukio Tomita
Nippon Steel Corp.

John C. Tompkins
SSC Laboratory

Mike Toner
Atlanta Journal-Constitution

Gerry Tool
SSC Laboratory

Rex E. Trekell
SSC Laboratory

Herbert D. Trenham
SSC Laboratory

Alvin Trivelpiece
Oak Ridge National Laboratory

Bob Troendly
Shaped Wire Inc.

Roman Turkevich
SSC Laboratory

William C. Turner
Lawrence Livermore National Laboratory

Alfred J. Unione
Jason Associates Corp.

George J. Urich
Everson Electric Co.

Charles Vacek
U.S. Department of Energy

William A. Van Dyke
U. S. Department of Energy

Robert L. Van Ness
SSC Laboratory

Voitto Vanhatalo
Outokumpu Copper

Pierre Vedrine
CEN Saclay

John T. Venard
Fermilab

Robert Viola
SSC Laboratory

Patrick M. Visintainer
Airco

Michael Vitale
Sumitomo Corporation of America

Richard J. Vopelak
Babcock & Wilcox

Duane Voy
SSC Laboratory

Eric Vransky
SSC Laboratory

Udo Wagner
Sulzer Bros., Ltd.

Yukichi Wakita
Mitsui & Co. (USA) Inc.

Ron Walker
Lockheed Missiles & Space Co.

Ian J. Walker
GMW Associates

Craig Walling
Hamamatsu Corp

Gully Walter
Tempel Steel Co.

Peter Wanderer
Brookhaven National Laboratory

Robert E. Warren
Hughes Aircraft Co.

Jerry M. Watson
SSC Laboratory

Joe Waynert
Babcock & Wilcox

Charlie M. Weber
Babcock & Wilcox

Marvin J. Weber
Lawrence Livermore National Laboratory

Mike Weiss
MDC Vacuum Products Corp.

Norman E. Wells
SSC Laboratory

Milton Werkema
Sheldahl

Mark W. West
Lawrence Berkeley Laboratory

James E. West
Cryogenic Consultants Inc.

J. T. West
IBM

Ed Whiting
SSC Laboratory

Colin Wilburn
Micron Semiconductor

Joe W. Wildmon Sr.
LTV Steel Corp.

Buddy Wilkinson
Wachs Technical Services Inc.

Erich Willen
Brookhaven National Laboratory

Bob Williamson
SSC Laboratory

Lola Williamson
SSC Laboratory

Mike Wilson
SSC Laboratory

Kenneth F. Wilson
CVI Inc.

Patricia A. Wilson
Hitachi America, Ltd.

Philip D. Winnie
Westinghouse Electric Corp.

Michael C. Winters
Fermilab

Kathryn Wirth
PB/MK Team

Lawrence J. Wolf
SSC Laboratory

Doug Wolfe
Hughes Aircraft Co.

Jack Woltz
SSC Laboratory

M. Edward Womble
C. S. Draper Laboratory, Inc.

John Womersley
Florida State University

Richard Woods
Los Alamos National Laboratory

Stuart Worley
Hughes Aircraft Co.

Jim Worth
Oxford Superconducting Tech

Russell L. Wylie
SSC Laboratory

Xun Xu
SSC Laboratory

Dan Yamamura
Teledyne Japan K.K.

Victor A. Yarba
Institute For High Energy Physics

Minoru Yokota
Sumitomo Electric Industries Ltd.

Richard York
SSC Laboratory

Neil E. Young
United Engineers & Constructors

Adnan Yucel
SSC Laboratory

Jon P. Zbasnik
SSC Laboratory

Bruce Zeitlin
IGC Advanced Superconductor

Burt X. Zhang
Gardner Cryogenics

Ren-yuan Zhu
Caltech

Bruno O. Ziegler
Sulzer Bros. Ltd.

Fred Zikas
Parker-ICD

Ewald Zimmer-Nixdorf
SSC Laboratory

George O. Zimmerman
Supersolder Technologies Inc.

Art Zinszer
General Dynamics

Exhibitors

Advanced Interconnection
Technology, Inc. (AIT)
AFA Industries
Air Products and Chemicals, Inc.
AISA
Alcatel Vacuum Products, Inc.
Allied Apical Company
American Magnetics, Inc.
APD Cryogenics Armco
Advanced Materials
AT&T
Babcock & Wilcox
Balzers
Best Southwest Partnership
Cabot Corporation
CEA
Ceramaseal
Climax Specialty Metals
Cryenco
Cryogenic Consultants, Inc.
Cryolab, Inc.
Cryomagnetics, Inc.
CVI Incorporated
Digital Corporation
Douglas Engineering Company
Dour Metal
Dynapower Corporation
Edwards High Vacuum International
Furukawa Electric Technologies, Inc.
General Dynamics Space Systems
GMW Associates
H & J Tool & Die
Hard Chrome Plating Consultants, Inc.
High Tech Services, Inc.

Hitachi America, Ltd.
Hitachi Cable America, Inc.
Hughes Aircraft Company
Hyspan Precision Products, Inc.
IASSC
IBM Engineering/Scientific Center
IBM-Federal Sector Division
Inland Steel Company
Intermagnetics General Corporation
Ishikawajima-Harima Heavy
Industries Co., Ltd. (IHI)
Iten Industries
Koch Process Systems, Inc.
Kurt J. Lesker Company
Kyser Company
Lake Shore Cryotronics, Inc.
LeCroy Corporation
Marcel M. Barbier, Inc.
Martin Marietta Astronautics Group
McDonnell Douglas
MDC Vacuum Products Corporation
Metallized Products, Inc.
Micron Semiconductor, Inc.
Mitsubishi Heavy Industries, Ltd.
MKS Instruments, Inc.
MPI International, Inc.
MTM Cryo Tech Lab
National SSC Coalition
New England Electric Wire Corporation
Nippon Steel USA, Inc.
Nor-Cal Products, Inc.
North Texas Commission
NRC, Inc.
Oracle Federal Division

Outokumpu Copper (USA), Inc.
Oxford Superconducting Technology
Parker Hannifin Coporation
Plainfield Stamping
PPB
Pressure Systems, Inc.
SAES Getter/U.S.A., Inc.
Scientific Instruments, Inc.
SciTech/COSI
Shaped Wire, Inc.

SSC Laboratory
Swagelok Company
Teledyne Wah Chang Albany
The Chambers of Ellis County, Texas
Thevenet - Clerjounie
Union Carbide Industrial Gases, Inc.
Vector Fields, Inc.
Von Roll Isola Inc.
Westinghouse Electric Corporation

Author Index

Subject Index

ANSYS analysis (continued)
of 50mm dipole vacuum vessel,
273
of ground motion, 1038, 1039
of ICR insulation system, 222,
223
of LAC, 242, 471
of support posts, 307
of thermal expansion joint, 1046,
1050, 1051
of ultra-low carbon steel, 147,
148, 152
ANSYS Mesh Tangential Load
Fixtures, 166
Anti-ovalized collars, 148, 326,
327, 554, 556, 566, 569,
570
Antiproton Source, 1022
Antiquarks, 533
Apple/Macintosh systems, 706
Applications Productivity Tool
(APT), 889
Argon
high-Mn non-magnetic steel and,
81
in LAC, 242
in liquid ionization calorimeter,
245
safety and, 989, 990
in warm liquid forward
calorimeter, 246
Argonne National Laboratory, 984,
985
detector development by, 937-943
scintillating calorimeter
development by, 945-950
ASCII text, 201, 1170, 1171
Asea Brown Boveri (ABB), 113
ASME, see American Society of
Mechanical Engineers
Associated Research Inc., 390
ASST program, see Accelerator
Systems String Test
program
ASTM method, on ultra-low carbon
steel, 140
Atoms, 530-532
Austenite, 129
Automated data collection (ADC),
253-254, 256-258
Azimuthal coil modulus, 165, 166,
169, 170, 171, 173

Backstop geometry, 249
Bar code systems, 256, 257
Barium fluoride crystal
calorimeter, 437-454
crystal production for, 451
electrical noise in, 445-446,
447

Barium fluoride crystal
calorimeter (continued)
energy resolution in, 438,
444-445, 447-448
geometry effect in, 447
intercalibration accuracy of,
447, 448-450
physics noise in, 446-447
position resolution in, 438, 444
radiation resistance of, 438,
451-453
readout design in, 443
slow component suppression in,
442-443
time resolution in, 438, 441
UV transmittance in, 451
Barrel Calorimeter, 233-234
Baskets, of electromagnetic
calorimeter, 397, 402, 403,
404, 405, 406
Battelle Memorial Institute,
ferromagnetic research at,
365-372
Beam diagnostics system, for UNK,
742, 747-748
Beam monitors
in HERA quadrupole, 315, 316,
321
in spool piece, 205-206
Beam tubes
bellows of, 1062, 1070-1076
CDM producibility and, 42
field uniformity and, 609
for 50mm dipole vacuum vessel,
272
ground motion and, 1039
in HERA quadrupoles, 315
high-Mn non-magnetic steel in,
84
non-magnetic stainless steel in,
117, 124
in spool piece, 203, 204
systems engineering in,
1104-1105
Beijing Glass Research Institute
(BGRI), 451
Bellows, 1059-1076, 1163, 1164
beam tube, 1062, 1070-1076
Euler buckling mode in,
1060-1062, 1065
large, 1060, 1063-1070, 1076
single-phase, 571
small, 1060, 1070, 1076
spring-supported, 1063
Berkeley SVXD readout chips, 903,
905, 907, 922
BGO crystals
barium fluoride compared with,
441, 444, 446, 448-450,
453

Finite element analysis (FEA)
(continued)
  of 50mm dipole vacuum vessel,
    273, 275-276, 282
  of ground motion, 1040, 1041,
    1042
  of HEB dipole, 628, 633
  of LAC, 478
  in magnetic field calculations,
    11, 12, 13, 14, 16,
    18-20
  of manufacturing tolerances, 49,
    51
  of support posts, 307
  three-dimensional, *see* Three-
    dimensional finite
    element analysis
  two-dimensional, *see* Two-
    dimensional finite
    element analysis
5-cm dipole, 615-624
  coil curing in, 620-621
  coil end in, 618-620
  coil winding in, 616, 620-621
  collars of, 616, 617, 621, 622
  design of, 616-618
  test results on, 622
  yokes of, 616, 618, 621
FLUKA program, 1017
Flux creep, 407, 412
FMECA, *see* Failure Modes Effects
    and Criticality Analysis
Formal Qualification Test, 1163,
    1166
FORTH, 201
Fourier coefficients, in magnetic
    field calculations, 25,
    27-28, 34
FRACAS, *see* Failure Reporting
    Analysis and Corrective
    Action System
Fracture
  in cold mass supports, 186
  cryogenic steels and, 91, 92,
    103
  ultra-low carbon steel and, 139,
    140-141, 142, 143, 144,
    145, 148
Free radicals, 1050
Freon 22, 490
Frequency response study, 295-299
FTAte TRAinSCC, 935-936
Furukawa, 616

Gamma radiation
  clinical applications of, 639,
    643
  insulation and, 108, 109
  integrated cryogenic sensors and,
    869, 870

Gamma radiation (continued)
  thermal expansion joint and,
    1045, 1050
Gauss-Legendre integration scheme,
    13, 30
GBFEL (Ground based free electron
    laser), 650
GEANT simulation, 446, 447, 448
General Dynamics, Space System
    Division (GDSSD), 967
  ALT program of, 1053-1058
  CDM producibility and, 39-47
  CDM program of, 539-547
  CDM reliability development by,
    283-285
  CDM test program of, 1159-1166
  manufacturing tolerance research
    by, 49-55
  SSP & AUD development by,
    253-258
  team-based organization in,
    1141-1148
General Dynamic Thermal Analyzer,
    222, 223
Global Positioning System (GPS),
    424
Global RF feedback systems,
    781-792
  computers in, 789-791
  description of, 782-784
  synchronization in, 783-784, 785,
    788, 792
Gold plating of wire, 375
Goodzeit gauges, 1077-1085
GPS (Global Positioning System),
    424
Ground based free electron laser
    (GBFEL), 650
Ground motion, 1037-1043
Grumman Corporation, spectrometer
    development by, 965-976
Guyan reduction, in ground motion
    analysis, 1040

Hadron calorimeter, 482
Hadronic modules
  in LAC, 234, 236, 473
  of SPACAL, 501-502
  in warm liquid calorimeter, 458,
    459, 462, 464
Hadrons, 532, 533
Hadron shielding, 1017, 1018-1019,
    1020, 1021, 1022
Hagedorn-Ranft model, in shielding
    calculations, 1018
Half cell
  helium venting model for, *see*
    Helium venting model
  systems engineering and, 1100,
    1105

Medium Energy Booster (MEB)
(continued)
  RF system for, 761, 762, 765,
      766
    global feedback in, 781, 783,
        784, 785, 788, 789,
        790
Mesons, 533
Metalforming, Incorporated, wire
      development by, 341-354
Micron Semi-Conductor, Ltd.,
      detector development by,
      903-926
Microsoft *Word*, 1170, 1171, 1172,
      1173
Midwest Research Institute (MRI),
      513, 514, 515
Minimum ionizing particle (MIP)
      calibration, 450
Mises-Hencky yield criterion, 150
MLI, *see* Multilayer insulation
MMS (MultiMuon Spectrometer), 741
Molybdenum, 93, 100, 132
Monoclonal antibodies, 643
Monte Carlo calculations
  for liquid xenon detector, 481,
      490, 491, 495
  for shielding, 1017-1024
  for TDC, 932
  for warm liquid forward
      calorimeter, 249
MOPS (Multiple Object Partitioned
      Structure), 804
MRI, *see* Magnetic resonance
      imaging
MTD 132 Octal Time Digitizer, *see*
      Time-to-Digital Converter
Multilayer insulation (MLI),
      849-859, 1032, 1036
  in ICR, 217-223
MultiMuon Spectrometer (MMS), 741
Multiple Object Partitioned
      Structure (MOPS), 804
Multipoles
  cryogenic magnetic field and,
      259, 260, 264
  defined, 50
  field uniformity and, 602, 604,
      605, 606, 608, 611, 612
  manufacturing tolerance and,
      49-55
  prestress performance and,
      330-332
  systems engineering and, 1102
  three-dimensional magnetic field
      and, 25, 26, 27, 35
  two-dimensional magnetic field
      and, 11, 12, 14, 16, 18,
      20

Muon detectors, 249, 250, 729, 730,
      741, *see also* Air core
      toroid muon spectrometer
Muons, 532
Muon shielding, 1017, 1020-1022

NASA, 177
NASTRAN analysis, of BGO crystals,
      406
National Academy of Engineering
      (NAE), 982
National Academy of Sciences (NAS),
      982
National Environmental Policy Act
      of 1969 (NEPA), 981, 982,
      983, 985
National Institute of Nuclear
      Physics (INFN), 397
National Institute of Standards and
      Technology (NIST), 95, 96,
      98
  ultra-low carbon steel
      manufacture by, 139-145,
      147
National Science Foundation, 515
National Technology Transfer Act of
      1989, 646
Neel temperature, 98, 122, 125
NEPA, *see* National Environmental
      Policy Act of 1969
Neumann boundary
  field uniformity and, 609, 610
  magnetic field and, 18, 20
Neutron radiation, 108, 109, 941,
      1045
  clinical applications of, 639,
      643, 646
Neutrons, 532, 938, 940, 1017,
      1018, 1020
Nevis Laboratories, 728-729, 730
New England Wire Corporation, 383
NFPA standards, 1009, 1012
Nickel
  in cryogenic steels, 91, 92, 93,
      95, 96, 98, 100, 101,
      102, 103, 105
  ferromagnetic effects of,
      365-372
  in non-magnetic stainless steel,
      117-118, 123, 124, 125
Nickel plating of wire, 376
Nickel wire, 375, 377
Niobium cable, 704
Niobium/tin conductor, 355-364
  tin-core process in, 356-360,
      361, 364
  tubular-tin-source process in,
      360-361, 362, 363, 364
Niobium/tin wire, 342

Radio Frequency Quadrupole (RFQ),
448-450, 454, 642, 644
RF system for, 771, 773-774
Random errors
field uniformity and, 601, 606,
612, 614
in 50mm dipole, 587, 596-597
Raster images, 1171, 1175
Ray Tracing Code, 459
Reflective memory, 803-804
in field bus, 193, 195-196, 197,
199
Refrigeration systems, *see*
Cryogenic systems
Relativistic Heavy Ion Collider
(RHIC)
conductor specifications for,
383, 384, 385, 387
FTAte TRAinSCC and, 935
support posts for, 175, 177,
178-180, 181, 183, 187,
188, 302, 311
Relay Ladder Logic (RLL), 889
Residual risk, 1009, 1014
Reversible Rack 2 Bus Bar System,
375
RF accelerator systems, 802
global feeback in, *see* Global RF
feedback system
for LINAC, *see* under LINAC
for synchrotrons, 761-769
for UNK, 742, 746-747
RFQ, *see* Radio Frequency
Quadrupole
R-glass, 113
RHIC, *see* Relativistic Heavy Ion
Collider
Riken, 482
Risk, 1009, 1014
Risk assessments, 1013
Risk management and control, 1133
Risk resolution, 1013-1014
RLL (Relay Ladder Logic), 889
Robert Bosch Corporation, 1147
Robotics, 431
Rockwell International, 535
Rugby Ball support structure, 397,
398, 403

Saclay, 710
HERA quadrupole development by,
313-324
Safan Co., 751
Safety, *see* System safety
Safety assessments, 1014-1015
Safety design criteria, 1009
SAIC, 644
Sausaging, 352, 353, 354, 360, 387,
680

Scalar potential approach, 26-32
Scanning, 1171
Science Applications International
Corporation, accelerators
of, 639-647
Scintillating calorimeters,
945-950, *see also* specific
types
Scintillating Fiber-Lead
Calorimeter (SPACAL),
497-505
EM module of, 501-502
eutectic module of, 497, 500-501
hadronic module of, 501-502
laminated module of, 497, 498
Scintillating/uranium calorimeters,
939
SDH (Synchronous Digital
Hierarchy), 802
Self-Describing Data Set (SDS),
804
Septum magnets, 745
Serpukhov, 1017
Sextupoles, 52, 205, 556, 563
ferromagnetic materials and,
365-372
field uniformity and, 604, 606,
607, 608, 609, 611
magnetic field calculations and,
13, 16, 25, 32, 33-34,
35-36
in UNK, 742-743
Shanghai Institute of Ceramics
(SIC), 451
Shell, 553, 570
of 50mm dipole, 273, 275, 276,
277, 576, 580
of 5-cm dipole, 618
prestress performance and, 328
thermal expansion joint and,
1046
Shell hoop prestress, 413-417
Shielding calculations, 1017-1024
Signal cable feedthru (SCFT),
233-244
Signal-to-noise ratio
in pixel detectors, 954, 956
in silicon microstrip detectors,
904, 905, 912, 918
Silicon, 132, 671
Silicon microstrip detector,
903-926
absolute calibration in, 922-926
advantages of, 904-905
data acquisition on, 910
diode signals in, 903, 905-907,
912-922, 923, 924, 925
leakage currents and, 910-911
microcode and triggering in,
909-910